Gene Biotechnology

Second Edition

Gene Biotechnology

Second Edition

William Wu
Michael J. Welsh
Peter B. Kaufman
Helen H. Zhang

CRC PRESS

Boca Raton London New York Washington, D.C.

Library of Congress Cataloging-in-Publication Data

Gene biotechnology / William Wu ... [et al.] -- 2nd ed.
p. cm.
First ed. published under title: Methods in gene biotechnology. Boca Raton : CRC Press, 1997.
Includes bibliographical references and index.
ISBN 0-8493-1288-4 (alk. paper)
1. Genetic engineering--Laboratory manuals. 2. Molecular biology--Laboratory manuals.
I. Wu, William. II. Methods in gene biotechnology.

QH442.M475 2003
572.8′6′072--dc21 2003053183

Direct all inquiries to CRC Press LLC, 2000 N.W. Corporate Blvd., Boca Raton, Florida 33431.

Visit the CRC Press Web site at www.crcpress.com

International Standard Book Number 0-8493-1288-4
Library of Congress Card Number 2003053183
Printed in the United States of America 1 2 3 4 5 6 7 8 9 0
Printed on acid-free paper

Preface

We are living in an era of genomics and biotechnology revolution. To many molecular and cellular biologists, the flood of information on human genomes and new methodologies is particularly overwhelming because biotechnology is forging ahead and bringing about rapid changes continuously. Because every organism depends on molecular action for survival, molecular biology research has become more dominant in multiple disciplines. In fact, it is a general tendency that, when the National Institutes of Health and other funding agencies award grants, they give high priority to research proposals that employ molecular biology approaches. How is it possible to catch up, receive updates on new biotechnology, and use the most recent, proven techniques for novel research? One of the aims of this book is to provide investigators with the tools needed for modern molecular and cellular biology research.

Another goal is to guide graduate students in their thesis research. In our experience, good graduate training mandates independent performance with minimum advice from a mentor. How is a novel research project for a thesis selected? What are the hypotheses, objectives and experimental designs? How can technical problems be grasped and current techniques mastered? Where does one begin and what are the predicted results? A graduate student needs some help with these questions; this book will provide the clues.

This book covers a wide range of current biotechnology methods developed and widely used in molecular biology, biochemistry, cell biology and immunology. The methods and protocols described in the appropriate chapters include:

- Strategies for novel research projects
- Rapid isolation of specific cDNAs or genes by PCR
- Construction and screening of cDNA and genomic DNA libraries
- Preparation of DNA constructs
- Nonisotopic and isotopic DNA or RNA sequencing
- Information superhighway and computer databases of nucleic acids and proteins
- Characterization of DNA, RNA or proteins by Southern, northern or western blot hybridization
- Gene overexpression, gene underexpression and gene knockout in mammalian systems
- Analysis of DNA or abundance of mRNA by radioactivity *in situ* hybridization (RISH)
- Localization of DNA or abundance mRNA by fluorescence *in situ* hybridization (FISH)
- *In situ* PCR hybridization of low copy genes and *in situ* RT-PCR detection of low abundance mRNAs

- New strategies for gene knockout
- Large-scale expression and purification of recombinant proteins in cultured cells
- High-throughput analysis of gene expression by real-time RT-PCR
- Gene expression profiling via DNA microarray
- Phage display
- siRNA technology

Each chapter covers the principles underlying the methods and techniques presented and a detailed step-by-step description of each protocol, as well as notes and tips. We have found that many of the currently available books in molecular biology contain only protocol recipes. Unfortunately, many fail to explain the principles and concepts behind the methods outlined or to inform the reader of possible pitfalls in the methods described. We intend to fill these gaps.

Although all four authors have worked as a team, for the information of the reader, the following table shows which authors wrote which chapters.

Chapter Number	Authors
1	W. Wu, P.B. Kaufman, M.J. Welsh
2, 3, 6, 10, 12–14	W. Wu
8, 9, 16, 17	W. Wu, M.J. Welsh
5, 7, 15	W. Wu, P.B. Kaufman
4, 11, 18	W. Wu, H.H. Zhang
19–22	W. Wu, M.J. Welsh, H.H. Zhang

William Wu

Michael J. Welsh

Peter B. Kaufman

Helen H. Zhang

Authors

William Wu, Ph.D., is currently an assistant research scientist in the Department of Cell and Developmental Biology at the University of Michigan Medical School, Ann Arbor. Dr. Wu is also a professor of biology at the Hunan Normal University, Changsha, Hunan, the People's Republic of China. He earned his M.S. degree in biology from the Hunan Normal University, Changsha, Hunan, in 1984. In 1992, he earned his Ph.D. degree in molecular and cellular biology at the Ohio University, Athens, Ohio. He has completed 3 years of postdoctoral training in molecular biology at the University of Michigan, Ann Arbor.

Dr. Wu is an internationally recognized expert in molecular biology. He has presented and published more than 35 research papers at scientific meetings and in national and international journals. As a senior author, Dr. Wu has contributed 34 chapters of molecular and cellular biology methodologies in 3 books. He has broad knowledge and profound understanding of molecular biology, biotechnology, protein biochemistry, cellular biology and molecular genetics. He is highly experienced and has extensive hands-on expertise in a variety of current molecular biology techniques in academic and bioindustrial settings. Dr. Wu is a member of the American Association for the Advancement of Science.

Michael J. Welsh, Ph.D., is a professor of cell biology in the Department of Cell and Developmental Biology at the University of Michigan Medical School, Ann Arbor. He is also a professor of toxicology in the School of Public Health at the University of Michigan, Ann Arbor. Dr. Welsh earned his Ph.D. degree in 1977 from the University of Western Ontario, London, Ontario, Canada.

Dr. Welsh is an internationally recognized scientist in molecular and cellular biology. He has published more than 120 research papers and has contributed several chapters to professional books. He is a leading authority on mammalian heat shock protein (HSP27) and has been awarded several major grants from the U.S. National Institutes of Health to examine the function and mechanisms of the HSP27 gene.

Peter B. Kaufman, Ph.D., is a professor of biology in the plant cellular and molecular biology program in the Department of Biology and a member of the faculty of the bioengineering program at the University of Michigan, Ann Arbor. He earned his Ph.D. degree in 1954 in plant biology at the University of California, Davis.

Dr. Kaufman is an internationally recognized scholar in molecular biology and physiology. He is a fellow of the American Association for the Advancement of Science and secretary–treasurer of the American Society for Gravitational and

Space Biology. He has served on the editorial board of *Plant Physiology* for 10 years and has published more than 190 research papers and 7 professional books. Dr. Kaufman teaches a popular course in plant biotechnology yearly at the University of Michigan, Ann Arbor. He has been awarded research grants from the National Science Foundation, the National Aeronautics and Space Administration, USDA and Parke–Davis Pharmaceutical Research Laboratories in Ann Arbor, Michigan.

Helen H. Zhang, M.S., is a research associate in the Department of Neurology at the University of Michigan Medical School, Ann Arbor. In 1996, she earned her M.S. degree in molecular and cellular biology at Eastern Michigan University, Ypsilanti.

Zhang has a strong background in molecular biology, microbiology, biochemistry and immunology. She is highly experienced in DNA recombination, PCR applications, cell transfection and gene expression in bacterial and animal systems. She also has extensive experience in protein/peptide purification, enzymatic assays and analysis of drug–protein binding.

Contents

1 Strategies for Novel Research Projects

CONTENTS

INTRODUCTION

We are living in the era of a technology revolution. To many molecular biologists, the flood of new information is particularly overwhelming because biotechnology is forging ahead and bringing about changes day after day. How can one catch up, update new biotechnology, and use the most recent, proven techniques for novel research? As indicated in the preface, one of the aims of this book is to provide investigators with tools for molecular biology research. Perhaps because every single organism depends on molecular actions for survival, molecular biology research has become more dominant in multiple disciplines.[1–20] In fact, it is a general tendency that the National Institutes of Health (NIH) and other funding agencies award grants giving high priority to those research proposals that use molecular biology approaches. Given that research funding resources are quite limited, funding budgets are decreasing and the number of research proposals rapidly increasing year after year, the fundamental question is how one can bring dollars to research and survive. One strategy is to write an excellent proposal using molecular biology tools. Nonetheless, what catches attention in view of research projects and funding resources? How does one grasp research problems and march in a new direction? How does one write an important and scientifically sound research proposal with new ideas, comprehensive designs and methodologies? What is the research plan and where does one start? These questions may be particularly obvious to an investigator who does not have much experience in molecular biology research. This chapter can provide some insights and ideas that may be helpful to novel research projects and to obtaining potential funding.

Another aim of this chapter is to guide graduate students in their thesis research. In our experience, good graduate training mandates independent performance with minimum advice from a mentor. How does one select a novel research project for a thesis? What are the hypotheses, objectives and experimental designs? How does one grasp technical problems and master current techniques? Where does one begin and what are the predicted results? How does one interpret research data and decide on the next experiments? A graduate student needs some help with these questions. This chapter will provide the clues.

The major objective of this book is to help the reader pursue research in molecular biology. The present chapter serves as a tour of how to use the current strategies and techniques in the book in order to approach novel research strategies. Several examples of 2 to 5 years' research proposals with specific aims, strategic designs and methods are illustrated below. The following examples are not the real proposal format; however, it is hoped that these examples will be a valuable guide for novel research, proposal funding, or thesis research. The examples can also be adapted and applied to other appropriate novel research projects.

PROPOSAL 1. IDENTIFICATION OF NEW DRUG-TARGETING PROTEINS AND ISOLATION OF NOVEL GENES

Drug discovery is one of the most interesting research projects pertaining to public health and is certainly invaluable to pharmaceutical companies. Once a new drug is produced, it will open up broad areas for new research. The fundamental question concerns its cure mechanism. Many kinds of research can be conducted using the drug. In our view, the most promising research proposal is the identification of the drug-targeting proteins and isolation of novel genes. Because the molecular interaction is the basis for the cure of a disease with a new drug, it is reasonable to hypothesize that one or more proteins or genes are potentially targeted by the drug. Because the discovery of a new drug is of greatest concern to the public, this plan would most likely be funded by NIH, pharmaceutical companies, or other private sectors. Figure 1.1 illustrates the strategies, research design, and molecular biology methodologies that one can use, along with references to the appropriate chapters in this book.

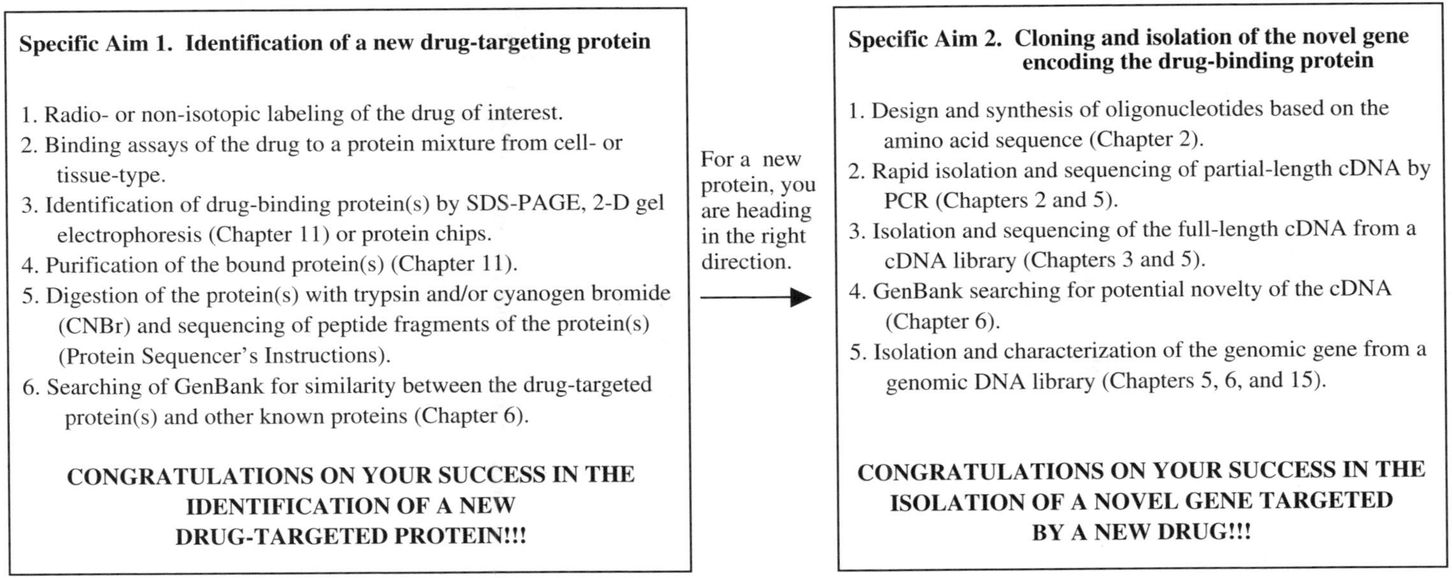

Specific Aim 1. Identification of a new drug-targeting protein

1. Radio- or non-isotopic labeling of the drug of interest.
2. Binding assays of the drug to a protein mixture from cell- or tissue-type.
3. Identification of drug-binding protein(s) by SDS-PAGE, 2-D gel electrophoresis (Chapter 11) or protein chips.
4. Purification of the bound protein(s) (Chapter 11).
5. Digestion of the protein(s) with trypsin and/or cyanogen bromide (CNBr) and sequencing of peptide fragments of the protein(s) (Protein Sequencer's Instructions).
6. Searching of GenBank for similarity between the drug-targeted protein(s) and other known proteins (Chapter 6).

CONGRATULATIONS ON YOUR SUCCESS IN THE IDENTIFICATION OF A NEW DRUG-TARGETED PROTEIN!!!

For a new protein, you are heading in the right direction. →

Specific Aim 2. Cloning and isolation of the novel gene encoding the drug-binding protein

1. Design and synthesis of oligonucleotides based on the amino acid sequence (Chapter 2).
2. Rapid isolation and sequencing of partial-length cDNA by PCR (Chapters 2 and 5).
3. Isolation and sequencing of the full-length cDNA from a cDNA library (Chapters 3 and 5).
4. GenBank searching for potential novelty of the cDNA (Chapter 6).
5. Isolation and characterization of the genomic gene from a genomic DNA library (Chapters 5, 6, and 15).

CONGRATULATIONS ON YOUR SUCCESS IN THE ISOLATION OF A NOVEL GENE TARGETED BY A NEW DRUG!!!

FIGURE 1.1 Research design for the identification of novel genes targeted by a new drug.

PROPOSAL 2. EXPLORATION OF FUNCTIONS OR ROLES FOR THE EXPRESSION OF A GENE TARGETED BY A NEW DRUG

Once a novel protein or gene targeted by a new drug or an important chemical has been identified, further research needs to be done. One logical and promising funding project is to determine the function of the targeted protein or gene. The information from the research will provide crucial evidence for the cure mechanism of a disease by the drug. This involves sophisticated skills, including generation and use of transgenic mice as animal models (Figure 1.2).

PROPOSAL 3. VERIFICATION OF POTENTIAL FUNCTION OF A SPECIFIC GENE BY THE GENE KNOCKOUT APPROACH

To verify the potential role of a novel, important gene, the best strategy is to knock out the expression of the gene. For example, if it is reasonable to believe that a gene plays a key role in heart development, a smart approach is to target the gene *in vivo* by knockout. If the gene becomes null, heart diseases such as failure of heart development would be predicted to occur based on the hypothesis. This is certainly a very sound proposal with promising funding for 3 to 4 years, assuming that no previous grants have been awarded for this type of proposal in other laboratories. The general approaches and methods are diagrammed in Figure 1.3.

PROPOSAL 4. IDENTIFICATION OF THE FUNCTIONAL DOMAIN OF A PROTEIN BY SITE-SPECIFIC MUTAGENESIS

Very often, the functions of a novel protein have been demonstrated but no one knows which specific fragment of the protein is the active domain or binding site. If the protein or enzyme is very important, there is a good reason to write a 2-year proposal on the identification of the functional domain of the novel protein or enzyme. The strategies for doing this are outlined in Figure 1.4.

PROPOSAL 5. IDENTIFICATION OF TOXICANT-BINDING PROTEINS AND ISOLATION OF A NOVEL GENE RELATED TO THE TOXICANT AND HEART HYPERTROPHY USING AN ANIMAL MODEL

The heart is the first organ formed in animals, and heart diseases such as heart hypertrophy, heart attack, or heart failure are of great concern to the public. Additionally, toxicity mediated by toxicants becomes more of an environmental concern. If there is good reason to hypothesize that a toxicant (e.g., chemicals, proteins, drugs, carbohydrates) may induce heart hypertrophy, a 4- to 5-year proposal would be

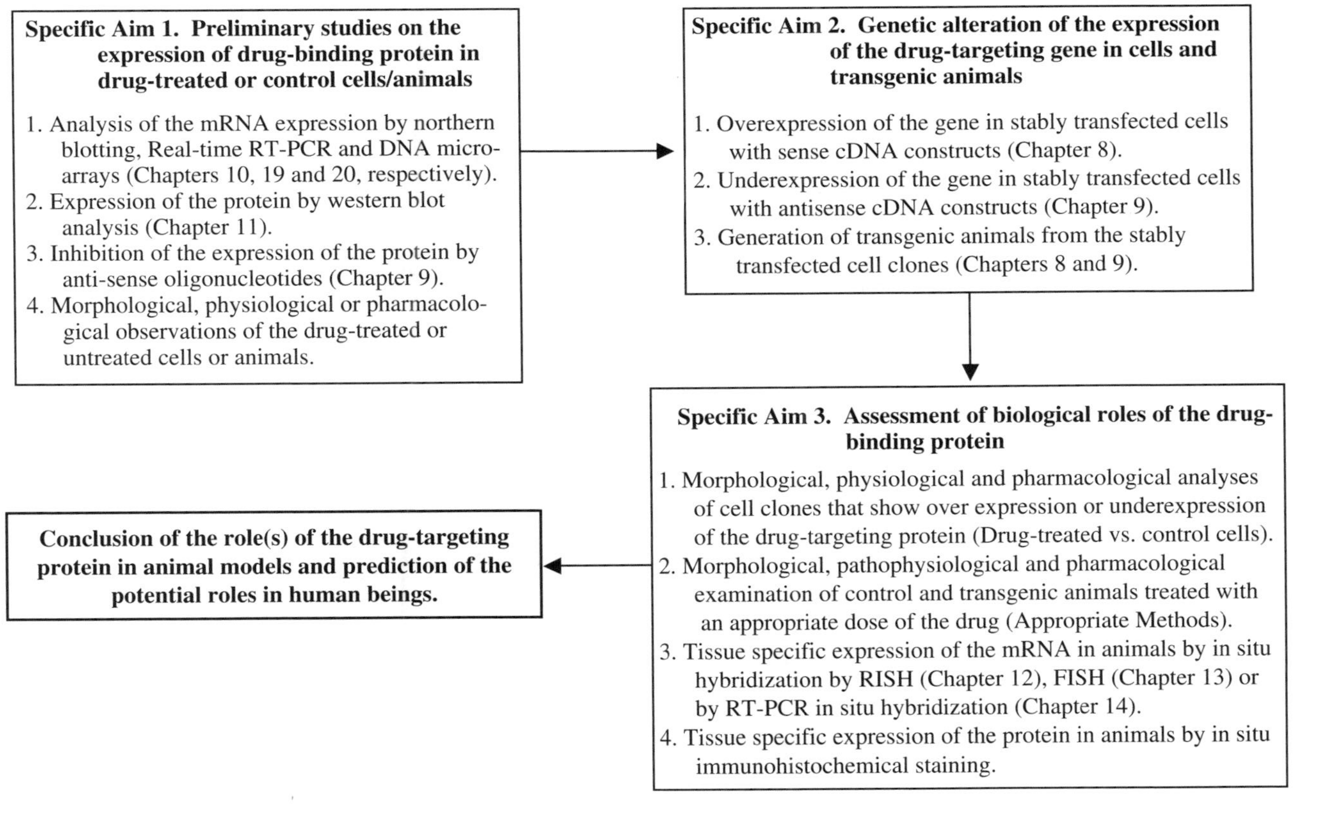

FIGURE 1.2 Research approaches and technologies for the exploration of the potential functions of a new drug-targeting protein via cultured cells and animal models.

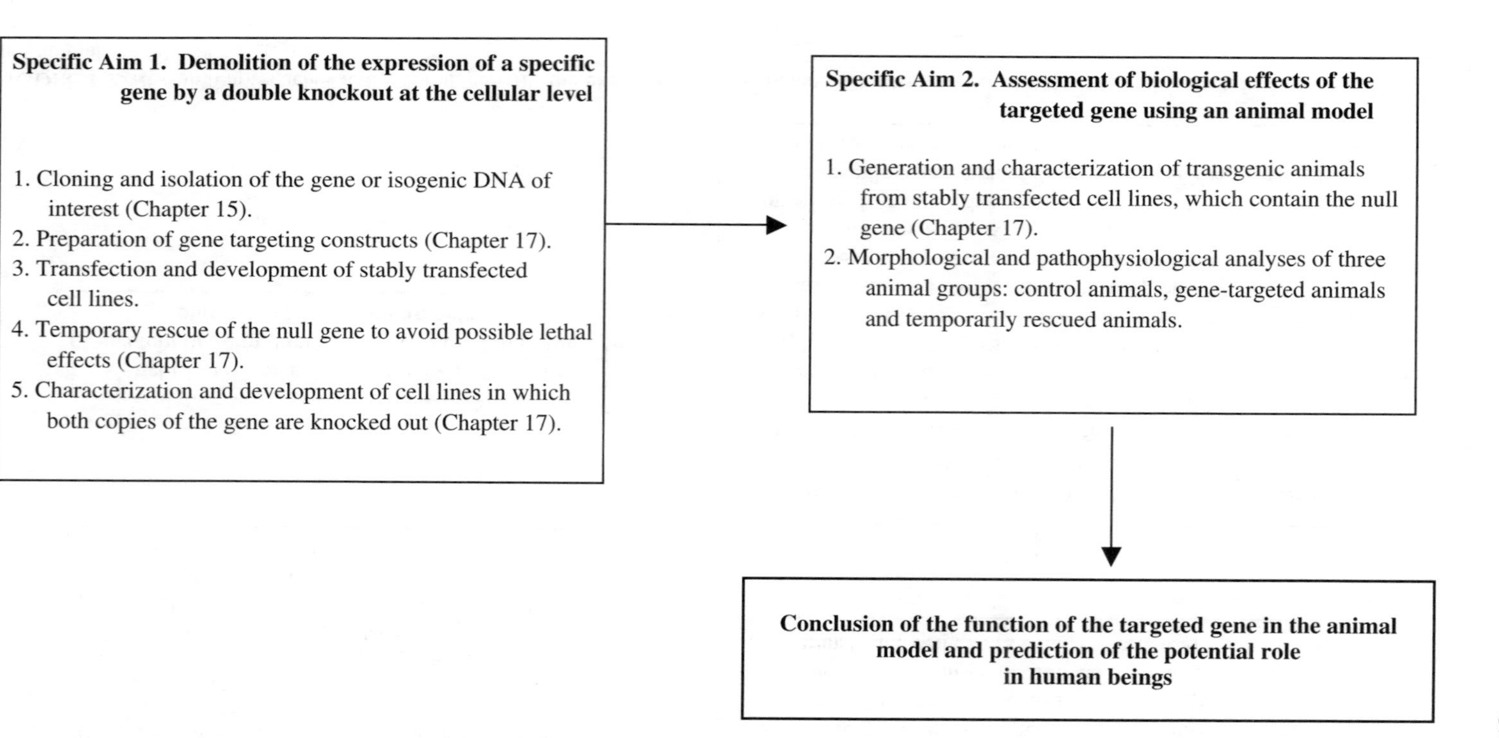

FIGURE 1.3 Research approaches and methods for exploration of the potential roles of a new drug-targeting protein via cultured cells and animal models.

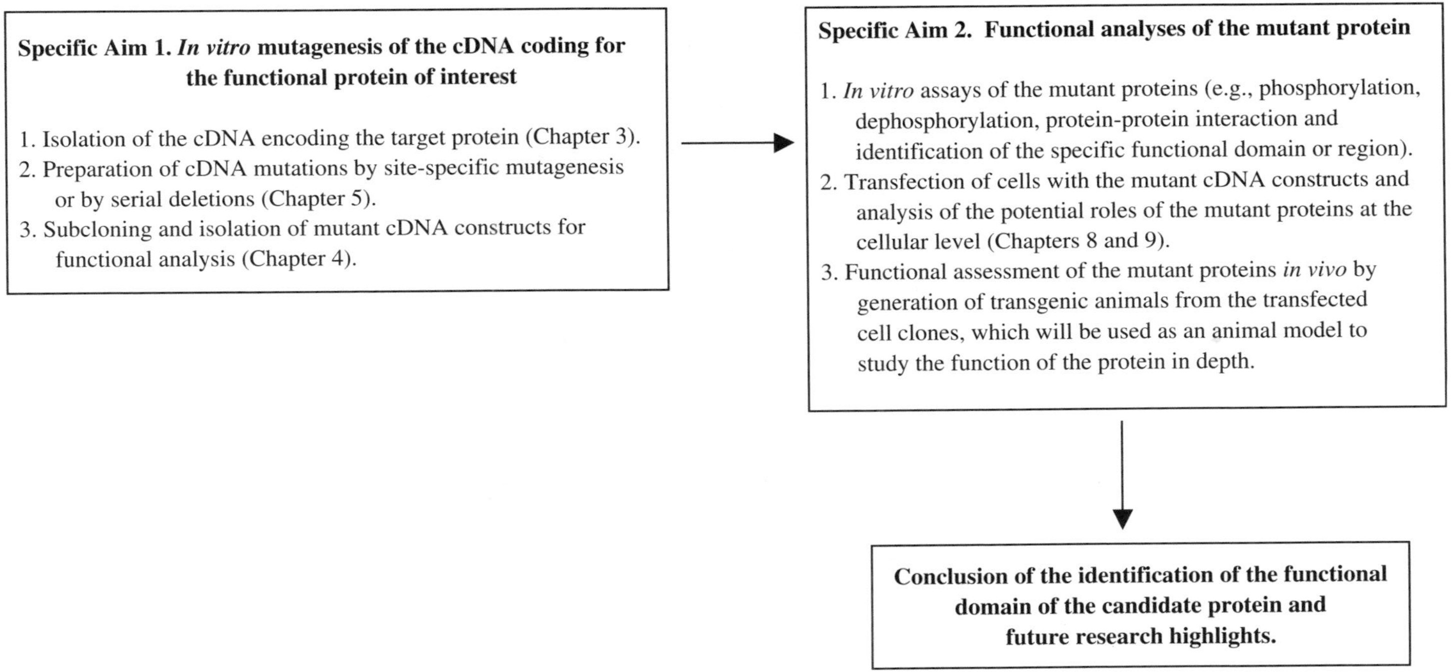

Specific Aim 1. *In vitro* mutagenesis of the cDNA coding for the functional protein of interest

1. Isolation of the cDNA encoding the target protein (Chapter 3).
2. Preparation of cDNA mutations by site-specific mutagenesis or by serial deletions (Chapter 5).
3. Subcloning and isolation of mutant cDNA constructs for functional analysis (Chapter 4).

→

Specific Aim 2. Functional analyses of the mutant protein

1. *In vitro* assays of the mutant proteins (e.g., phosphorylation, dephosphorylation, protein-protein interaction and identification of the specific functional domain or region).
2. Transfection of cells with the mutant cDNA constructs and analysis of the potential roles of the mutant proteins at the cellular level (Chapters 8 and 9).
3. Functional assessment of the mutant proteins *in vivo* by generation of transgenic animals from the transfected cell clones, which will be used as an animal model to study the function of the protein in depth.

↓

Conclusion of the identification of the functional domain of the candidate protein and future research highlights.

FIGURE 1.4 Design and methods for identification of functional domains for a given protein by site-directed mutagenesis.

profound and promising for funding. This is a long-term research project that may need collaboration between laboratories. Specific aims, strategies, designs, expected results, and methodologies, as found in the appropriate chapters, are outlined in Figure 1.5.

PROPOSAL 6. RESCUE OF AN IMMUNE-DEFICIENT SYSTEM VIA GENE THERAPY

It is well known that immune deficiency, such as HIV, is the disease of greatest concern to the public at the present time. Many important proteins, such as the CD families, have been discovered to play a significant role in the deficient immune system. In order to increase the immune capabilities of patients, gene therapy has been established. Due to the great public interest in this type of therapy, research funding is virtually unlimited. Therefore, one may switch to this new research avenue that involves several collaborators. In view of recent advances in molecular biology, a proposal regarding gene therapy will be a very good approach. Specifically, one may transfect human cells such as bone marrow stem cells with overexpression of sense cDNA constructs. Stably transfected cells that constitutively express proteins (e.g., CD4) can be directly injected into the immune system of the patient. An alternative is to overexpress and purify a large amount of the proteins in bacteria or yeast, and then inject an appropriate dose of the protein molecules into the patient to activate the immune system using appropriate methodologies. Figure 1.6 illustrates the relevant research design strategies and methods covered in this book.

PROPOSAL 7. DISCOVERY OF IAA- OR GA_3-BINDING PROTEINS AND ISOLATION OF NOVEL GENES IN PLANTS

GA_3 and IAA are two well-known growth hormones that have been well studied in view of their actions on the physiology and biochemistry of plants. In spite of the fact that some laboratories have recently been working on the molecular biology of these hormones, the mechanisms by which GA_3 and IAA can promote the growth of plants, especially in mutant phenotypes, are not really understood well. It appears that not much progress has been made so far. However, these two chemicals are very important hormones for normal plant development. From this point of view, and in order to promote new studies on the models of actions of these two hormones, one may launch a new proposal to the Department of Energy (DOE) or the U.S. Department of Agriculture (USDA) for the identification and isolation of GA_3- and IAA-targeting proteins and novel genes, based on the hypothesis that the mechanisms of these hormones depend on hormone–protein interactions. This is an excellent research project for a Ph.D. thesis, and is most likely to get funded. The suggested strategies and techniques for doing this are shown in Figure 1.7.

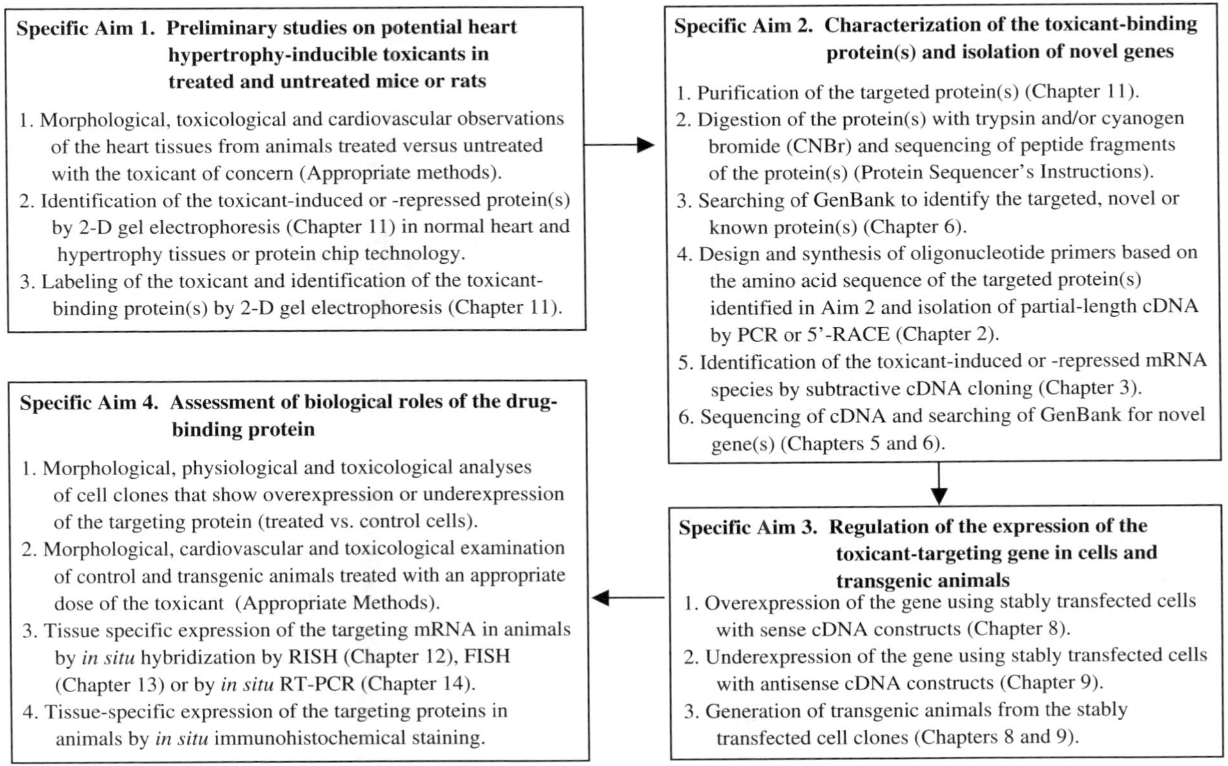

Specific Aim 1. Preliminary studies on potential heart hypertrophy-inducible toxicants in treated and untreated mice or rats

1. Morphological, toxicological and cardiovascular observations of the heart tissues from animals treated versus untreated with the toxicant of concern (Appropriate methods).
2. Identification of the toxicant-induced or -repressed protein(s) by 2-D gel electrophoresis (Chapter 11) in normal heart and hypertrophy tissues or protein chip technology.
3. Labeling of the toxicant and identification of the toxicant-binding protein(s) by 2-D gel electrophoresis (Chapter 11).

Specific Aim 2. Characterization of the toxicant-binding protein(s) and isolation of novel genes

1. Purification of the targeted protein(s) (Chapter 11).
2. Digestion of the protein(s) with trypsin and/or cyanogen bromide (CNBr) and sequencing of peptide fragments of the protein(s) (Protein Sequencer's Instructions).
3. Searching of GenBank to identify the targeted, novel or known protein(s) (Chapter 6).
4. Design and synthesis of oligonucleotide primers based on the amino acid sequence of the targeted protein(s) identified in Aim 2 and isolation of partial-length cDNA by PCR or 5'-RACE (Chapter 2).
5. Identification of the toxicant-induced or -repressed mRNA species by subtractive cDNA cloning (Chapter 3).
6. Sequencing of cDNA and searching of GenBank for novel gene(s) (Chapters 5 and 6).

Specific Aim 3. Regulation of the expression of the toxicant-targeting gene in cells and transgenic animals

1. Overexpression of the gene using stably transfected cells with sense cDNA constructs (Chapter 8).
2. Underexpression of the gene using stably transfected cells with antisense cDNA constructs (Chapter 9).
3. Generation of transgenic animals from the stably transfected cell clones (Chapters 8 and 9).

Specific Aim 4. Assessment of biological roles of the drug-binding protein

1. Morphological, physiological and toxicological analyses of cell clones that show overexpression or underexpression of the targeting protein (treated vs. control cells).
2. Morphological, cardiovascular and toxicological examination of control and transgenic animals treated with an appropriate dose of the toxicant (Appropriate Methods).
3. Tissue specific expression of the targeting mRNA in animals by *in situ* hybridization by RISH (Chapter 12), FISH (Chapter 13) or by *in situ* RT-PCR (Chapter 14).
4. Tissue-specific expression of the targeting proteins in animals by *in situ* immunohistochemical staining.

FIGURE 1.5 Comprehensive project design and methods for identification of novel proteins and genes in heart hypertrophy induced by a toxicant via an animal model.

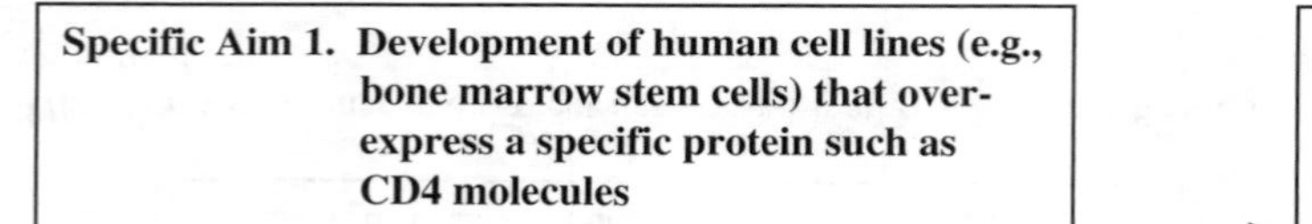

FIGURE 1.6 Research projects and methods for rescue of patients with immune deficiency by gene therapy.

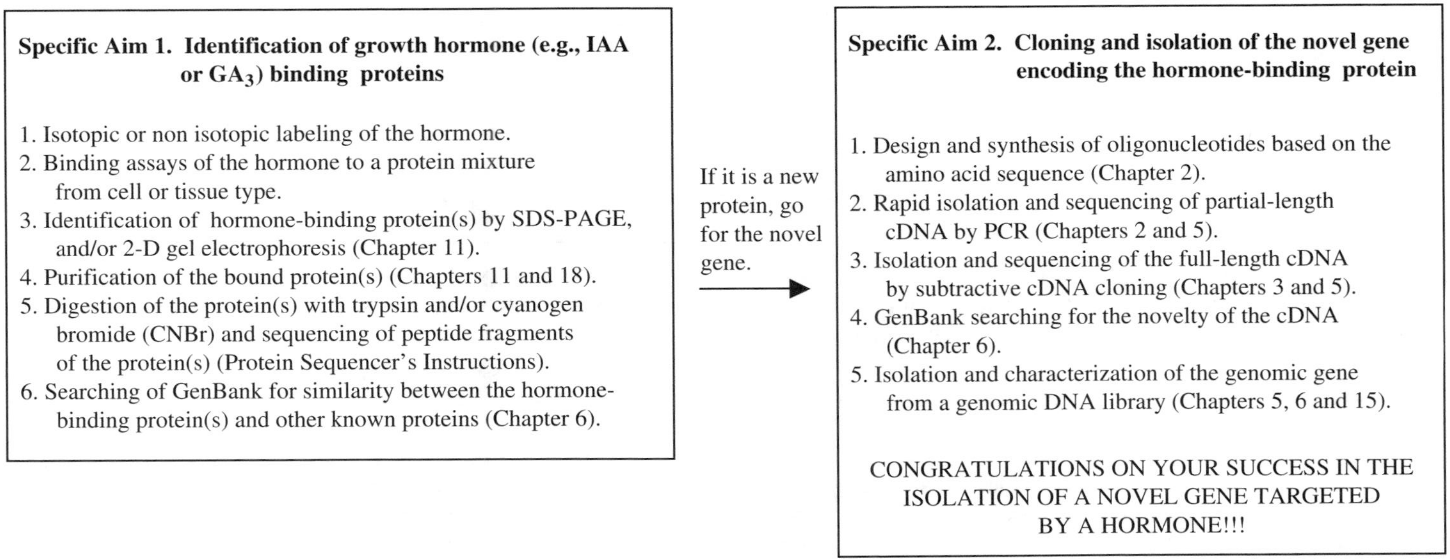

Specific Aim 1. Identification of growth hormone (e.g., IAA or GA_3) binding proteins

1. Isotopic or non isotopic labeling of the hormone.
2. Binding assays of the hormone to a protein mixture from cell or tissue type.
3. Identification of hormone-binding protein(s) by SDS-PAGE, and/or 2-D gel electrophoresis (Chapter 11).
4. Purification of the bound protein(s) (Chapters 11 and 18).
5. Digestion of the protein(s) with trypsin and/or cyanogen bromide (CNBr) and sequencing of peptide fragments of the protein(s) (Protein Sequencer's Instructions).
6. Searching of GenBank for similarity between the hormone-binding protein(s) and other known proteins (Chapter 6).

If it is a new protein, go for the novel gene. →

Specific Aim 2. Cloning and isolation of the novel gene encoding the hormone-binding protein

1. Design and synthesis of oligonucleotides based on the amino acid sequence (Chapter 2).
2. Rapid isolation and sequencing of partial-length cDNA by PCR (Chapters 2 and 5).
3. Isolation and sequencing of the full-length cDNA by subtractive cDNA cloning (Chapters 3 and 5).
4. GenBank searching for the novelty of the cDNA (Chapter 6).
5. Isolation and characterization of the genomic gene from a genomic DNA library (Chapters 5, 6 and 15).

CONGRATULATIONS ON YOUR SUCCESS IN THE ISOLATION OF A NOVEL GENE TARGETED BY A HORMONE!!!

FIGURE 1.7 Approaches of identification of the novel genes targeted by a hormone.

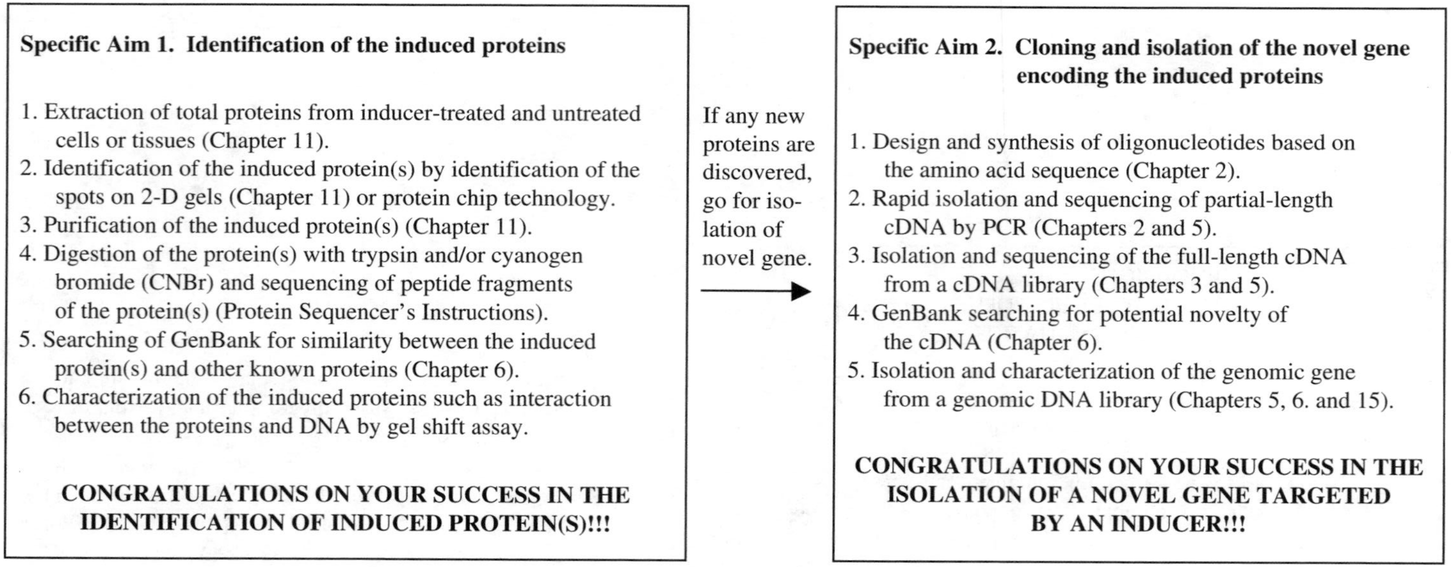

Specific Aim 1. Identification of the induced proteins

1. Extraction of total proteins from inducer-treated and untreated cells or tissues (Chapter 11).
2. Identification of the induced protein(s) by identification of the spots on 2-D gels (Chapter 11) or protein chip technology.
3. Purification of the induced protein(s) (Chapter 11).
4. Digestion of the protein(s) with trypsin and/or cyanogen bromide (CNBr) and sequencing of peptide fragments of the protein(s) (Protein Sequencer's Instructions).
5. Searching of GenBank for similarity between the induced protein(s) and other known proteins (Chapter 6).
6. Characterization of the induced proteins such as interaction between the proteins and DNA by gel shift assay.

CONGRATULATIONS ON YOUR SUCCESS IN THE IDENTIFICATION OF INDUCED PROTEIN(S)!!!

If any new proteins are discovered, go for isolation of novel gene. →

Specific Aim 2. Cloning and isolation of the novel gene encoding the induced proteins

1. Design and synthesis of oligonucleotides based on the amino acid sequence (Chapter 2).
2. Rapid isolation and sequencing of partial-length cDNA by PCR (Chapters 2 and 5).
3. Isolation and sequencing of the full-length cDNA from a cDNA library (Chapters 3 and 5).
4. GenBank searching for potential novelty of the cDNA (Chapter 6).
5. Isolation and characterization of the genomic gene from a genomic DNA library (Chapters 5, 6. and 15).

CONGRATULATIONS ON YOUR SUCCESS IN THE ISOLATION OF A NOVEL GENE TARGETED BY AN INDUCER!!!

FIGURE 1.8 Research outline for discovery of new proteins, cDNAs, and novel genes induced by a chemical, toxicant, hormone, alcohol or drug.

PROPOSAL 8. IDENTIFICATION OF NOVEL PROTEINS, cDNA AND GENES INDUCED OR REPRESSED BY A SPECIFIC TREATMENT

It is reasonable to hypothesize that treatment of cultured cells or organisms with an important chemical, drug, alcohol or hormone may induce or repress the expression of certain proteins. Therefore, a research proposal on the identification of these proteins and isolation of the cDNAs or genes coding for the proteins, especially the novel genes, would provide fundamental information about the action or mechanisms involved as a result of the treatment. Once a novel gene is isolated, it allows one to open a broad research area concerning the treatment, which has promise for funding. Meanwhile, this is an excellent research project for a master's or Ph.D. thesis. Figure 1.8 gives a guide for this type of research.

REFERENCES

1. Welsh, M.J., Wu, W., Parvinen, M., and Gilmont, R.R., Variation in expression of HSP27 messenger RNA during the cycle of the seminiferous epithelium and co-localization of HSP27 and microfilaments in Sertoli cells of the rat, *Biol. Reprod.,* 55, 141–151, 1996.
2. Mehlen, P., Preville, X., Chareyron, P., Briolay, J., Klemenz, R., and Arrigo, A-P., Constitutive expression of human hsp27, *Drosophila* hsp27, or human αB-crystallin confers resistance to TNF- and oxidative stress-induced cytotoxicity in stably transfected murine L929 fibroblasts, *J. Immunol.*, 154, 363–374, 1995.
3. Iwaki, T., Iwaki, A., Tateishi, J., and Goldman, J.E., Sense and antisense modification of glial αB-crystallin production results in alterations of stress fiber formation and thermoresistance, *J. Cell Biol.*, 15, 1385–1393, 1994.
4. Wu, W. and Welsh, M.J., Expression of HSP27 correlates with resistance to metal toxicity in mouse embryonic stem cells transfected with sense or antisense HSP27 cDNA, *Appl. Toxicol. Pharmacol.*, 141, 330, 1996.
5. Yang, N.S., Burkholder, J., Roberts, B., Martinell, B., and McCabe, D., *In vitro* and *in vivo* gene transfer to mammalian somatic cells by particle bombardment, *Proc. Natl. Acad. Sci., USA,* 87, 9568, 1990.
6. Carmeliet, P., Schoonjans, L., Kiechens, L., Ream, B., Degen, J., Bronson, R., Vos, R.D., Oord, J. Jvan den, Collen, D., and Mulligan, R.C., Physiological consequences of loss of plasminogen activator gene function in mice, *Nature,* 368, 419, 1994.
7. Johnston, S.A., Biolistic transformation: microbes to mice, *Nature,* 346, 776, 1990.
8. Wu, W., Gene transfer and expression in animals, in *Handbook of Molecular and Cellular Methods in Biology and Medicine,* Kaufman, P.B., Wu, W., Kim, D., and Cseke, L.J., CRC Press, Boca Raton, FL, 1995.
9. Capecchi, M.R., Targeted gene replacement, *Sci. Am.*, 52, 1994.
10. Capecchi, M.R., Altering the genome by homologous recombination, *Science*, 244, 1288, 1989.
11. Mansour, S.L., Thomas, K.R., and Capecchi, M.R., Disruption of the proto-oncogene *int-2* in mouse embryo-derived stem cells: a general strategy for targeting mutations to nonselectable genes, *Nature*, 336, 348, 1988.

12. Riele, H.T., Maandag, E.R., and Berns, A., Highly efficient gene targeting in embryonic stem cells through homologous recombination with isogenic DNA constructs, *Proc. Natl. Acad. Sci. USA,* 89, 5128, 1992.
13. Schorle, H., Holtschke, T., Hunig, T., Schimpl, A., and Horak, I., Development and function of T cells in mice rendered interleukin-2 deficient by gene targeting, *Nature,* 352, 621, 1991.
14. Weinstock, P.H., Bisgaier, C.L., Setala, K.A., Badner, H., Ramakrishnan, R., Frank, S.L., Essenburg, A.D., Zechner, R., and Breslow, J.L., Severe hypertriglyceridemia, reduced high density lipoprotein, and neonatal death in lipoprotein lipase knockout mice, *J. Clin. Invest.*, 96, 2555, 1995.
15. Michalska, A.E. and Choo, K.H.A., Targeting and germ-line transmission of a null mutation at the metallothionein I and II loci in mouse, *Proc. Natl. Acad. Sci. USA*, 90, 8088, 1993.
16. Paul, E.L., Tremblay, M.L., and Westphal, H., Targeting of the T-cell receptor chain gene in embryonic stem cells: strategies for generating multiple mutations in a single gene, *Proc. Natl. Acad. Sci. USA*, 89, 9929, 1992.
17. Robert, L.S., Donaldson, P.A., Ladaigue, C., Altosaar, I., Arnison, P.G., and Fabijanski, S.F., Antisense RNA inhibition of β-glucuronidase gene expression in transgenic tobacco can be transiently overcome using a heat-inducible β-glucuronidase gene construct, *Biotechnology,* 8, 459, 1990.
18. Rezaian, M.A., Skene, K.G.M., and Ellis, J.G., Anti-sense RNAs of cucumber mosaic virus in transgenic plants assessed for control of the virus, *Plant Mol. Biol.,* 11, 463, 1988.
19. Hemenway, C., Fang, R.-X., Kaniewski, W.K., Chua, N.-H., and Tumer, N.E., Analysis of the mechanism of protection in transgenic plants expressing the potato virus X coat protein or its antisense RNA, *EMBO J.,* 7, 1273, 1988.
20. Dag, A.G., Bejarano, E.R., Buck, K.W., Burrell, M., and Lichtenstein, C.P., Expression of an antisense viral gene in transgenic tobacco confers resistance to the DNA virus tomato golden mosaic virus, *Proc. Natl. Acad. Sci. USA,* 88, 6721, 1991.

2 Rapid Isolation of Specific cDNAs or Genes by PCR

CONTENTS

INTRODUCTION

The polymerase chain reaction (PCR) is a powerful technique that is widely used for amplification of specific DNA sequences *in vitro* using appropriate primers.[1–4] PCR is a major breakthrough technology and is a relatively rapid, sensitive, and inexpensive procedure for amplification and cloning of the cDNA or genomic DNA of interest. It is also invaluable for analysis of RNA expression, genetic diagnosis, detection of mutations and genetic engineering.[2–11] The general principles of PCR start from a pair of oligonucleotide primers designed so that forward or sense primer directs the synthesis of DNA towards reverse or antisense primer, and vice versa. During the PCR, *Taq* DNA polymerase, which is purified from bacterial *Thermus aquaticus* and is a heat-stable enzyme, catalyzes the synthesis of a new DNA strand complementary to a template DNA from 5′→3′ direction by primer extension reaction. This results in production of the DNA region flanked by the two primers. Because the *Taq* DNA polymerase is high-temperature (95°C) stable, it is possible for target sequences to be amplified for many cycles using excess primers in a commercial thermocycler apparatus. Recently, high-quality *Taq* polymerases, such as recombinant polymerase *Tth*, long-spand polymerase and high-fidelity PCR polymerase, have been developed. Due to their capability of proofreading, these polymerases are more advanced enzymes compared with traditional *Taq* DNA polymerase.

The present chapter focuses on rapid amplification and isolation of specific cDNAs or genomic genes by PCR strategies. Traditionally, cDNA or the gene is isolated from cDNA or genomic DNA libraries, which involves construction and screening of cDNA or genomic DNA libraries. The procedures are complicated, time-consuming, and costly. Besides, it may be impossible to "fish" out the low copy cDNAs transcribed from rare mRNAs in cDNA libraries. In contrast, the rare cDNA can be rapidly amplified and isolated by the reverse transcription polymerase chain reaction (RT-PCR). This chapter describes detailed protocols for fast isolation and purification of the cDNAs or genes of interest. These protocols have been successfully used in our laboratories.

ISOLATION OF SPECIFIC FULL-LENGTH cDNAs BY THE RT-PCR METHOD

A full-length cDNA can be amplified by reverse transcription PCR or RT-PCR. To achieve this objective, a forward or sense primer and a reverse or antisense primer should be designed in the 5′-UTR (untranslation region) and the 3′-UTR regions of a known cDNA sequence or based on known amino acid sequence.

ISOLATION OF RNAs

RNA isolation is described in Chapter 3.

DESIGN AND SYNTHESIS OF SPECIFIC FORWARD AND REVERSE PRIMERS

A pair of forward (sense or upstream) primer and reverse (antisense or downstream) primer should be designed based on the 5′-UTR or 3′-UTR sequences of a known cDNA to be isolated (Figure 2.1). These sequences can be found from published cDNA from the same species or may be from different organisms. Specifically, the forward primer can be designed from the 5′-UTR in the 5′→3′ direction with 20 to 30 bases, which should be complementary to the first or (–)strand of the cDNA template. The reverse primer is designed from the 3′-UTR region in the 5′→3′ direction with 20 to 30 bases, which should be complementary to the second or (+)strand of the cDNA template.

Alternatively, forward and reverse primers can be designed based on the very N-terminal and C-terminal amino acid sequences if the cDNA sequence is not available. As a result, if all goes well, a cDNA including the entire open reading frame (ORF) can be isolated with ease.[2–4] For example, assuming that the very N-

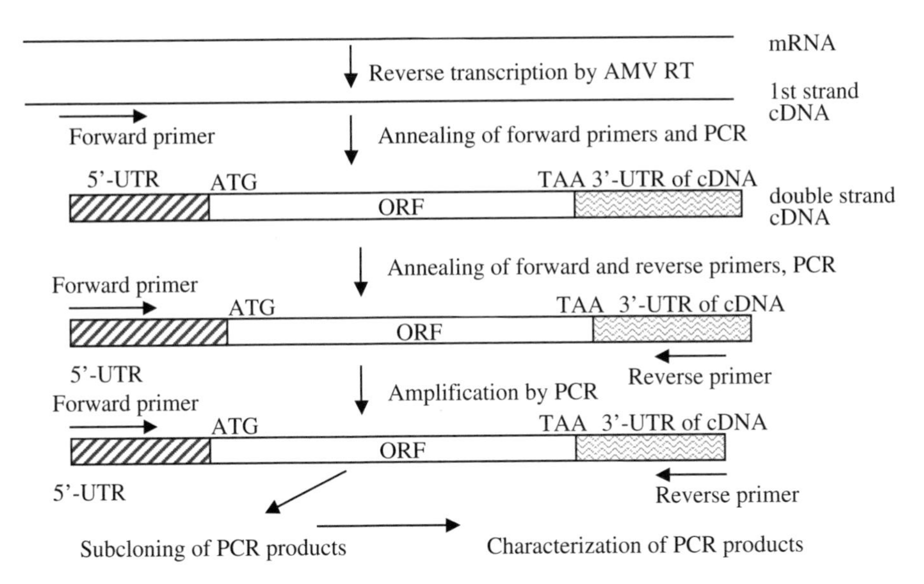

FIGURE 2.1 Diagram of isolation of full-length cDNA by RT-PCR.

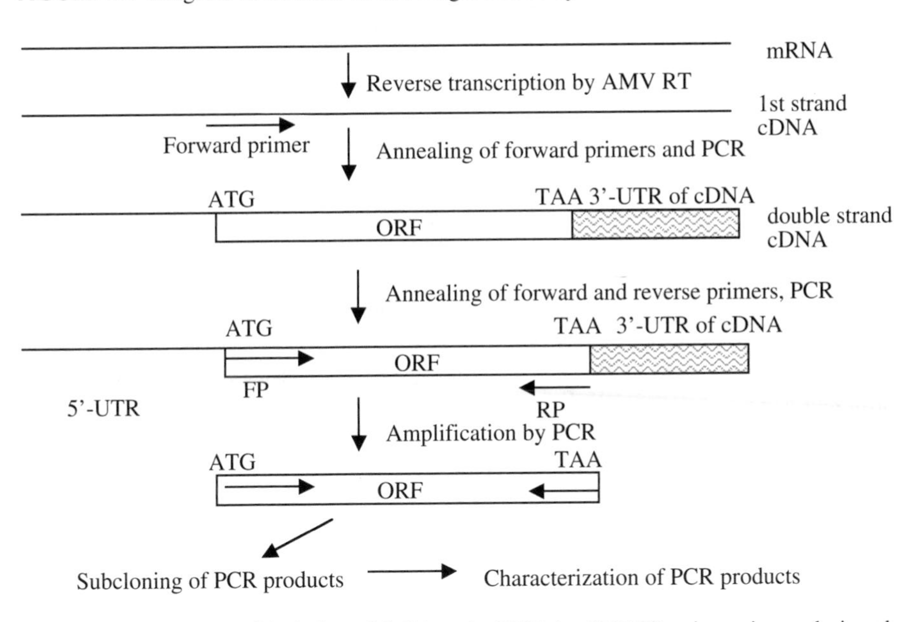

FIGURE 2.2 Diagram of isolation of full-length cDNA by RT-PCR using primers designed from known amino acid sequences.

terminal and C-terminal amino acid sequences are NDPNG and DPCEW, respectively, the forward and reverse primers can be designed accordingly to be 5′-AAC(T)GAT(C)CCIAA(C)TGGI-3′ and 3′-GGTGAGCGTCCCTAG-5′ (Figure 2.2). If only the N-terminal amino acid sequence is available, a reverse primer may be designed as oligo(dT) (Figure 2.3).

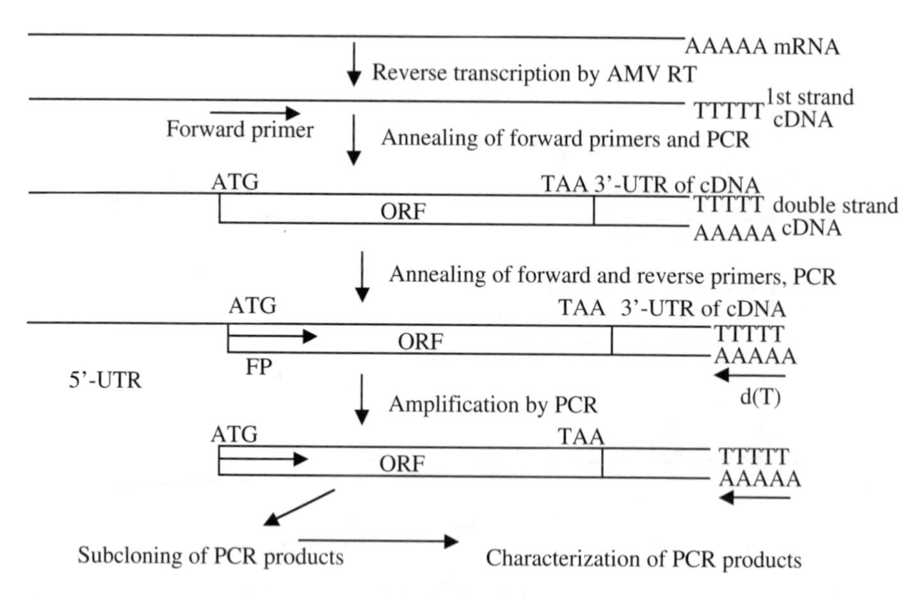

FIGURE 2.3 Diagram of isolation of full-length cDNA by RT-PCR using primers designed from known amino acid sequence and poly(A) tail. FP: forward primer.

It is extremely important that each pair of primers should be carefully designed to anneal with two DNA strands. If two primers are annealed with the same DNA strand of the template, no PCR products will be produced. The primers should be analyzed by an appropriate computer program prior to being used for PCR. We routinely use the Oligo Version 4.0 or CPrimer f program to check the quality of primers, including GC contents, *Tm* value, potential formation of intraloops/dimers and interduplex of the primers, and potential annealing between the primers and the region flanking the primers. In general, primers should be >17 bases with <60% GC contents. The first base at 5′ or 3′ end should be G or C followed by A or T for an efficient annealing. No more than four bases of introprimer or interprimer complementary are allowed.

In experience, the author recommends that two different restriction enzyme sites be designed at the 5′ end of forward primer and the 3′ end of reverse primer. This is invaluable to facilitate efficient subcloning of the PCR products for characterization, such as DNA sequencing and *in vitro* transcription. The enzyme cutting sites depend on the particular multiple cloning site (MCS) of an appropriate vector to be used for subcloning of the PCR products, for example, 5′ GTGGATCCAACGATCCCIAATGGTATTC 3′ (forward primer containing *BamH* I site) and 3′ GGTGAGCGTCCCTAGTTCGAATG 5′ (reverse primer containing a *Hind* III site). After PCR amplification, the PCR products can be digested with *BamH* I and *Hind* III and be subcloned at the *BamH* I and *Hind* III sites of appropriate vectors are digested with *BamH* I and *Hind* III. The general procedures are outlined in Figure 2.1 through Figure 2.3.

AMPLIFICATION OF cDNA OF INTEREST BY RT-PCR

Target mRNA can be selectively transcribed into cDNA using a specific primer that anneals to the 3′portion near the target region of the mRNA. The author recommends that oligo(dT) primers be used to transcribe all the mRNA species into cDNAs. Oligo(dT) primers anneal to the 3′ poly(A) tails of mRNA molecules and facilitate the synthesis of the first stranded cDNAs. Avian myeloblastosis virus (AMV) reverse transcriptase and Moloney murine leukemia virus (MoMuLV) reverse transcriptase are commonly used RT reaction. RT kits are commercially available; a standard reaction volume is 25 µl.

Perform Synthesis of the First-Stranded cDNAs from mRNA

1. Anneal 10 µg of total RNAs or 1 µg of mRNA template with 1 µg of oligo(dT) primers in a sterile RNase-free microcentrifuge tube. Add nuclease-free dd.H_2O to a total volume of 15 µl. Heat the reaction at 70°C for 5 min and allow it to slowly cool to room temperature to finish annealing. Briefly spin down the mixture to the bottom of the microcentrifuge tube.
2. To the annealed primer–template mixture, add the following in the order shown below:
   First strand 5X buffer, 5 µl
   rRNasin ribonuclease inhibitor, 50 units (25 units/µg mRNA)
   40 m*M* sodium pyrophosphate, 2.5 µl
   AMV reverse transcriptase, 30 units (15 units/µg mRNA)
   Nuclease-free dd.H_2O to a total volume of 25 µl
3. Incubate the reaction at 42°C for 60 min. At this point, the synthesis of the first-strand cDNAs is complete.
4. Stop the reaction by adding 2 µl of 0.2 *M* EDTA to the mixture and place it on ice.
5. Precipitate the cDNAs by adding 2.5 volumes of chilled (–20°C) 100% ethanol to the tube. Gently mix and allow precipitation to occur at –20°C for 2 h.
6. Centrifuge at the top speed for 5 min. Carefully remove the supernatant, briefly rinse the pellet with 1 ml of cold 70% ethanol and gently drain away the ethanol.
7. Dry the cDNA pellet under vacuum for 10 min, and dissolve the cDNA pellet in 10 µl of TE buffer. Take 2 µl of the sample to measure the concentration of cDNAs prior to PCR. Store the sample at –20°C until use.

Reagents Needed

10X AMV or MoMuLV Buffer

0.1 *M* Tris-HCl, pH 8.3
0.5 *M* KCl
25 m*M* $MgCl_2$

Stock dNTPs

dATP, 10 m*M*
dTTP, 10 m*M*
dATP, 10 m*M*
dTTP, 10 m*M*

Working dNTPs Solution

dATP (10 m*M*), 10 µl
dTTP (10 m*M*), 10 µl
dATP (10 m*M*), 10 µl
dTTP (10 m*M*), 10 µl

Carry Out Amplification of Specific cDNAs by PCR in a 0.5-ml Microcentrifuge Tube

1. Prepare a PCR cocktail as follows:
   Forward primer, 5 to 8 pmol (30 to 40 ng) depending on the size of primer (36- to 43-mer, 10 to 30 ng/µl)
   Reverse primer, 5 to 8 pmol (30 to 40 ng) depending on the size of primer (36- to 43-mer, 10 to 30 ng/µl)
   cDNAs, 1 to 2 µg
   10X amplification buffer, 4 µl
   dNTPs (2.5 m*M* each), 4 µ
   Taq or *Tth* DNA polymerase, 5 to 10 units
   Add dd.H_2O to a final volume of 40 µl.

Note: The DNA polymerase should be high fidelity and long expand, which is commercially available.

2. Carefully overlay the mixture with 30 µl of mineral oil to prevent evaporation of the samples during the PCR amplification. Place the tubes in a thermal cycler and perform PCR.

Profile	Cycling (30 cycles)				
	Predenaturation	Denaturation	Annealing	Extension	Last
Template (<4 Kb) Primer is <24 bases or with <40% G-C content	94°C, 3 min	94°C, 1 min	60°C, 1 min	70°C, 1.5 min	4°C
Template is >4 Kb Primer is >24-mer or <24 bases with >50% G-C content	95°C, 3 min	95°C, 3 min	60°C, 1 min	72°C, 2 min	4°C

Purification of PCR Products by High-Speed Centrifugation of Agarose Gel Slices

1. Load the amplified PCR mixture into a 1 to 1.4% agarose gel, depending on the sizes of the PCR products, and carry out electrophoresis.
2. When electrophoresis is complete, quickly locate the DNA band of interest by illuminating the gel on a long wavelength (>300 nm) UV transilluminator. Quickly slice out the band of interest using a sharp, clean razor blade.

Note: to avoid potential damage to the DNA molecules, the UV light should be turned on as briefly as possible.

3. To enhance the yield of DNA, trim away extra agarose gel outside the band and cut the gel slice into tiny pieces with a razor blade.
4. Transfer the fine slices into a 1.5 ml microcentrifuge tube (Eppendorf).

Notes: (1) If one does not need to elute DNA out of the gel slices, they need not be sliced into tiny pieces. They can be directly placed in a tube. (2) At this point, there are two options for eluting DNA from the agarose gel pieces. The first is to carry out high-speed centrifugation (Step 5) immediately. The second option is to elute the DNA as high yield as possible (see below).

5. Centrifuge at 12,000 to 14,000 × *g* or at the highest speed using an Eppendorf centrifuge 5415C (Brinkmann Instruments, Inc.) for 15 min at room temperature.

Principles: With high-speed centrifugation, the agarose matrix is compressed or even partially destroyed by the strong force of centrifugation. The DNA molecules contained in the matrix are released into the supernatant fluid.

6. Following centrifugation, carefully transfer the supernatant fluid containing DNA into a clean microcentrifuge tube. The DNA can be used directly for ligation, cloning, and labeling as well as restriction enzyme digestion without ethanol precipitation. Store the DNA solution at 4°C or –20°C until use.

Tips: (1) In order to confirm that the DNA is released from the gel pieces, the tube containing the fluid can be briefly illuminated with long wavelength UV light after centrifugation. An orange–red fluid color indicates the presence of DNA in the fluid. (2) The supernatant fluid should be immediately transferred from the agarose pellet; within minutes, the temporarily compressed agarose pellet may swell, absorbing the supernatant fluid.

High-Yield and Cleaner Elution of DNA

1. Add 100 to 300 µl of distilled deionized water (dd.H_2O) or TE buffer (10 m*M* Tris-HCl, 1 m*M* EDTA, pH 8.0) to the gel pieces at the preceding Step 4.

Function: TE buffer or dd.H_2O is added to the gel pieces to help elute the DNA and to simultaneously wash EtBr from the DNA during high-speed centrifugation.

2. Centrifuge at 12,000 to 14,000 × *g* at room temperature for 20 min. When the centrifugation is complete, briefly visualize the eluted DNA in the fluid by illumination using a long wavelength UV light.
3. Immediately transfer the supernatant that contains the eluted DNA into a fresh tube.
4. To increase the recovery of DNA, extract the agarose pellet by resuspending the pellet in 100 μl of dd.H_2O or TE buffer followed by another cycle of centrifugation.
5. Pool the DNA supernatant and precipitate the DNA with ethanol. Dissolve DNA 10 μl of dd.H_2O or TE buffer. Store the DNA at 4 or –20°C until use. Proceed to subcloning and characterization of the PCR products.

AMPLIFICATION AND ISOLATION OF cDNA ENDS BY 5′-RACE

5′-RACE stands for rapid amplification of 5′-cDNA ends and is widely utilized to amplify or isolate partial-length cDNA transcribed from some rare mRNAs. As shown in Figure 2.4, the procedures and principles are quite similar to those described previously except for the addition of oligo(dC) and the use of oligo(dG) as an anchor primer or forward primer. A reverse primer can be designed according to the known sequence in the 5′ portion of the cDNA to be amplified or the N-terminal amino acid of the protein of interest. Addition of oligo(dC) to the 3′ end of the first strand cDNA can be facilitated using terminal deoxynucleotidyl transferase. The drawback of this approach is the potential nonspecific priming if the target cDNA sequence is GC rich. Nonetheless, this strategy provides a powerful tool for analysis of the expression of rare messages. If the 5′-UTR sequence is available or the N-terminal amino acid sequence is known, the addition of oligo(dC) and synthesis of (dG) anchor primer is not necessary. A forward primer can be designed in the 5′-UTR sequence or N-terminal amino acid sequence.

AMPLIFICATION AND ISOLATION OF cDNA ENDS BY 3′-RACE

Similar to 5′-RACE, 3′-RACE or rapid amplification of 3′-cDNA ends is applied to amplify and to isolate the 3′ portion of rare cDNAs made from rare mRNAs. It is simpler than 5′-RACE because oligo(dC) does not need to be added. A forward primer can be designed according to the 3′ portion of cDNA to be amplified or based on known C-terminal amino acid sequence. A reverse sequence may be universal primer or poly(T) primer. The procedure is outlined in Figure 2.5, which is similar to that described earlier.

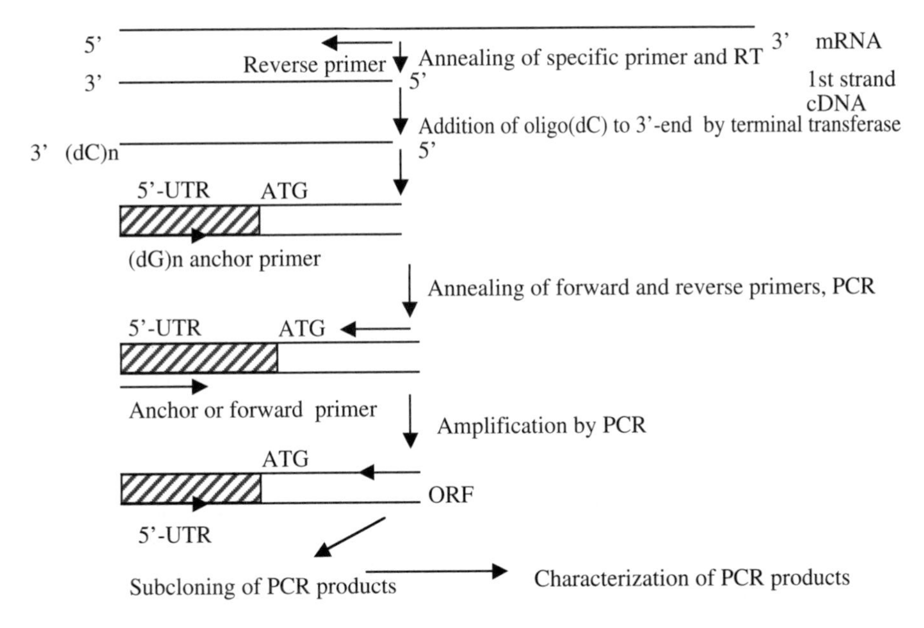

FIGURE 2.4 Diagram of 5′-RACE PCR.

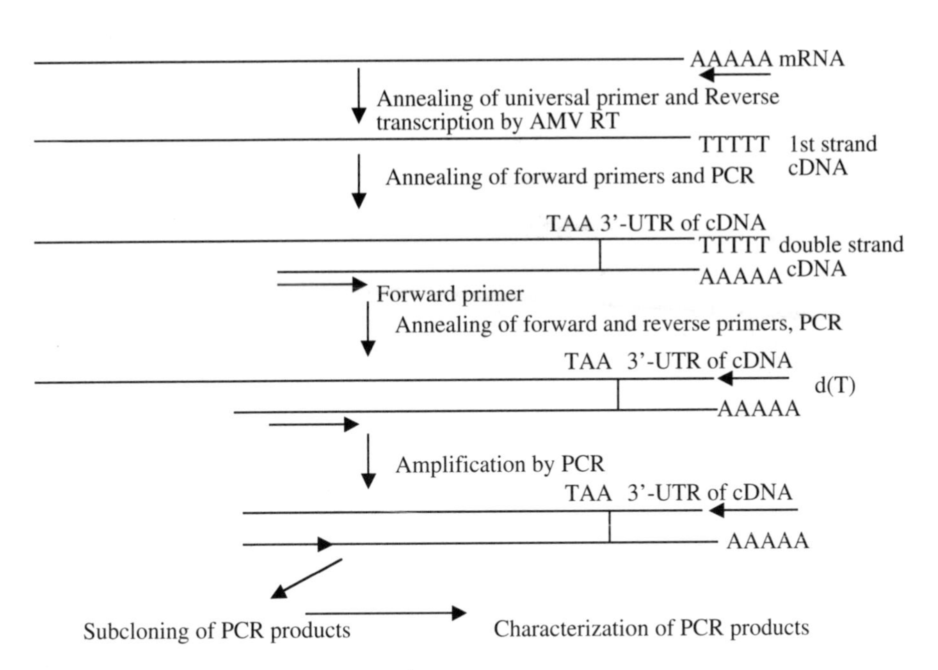

FIGURE 2.5 Diagram of isolation of 3′-RACE PCR.

ISOLATION OF THE GENE OF INTEREST BY PCR

Isolation of Genomic DNA

1. For cultured cells, aspirate culture medium and rinse the cells twice in PBS. Add 10 ml/60 mm culture plate (with approximately 10^5 cells) of DNA isolation buffer to the cells and allow cells to be lysed for 5 min at room temperature with shaking at 60 rpm. For tissues, grind 1 g tissue in liquid N_2 to fine powder and transfer the powder into a tube containing 15 ml of DNA isolation buffer. Allow cells to be lysed for 10 min at room temperature with shaking at 60 rpm.
2. Incubate at 37°C for 5 h or overnight to degrade proteins, and extract genomic DNA.
3. Allow the lysate to cool to room temperature and add three volumes of 100% ethanol precooled at –20°C to the lysate. Gently mix it and allow DNA to precipitate at room temperature with slow shaking at 60 rpm. Precipitate DNA in appearance of white fibers should be visible in 20 min.
4. "Fish" out the DNA using a sterile glass hook or spin down the DNA. Rinse the DNA twice in 70% ethanol and partially dry the DNA for 40 to 60 min at room temperature to evaporate ethanol. The DNA can be directly subjected to restriction enzyme digestion without being dissolved in TE buffer or dd.H_2O. We usually overlay the precipitated DNA with an appropriate volume of restriction enzyme digestion cocktail and incubate the mixture at the appropriate temperature for 12 to 24 h. The digested DNA is mixed with an appropriate volume of DNA loading buffer and is ready for agarose gel electrophoresis.

Reagents Needed

DNA Isolation Buffer

75 m*M* Tris-HCl, pH 6.8
100 m*M* NaCl
1 mg/ml Protease K (freshly added)
0.1% (w/v) *N*-Lauroylsarcosine

Partial Digestion of Genomic DNA Using *Sau3A I*

It is necessary to partially cut the high molecular weight of genomic DNA with a four-base cutter, *Sau3A I*, to increase the efficiency of PCR.

Optimization of Partial Digestion of Genomic DNA with *Sau3A I*

In order to determine the amount of enzyme used to digest the high molecular weight DNA into 20- to 30-Kb fragments, small-scale reactions, or pilot experiments, should be carried out.

1. Prepare 1X *Sau3A I* buffer on ice:
   10X *Sau3A I* buffer, 0.2 ml
   Add dd.H_2O to a final volume of 2.0 ml.
2. Perform *Sau3A I* dilutions in 10 individual microcentrifuge tubes on ice:

Tube #	*Sau3A I* (3 u /μl) dilution	Dilution factors
1	2 μl *Sau3A I* + 28 μl 1X *Sau3A I* buffer	1/15
2	10 μl of 1/15 dilution + 90 μl 1X *Sau3A I* buffer	1/150
3	10 μl of 1/150 dilution + 10 μl 1X *Sau3A I* buffer	1/300
4	10 μl of 1/150 dilution + 30 μl 1X *Sau3A I* buffer	1/600
5	10 μl of 1/150 dilution + 50 μl 1X *Sau3A I* buffer	1/900
6	10 μl of 1/150 dilution + 70 μl 1X *Sau3A I* buffer	1/1200
7	10 μl of 1/150 dilution + 90 μl 1X *Sau3A I* buffer	1/1500
8	10 μl of 1/150 dilution + 110 μl 1X *Sau3A I* buffer	1/1800
9	10 μl of 1/150 dilution + 190 μl 1X *Sau3A I* buffer	1/3000
10	10 μl of 1/150 dilution + 290 μl 1X *Sau3A I* buffer	1/4500

3. Carry out 10 individual, small-scale digestion reactions on ice in the order listed below:

Components	Tube Number									
	1	2	3	4	5	6	7	8	9	10
Genomic DNA (1 μg/μl)	1(μl)	1	1	1	1	1	1	1	1	1
10X *Sau3A I* buffer	5(μl)	5	5	5	5	5	5	5	5	5
dd.H_2O (μl)	39	39	39	39	39	39	39	39	39	39
Sau3A I dilution in the same order as in (b) above	5(μl)	5	5	5	5	5	5	5	5	5
Final volume (μl)	50	50	50	50	50	50	50	50	50	50

Note: The final *Sau3A I* concentration used in tube 1 to tube 10 should be 1, 0.1, 0.05, 0.025, 0.015, 0.0125, 0.01, 0.0085, 0.005, and 0.0035 unit/μg DNA, respectively.

4. Incubate the reactions at the same time at 37°C for 30 min. Place the tubes on ice and add 2 μl of 0.2 *M* EDTA buffer (pH 8.0) to each tube to stop the reaction.
5. While the reactions are performed, prepare a large size of 0.4% agarose gel in 1X TBE buffer.
6. Add 10 μl of 5X DNA loading buffer to each of the 10 tubes containing the digested DNA prepared in Step (d).
7. Carefully load 30 μl of each sample to the wells in the order of 1 to 10. Load DNA markers (e.g., λDNA *Hind III* markers) to the left or the right well of the gel to estimate the sizes of digested DNA.
8. Carry out electrophoresis of the gel at 2 to 5 V/cm until the bromphenol blue reaches the bottom of the gel.

9. Photograph the gel under UV light and find the well that shows the maximum intensity of fluorescence in the desired DNA size range of 20 to 30 Kb.

Large-Scale Preparation of Partially Digested Genomic DNA

1. Based on the optimal conditions established above, carry out a large-scale digestion of 50 μg of high-molecular-weight genomic DNA using half units of *Sau3A I*/μg DNA that produced the maximum intensity of fluorescence in the DNA size range of 20 to 30 Kb. The DNA concentration, time and temperature should be exactly the same as those used for the small-scale digestion. For instance, if tube 7 (0.01 unit of *Sau3A I* per microgram DNA) in the small-scale digestion of the DNA shows a maximum intensity of fluorescence in the size range of 20 to 30 Kb, the large-scale digestion of the same DNA can be carried out as follows.
   Genomic DNA (1μg/μl), 50 μl
   10X *Sau3A I* buffer, 250 μl
   dd.H_2O, 1.95 ml
   Diluted *Sau3A I,* 250 μl
   (0.005 u /μg) prepared as in Step 1 (b), tube 9
   Final volume of 2.5 ml
2. Incubate the reactions at 37°C in a water bath for 30 min. Stop the reaction by adding 20 μl of 0.2*M* EDTA buffer (pH 8.0) and place the tube on ice until use.
3. Add 2 to 2.5 volumes of chilled 100% ethanol to precipitate the DNA at –70°C for 30 min.
4. Centrifuge at 12,000 × *g* for 10 min at room temperature. Carefully decant the supernatant and briefly rinse the DNA pellet with 5 ml of 70% ethanol. Dry the pellet under vacuum for 8 min. Dissolve the DNA in 50 μl of TE buffer. Take 5 μl of the sample to measure the concentration at A_{260} nm and store the DNA sample at –20°C prior to use.

Reagents Needed

10X Sau3A I Buffer

0.1 *M* Tris-HCl, pH 7.5
1 *M* NaCl
70 m*M* $MgCl_2$

5X Loading Buffer

38% (w/v) sucrose
0.1% Bromphenol blue
67 m*M* EDTA

5X TBE Buffer

Tris base54 g
Boric acid27.5 g
20 ml of 0.5 *M* EDTA, pH 8.0

TE Buffer

10 m*M* Tris-HCl, pH 8.0
1 m*M* EDTA, pH 8.0
Ethanol (100%, 70%)

Design and Synthesis of Specific Forward and Reverse Primers

Amplification and Isolation of Exon and Intron Sequences

A pair of forward and reverse primers can be designed based on the 5′-UTR or 3′-UTR sequences of a known cDNA. The forward primer can be designed in the 5′-UTR region in the 5′→3′ direction with 20 to 30 bases, which should be complementary to the first or (–)strand of cDNA template. The reverse primer is designed in the 3′-UTR region in the 5′→3′ direction with 20 to 30 bases, which should be complementary to the second or (+)strand of cDNA template (Figure 2.6A). Alternatively, forward and reverse primers can be designed based on the very N-terminal and C-terminal amino acid sequences if the cDNA sequence is not known (Figure 2.6B).

Amplification and Isolation of Promoter Sequence

To obtain the promoter sequence of interest, a reverse primer can be designed according to the 5′ end sequence of a known cDNA or based on the N-terminal amino acid sequence. This primer will be employed to anneal with a DNA strand of the denatured genomic DNA fragments. The primer facilitates the synthesis of a new strand, which is complementary to the template and includes the promoter

A.

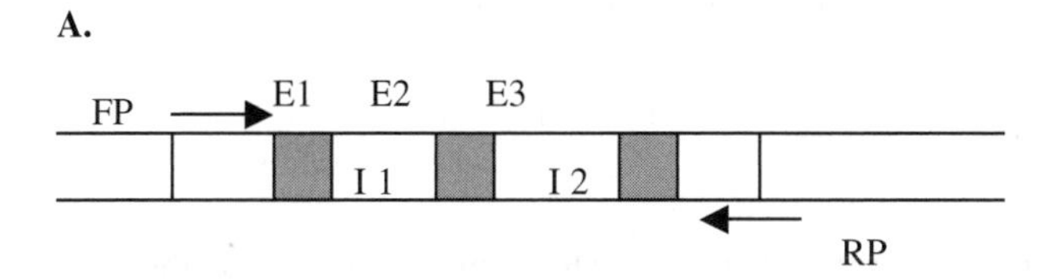

B.

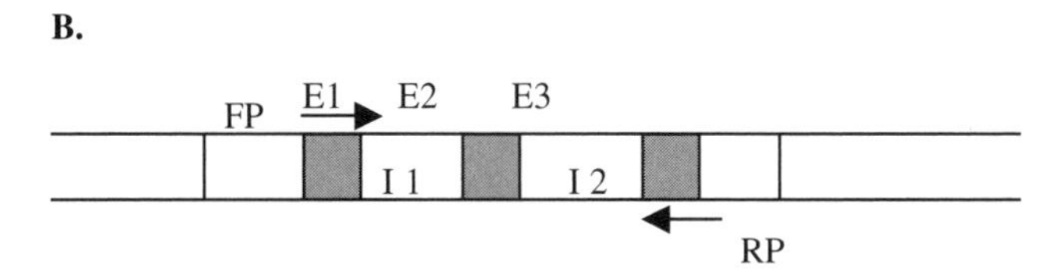

FIGURE 2.6 Design of primers for amplification and isolation of exon and intron sequences. (A) Forward primer (FP) and reverse primer (RP) are designed from the 5′-UTR and 3′-UTR sequences of a known cDNA. (B) FP and RP are designed based on the very N-terminal or C-terminal amino acid sequence. E: exon; I: intron.

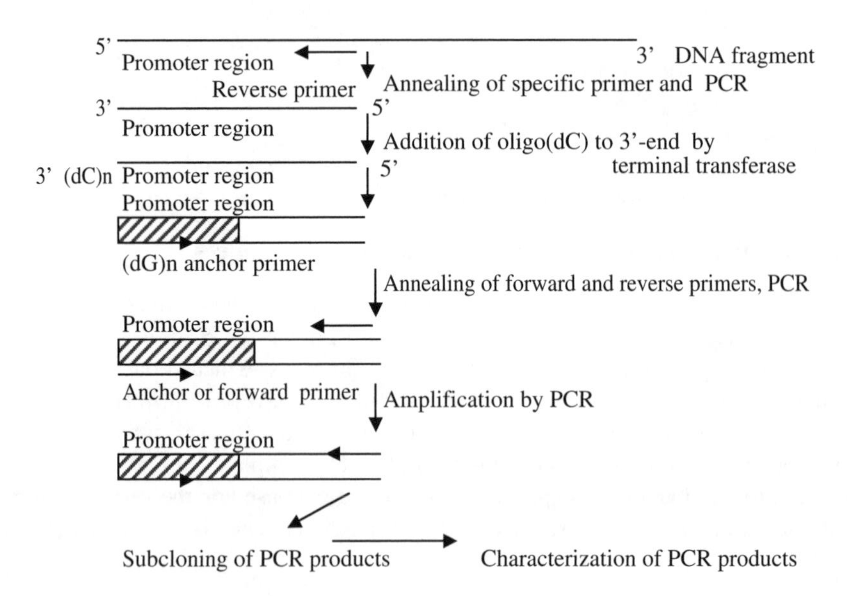

FIGURE 2.7 Diagram of amplification and isolation of promoter region of interest by PCR.

sequence. Following addition of oligo(dC) to the 3′ end of the new strand, an oligo(dG) anchor primer or forward primer in the cloning site of the vector will facilitate the synthesis of the DNA strand complementary to the new strand. As a result, major PCR products including the promoter sequence will be obtained (Figure 2.7).

Amplification of Specific DNA Fragments by PCR

1. Prepare a PCR cocktail in a 0.5-ml microcentrifuge tube.
   Forward primer, 5 to 8 pmol (30 to 40 ng) depending on the size of primer (20- to 26-mer, 10 to 30 ng/μl)
   Reverse primer, 5 to 8 pmol (30 to 40 ng) depending on the size of primer (20- to 26-mer, 10 to 30 ng/μl)
   Digested genomic DNA, 2 to 3 μg
   10X amplification buffer, 4 μl
   dNTPs (2.5 m*M* each), 4 μl
   Taq or *Tth* DNA polymerase, 5 to 10 units
   Add dd.H_2O to a final volume of 40 μl.
   Note: the DNA polymerase should be of high fidelity and long expand.
2. Carefully overlay the mixture with 30 μl of mineral oil to prevent evaporation of the samples during the PCR amplification. Place the tubes in a thermal cycler and perform PCR.

	Cycling (35 cycles)			
Predenaturation	**Denaturation**	**Annealing**	**Extension**	**Last**
95°C, 3 min	95°C, 1 min	60°C, 1 min	72°C, 2 min	4°C

Purification of PCR Products by Agarose Gels

Carry out electrophoresis and elution of PCR products as described earlier in this chapter. Due to different sizes of genomic DNA fragments, multiple PCR bands may be shown in agarose gels. These bands should be individually isolated for characterization.

Subcloning of cDNA or Gene of Interest

Detailed protocols are described in Chapter 4.

CHARACTERIZATION OF PCR PRODUCTS

To verify whether DNA of interest is obtained, PCR products should be characterized by Southern blotting using a specific probe (Chapter 7) and by DNA sequencing (Chapter 5). The sequence can be compared with known or expected DNA sequences in GenBank databases (Chapter 6).

REFERENCES

1. Wu, W., PCR techniques and applications, in *Handbook of Molecular and Cellular Methods in Biology and Medicine,* Kaufman, P.B., Wu, W., Kim, D., and Cseke, L., pp. 243–262, CRC Press, Boca Raton, FL, 1995.
2. Ausubel, F.M., Brent, R., Kingston, R.E., Moore, D.D., Seidman, J.G., Smith, J.A., and Struhl, K., *Current Protocols in Molecular Biology,* Greene Publishing Associates and Wiley-Interscience, John Wiley & Sons, New York, 1995.
3. Erlich, H.A., *PCR Technology: Principles and Applications for DNA Amplification,* Stockton Press, New York, 1989.
4. Wu, L.-L., Song, I., Karuppiah, N., and Kaufman, P.B., Kinetic induction of oat shoot pulvinus invertase mRNA by gravistimulation and partial cDNA cloning by the polymerase chain reaction, *Plant Mol. Biol.,* 21, 1175, 1993.
5. Innis, M.A., Gelfand, D.H., Sninsky, J.J., and White, T.J., *PCR Protocols: A Guide to Methods and Applications,* Academic Press, New York, 1989.
6. Wang, A.M., Doyle, M.V., and Mark, D.F., Quantitation of mRNA by the polymerase chain reaction, *Proc. Natl. Acad. Sci., USA,* 86, 9717, 1989.
7. Arnold, C. and Hodgson, I.J., Vectorette PCR: a novel approach to genomic walking, *PCR Methods Appl.*, 1, 39–42, 1991.
8. Barany, F., Genetic disease detection and DNA amplification using cloned thermostable ligase, *Proc. Natl. Acad. Sci., USA*, 88, 189, 1991.
9. Lu, W., Han, D.-S., Yuan, J., and Andrieu, J.-M., Multi-target PCR analysis by capillary electrophoresis and laser-induced fluorescence, *Nature,* 368, 269, 1994.
10. Bloomquist, B.T., Johnson, R.C., and Mains, R.E., Rapid isolation of flanking genomic DNA using biotin-RAGE a variation of single-sided polymerase chain reaction, *DNA Cell Biol.*, 10, 791, 1992.
11. Edwards, J.B.D.M., Delort, J., and Mallet, J., Oligodeoxyribonucleotide ligation to single-stranded cDNAs: a new tool for cloning 5′ ends of mRNAs and for constructing cDNA libraries by *in vitro* amplification, *Nucleic Acids Res.*, 19, 5227–5232, 1991.

3 Construction and Screening of Subtracted and Complete Expression cDNA Libraries

CONTENTS

INTRODUCTION

cDNA cloning plays a major role in current molecular biology.[1–7] Construction of a cDNA library is a highly sophisticated technology that involves a series of enzymatic reactions. The quality and integrity of a cDNA library greatly influence the success or failure in isolation of the cDNAs of interest.[2] General principles begin with an mRNA transcribed into the first-strand DNA, called a complementary DNA or cDNA, which is based on nucleotide bases complementary to the mRNA template. This step is catalyzed by AMV reverse transcriptase using oligo(dT) primers. The second-strand DNA is copied from the first-strand cDNA using DNA polymerase I, thus producing a double-stranded cDNA molecule. Subsequently, the double-stranded cDNA is ligated to an adapter and then to an appropriate vector via T4 DNA ligase. The recombinant vector–cDNA molecules are then packaged *in vitro* and cloned in a specific host, generating a cDNA library. Specific cDNA clones can be "fished" out by screening the library with a specific probe.[1,4–10]

Traditionally, cDNA cloning takes advantage of the 3′ hairpins produced by AMV reverse transcriptase during the synthesis of the first-strand cDNAs. The hairpins are utilized to prime the second-strand cDNA that is catalyzed by Klenow

DNA polymerase and reverse transcriptase. Hence, *S1* nuclease is needed to cleave the hairpin loops. However, the digestion is difficult and causes a low cloning efficiency and the loss of significant sequence information corresponding to the 5' end of the mRNA.[2] To solve this problem, 4 m*M* sodium pyrophosphate is utilized to suppress the formation of hairpins during the synthesis of the first-stranded cDNA. The second-strand cDNA is then produced by RNase H to create nicks and gaps in the hybridized mRNA template, generating 3′ OH priming sites for DNA synthesis using DNA polymerase I. Following treatment with T4 DNA polymerase to remove any remaining 3′ protruding ends, the blunt-ended, double-stranded cDNAs are ready for ligation with adaptors or linkers. Therefore, *SI* digestion is eliminated and the cloning efficiency is much higher than that obtained with the classic method. Besides, the sequence information is optimally reserved.

The general strategies for cDNA cloning can be grouped into two classes. One is random cloning; the other is orientation-specific cloning.[1] The random or classical cDNA cloning uses oligo(dT) as a primer and λgt10 as cloning vectors (Figure 3.1A). The cDNA is cloned at the unique *EcoR* I site. Because of the single *EcoR* I site for cloning, the cDNAs are cloned in a random manner in sense and in antisense orientations. The constructed cDNA library can be screened using a DNA or RNA probe but not antibodies. In random cDNA cloning, if the expression vector λgt11 (Figure 3.1B) is used for a cDNA library, the library can be screened using a specific antibody as well as a DNA or RNA probe. The disadvantage of using antibodies is that the possibility of obtaining the positive clones is at least 50% less than it actually is. This is because approximately 50% of cloned cDNAs are expressed as antisense RNA that can interfere with sense RNA by inhibiting the translation of the sense RNA to proteins. This shortcoming is eliminated in the orientation-specific or directional cDNA cloning using a primer–adapter to prime the synthesis of the first-strand cDNA and the use of λgt11 as cloning vectors (see Construction and Screening of a Complete Expression cDNA library in Part B).

The primer–adapter consists of oligo(dT) adjacent to a unique restriction site (*Xba* I or *Not* I). The double-stranded cDNAs with *EcoR* I adaptors at the ends are digested with *Xba* I or *Not* I. As a result, the digested cDNAs contain two incompatible termini: one *EcoR* I and one *Xba* I or *Not* I. In this way, self-ligation of adapter–cDNA molecules is greatly eliminated. These molecules are readily ligated with vectors containing the same restriction enzyme termini (see Figure 3.6 for details). Compared with the random cloning method, directional cDNA cloning is much more powerful and valuable. A factor of two is the likelihood of expressing the cDNA inserts as the correct polypeptides using λgt11 *Sfi-Not* I. In theory, no antisense RNA interference will occur in the directional cDNA cloning approach.

More advanced lambda vectors are available, in which promoters of T7, T3 or SP6 RNA polymerase are included adjacent to the multiple cloning sites (Figure 3.1C). All cDNAs can be cloned in the same direction downstream from T7, T3 or SP6 promoter. Sense or antisense RNA probes can be produced *in vitro,* and can be utilized as RNA probes for screening a cDNA or genomic DNA library or for northern blot analysis.

Based on the author's experience, three steps are critical for the success or failure in constructing a cDNA library. The first is the purity and integrity of the mRNAs

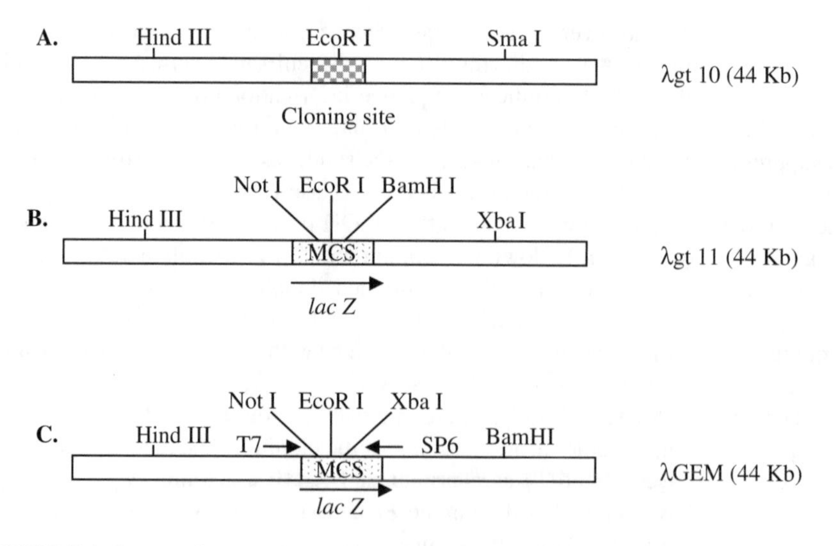

FIGURE 3.1 Structural maps of vectors used for cDNA cloning.

used for the synthesis of the first-strand cDNAs. Any degradation or absence of specific mRNAs will result in partial-length cDNAs or complete loss of the specific cDNAs, especially for some rare mRNAs. The second important step is to obtain full-length cDNAs. If this procedure is not performed well, even with a very good mRNA source, the cDNA library is not so good. In that case, only partial-length cDNAs or no positive clones at all may be "fished" out. Once double-strand cDNAs are obtained, they are much more stable compared with mRNAs. A third essential step in cDNA cloning is the ligation of cDNAs with adaptors to vectors. If the ligation fails or is of low efficiency, *in vitro* packaging of recombinant λDNAs cannot be carried out effectively. As a result, the plaque-forming units' (pfu) number or efficiency will be very low. In that case, a poor cDNA library is not recommended for screening specific cDNA clones. In order to construct an excellent cDNA library, the following precautions should be taken. Elimination of RNase contamination must be carried out whenever possible. Procedures for creating an RNase-free laboratory environment for mRNA and cDNA synthesis include:

1. Disposable plastic test tubes, micropipet tips and microcentrifuge tubes should be sterilized.
2. Gloves should be worn at all times and changed often to avoid finger-derived RNase contamination.
3. All glassware and electrophoresis apparatus used for cDNA cloning should be separated from other labware.
4. Glassware should be treated with 0.1% diethyl pyrocarbonate (DEPC) solution, autoclaved to remove the DEPC and baked overnight at 250°C before being used.

5. Solutions to be used with mRNA and cDNA synthesis should be treated with 0.1% DEPC (v/v) to inhibit RNase by acylation.
6. It is recommended to add DEPC to solutions with vigorous stirring for 20 min. The solutions should be autoclaved or heated at 70°C for 1 h to remove the DEPC. It should be noted that because DEPC reacts with amines and sulfhydryl groups, reagents such as Tris and DTT cannot be treated directly with DEPC. These solutions can be made using DEPC-treated and sterile water. If solutions such as DTT cannot be autoclaved, the solution should be sterile-filtered.

Construction of a cDNA library is a highly sophisticated procedure. One should have a strong molecular biology background and intensive laboratory experience in molecular biology when working on this type of library. In spite of the fact that some commercial kits or cDNA libraries are available, often, many laboratories need to construct a specific library with the mRNAs expressed in a specific cell or tissue type in the spatial and temporal manner. The present chapter provides detailed and successful protocols for constructing and screening a cDNA library. Two major strategies are described. One is the subtracted cDNA library in which cDNAs derived from mRNAs are expressed in a specific cell or tissue type but not in another type. The cell or tissue type-specific cDNA clones are greatly enriched in the library, which allows specific cDNAs copied from rare mRNAs to be readily isolated. The other is the complete expression cDNA library that includes all cDNA clones from all mRNAs in a specific cell or tissue type. The step-by-step protocols described in this chapter allow experienced workers as well as beginners to construct their own cDNA libraries.

PART A. CONSTRUCTION AND SCREENING OF A SUBTRACTED cDNA LIBRARY

Construction of a subtracted cDNA library can allow identification of specific cDNA clones corresponding to a specific class of mRNAs expressed in one cell or tissue type but not in another type. For instance, treatment of cultured cells or organisms with an important chemical, drug, alcohol or hormone may induce or repress the expression of certain genes. In some cases, some novel genes encoding new mRNAs and proteins can be identified corresponding to a specific treatment, which may provide fundamental information about the action or mechanisms of the treatment. Once a novel gene is isolated, it allows opening a broad research area concerning the treatment with promising results. Additionally, a subtracted cDNA library consists of enriched cDNAs that provide a powerful tool to isolate some rare cDNAs transcribed from rare mRNAs, which may be difficult to be identified via screening of a complete cDNA library.[1,2]

The general procedure is outlined in Figure 3.2. The first step is to isolate total RNAs or mRNAs and to make the first-strand cDNAs representing all of the mRNAs expressed in each of two cell or tissue types of interest. The single-stranded cDNAs

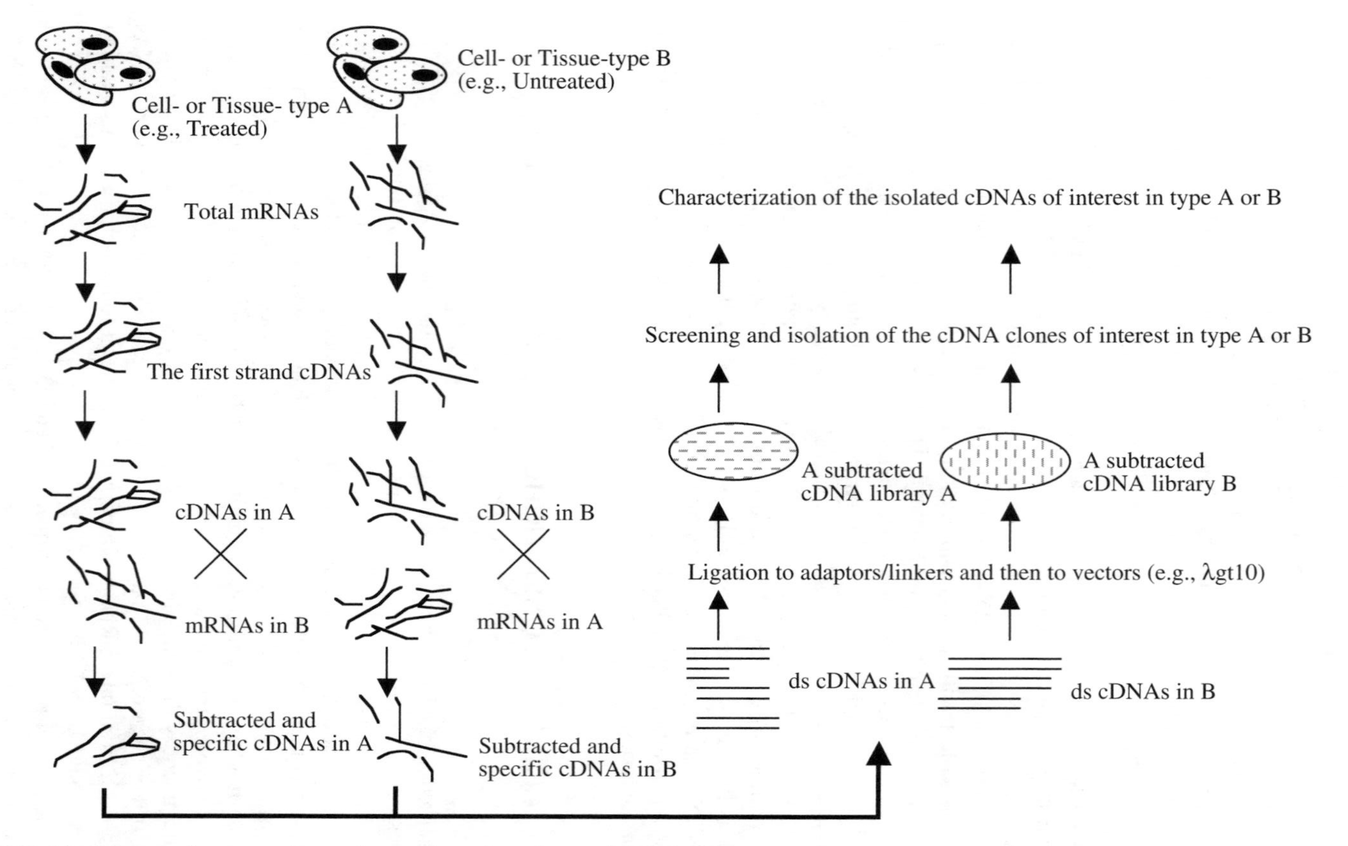

FIGURE 3.2 Scheme of construction and screening of a subtracted cDNA library.

in one cell or tissue type are subjected to hybridization with total mRNAs from another type. For example, in cell or tissue type A, any cDNAs that represent sequences expressed in type B will form DNA/mRNA hybrids with the corresponding mRNAs in cell or tissue type B, or vice versa. Therefore, the unhybridized cDNAs are specific in type A, but not in type B, or vice versa. Single-stranded cDNAs and cDNA/mRNA hybrids can be separated by chromatography on hydroxyapatite columns. The unhybridized cDNAs are then used to synthesize double-stranded (ds) cDNAs, which are then ligated to appropriate adaptors and vectors and utilized to construct a subtracted cDNA library in which the sequences specific to one cell or tissue type are greatly enriched. Compared with a complete cDNA library, a subtracted cDNA library contains an enriched set of cDNA clones, which simplifies the screening procedures, using much smaller numbers of pfu in the primary screening. As a result, the purification process of the cDNA of interest can be speeded up a great deal.[1]

ISOLATION OF TOTAL RNAs FROM CELL OR TISSUE TYPES A AND B OF INTEREST

The quality and integrity of RNAs play a crucial role in the successful construction of a cDNA library. Compared with DNA, RNAs are very mobile molecules due to their degradation by RNases. In order to obtain high yield and high quality of RNAs, four procedures should be handled properly and effectively: first, optimal disruption of cells or tissues; second, effective denaturation of nucleoprotein complexes; third, maximum inhibition of exogenous RNases and endogenous RNases released from cells upon cell disruption; and finally, effective purification of RNA from DNA and proteins. The most difficult thing, however, is to inactivate RNase activity. In our experience, the following procedures are very effective in inactivation of RNases.

To inactivate exogenous RNase: Two common sources of RNase contamination are the user's hands and bacteria and fungi in airborne dust particles. Therefore, (1) gloves should be worn and changed frequently; (2) whenever possible, disposable plasticware should be autoclaved; (3) nondisposable glassware and plasticware should be treated with 0.1% Diethyl pyrocarbonate (DEPC) in dd.H_2O and be autoclaved prior to use; and (4) glassware should be baked at 250°C overnight. Disposable and sterile polypropylene centrifuge tubes are strongly recommended for use in RNA isolation. Glass Corex tubes must be thoroughly cleaned, DEPC treated and autoclaved, followed by baking at 250°C overnight before use.

To inactivate endogenous RNase: 4M guanidine thiocyanate and β-mercaptoethanol are strongly recommended in the extraction buffer. These chemicals are strong inhibitors of RNases. Whenever possible, solutions should be treated with 0.05% DEPC at 37°C for 2 h or overnight at room temperature and then autoclaved for 20 min to remove any trace of DEPC. Tris buffers that cannot be treated with DEPC should be made from DEPC-treated dd.H_2O.

Protocol A. Rapid Isolation of total RNA by Acid Guanidinium Thiocyanate-Phenol-Chloroform Method

1. **From tissue**: Harvest tissue and immediately drop it into liquid nitrogen. Store at –80°C until use. Grind 1 g tissue in liquid nitrogen in a clean blender. Keep adding liquid nitrogen while grinding the tissues until a fine powder is obtained. Warm the powder at room temperature for a few minutes and transfer the powder with a sterile spatula into a clean 14- or 50-ml polypropylene tube. Add 10 ml of solution B, mix and keep the tube on ice. Proceed to step 2. Alternatively, homogenize the tissue in 10 ml of solution B in a sterile polypropylene tube on ice using a glass-Teflon homogenizer or an equivalent polytron at top speed for three 30-s bursts. Transfer the homogenate to a fresh polypropylene tube and proceed to step 2.

   From cultured cells grown in suspension culture or in a monolayer: Collect 1 to 2 × 10^8 cells in a sterile 50-ml polypropylene tube by centrifugation at 400 × *g* for 5 min at 4°C. Wash the cell pellet with 25 ml of ice cold, sterile 1X PBS or serum-free medium (e.g., DMEM) and centrifuge at 400 × *g* for 5 min at 4°C. Repeat washing and centrifugation once more to remove all traces of serum containing RNases. Aspirate the supernatant and resuspend the cells in 10 ml solution B. Keep the tube on ice and homogenize the cell suspension using a microtip on a polytron for two 0.5- to 1-min bursts at the top speed. Carefully transfer homogenate to a fresh tube. Proceed to step 2. Alternatively, wash and resuspend the cells as described above. Keep the suspension of cells on ice. In order to shear the DNA and lower the viscosity, sonicate the suspension of cells at the maximum power for a clean microtip (60%) using two 0.5- to 1-min bursts. A good sonicated solution should be thin enough to drop freely from the end of a Pasteur pipette. Proceed to step 2.

Note: As long as the sample is in the solution B, RNase released from the tissue may be inhibited by guanidinium thiocyanate.

Caution: Guanidine thiocyanate is a potent chaotropic agent and irritant.

2. Add the following components to the sample in the order shown:
   1 ml of 2 *M* sodium acetate buffer (pH 4.0)
   10 ml of water-saturated phenol
   2 ml of chloroform:isoamyl alcohol
   Cap the tube and mix by inversion after each addition. Vigorously shake the tube for 20 s to shear genomic DNA.
   Caution: Phenol is poisonous and can cause severe burns. Proper laboratory clothing including gloves and goggles should be worn when handling these reagents. If phenol contacts skin, wash the area immediately with large volumes of water, but DO NOT RINSE WITH ETHANOL!
3. Centrifuge at 12,000 × *g* for 20 min at 4°C using Corex tube or at 7000 to 8000 rpm for 30 to 40 min at 4°C using a 14-ml polypropylene tube

in the swing-out rotor. Carefully transfer the top, aqueous phase to a fresh tube. Decant the phenol phase into a special container for disposal.

4. Add 10 ml of isopropanol, mix, and precipitate RNAs at –20°C for 2 h.
5. Centrifuge at 12,000 × *g* for 15 min at 4°C in a Corex tube or at 7000 to 8000 rpm for 30 min at 4°C using a 14-ml polypropylene tube in the swing-out rotor. Resuspend the RNA pellet in 3 ml of solution B and add 3 ml of isopropanol. Mix and then place at –20°C for 2 h or overnight.
6. Centrifuge at 12,000 × *g* for 15 min at 4°C or at 7000 to 8000 rpm for 30 to 40 min at 4°C using a 14-ml polypropylene tube in the swing-out rotor. Briefly rinse the total RNA pellet with 4 ml of 75% ethanol. Dry the RNA pellet under vacuum for 15 min.
7. Dissolve the total RNAs in 100 to 150 μl of 0.5% SDS solution. Take 5 μl to measure the concentration with the spectrophotometer at 260 and 280 nm.

Tips: The ratio of A_{260}/A_{280} should be 1.8:2.0 to be judged a good RNA preparation. The RNA isolated may be checked by 1% agarose–formaldehyde denaturing gel electrophoresis. Two strong rRNA bands should be visible with some smear through the well (Figure 3.3). At this point, the RNA can be used for northern blot analysis, dot blot hybridization or mRNA purification. The RNA sample may be stored at –20°C until use.

Protocol B. Rapid Isolation of Total RNAs Using Trizol Reagent™ (Gibco BRL Life Technologies)

1. **From tissue:** Harvest tissue and immediately drop it into liquid nitrogen. Store at –80°C until use. Homogenize 0.3 g tissue in 1 ml of Trizol reagent solution using a polytron at the top speed for three 30-s bursts. Allow the cells to be lysed/solublized for 5 min at room temperature. Proceed to step 2.

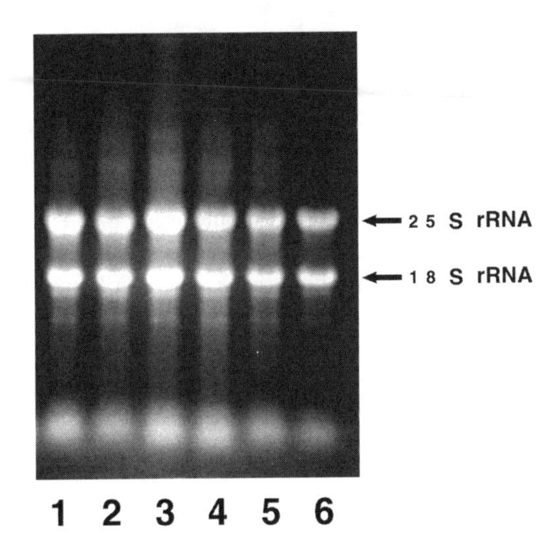

FIGURE 3.3 Electrophoresis of total RNA.

From cultured cells grown in suspension culture or in a monolayer: Collect 1×10^7 cells in a sterile 50-ml polypropylene tube by centrifugation at $400 \times g$ for 5 min at 4°C. Wash the cell pellet with 25 ml of ice cold, sterile 1X PBS or serum-free medium (e.g., DMEM) and centrifuge at $400 \times g$ for 5 min at 4°C. Repeat washing and centrifugation once more to remove all traces of serum that contain RNases. Aspirate the supernatant and lyse the cells in 1 ml of Trizol reagent solution for 3 to 5 min at room temperature. Proceed to step 2.

Note: Isolation of RNAs can be scaled up using 0.3 g tissue or 1×10^7 cells per 1 ml of Trizol reagent.

Caution: Trizol reagent contains toxic phenol.

2. Add 0.2 ml of chloroform per 1 ml of the Trizol reagent used. Cap the tube and vigorously shake it for 20 s to shear genomic DNA. Incubate at room temperature for 3 min for phase separation.
3. Centrifuge at $12{,}000 \times g$ for 20 min at 4°C in an Eppendorf tube or at 7000 to 8000 rpm for 30 to 40 min at 4°C using a 14-ml polypropylene tube in a swing-out rotor. Carefully transfer the top, aqueous phase to a fresh tube. Decant the bottom, phenol phase into a special container for disposal.
4. Add 1 ml of isopropanol, mix, and precipitate RNAs at –20°C for 2 h.
5. Centrifuge at $12{,}000 \times g$ for 15 min at 4°C. Briefly rinse the total RNA pellet with 1 ml of 75% ethanol. Partially dry the RNA pellet at room temperature for 15 min.
6. Dissolve the total RNAs in 20 to 50 µl of 0.5% SDS solution. Take 2 µl to measure the concentration with the spectrophotometer at 260 and 280 nm.

Protocol C. Measurement of RNAs

1. Turn on an ultraviolet (UV) spectrophotometer and set the wavelength at 260 and 280 nm according to the manufacturer's instructions. Some spectrophotometers (e.g., UV 160U, Shimadzu) have a computer unit that can automatically and simultaneously measure RNA at A_{260} and A_{280} nm, calculate and display RNA concentration together with the ratio of A_{260}/A_{280} reading numbers on the computer screen. If such a spectrophotometer is not available, RNA concentration can be measured by other relatively simple UV-spectrophotometers such as Hitachi model 100-10 UV-visible spectrophotometer. The disadvantages, however, are that RNAs are first measured and recorded at 260 nm and then the machine switches to 280 nm and records the reading, and that RNA concentration and the ratio of A_{260}/A_{280} must be calculated manually.
2. Set up a reference using blank solution, depending on the solution in which RNAs are dissolved. Add an appropriate volume (0.1 or 1 ml) of the solution (TE buffer, pH 8.0 or dd.H_2O) to a clean cuvette. Insert the cuvette into the cuvette holder in the sample compartment with the optical (clear) sides of the cuvette facing the light path. Close the sample com-

partment cover and adjust the number to 0.000 manually or by pressing the auto zero button, depending on the specific spectrophotometer.

Note: Gloves should be worn when handling RNA measurement. Cuvettes should be rinsed with 95% ethanol followed by dd.H_2O, and wiped dry using Kimwipe paper prior to reuse. Unclean cuvettes may contain contaminated materials that may affect readings between samples.

3. Check for mismatching of cuvettes by filling cuvettes with blank solution and recording the absorbency (positive or negative number). Make appropriate additive (for negative number) or subtractive (for positive number) corrections to future RNA sample readings. In most cases, however, this step is optional.
4. Dilute the RNA sample to be measured in blank solution appropriately in a total volume of 0.1 or 1 ml, depending on the size of the cuvette. For example, if 2, 5, 10, 15, 20, or 25 μl of the RNA sample is taken, the blank solution should be 998, 995, 990, 985, 980 or 975 μl, respectively, in a clean cuvette. Tightly cover the cuvette with a piece of parafilm and mix well by inverting the cuvette two or three times. Alternatively, the sample can be mixed well using a pipette tip.
5. Insert the sample cuvette in place and read its absorbency against the blank cuvette according to the instructions of spectrophotometer. If only one cell is available, remove the reference cuvette after adjusting the number at 0.000 and insert the sample cuvette in its place.
6. Following measurement of RNA samples at A_{260} and A_{280} nm respectively, calculate the ratio of A_{260}/A_{280}, reading numbers and concentration for each RNA sample. Some spectrophotometers (e.g., UV 160U, Shimadzu) can automatically calculate, display and print out the ratio and concentration for each RNA sample. A pure RNA should have a ratio of 1.85:2.0 of A_{260}/A_{280} readings. The concentration of an RNA sample can be manually calculated as follows:

$$\text{Total RNA } (\mu g/\mu l) = \text{OD at } A_{260} \times 40\ \mu g/ml \times \text{dilution factor}$$

For example, if 20 μl of RNA is diluted to 1000 μl for measuring and its A_{260} reading number is 0.5000, then total RNA (μg/μl) = A_{260} reading number × 40 μg/ml × dilution factor = 0.5000 × 40 μg/ml × 1000/20 = 1000 μg/ml = 1 μg/μl.

Reagents Needed

Solution A

4 *M* Guanidinium thiocyanate
25 m*M* Sodium citrate, pH 7.0
0.5% Sarcosyl
Sterile-filter and store in a light-tight bottle at room temperature up to 3 months.

Solution B

100 ml Solution A
0.72 ml β-Mercaptoethanol
Store at room temperature up to 1 month.

Sodium Acetate Buffer (pH 4.0)

2 *M* Sodium acetate
Adjust pH to 4.0 with 2 *N* acetic acid. Autoclave.

Sodium Acetate Solution

3 *M* Sodium acetate
Autoclave.

Water-Saturated Phenol

Thaw crystals of phenol at 65°C in a water bath and mix one part phenol and one part sterile dd.H_2O. Mix well and allow the two phases to separate. Store at 4°C.

Caution: Phenol is a dangerous reagent; gloves should be worn and work should be performed while wearing a fume hood.

Chloroform:Isoamyl Alcohol (49:1)

98 ml Chloroform
2 ml Isoamyl alcohol
Mix and store at 4°C.

Ethanol

70%: 70 ml of 100% ethanol and 30 ml dd.H_2O
75%: 75 ml of 100% ethanol and 25 ml dd.H_2O
100%: store at –20°C.

20% SDS

20 g SDS (sodium dodecyl sulfate) in 100 ml dd.H_2O

PURIFICATION OF mRNAs FROM TOTAL RNAs

Protocol A. Purification of poly(A)+RNAs Using Oligo(dT)-Cellulose Column

1. Add 4 ml of binding buffer in a clean tube containing 0.5 g of oligo(dT)-cellulose type 7 (Collaborative Research) and mix well. Transfer the slurry into a 10-ml poly-prep chromatography column (Bio-Rad) or its equivalent. Vertically fix the column in a holder with a clamp and place the column at 4°C for 2 h to equilibrate the cellulose resin prior to use.
2. Transfer the column to room temperature and wash it with 2 ml of 0.1 *N* NaOH solution through the top of the resin. Drain away the fluid by

gravity and repeat washing the column five times with 2 ml of 0.1 *N* NaOH solution per washing.

Note: The color of the resin is changed from white to yellow.

3. Neutralize the column by adding 4 ml of binding buffer to the top of the resin followed by draining away the binding buffer. Repeat this step five times and store at 4°C until use.

Note: The color of the resin returns to white.

4. Add 1 volume of loading buffer to the total RNA sample and heat the mixture for 10 min to 65°C to denature the secondary structure of RNA. Cool the sample to room temperature.

Note: The concentration of SDS in the loading buffer should never be >0.5%; otherwise, the SDS will precipitate and block the column. In our experience, 0.2% SDS in the binding buffer works well.

5. Carefully load the total RNA sample to the top of the column capped at the bottom and allow the sample to run through the column by gravity. Gently loosen the resin in the column using a clean needle and let the column stand for 5 min. Drain away the fluid from the bottom of the column and collect the fluid containing some unbound mRNA. Reload the fluid onto the column to allow the unbound mRNA to bind to the oligo(dT)-cellulose column. Collect the eluate and repeat this procedure twice.
6. Wash the column with 2 ml of washing buffer and drain away the buffer. Repeat washing three to four times.
7. Cap the bottom of the column and add 1 ml of elution buffer to the column to elute the bound poly(A)+RNA. Gently loosen the resin using a clean needle and allow elution to proceed for 4 min at room temperature. Remove the bottom cap and collect the eluate in a sterile tube. Elute the column twice with 1 ml of elution buffer and pool the eluate.
8. To the pooled eluates, add 0.15 volume of 3 *M* sodium acetate and 2.5 volumes of 100% chilled ethanol to precipitate the mRNA. Mix and place at –80°C for 1 h or at –20°C for 2 h or overnight.
9. Centrifuge at 12,000 × *g* for 15 min at 4°C. Aspirate the supernatant and briefly wash the mRNA pellet with 2 ml of 70% ethanol. Dry the pellet under vacuum for 15 min and dissolve the mRNA in 20 to 40 μl of TE buffer. Store the sample at –80°C until use.
10. Take 2 to 4 μl of the sample to measure the concentration of the mRNA as described previously.

Notes: (1) At this point, the sample should have a ratio of A_{260}/A_{280} *of 1.9:2.0. If the ratio is <1.75, the sample is not recommended for synthesis of cDNAs. (2) The quality of the purified mRNA may be checked by 1% agarose–formadehade denaturing gel electrophoresis. A broad range of smear should be visible from the top to the bottom of the well with or without two weak rRNA bands (Figure 3.3B). (3) The yield of poly(A)+RNA may be expected to be approximately 6% of total RNA. The calculation is as follows:*

Poly(A) + RNA (μg/μl) = A_{260} reading number × 42 μg/ml × dilution factor

For example, if 20 μl of mRNA is diluted to 1000 μl for measuring and its A_{260} reading number is 0.4000, then poly(A) + RNA (μg/μl) = A_{260} reading number × 42 μg/ml × dilution factor = 0.4000 × 42 μg/ml × 1000/20 = 840 μg/ml = 0.84 μg/μl.

Protocol B. Minipurification of mRNAs using Oligo(dT)-Cellulose Resin

1. Add 1 ml of binding buffer to 0.3 g of oligo(dT)–cellulose type 7 (Collaborative Research) in an Eppendorf tube and mix well. Place the tube at 4°C for 30 min.
2. Wash the resin with 1 ml of 0.1 *N* NaOH solution and centrifuge at 1500 × *g* for 2 min. Carefully aspirate the supernatant and repeat this step eight times.

Note: Centrifugation greater than 1500 × g may cause damage to the oligo(dT) beads.

3. Neutralize the resin with 1 ml of binding buffer and centrifuge at 1500 × *g* for 2 min. Carefully decant the supernatant and repeat this step eight times. Resuspend the resin in 1 ml of binding buffer.
4. Add 1 volume of loading buffer to the total RNA sample (maximum 100 μg, about 2 μg/μl). Heat the mixture for 10 min at 65°C and cool it to room temperature.
5. Add the total RNA sample to the 1 ml slurry of oligo(dT)–cellulose prepared at step 3. Gently mix and place at room temperature for 15 min to allow the RNA to bind to the resin.
6. Centrifuge at 1500 × *g* for 5 min and carefully aspirate the supernatant.
7. Add 0.5 ml of wash buffer to the resin and centrifuge at 1500 × *g* for 2 min. Carefully aspirate the supernatant. Repeat this step three times.
8. Add 0.2 ml of elution buffer to the resin and gently shake for 5 min. Centrifuge at 1500 × *g* for 5 min and transfer the supernatant containing mRNAs to a fresh tube. Repeat this step two times.
9. Pool the supernatant and add 0.1 volume 3 *M* sodium acetate and 2.5 volumes of 100% chilled ethanol to precipitate the mRNA as described above.

Reagents Needed

Ethanol

70%: 70 ml of 100% ethanol and 30 ml dd.H_2O
75%: 75 ml of 100% ethanol and 25 ml dd.H_2O
100%: store at –20°C.

*0.1*N *NaOH*

4 g NaCl in 100 ml dd.H_2O

RNA Buffer

10 m*M* Tris-HCl, pH 7.5
1 m*M* EDTA, pH 8.0
0.2% SDS
Sterile-filter.

Binding Buffer

10 m*M* Tris-HCl, pH 7.5
0.5 *M* NaCl
1 m*M* EDTA
0.5% SDS
Sterile-filter.

Loading Buffer

20 m*M* Tris-HCl, pH 7.5
1 *M* NaCl
2 m*M* EDTA
0.2% SDS
Sterile-filter.

Wash Buffer

10 m*M* Tris-HCl, pH 7.5
100 m*M* NaCl
1 m*M* EDTA
Sterile-filter.

Elution Buffer

10 m*M* Tris-HCl, pH 7.5
1 m*M* EDTA

TE Buffer

10 m*M* Tris-HCl, pH 8.0
1 m*M* EDTA

SYNTHESIS OF FIRST-STRAND cDNAs FROM CELL OR TISSUE TYPE A OR B

Protocol A. Synthesis of First-Strand cDNAs from mRNAs

It is recommended that two reactions that will be carried out separately all the way to the end be set up in two microcentrifuge tubes. The benefit is that if one reaction is stopped by an accident during the experiments, the other reaction can be used as a back-up. Otherwise, it is necessary to start over again, thus wasting time and money. A standard reaction is total 25 μl using 0.2 to 2 μg mRNA or 5 to 10 μg total RNAs. For each additional 1 μg mRNA, increase the reaction volume by 10 μl. A control reaction is recommended to be carried out whenever possible under the same conditions.

1. Anneal 2 μg of mRNAs as templates with 1 μg of primer of oligo(dT), or with 0.7 μg of primer–adapter of oligo(dT)-*Not I*/*Xba I* in a 1.5-ml sterile, RNase-free microcentrifuge tube. Add nuclease-free dd.H_2O to 15 μl. Heat the tube at 70°C for 5 min to denature the secondary structures of mRNAs and allow it to slowly cool to room temperature to complete the annealing. Briefly spin down the mixture.
2. To the annealed primer–template, add the following in the order listed. To prevent precipitation of sodium pyrophosphate when adding it to the reaction, the buffer should be preheated at 40 to 42°C for 4 min prior to the addition of sodium pyrophosphate and AMV reverse transcriptase. Gently mix well following each addition.
   First-strand 5X buffer, 5 μl
   rRNasin ribonuclease inhibitor, 50 units (25 units/μg mRNA)
   40 m*M* sodium pyrophosphate, 2.5 μl
   AMV reverse transcriptase, 30 units (15 units/μg mRNA)
   Add nuclease-free dd.H_2O to a final 25 μl

Function of components*: The 5X reaction buffer contains components for the synthesis of cDNAs. Ribonuclease inhibitor serves to inhibit RNase activity and to protect the mRNA templates. Sodium pyrophosphate suppresses the formation of hairpins that are commonly generated in traditional cloning methods, thus avoiding the S1 digestion step that is difficult to carry out. AMV reverse transcriptase catalyzes the synthesis of the first-strand cDNAs from mRNA templates according to the rule of complementary base pairing.*

3. Transfer 5 μl of the mixture to a fresh tube, as a tracer reaction, which contains 1 μl of 4 μCi of [α-^{32}P]dCTP (>400 Ci/mmol, less than 1 week old). The synthesis of the first-strand cDNAs will be measured by trichloroacetic acid (TCA) precipitation and alkaline agarose gel electrophoresis using the tracer reaction.

Caution: [α-^{32}P]dCTP is a toxic isotope. Gloves should be worn when carrying out the tracer reaction, TCA assay and gel electrophoresis. Waste materials such as contaminated gloves, pipette tips, solutions and filter papers should be put in special containers for disposal of waste radioactive materials.

4. Incubate both reactions at 42°C for 1 to 1.5 h, and then place the tubes on ice. At this point, the synthesis of the first-strand cDNAs has been completed. To the tracer reaction mixture, add 50 m*M* EDTA solution up to a total volume of 100 μl and store on ice until used for TCA incorporation assays and analysis by alkaline agarose gel electrophoresis.
5. To remove RNAs completely, add RNase A (DNase-free) to a final concentration of 20 μg/ml or 1 μl of RNase (Boehringer Mannheim, 500 μg/ml) per 10 μl of reaction mixture. Incubate at 37°C for 30 min.
6. Extract the cDNAs from the unlabeled reaction mixture with one volume of TE-saturated phenol/chloroform. Mix well and centrifuge at 11,000 × *g* for 5 min at room temperature.

7. Carefully transfer the top, aqueous phase to a fresh tube. Do not take any white materials at the interphase between the two phases. To the supernatant, add 0.5 volume of 7.5 *M* ammonium acetate or 0.15 volume of 3 *M* sodium acetate (pH 5.2), mix well and add 2.5 volumes of chilled (at –20°C) 100% ethanol. Gently mix and precipitate cDNAs at –20°C for 2 h.
8. Centrifuge in a microcentrifuge at 12,000 × *g* for 5 min. Carefully remove the supernatant, briefly rinse the pellet with 1 ml of cold 70% ethanol. Centrifuge for 4 min and carefully aspirate the ethanol.
9. Dry the cDNAs under vacuum for 15 min. Dissolve the cDNA pellet in 15 to 25 μl of TE buffer. Take 2 μl of the sample to measure the concentration of cDNAs prior to the next reaction. Store the sample at –20°C until use.

Protocol B. Trichloroacetic Acid (TCA) Assay and Calculation of cDNA-Yield

1. Spot 4 μl of the first-strand cDNA tracer reaction mixture on glass fiber filters and air-dry. These will be used to obtain the total counts per minute (cpm) in the sample.
2. Add another 4 μl of the same reaction mixture to a tube containing 100 μl of carrier DNA solution (1 mg/ml) and mix well. Add 0.5 ml of 5% TCA and mix by vortexing and precipitate the DNA on ice for 30 min.
3. Filter the precipitated cDNA through glass fiber filters and wash the filters four times with 6 ml cold 5% TCA. Rinse the filters once with 6 ml of acetone or ethanol and air dry. This sample represents incorporated cpm in the reactions.
4. Place the filters at step 1 and step 3 in individual vials and add 10 to 15 ml of scintillation fluid to cover each filter. Count both total and incorporated cpm samples according to the instructions of an appropriate scintillation counter. The cpm can also be counted by Cerenkov radiation (without scintillant).
5. Calculate the yield of first-strand cDNA as follows:

   Percentage incorporated = incorporated cpm/total cpm × 100 dNTP incorporated (nmol) = 4 nmol dNTP/ml × reaction vol.(ml) × % incorporation/100 cDNA synthesized (ng) = nmol dNTP incorporated × 330 ng/nmol % mRNA converted to cDNA = ng cDNA synthesized/ng mRNA in reaction.

Protocol C. Analysis of cDNAs by Alkaline Agarose Gel Electrophoresis

The quality and the size range of the synthesized cDNAs in the first-strand cDNA reactions should be checked by 1.4% alkaline agarose gel electrophoresis using the tracer reaction samples.

1. Label λ*Hind III* fragments with ^{32}P in a fill-in reaction that will be used as DNA markers to estimate the sizes of cDNAs. Add the following components to a sterile Eppendorf tube in the order shown:
   Hind III 10X buffer, 2 μl
   dGTP, 0.2 m*M*
   dATP, 0.2 m*M*
   [α-^{32}P]dCTP, 2 μCi
   λ*Hind III* markers, 1 μg
   Klenow DNA polymerase, 1 unit
   Add dd.H_2O to a final volume of 20 μl.
2. Mix well after each addition and incubate the reaction for 15 min at room temperature. Add 2 μl of 0.2 *M* EDTA to stop the reaction. Transfer 6 μl of the sample directly to 6 μl of 2X alkaline buffer and store the remainder at –20°C.
3. Dissolve 1.4% (w/v) agarose in 50 m*M* NaCl, 1 m*M* EDTA solution and melt it in a microwave for a few minutes. Allow the gel mixture to cool to 50°C and pour the gel into a mini-electrophoresis apparatus. Allow the gel to harden and equilibrate it in alkaline gel running buffer for 1 h before electrophoresis.
4. Transfer the same amount of each sample (50,000 cpm) to separate tubes and add an equal volume of TE buffer to each tube. Add one volume of 2X alkaline buffer to each tube.
5. Carefully load the samples into appropriate wells and immediately run the gel at 7 V/cm until the dye has migrated to about 2 cm from the end.
6. When electrophoresis is complete, soak the gel in five volumes of 7% TCA at room temperature until the dye changes from blue to yellow. Dry the gel on a piece of Whatman 3MM paper in a gel dryer or under a weighted stack of paper towels for 6 h. The 7% TCA is used here to neutralize the denatured gel.
7. Wrap the gel with a piece of SaranWrap™ and expose to x-ray film at room temperature or at –70°C with an intensifying screen for 1 to 4 h.

Note: A successful synthesis of cDNAs should have a signal range from 0.15 to 10 kb with a sharp size range of 1.5 to 4 kb (Figure 3.4).

Materials Needed

[α-^{32}P]dCTP (>400 Ci/mmol)
50 m*M* EDTA
7.5 *M* Ammonium acetate
Ethanol (100% and 70%)
Chloroform:isoamyl alcohol (24:1)
2 *M* NaCl
0.2 *M* EDTA
1 mg/ml Carrier DNA (e.g., salmon sperm)
Trichloroacetic acid (5% and 7%)

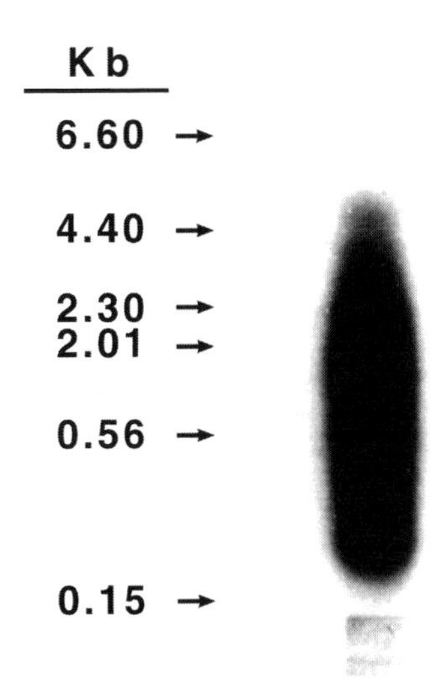

FIGURE 3.4 Analysis of first-strand cDNAs using agarose electrophoresis.

First-Strand 5X Buffer

250 m*M* Tris-HCl, pH 8.3 (at 42°C)
250 m*M* KCl
2.5 m*M* Spermidine
50 m*M* $MgCl_2$
50 m*M* DTT
5 m*M* Each of dATP, dCTP, dGTP and dTTP

TE Buffer

10 m*M* Tris-HCl, pH 8.0
1 m*M* EDTA

TE-Saturated Phenol/Chloroform

Thaw phenol crystals at 65°C and mix an equal part of phenol and TE buffer. Mix well and allow the phases to separate at room temperature for 30 min. Take 1 part of the lower, phenol phase to a clean beaker or bottle and mix with 1 part of chloroform:isoamyl alcohol (24:1). Mix well and allow phases to be separated, and store at 4°C until use.

Alkaline Gel Running Buffer (Fresh)

30 m*M* NaOH
1 m*M* EDTA

2X Alkaline Buffer

20 m*M* NaOH
20% Glycerol
0.025% Bromophenol blue (use fresh each time)

HYBRIDIZATION OF cDNAs FROM CELL OR TISSUE TYPE A OR B WITH mRNAs FROM CELL OR TISSUE TYPE B OR A, OR VICE VERSA

1. Perform a hybridization reaction in a microcentrifuge tube as follows:
   4 to 6 μg First-strand cDNAs from tissue or cell type A
   15 μl of 2X Hybridization buffer
   5 to 15 μg Poly(A)±RNA or mRNAs from tissue or cell type B
   Add dd.H_2O to a final volume of 30 μl

or

   4 to 6 μg First-strand cDNAs from tissue or cell type B
   15 μl of 2X Hybridization buffer
   5 to 15 μg mRNAs from tissue or cell type A
   Add dd.H_2O to a final volume of, 30 μl.
2. Cap the tubes to prevent evaporation and incubate at 65 to 68°C overnight to 16 h.

SEPARATION OF cDNA/mRNA HYBRIDS FROM SINGLE-STRANDED cDNAs BY HYDROXYAPATITE (HAP) CHROMATOGRAPHY

1. Carry out preparation of an HAP column: add 1g of HAP (DNA Grade Bio-Gel, BioRad, Cat.# 130-0520) in 0.1 *M* phosphate buffer (PB), pH 6.8 for 5 ml of slurry and mix well. Heat in boiling water for 5 min. Place a 5-ml plastic syringe closed at the bottom in a water bath to equilibrate at 60°. Place some sterile glass wool at the bottom of the syringe and slowly add 0.5 to 1 ml of HAP slurry to the bottom-closed column and allow it to set for 5 min prior to opening the column. Wash the column twice with 5 volumes of 0.1 *M* PB (60°).
2. Dilute the hybridized sample 10-fold in prewarmed (60°C) 0.1 *M* PB containing 0.15 *M* NaCl. Gently load the sample onto the prepared HAP column with the bottom closed. Gently loosen the mixture in the column with a needle (avoid bubbles) and allow it to set for 10 min.
3. Open the column from the bottom and collect the effluent containing the single-strand cDNAs. Wash the column twice with 0.5 ml of 0.1 *M* PB containing 0.15 *M* NaCl and collect the effluent. The cDNA/mRNA hybrids in the column are then eluted with 0.5 *M* PB.
4. Pool the effluent together and load onto a fresh HAP column as in step 3 and step 4 to obtain the maximum of any cDNA/mRNA hybrids that potentially remain in the effluent.
5. Pool the effluent together and dialyze against 500 ml dd.H_2O overnight to remove the PB salts.

6. Add 0.15 volume of 3 *M* sodium acetate buffer (pH 5.2) and 2.5 volumes of chilled 100% ethanol to the dialyzed sample and place at –80°C for 30 min.
7. Centrifuge at 12,000 × *g* for 5 min and briefly rinse the cDNA pellet with 1 ml of 70% cold ethanol. Dry the pellet under vacuum for 15 min and dissolve the cDNAs in 20 to 25 µl TE buffer. Take 2 µl of the sample to measure the concentration of cDNA. Store the sample at –20°C until use.

SYNTHESIS OF DOUBLE-STRANDED cDNAs FROM SUBTRACTED cDNAs

1. Set up the following reaction:
   Subtracted first-strand cDNAs, 1 µg
   Second-strand 10X buffer, 50 µl
   Poly(A)$_{12\text{ to }20}$ primers, 0.2 µg
   Escherichia coli DNA polymerase I, 20 units
   Nuclease-free dd.H_2O to a final volume of 50 µl
2. Set up a tracer reaction for the second-strand DNA by transferring 5 µl of the mixture to a fresh tube containing 1 µl 4 µCi [α-^{32}PJdCTP (>400 Ci/mmol). The tracer reaction will be used for TCA assay and alkaline agarose gel electrophoresis to monitor the quantity and quality of the synthesis of the second-strand DNA, as described previously.
3. Incubate both reactions at 14°C for 3.5 h and then add 95 µl of 50 m*M* EDTA to the 5-µl tracer reaction.
4. Heat the unlabeled (nontracer portion) double-strand DNA sample at 70°C for 10 min to stop the reaction. At this point, the double-strand DNAs should be produced. Briefly spin down to collect the contents at the bottom of the tube and place on ice until use.
5. Add 4 units of T4 DNA polymerase (2 u/µg input cDNA) to the mixture and incubate at 37°C for 10 min. This step functions to make the blunt-end double-stranded cDNAs by T4 polymerase. The blunt ends are required for later adapter ligations.
6. Stop the T4 polymerase reaction by adding 3 µl of 0.2 *M* EDTA and place on ice.
7. Extract the cDNAs with one volume of TE-saturated phenol/chloroform. Mix well and centrifuge at 11,000 × *g* for 4 min at room temperature.
8. Carefully transfer the top aqueous phase to a fresh tube and add 0.5 volume of 7.5 *M* ammonium acetate or 0.15 volume of 3 *M* sodium acetate (pH 5.2). Mix well and add 2.5 volumes of chilled (–20°C) 100% ethanol. Gently mix and precipitate the cDNAs at –20°C for 2 h.
9. Centrifuge at 12,000 × *g* for 5 min and carefully remove the supernatant. Briefly rinse the pellet with 1 ml of cold 70% ethanol, centrifuge at 12,000 × *g* for 5 min and aspirate the ethanol.

10. Dry the cDNA pellet under vacuum for 15 min. Dissolve it in 20 μl of TE buffer. Measure the concentration of the cDNAs as described previously. Store the sample at –20°C until use.

Note: Prior to adapter ligation, it is strongly recommended that the quantity and quality of the double-stranded (ds) cDNAs should be checked by TCA precipitation and gel electrophoresis as described previously.

Materials Needed

[α-^{32}P]dCTP (>400 Ci/mmol)
50 m*M* EDTA
3 *M* Sodium acetate buffer, pH 5.2
Ethanol (100% and 70%)
Chloroform:isoamyl alcohol (24:1)
0.2*M* EDTA
Trichloroacetic acid (5% and 7%)

Hybridization Buffer (2X)

40 m*M* Tris-HCl, pH 7.7
1.2 *M* NaCl
4 m*M* EDTA
0.4% SDS
1 μg/μl Carrier yeast tRNA (optional)

Phosphate Buffer (Stock)

0.5 *M* Monobasic sodium phosphate
Adjust the pH to 6.8 with 0.5 *M* dibasic sodium phosphate.

LIGATION OF cDNAs TO ΛGT10 OR APPROPRIATE VECTORS

LIGATION OF *EcoR* I LINKERS OR ADAPTORS TO DOUBLE-STRANDED, BLUNT-END cDNAs

The *EcoR* I linker or *EcoR* I adapter ligation system is designed to generate *EcoR* I sites at the termini of blunt-ended cDNA that are commercially available. If *EcoR* I linkers are used, digestion with *EcoR* I is required after ligation. However, there is no need to digest with *EcoR* I when using the *EcoR* I adaptors if each adapter has one sticky end. The ligation of *EcoR* I linker or adapter allows cDNAs to be cloned at the unique *EcoR* I sites of vectors λgt10 or λgt11. If cDNAs are synthesized with an *Xba* I or *Not* I primer–adapter, they can be directionally cloned between the EcoR I and *Xba* I or *Not* I sites of λgt11 *Sfi-Not* I vector or other vectors such as λGEM-2 and λGEM-4.

Carry Out Size Fractionation of Double-Stranded cDNAs with an Appropriate Column (e.g., Sephacryl S-400 Spin Columns) (Promega Corporation)

Note: It is strongly recommended that cDNA samples be size-fractionated prior to ligation with linkers to eliminate <200 bp cDNAs.

1. Add approximately 1.2 ml of Sephacryl S-400 slurry to a vertical column and drain the buffer completely. Continue adding the slurry until the final height of the gel bed reaches the neck of the column at the lower "ring" marking. Rehydrate the column with buffer, cap the top and seal the bottom. Store the column at 4°C until use.
2. Prior to use, drain away the buffer and place the column in a collecting tube provided or a microcentrifuge tube. Briefly centrifuge in a swinging bucket for 4 min at 800 × *g* to remove excess buffer.
3. Place the column in a fresh collecting tube and slowly load the cDNA samples (30 to 60 µl), dropwise, to the center of the gel bed. Centrifuge in a swinging bucket at 800 × *g* for 5 min and collect the eluate. The eluted cDNAs can be directly used or precipitated with ethanol and dissolved in TE buffer.

Protection of Internal *EcoR* I Sites of cDNAs by Methylation

1. Set up the *EcoR* I methylation reaction as follows:
   1 m*M* S-Adenosyl-L-methionine, 2 µl
   EcoR I Methylase 10X buffer, 2 µl
   DNA (1 µg/µl), 2 µl
   EcoR I Methylase, 20 units
   Add dd.H_2O to a final volume of 20 µl.
2. Incubate the reaction at 37°C for 20 min and inactivate the enzyme at 70°C for 10 min.
3. Extract the enzyme with 20 µl of TE-saturated phenol/chloroform, mix well by vortexing and centrifuge at 11,000 × *g* for 4 min.
4. Carefully transfer the top, aqueous phase to a fresh tube and add an equal volume of chloroform:isoamyl alcohol (24:1). Vortex to mix and centrifuge as described previously.
5. Carefully transfer the top, aqueous phase to a fresh tube. Do not take any white materials. To the supernatant, add 0.15 volume of 3 *M* sodium acetate buffer, pH 5.2 and 2.5 volumes of chilled 100% ethanol. Precipitate the DNA at –70°C for 30 min or at –20°C for 2 h.
6. Perform centrifugation and dissolving the DNA as described previously.

Ligation of *EcoR I* Linkers to Methylated cDNAs

1. Set up the reaction:
   cDNA,0.2 μg (0.1 μg/μl)
   Ligase 10X buffer, 2 μl
   Phosphorylated *EcoR* I linkers, 60-fold molar excess
   (diluted stock linkers, 0.001 A_{260} unit of the linker = 10 pmol)
   T4 DNA ligase (Weiss units), 5 units
   Add dd.H_2O to a final volume of 20 μl.
2. Incubate the reaction at 15°C overnight and stop the reaction at 70°C for 10 min.
3. Cool the tube on ice for 2 min and carry out *EcoR I* digestion:
   Ligated cDNAs, 20 μl
   EcoR I 10X buffer, 5 μl
   EcoR I, 1 unit/pmol of linker used in step 1
   Add dd.H_2O to a final volume of 50 μl.
4. Incubate the reaction at 37°C for 2 h and add 4 μl of 0.2 *M* EDTA to stop the reaction.
5. Extract enzymes and precipitate DNA as described earlier.
6. Set up the *Xba* I or *Not* I digestion reaction:
   cDNAs from step a, 40 μl
   Xba I or *Not* I 10X buffer, 8 μl
   Xba I or *Not* I, 1 unit/pmol of linker used in step 1
   Add dd.H_2O to a final volume of 80 μl.

Incubate the reaction at 37°C for 2 h and add 8 ml of 0.2 *M* EDTA to stop the reaction. Extraction of DNA as described earlier.

Note: The following steps are designed for directional cloning of cDNAs synthesized with an Xba I or Not I primer–adapter. The ligated cDNAs with linkers must be digested with Xba I or Not I following the EcoR I digestion. Alternatively, a double enzyme digestion with EcoR I and Xba I or Not I can be executed in the same reaction using 10X digestion buffer for Xba I or Not I in the reaction. However, double digestions are usually of lower efficiency.

7. To check the efficiency of ligation, load 4 μl (about 0.3 μg) of the ligated sample on a 0.8% agarose gel containing EtBr and an equal amount of the unligated, blunt-ended cDNAs next to the ligated sample as a control. After electrophoresis, take a photograph under UV light. An efficient ligation will cause a shift in the migration rate because of the ligation of linker concatemers that run more slowly than the unligated sample.

Reagents Needed

7.5 *M* Ammonium acetate
Ethanol (100% and 70%)
Chloroform:isoamyl alcohol (24:1)
2 *M* NaCl
0.2 *M* EDTA

EcoR I Methylase 10X Buffer
1 *M* Tris-HCl, pH 8.0
0.1 *M* EDTA

TE Buffer
10 m*M* Tris-HCl, pH 8.0
1 m*M* EDTA

Sample Buffer
50% Glycerol
0.2% SDS
0.1% Bromophenol blue
10 m*M* EDTA

Ligase 10X Buffer
0.3 *M* Tris-HCl, pH 7.5
0.1 *M* DTT
0.1 *M* $MgCl_2$
10 m*M* ATP

EcoR I 10X Buffer
0.9 *M* Tris-HCl, pH 7.5
0.5 *M* NaCl
0.1 *M* $MgCl_2$

Not I or Xba I 10X Buffer
0.1 *M* Tris-HCl, pH 7.5
1.5 *M* NaCl
60 m*M* $MgCl_2$
10 m*M* DTT

Ligation of *EcoR I* Adaptors to cDNAs

An *EcoR* I adapter is a duplex DNA molecule with one *EcoR* I sticky end and one blunt end for ligation to cDNA. Use of adaptors can eliminate the *EcoR* I methylation as described previously. The adapter–cDNA does not need to be digested with *EcoR* I after ligation, which is necessary for *EcoR* I linker ligation methodology. However, a phosphorylation reaction should be performed if the cloning vector is dephosphorylated.

1. Set up ligation of *EcoR* I adaptors to cDNAs in a microcentrifuge tube.
   cDNA, 0.2 μg
   Ligase 10X buffer, 3 μl
   EcoR I adaptors, 25-fold molar excess
   T4 DNA ligase (Weiss units), 8 units
   Add dd.H_2O to a final volume of 30 μl.

2. Incubate the reaction at 15°C overnight and stop the reaction at 70°C for 10 min.
3. Carry out a kinase reaction and *Xba* I or *Not* I digestion.

Note: This reaction is necessary for the dephosphorylated vectors but is not required for the phosphorylated vectors. For directional cDNA cloning, Xba I or Not I digestion should be carried out. This can be done simultaneously with the kinase reaction using restriction enzyme 10X buffer instead of the kinase 10X buffer.

Place the tube at step 2 on ice for 1 min and add the following in the order shown:

For kinase reaction only:
Kinase 10X buffer, 5 μl
0.1 m*M* ATP (diluted from stock), 2.5 μl
T4 Polynucleotide kinase, 10 units
Add dd.H_2O to a final volume of 50 μl.

For kinase reaction and *Xba* I or *Not* I digestion:
Xba I or *Not* I 10X buffer, 5 μl
0.1 m*M* ATP (diluted from stock), 2.5 μl
T4 Polynucleotide kinase, 10 units
Either *Xba* I or *Not* I, 8 units
Add dd.H_2O to a final volume of 50 μl.

4. Incubate the reaction at 37°C for 60 min and extraction of DNA as described earlier.
5. Remove unligated adaptors:
   Place a spin column in the collection tube and slowly load the supernatant onto the top center of the gel bed. Centrifuge in a swinging bucket at 800 × *g* for 5 min and collect the eluate containing the cDNA–adapter.
6. Add 0.5 volume of 7.5 *M* ammonium acetate to the eluate, mix and add two volumes of chilled 100% ethanol. Precipitate DNA-linker/adapter at –20°C for 2 h.
7. Centrifuge at 12,000 × *g* for 10 min and briefly rinse the pellet with 1 ml of cold 70% ethanol. Dry the cDNA-linker/adapter under vacuum for 10 min and resuspend the pellet in 20 μl of TE buffer. Store the sample at –20°C until use.

Reagents Needed

Ligase 10X Buffer

300 m*M* Tris-HCl, pH 7.8
0.1 *M* $MgCl_2$
0.1 *M* DTT
10 m*M* ATP

Kinase 10X Buffer

0.7 *M* Tris-HCl, pH 7.5
100 m*M* $MgCl_2$
50 m*M* DTT

Xba I or Not I 10X Buffer

300 mM Tris-HCl, pH 7.8
100 mM $MgCl_2$
500 mM NaCl
100 mM DTT
7.5 M ammonium acetate
Ethanol (100% and 70%)

Restriction Enzyme Digestion of Vectors

If it is necessary, the vectors should be digested with appropriate restriction enzymes prior to being ligated with cDNA-linkers or adaptors, as described previously.

Ligation of cDNAs to Vectors

Determination of Optimal Ligation of Vectors with cDNA Inserts

Prior to a large-scale ligation, several ligations of vector arms and cDNA inserts should be carried out to determine the optimal molar ratio and ligation conditions. The molar concentration of the vectors should be constant while adjusting the amount of DNA inserts. The molar ratio range of 43-kb lambda arms to average 1.4-kb cDNA inserts is usually from 1:1 to 1:0.5. Two control ligations should be carried out. One is the ligation of vectors and positive DNA inserts to check the efficiency; the other is ligation of the vector arms only to determine the background level of religated arms.

1. Set up the ligation reactions as follows and carry them out separately:

Sample Ligations

Components	Microcentrifuge Tubes			
	A	B	C	D
Ligase 10X buffer	0.5	0.5	0.5	0.5 (μl)
λDNA vectors (0.5 μg/μl; 1 μg = 0.04 pmol)	1	1	1	1 (μg)
Linker/adapter–cDNA inserts				
(0.1 μg/μl; 10 ng = 0.01 pmol)	10	20	30	40 (ng)
T4 DNA ligase (Weiss unit)	2	2	2	2 (units)
Add dd.H_2O to a final volume of	5	5	5	5 (μl)

Control Ligation 1

Ligase 10X buffer, 0.5 μl
λDNA vector (0.5 μg/μl), 1.0 μg
Positive control insert DNA, 0.1 μg
T4 DNA ligase (Weiss unit), 2.0 units
Add dd.H_2O to a final volume of 5 μl.

Control Ligation 2

Ligase 10X buffer, 0.5 μl
λDNA vector (0.5 μg/μl), 1.0 μg
T4 DNA ligase (Weiss unit), 2.0 units
Add dd.H_2O to a final volume of 5 μl.

2. Incubate the reactions at room temperature (22 to 24°C) for 1 to 4 h.
3. Check the efficiency of ligation by 0.8% agarose electrophoresis and determine the optimal conditions for a large-scale ligation.
4. Carry out a large-scale ligation of vector arms and cDNA inserts using the optimal conditions determined earlier. Proceed to *in vitro* packaging.

GENERATION OF A SUBTRACTED cDNA LIBRARY

In Vitro Packaging

Performance of packaging *in vitro* is to use a phage-infected *E. coli* cell extract to supply the mixture of proteins and precursors required for encapsulating the recombinant λDNA (vector–cDNA inserts). Packaging systems are commercially available with the specific bacterial strain host and control DNA.

1. Thaw a packaging extract (50 μl/extract) on ice.

Note: Do not thaw the extract at room temperature or 37°C, and do not freeze the extract once it has thawed.

2. Once the extracts have thawed, immediately add 5 to 10 μl of the ligated mixture to the extract and mix gently.
3. Incubate at 22 to 24°C for 2 to 4 h and add 0.5 ml of phage buffer and 20 μl of chloroform to the mixture. Gently mix well and allow the chloroform to settle to the bottom of the tube. Store the packaged phage at 4°C for up to 5 weeks, although the titer may drop. Proceed to titration of lambda phage.

Titration of Packaged Phage

1. Briefly thaw the specific bacterial strain such as *E. coli* DH5α, *E. coli* 392, Y 1090, Y1089 and KW 251 on ice. The bacterial strain is usually kept in 20% glycerol and stored at –70°C until use. Pick up a small amount of the bacteria using a sterile wire transfer loop and immediately inoculate the bacteria on the surface of an LB plate (Figure 3.5). Invert the inoculated LB plate and incubate it in an incubator at 37°C overnight. Multiple bacterial colonies will form.

Note: The LB plate should be freshly prepared and dried at room temperature for a couple of days prior to use.

2. Prepare fresh bacterial culture by inoculating a single colony from the streaked LB plate using a sterile wire transfer loop in 5 ml of LB medium supplemented with 50 ml of 20% maltose and 50 ml of 1 *M* $MgSO_4$

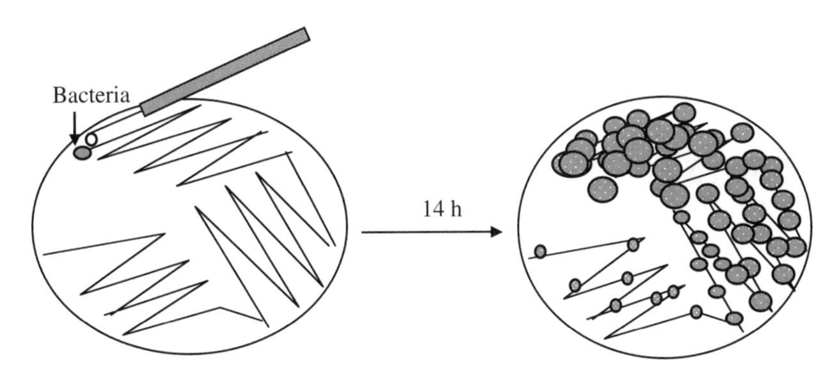

FIGURE 3.5 Streaking of bacterial stock in a zig-zag pattern on an LB plate. Well-isolated bacterial colonies are generated 14 h later in the last streaking area in the plate.

solution. Shake at 160 rpm at 37°C for 6 to 9 h or until the OD_{600} has reached 0.6.

3. Dilute each of the packaged recombinant phage samples at 1000X, 5000X, 10,000X and 100,000X in phage buffer.
4. Add 20 µl of 1 *M* $MgSO_4$ solution and 2.8 ml of melted top agar to 36 sterile glass tubes in a sterile laminar flow hood. Cap the tubes and immediately place them in 50°C water bath for at least 30 min prior to use.
5. Mix 0.1 ml of the diluted phage with 0.1 ml of freshly cultured bacterials in a microcentrifuge tube. Cap the tube and allow the phage to adsorb the bacteria in an incubator at 37°C for 30 min.
6. Add the phage–bacterial mixture into specified tubes in the water bath. Vortex gently and immediately pour onto the center of the LB plates, quickly overlaying the complete surface of the plate with the mixture by gently tilting the plate. Cover the plates and allow the top agar to harden for 10 min. Invert the plates and incubate them in an incubator at 37°C overnight or for 15 h.

Note: The top agar mixture should be evenly distributed over the surface of the LB plate. Otherwise, the growth of bacteriophage will be affected by decreasing the pfu number.

7. Count the number of plaques for each plate and calculate the titer of the phage (pfu).

Plaque forming units (pfu) per milliliter = number of plaques/plate × dilution times × 10

The last "10" of the calculation refers to the 0.1 ml, per milliliter basis, of the packaging extract used for one plate. For instance, if are 100 plaques are on a plate made from a 1/5000 dilution, the pfu per milliliter of the original packaging extract = $100 \times 5000 \times 10 = 5 \times 10^6$.

Reagents Needed

Ligase 10X Buffer

300 m*M* Tris-HCl, pH 7 8
0.1 *M* $MgCl_2$
0.1 *M* DTT
10 m*M* ATP

Phage Buffer

20 m*M* Tris-HCl, pH 7.4
0.1 *M* NaCl
10 m*M* $MgSO_4$

LB (Luria-Bertaini) Medium (per Liter)

10 g Bacto-tryptone
5 g Bacto-yeast extract
5 g NaCl
Adjust to pH 7.5 with 0.2 *N* NaOH and autoclave.
20% (v/v) maltose
1 *M* $MgSO_4$

LB Plates

Add 15 g of Bacto-agar to 1 l of prepared LB medium and autoclave. Allow the mixture to cool to about 60°C and pour 30 ml/dish into 85- or 100-mm Petri dishes in a sterile laminar flow hood with filtered air flowing. Remove any bubbles with a pipette tip and let the plates cool for 5 min prior to their being covered. Allow the agar to harden for 1 h and store the plates at room temperature for up to 10 days or at 4°C in a bag for 1 month. The cold plates should be placed at room temperature for 1 to 2 h before use.

TB Top Agar

Add 3.0 g agar to 500 ml prepared LB medium and autoclave. Store at 4°C until use. For plating, melt the agar in a microwave. When the solution has cooled to 60°C, add 0.1 ml of 1 *M* $MgSO_4$ per 10 ml of the mixture. If color selection methodology is used to select recombinants in conjunction with bacterial host strain Y1090, 0.1 ml IPTG (20 mg/ml in water, filter-sterilized) and 0.1 ml X-Gal (50 mg/ml in dimethylformamide) should be added to the cooled (55°C) 10 ml of the top agar mixture.

S Buffer

Phage buffer + 2% (w/v) gelatin

Amplification of cDNA Library (Optional)

1. For each 50 ml of bacterial culture, add 10 to 15 µl of 1 x 10^5 pfu of packaged bacteriophage or total plaque eluate into a microcentrifuge tube

containing 240 to 235 μl of phage buffer. Mix with 0.25 ml of cultured bacteria and allow the bacteria and phage to adhere to each other by incubating at 37°C for 30 min.

2. Add this mixture to a 250- or 500-ml sterile flask containing 50 ml of LB medium prewarmed at 37°C, which is supplemented with 1 ml of 1 *M* $MgSO_4$. Incubate at 37°C with shaking at 260 rpm until lysis occurs.

Tips: It usually takes 9 to 11 h for lysis to occur. The medium should be cloudy after several h of culture and then be clear upon cell lysis. Cellular debris also becomes visible in the lysed culture. There is a density balance between bacteria and bacteriophage; if the bacteria density is much over that of bacteriophage, it takes longer for lysis to occur, or no lysis takes place at all. In contrast, if bacteriophage concentration is much over that of bacteria, lysis is too quick to be visible at the beginning of the incubation and, later on, no lysis will happen. The proper combination of bateria and bacteriaphage used in step 1 will assure success. In addition, careful observations should be made after 9 h of culture because the lysis is usually quite rapid after that time. Incubation of cultures should stop once lysis occurs; otherwise, the bacteria grow continuously and the cultures become cloudy again. Once that happens, it will take long time to see lysis, or no lysis will take place.

3. Immediately centrifuge at 9000 × *g* for 10 min at 4°C to spin down the cellular debris. Transfer the supernatant containing the amplified cDNA library in bacteriophage particles into a fresh tube. Aliquot the supernatant, add 20 to 40 μl of chloroform and store at 4°C for up to 5 weeks.

ISOLATION OF SPECIFIC cDNA FROM THE SUBTRACTED cDNA LIBRARY

Detailed screening protocols are described in Part B.

CHARACTERIZATION OF cDNA

DNA sequencing and sequence analysis are described in detail in Chapter 5 and Chapter 6.

PART B. CONSTRUCTION AND SCREENING OF A COMPLETE EXPRESSION cDNA LIBRARY

GENERAL PRINCIPLES AND CONSIDERATIONS OF AN EXPRESSION cDNA LIBRARY

Unlike the subtracted cDNA library described previously, a complete cDNA library theoretically contains all cDNA clones corresponding to all mRNAs expressed in a cell or tissue type. An expression cDNA library refers to one in which all cDNAs are cloned in the sense orientation so that all the cDNA clones in the library can be induced to express their mRNAs and proteins. As a result, this cDNA library can be screened with specific antibodies against the expressed protein of interest or with a specific nucleic acid probe.

The general principles and procedures are similar to those of the subtracted cDNA library described earlier. Briefly, mRNAs are transcribed into the first-stranded DNAs, or called complementary DNA (cDNA), based on nucleotide bases complementary to each other. This step is catalyzed by AMV reverse transcriptase using oligo(dT) adapter–primers. The second-stranded DNA is copied from the first-stranded DNA using RNase H that creates nicks and gaps in the hybridized mRNA template, generating 3′-OH priming sites for DNA synthesis by DNA polymerase I. Following treatment with T4 DNA polymerase to remove any remaining 3′ protruding ends, the blunt-ended, double-stranded cDNAs are ready for the ligation with appropriate adaptors or linkers. The resultants are then packaged *in vitro* and cloned in a specific host, generating a complete express cDNA library. In theory, this type of cDNA library preserves as much of the original cDNA as possible, which can allow "fishing" out any possible cDNA clones by screening the cDNA library as long as a specific probe or specific antibodies are available (Figure 3.6).

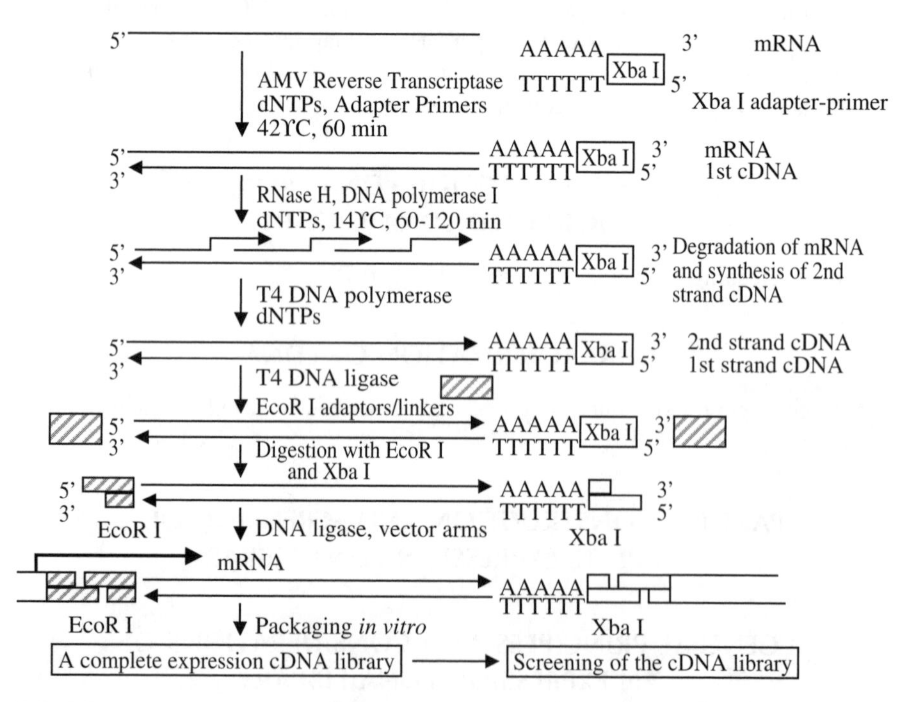

FIGURE 3.6 Construction of a complete expression cDNA library using a directional cloning strategy.

ISOLATION OF TOTAL RNAs AND PURIFICATION OF mRNAs FROM CELLS OR TISSUES OF INTEREST

Detailed protocols are described in Part A.

SYNTHESIS OF cDNAs FROM mRNAs

See the section entitled Subtracted cDNA Library for detailed protocols.

LIGATION OF cDNAs TO ΛGT11 EXPRESSION VECTORS

Λgt11 or equivalent vectors should be used for construction of an expression cDNA library. Restriction enzyme digestion of cDNAs and vector arms, as well as their ligation, are described in detail in Part A: Construction of a Subtracted cDNA Library.

GENERATION OF AN EXPRESSION cDNA LIBRARY

Packaging *in vitro* and generation of a library are the same as that described in Part A.

SCREENING OF THE EXPRESSION LIBRARY AND ISOLATION OF THE cDNA OF INTEREST

METHOD A. IMMUNOSCREENING OF EXPRESSION cDNA LIBRARY USING SPECIFIC ANTIBODIES

cDNAs cloned at the multiple cloning site (MCS) in the *lac* Z gene of vector λgt11 *Sfi-Not* I can be expressed as a part of a β-galactosidase fusion protein. The expression cDNA library can be screened with specific antibodies of interest. The primary positive clones are then "fished" out from the library. Following several rounds of retitration and rescreening of the primary positive clones, putative cDNA clones of interest will be isolated.

1. Prepare LB medium, pH 7.5, by adding 15 g of Bacto-agar to 1 l of the LB medium followed by autoclaving. Allow the mixture to cool to about 55°C and add ampicillin (100 μg/ml) or tetracycline (15 μg/ml), depending on the antibiotic-resistant gene contained in the vectors. Mix well and pour 30 ml/dish of the mixture into 85-or 100-mm Petri dishes in a sterile laminar flow hood with filtered air flowing. Remove any bubbles with a pipette tip and allow the plates to cool for 5 min prior to being covered. Let the agar harden for 1 h and store the plates at room temperature for up to 10 days or at 4°C in a bag for 1 month. The cold plates should be placed at room temperature for 1 to 2 h prior to use.
2. Partially thaw a specific bacterial strain such as *E. coli* Y 1090 on ice. Pick up a tiny bit of the bacteria using a sterile wire transfer loop and

immediately streak out *E. coli* Y1090 on a LB plate by gently drawing several lines on the surface of an LB plate (Figure 3.5). Invert the inoculated LB plate and incubate in an incubator at 37°C overnight. Multiple bacterial colonies will be generated.

3. Prepare fresh bacterial cultures by picking up a single colony from the streaked LB plate using a sterile wire loop and inoculate into a culture tube containing 50 ml of LB medium supplemented with 500 μl of 20% maltose and 500 μl of 1 *M* $MgSO_4$ solution. Shake at 160 rpm at 37°C overnight or until the OD_{600} has reached 0.6. Store the culture at 4°C until use.
4. Add 20 μl of 1 *M* $MgSO_4$ solution and 2.8 ml of melted LB top agar to each of sterile glass test tubes in a sterile laminar flow hood. Cap the tubes and immediately place the tubes in 50°C water bath for at least 30 min before use.
5. According to the titration data, dilute the λcDNA library with phage buffer. Set up 20 to 25 plates for primary screening of the cDNA library. For each of the 100-mm plates, mix 0.1 ml of the diluted phage containing 2×10^5 pfu of the library with 0.15 ml of fresh bacteria in a microcentrifuge tube. Cap the tube and allow the phage to adsorb the bacteria at 37°C for 30 min.

Note: The number of clones needed to detect a given probability that a low-abundance mRNA is converted into a cDNA in a library is:

$$N = \ln (1 - P)/\ln (1 - 1/n)$$

where N is the number of clones needed, P is the possibility given (0.99), and 1/n is the fractional portion of the total mRNA, which is represented by a single low-abundant mRNA molecule. For example, if P = 0.99, 1/n = 1/37,000 and the N = 1.7×10^5.

6. Add the incubated phage–bacterial mixture to the tubes from the water bath. Vortex gently and quickly pour onto the center of the LB plate, quickly spreading the mixture over the entire surface of an LB plate by tilting the plate. Cover the plates and allow the top agar to harden for 5 to 10 min in a laminar flow hood. Invert the plates and incubate them in an incubator at 42°C for 4 h.

Note: The top agar mixture should be evenly distributed over the surface of an LB plate. Otherwise, the growth of the bacteriophage may be affected.

7. Saturate nitrocellulose filter disks in 10 m*M* IPTG in water for 30 min and air-dry the filters at room temperature for 30 min.
8. Carefully overlay each plate with a dried nitrocellulose filter disk from one end and slowly lower it to the other end of the plate. Avoid any air bubbles underneath the filter. Quickly and carefully mark the top side and position of each filter in a triangle fashion by punching the filter through the bottom agar using a 20-gauge needle containing India ink. Cover the plates and incubate at 37°C for 4 to 6 h. If a duplicate is needed, a second

filter may be overlaid on the plate and incubated for another 5 h at 37°C. However, the signal will be relatively weaker.

Note: IPTG induces the expression of the cDNA library and the expressed proteins are transferred onto the facing side of the membrane filters.

9. Place the plates at 4°C for 30 min to chill the top agar and to prevent it from sticking to the filters. Transfer the plates to room temperature and carefully remove the filters using forceps. Label and wrap each plate with Parafilm and store at 4°C until use. The filters can be rinsed briefly in TBST buffer to remove any agar.

Tips: Due to the possible diffusion of proteins, it is strongly recommended to process the filters according to the following steps. Storage at 4°C for a couple of days before processing may cause difficulty in the identification of the actual positive plaque in the plates. The filters must not be allowed to dry out during any of the subsequent steps carried out at room temperature. Based on our experience, high background and strange results may appear, even partial drying. If time is limiting, the damp filters may be wrapped in SaranWrap, then in aluminum foil, and stored at 4°C for up to 12 h. It is recommended to process the filters individually during the following steps to obtain an optimal detection signal.

10. Incubate a filter in 10 ml of TBST buffer containing 2 to 3% BSA or 1% gelatin or 5% nonfatty milk or 20% calf serum to block nonspecific protein-binding sites on the filter. Treat the filters for 60 min with slow shaking at 60 rpm.

Note: It is recommended that one dish be used for one filter only and that the facing side of the filter be down. Each filter is incubated with about 10 ml of blocking buffer. Less than 5 ml will cause the filter to dry but more than 15 ml for an 82-mm filter may prevent obtaining optimal results.

11. Carefully transfer the filters to fresh dishes, each containing 10 ml of TBST buffer with primary antibody.

Tip: The antibodies should be diluted at 200 to 10,000X, depending on the concentration of the antibody stock. The antibodies purified with an IgG fraction or with an affinity column usually produce better signals. The diluted primary antibody can be reused several times if stored at 4°C.

12. Wash each filter in 20 ml of TBST buffer containing 0.5% BSA for 10 min with shaking at 60 rpm. Repeat washing twice.
13. Transfer the filter to a fresh dish containing 10 ml of TBST buffer with the second antibody–alkaline phosphatase conjugate at 1:5000 to 10,000 dilution. Incubate for 40 min with shaking at 60 rpm.
14. Wash the filters as described in step 12.
15. Briefly damp-dry the filter on a filter paper and place it in 10 ml of freshly prepared AP color development substrate solution. Allow the color to develop for 0.5 to 5 h or overnight. Positive clones should appear as purple circle spots on the white filter.

16. Stop the color development when the color has developed to a desired intensity in 15 ml of stop solution. The filter can be stored in the solution or stored dry. The color will fade after drying but can be restored in water.
17. In order to locate the positive plaques, carefully match the filter to the original plate by placing the filter, facing side down, underneath the plate with the help of marks made previously. This can be done by placing a glass plate over a lamp and putting the matched filter and plate on the glass plate. Turn on the light so that the positive clones can be easily identified. Transfer individual positive plugs containing phage particles from the plate using a sterile pipette tip with the tip cut off. Expel the plug into a microcentrifuge tube containing 1 ml of elution buffer. Elute phage particles from the plug at room temperature for 4 h with occasional shaking.
18. Transfer the supernatant containing phage particles into a fresh tube and add 20 μl of chloroform. Store at 4°C for up to 5 weeks.
19. Determine the pfu of the eluate, as described in the section entitled Subtracted cDNA Library. Replate the phage and repeat the screening procedure with the antibody probe several times until 100% of the plaques on the plate are positive. (Figure 3.7)

Tips: During the rescreening process, the plaque number used for each plate should be gradually reduced. In my experience, the plaque density for one 100-mm plate in the rescreening procedure is 1000 to 500 to 300, and to 100.

20. Amplify the putative cDNA clones and isolate the recombinant λDNAs by either plate or liquid methods.

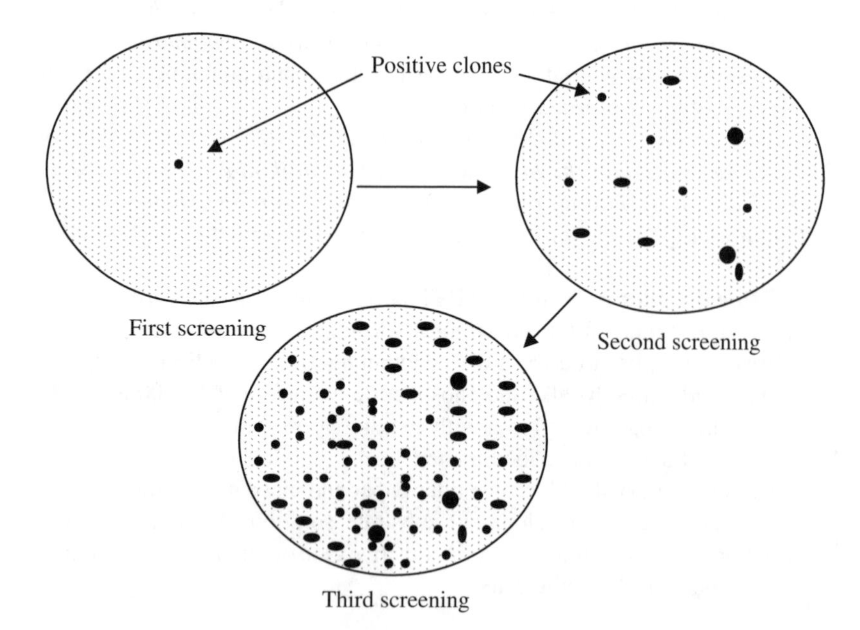

FIGURE 3.7 Diagram of rescreening and purification of positive clones from a cDNA library.

Reagents Needed

LB (Luria-Bertaini) Medium (per Liter)

10 g Bacto-tryptone
5 g Bacto-yeast extract
5 g NaCl
Adjust pH to 7.5 with 0.2 *N* NaOH. Autoclave.

LB Top Agar (500 ml)

Add 4 g agar to 500 ml of LB medium and autoclave.

Elution Buffer

10 m*M* Tris-HCl, pH 7.5
10 m*M* $MgCl_2$

IPTG Solution

10 m*M* Isopropyl β-D-thiogalactopyranoside in dd.H_2O

Phage Buffer

20 m*M* Tris-HCl, pH 7.4
0.1 *M* NaCl
10 m*M* $MgSO_4$

TBST Buffer

10 m*M* Tris-HCl, pH 8.0
150 m*M* NaCl
0.05% Tween 20

Blocking Solution

3% BSA, or 1% gelatin, or 5% nonfat milk in TBST buffer

AP Buffer

0.1 *M* Tris-HCl, pH 9.5
0.1 *M* NaCl
5 m*M* $MgCl_2$

Color Development Substrate Solution

50 ml AP buffer
0.33 ml NBT (nitroblue tetrazolium) stock solution
0.165 ml BCIP (5-bromo-4-chloro-3-indolyl phosphate) stock solution. Mix well after each addition and protect the solution from strong light. Warm the substrate solution to room temperature to prevent precipitation.

Stop Solution

20 m*M* Tris-HCl, pH 8.0
5 m*M* EDTA

Method B. Screening a cDNA Library Using ^{32}P-Labeled DNA as a Probe

In addition to the immunoscreening methodology described previously, the cDNA library can also be screened with labeled DNA, which can be labeled by isotopic or nonisotopic methods (see Chapter 7 for detailed protocols), as a probe. The library is screened by *in situ* plaque hybridization with the probe. Identification of cDNA clones of a potential multigene family can be obtained at relatively low stringency of hybridization. This will be achieved, based on our experience, by reducing the percentage of formamide (30 to 40%) in hybridization solution and washing temperature (room temperature followed by 50 to 55°C) and keeping the temperature of hybridization constant and the salt in the wash solution constant at 2X SSC in 0.1% SDS. To prevent nonspecific cross-linking hybridization, however, a high-stringency condition should be applied. Positive plaques will then be isolated to homogeneity by successive rounds of phage titration and rescreening. The resulting phage lysates will be used in the large-scale preparation of phage DNAs.

The general procedures are the same as those described in the section entitled Screening of a Genomic Library in Chapter 19.

Method C. Screening a cDNA Library Using a Nonradioactive Probe

These general procedures are described in Chapter 19.

Method D. Isolation of λPhage DNAs by the Liquid Method

1. For each 50 ml of bacterial culture, add 10 to 15 μl of 1 x 10^5 pfu of plaque eluate into a microcentrifuge tube containing 240 to 235 μl of phage buffer. Mix with 0.25 ml of cultured bacteria and allow the bacteria and phage to adhere to each other by incubating at 37°C for 30 min.
2. Add this mixture to a 250- or 500-ml sterile flask containing 50 ml of LB medium prewarmed at 37°C, which is supplemented with 1 ml of 1 *M* $MgSO_4$. Incubate at 37°C with shaking at 260 rpm until lysis occurs.

Tips: It usually takes 9 to 11 h for lysis to occur. The medium should be cloudy after several h of culture and then be clear upon cell lysis. Cellular debris also becomes visible in the lysed culture. There is a density balance between bacteria and bacteriophage; if the bacteria density is much over that of bacteriophage, it takes longer for lysis to occur, or no lysis takes place at all. In contrast, if bacteriophage concentration is much over that of bacteria, lysis is too quick to be visible at the beginning of the incubation and, later on, no lysis will happen. The proper combination of bateria and bacteriophage used in step 1 will assure success. In addition, careful observations should be made after 9 h of culture because the lysis is usually quite rapid after that time. Incubation of cultures should stop once lysis occurs. Otherwise, the bacteria grow continuously and the cultures become cloudy again. Once that happens, it will take a long time to see lysis, or no lysis will take place.

3. Immediately centrifuge at 9000 g for 10 min at 4°C to spin down the cellular debris. Transfer the supernatant containing the amplified cDNA library in bacteriophage particles into a fresh tube. Aliquot the supernatant, add 20 to 40 µl of chloroform and store at 4°C for up to 5 weeks.
4. Add RNase A and DNase I to the lambda lysate supernatant, each to a final concentration of 2 to 4 µg/ml. Place at 37°C for 30 to 60 min.

Note: RNase A functions to hydrolyze the RNAs. DNase I will hydrolyze chromosomal DNA but not phage DNA that is packed.

5. Precipitate the phage particles with one volume of phage precipitation buffer and incubate for at least 1 h on ice.
6. Centrifuge at 12,000 × *g* for 15 min at 4°C and allow the pellet to dry at room temperature for 5 to 10 min.
7. Resuspend the phage particles with 1 ml of phage release buffer per 10 ml of initial phage lysate and mix by vortexing.
8. Centrifuge at 5000 × *g* for 4 min at 4°C to remove debris and carefully transfer the supernatant into a fresh tube.
9. Extract the phage particle proteins using an equal volume of TE-saturated phenol/chloroform. Mix for 1 min and centrifuge at 10,000 × *g* for 5 min.
10. Transfer the top, aqueous phase to a fresh tube and extract the supernatant one more time as in step 9.
11. Transfer the top, aqueous phase to a fresh tube and extract once with 1 volume of chloroform:isoamyl alcohol (24:1). Vortex and centrifuge as in step 9.
12. Carefully transfer the upper, aqueous phase containing λDNAs to a fresh tube and add one volume of isopropanol or two volumes of chilled 100% ethanol. Mix well and precipitate the DNAs at –80°C for 30 min or –20°C for 2 h.
13. Centrifuge at 12,000 g for 10 min at room temperature and aspirate the supernatant. Briefly wash the pellet with 2 ml of 70% ethanol and dry the pellet under vacuum for 10 min. Resuspend the DNA in 50 to 100 µl of TE buffer. Measure the DNA concentration (similar to RNA measurement described previously) and store at 4 or –20°C until use.

Reagents Needed

SM Buffer (per Liter)

50 m*M* Tris-HCl, pH 7.5
0.1 *M* NaCl
8 m*M* $MgSO_4$
0.01% Gelatin (from 2% stock)
Sterilize by autoclaving.

Phage Buffer

20 m*M* Tris-HCl, pH 7.4
100 m*M* NaCl
10 m*M* $MgSO_4$

Release Buffer

10 m*M* Tris-HCl, pH 7.8
10 m*M* EDTA

Phage Precipitation Solution

30% (w/v) Polyethylene glycol (MW 8000) in 2 *M* NaCl solution

TE Buffer

10 m*M* Tris-HCI, pH 8.0
1 m*M* EDTA

TE-Saturated Phenol/Chlorolorm

Thaw crystals of phenol in a 65°C water bath and mix with an equal volume of TE buffer. Allow the phases to separate for about 10 to 20 min. Mix one part of the lower phenol phase with one part of chloroform:isoamyl alcohol (24:1). Allow the phases to separate and store at 4°C in a light-tight bottle.

DNase-Free RNase A

To make DNase-free RNase A, prepare a 10 mg/ml solution of RNase A in 10 m*M* Tris-HCl, pH 7.5 and 15 m*M* NaCl. Boil for 15 min and slowly cool to room temperature. Filter if necessary. Alternatively, RNase (DNase-free) from Boehringer Mannheim can be utilized.

CHARACTERIZATION OF cDNA

Isolated cDNA clones should be carefully characterized. For DNA sequencing and sequence analysis, detailed methods are described in Chapter 5 and Chapter 6.

TROUBLESHOOTING GUIDE

1. **Low yield of the first-strand cDNAs**. The success of the synthesis of first- and second-strand cDNAs is extremely important and is considered to be halfway done in constructing a cDNA library. If the yield of the first-strand cDNA conversion is less than 8%, the intensity of the different cDNA sizes shown on the x-ray film is very weak or the size range is very small and of lower molecular weight, these problems are likely caused by poor mRNA preparation or the failure of reverse transcriptase. The quality of the mRNA templates is the most important factor for successful cDNA synthesis. If the poly(A)+RNA is degraded during isolation or purification procedures, the first-strand cDNA synthesized from the template is only partial-length cDNA. To avoid this problem, check the quality of mRNA

templates prior to the synthesis of first-strand cDNAs by analyzing 2 μg mRNAs in 1% agarose–formaldehyde denaturing gel containing ethidium bromide (EtBr). Following electrophoresis, take a photograph of the gel under UV light. If the smear range is from 0.65 up to 10 kb, the integrity of the mRNA is very good for the synthesis of cDNA. Another possible cause is that AMV reverse transcriptase for the synthesis of first-strand cDNAs fails in the middle of the procedure. In that case, the cDNAs produced are likely to be only partial length. Try to use positive mRNA templates as a control to check the activity of reverse transcriptase. If mRNA templates and reverse transcriptase are very good, then use control RNA to check for the presence of inhibitors such as SDS, EDTA, and salts.

2. **Low yield of the second-strand cDNAs**. A yield of the second-strand DNA conversion of less than 40% or a size range that is very small and of abundant lower molecular weight may be due to RNase H or DNA polymerase I that does not function well. To check the activity of RNase H, treat control mRNA with the enzyme and assayed by electrophoresis. If the RNase H works well, the mRNA is nicked into fragments. This will be indicated by gel electrophoresis as compared with untreated mRNA. On the other hand, if DNA polymerase I does not function well, the synthesis of second-strand cDNAs or the repair of gaps will not be completed, thus generating partial-length cDNAs. In that case, set up another reaction using a control experiment or use fresh DNA polymerase I. If low incorporation occurs even with the positive control, it is likely that the radioisotope used as a labeled probe is not good. Check the specificity and half-life of the isotope and use fresh [α-^{32}P]dCTP as a control.
3. **Plaques are small in one area and large in other areas.** This is due to uneven distribution of the top agar mixture. Make sure that the top agar mixture is spread evenly.
4. **Too many positive clones in the primary screening**. The specificity of the primary antibodies is low, the concentration is too high, or the blocking efficiency is low. Try to use IgG or an affinity column-purified antibodies and carry out different dilutions of the antibodies, or increase the percentage of the blocking reagent in the TBST buffer.
5. **Purple background on the filter.** Color development is too long. Try to stop the color reaction as soon as desired signals appear.
6. **Unexpected larger purple spots.** Air bubbles may be produced when the filters are overlaid on the LB plates. Make sure that no air bubbles are generated underneath the filters.
7. **No signals appear at all in any of the primary screening plates**. The pfu numbers used for each plate may be too low. For rare proteins, 2×10^6 pfu should be used in primary screening of the library.
8. **Signals are weak on the filters of subsequent screening**. The antibodies have low activity or color development was stopped too early. Try to use an immunoassay to check the quality of the antibodies, allow the color development to proceed for a longer time, or try to use high-quality nitrocellulose filters.

REFERENCES

1. Wu, L.-L., Mitchell, J.P., Cohn, N.S., and Kaufman, P.B., Gibberellin (GA_3) enhances cell wall invertase activity and mRNA levels in elongating dwarf pea (*Pisum sativum*) shoots, *Int. J. Plant Sci.,* 154, 280, 1993.
2. Wu, W. and Welsh, M.J. Expression of the 25-kDa heat-shock protein (HSP27) correlates with resistance to the toxicity of cadmium chloride, mercuric chloride, *cis*-platinum(II)-diammine dichloride, or sodium arsenite in mouse embryonic stem cells transfected with sense or antisense HSP27 cDNA, *Toxicol. Appl. Pharmacol.*, 141, 330, 1996.
3. Welsh, M.J., Wu, W., Parvinen, M., and Gilmont, R.R., Variation in expression of HSP27 messenger ribonucleic acid during the cycle of the seminiferous epithelium and co-localization of HSP27 and microfilaments in Sertoli cells of the rat, *Biol. Reprod.*, 55, 141–151, 1996.
4. Wu, W., cDNA and genomic DNA libraries, in *Handbook of Molecular and Cellular Methods in Biology and Medicine,* Kaufman, P.B., Wu, W., Kim, D., and Leland, C., pp. 123–210, CRC Press, Boca Raton, FL, 1995.
5. Okayama, H. and Berg, P., High-efficiency cloning of full-length cDNA, *Mol. Cell Biol.,* 2, 161, 1982.
6. Wu, L.-L., Song, I., Karuppiah, N., and Kaufman, P.B., Kinetic induction of oat shoot *pulvinus* invertase mRNA by gravistimulation and partial cDNA cloning by the polymerase chain reaction, *Plant Mol. Biol.,* 21, 1175, 1993.
7. Mignery, G.A., Pikaard, C.S., Hannapel, D.J., and Park, W.D., Isolation and sequence analysis of cDNAs for the major tuber proteun patatin, *Nucl. Acids Res.,* 12, 7987–8000, 1984.
8. Frohman, M.A., Dush, M.K., and Martin, G.R., Rapid production of full-length cDNAs from rare transcripts: amplification using a single gene-specific oligonucleotide primer, *Proc. Natl. Acad. Sci. USA,* 85, 8998, 1988
9. Sambrook, J., Fritsch, E.F., and Maniatis, T., *Molecular Cloning, A Laboratory Manual,* 2nd ed., Cold Spring Harbor Laboratory Press, Cold Spring Harbor, NY, 1989.
10. Chomczynski, P. and Sacchi, N., Single-step method of RNA isolation by acid guanidinium thiocyanate phenol-chloroform extraction, *Anal. Biochem.,* 162, 156, 1987.

4 Subcloning of Genes or DNA Fragments

CONTENTS

INTRODUCTION

Subcloning of a gene or DNA fragment into an appropriate vector is an essential technique that is widely used in current molecular biology. Major advantages include: (1) DNA fragments of interest can be subcloned into a high-copy number of plasmid vectors and be amplified up to 300-fold by plasmid replication in *E. coli*, (2) vectors used for subcloning usually contain SP6, T7 or T3 promoters upstream from the polycloning sites, which allows preparing sense RNA or antisense RNA of the inserted cDNA as hybridization probes, (3) the SP6, T7 and T3 promoters can enable an investigator to sequence the DNA insert on both strands in opposite directions, and (4) DNA constructs can be prepared by inserting a gene or cDNA of interest into an appropriate vector for gene transfer and expression in cultured cells and animals.

The question is what the principles are and how to insert a gene or DNA fragment of interest into the right place in a vector. The general procedures and principles start with restriction enzyme digestion of vector and insert DNA. Following digestion, sticky or blunt ends are generated. Compatible sticky ends of vectors and DNA inserts will be ligated together using T4 DNA ligase, forming recombined constructs. In case of incompatible ends between vectors and inserts, the ends can be blunt-ended by 5′ or 3′ end filling. DNA constructs can be made by blunt-end ligation. The constructs are then transformed into an appropriate bacterial strain in which plasmids are replicated up to 300 copies per cell. Transformed bacterial colonies will be selected using appropriate antibiotics. This chapter describes the well-established, detailed methods routinely used in our laboratories.

RESTRICTION ENZYME DIGESTION OF VECTOR OR DNA INSERT FOR SUBCLONING

SELECTION OF RESTRICTION ENZYMES

Restriction enzymes or endonucleases recognize specific base sequences in DNA and make a cut on both strands. In general, these enzymes can be divided into two types. Type A refers to endonucleases that can produce sticky ends after digestion. Type B enzymes generate blunt ends following digestion. Hundreds of restriction enzymes are commercially available. Typical type A enzymes include *BamH* I, *Hind* III, *EcoR* I, *Kpn* I, *Not* I, *Sac* I, *Xho* I and *Xba* I. They can recognize and produce sticky ends as follows.

```
BamH I:     5′---G↓GATCC---3′   →    5′---G↓      GATCC---3′
            3′---CCTAA↑G---5′        3′---CCTAA      ↑G---5′
Hind III:   5′---A↓AGCTT---3′   →    5′---A↓      AGCTT---3′
            3′---TTCGA↑A---5′        3′---TTCGA      ↑A---5′
EcoR I:     5′---G↓AATTC---3′   →    5′---G↓      AATTC---3′
            3′---CTTAA↑G---5′        3′---CTTAA      ↑G---5′
Kpn I:      5′---GGTAC↓C---3′   →    5′---GGTAC      ↓C---3′
            3′---C↑CATGG---5′        3′---C↑      CATGG---5′
Not I:      5′---GC↓GGCCGC---3′ →    5′---GC↓    GGCCGC---3′
            3′---CGCCGG↑CG---5′      3′---CGCCGG    ↑CG---5′
Sac I:      5′---GAGCT↓C---3′   →    5′---GAGCT      ↓C---3′
            3′---C↑TCGAG---5′        3′---C↑      TCGAG---5′
```

```
Xho I:   5'---C↓TCGAG---3'  →  5'---C↓      TCGAG---3'
         3'---GAGCT↑C---5'     3'---GAGCT      ↑C---5'
Xba I:   5'---T↓CTAGA---3'  →  5'---T↓      CTAGA---3'
         3'---AGATC↑T---5'     3'---AGATC      ↑T---5'
```

The common type B enzymes that generate blunt ends are as follows:

```
Dra I:   5'---TTT↓AAA---3'  →  5'---TTT↓     AAA---3'
         3'---AAA↑TTT---5'     3'---AAA     ↑TTT---5'
EcoR V:  5'---GAT↓ATC---3'  →  5'---GAT↓     ATC---3'
         3'---CTA↑TAG---5'     3'---CTA     ↑TAG---5'
Sma I:   5'---CCC↓GGG---3'  →  5'---CCC↓     GGG---3'
         3'---GGG↑CCC---5'     3'---GGG     ↑CCC---5'
```

Tips: (1) Selection of restriction enzymes depends on the particular restriction enzyme cutting site in the vectors and on the DNA to be inserted. To make the best choice, it is necessary to find out which enzyme has a unique cut at the right place by using a computer program such as GCG or Eugene (see Chapter 6). In our experience, several selections can be made. Choice A: two incompatible, sticky ends in the vector and the same sticky ends in the DNA insert will be the best choice for high efficiency of ligation and transformation. For example, for the DNA insert to have Xho I *and* Hind III *sites at both ends of the DNA to be subcloned, and if the vector also has unique* Xho I *and* Hind III *sites at the multiple cloning site, is very fortunate. The vector and insert DNA can be easily digested with these two enzymes. After purification of the digested DNA, vectors or inserts cannot undergo self-ligation because of the two incompatible sticky ends produced by these two enzymes. During ligation, only insert and vector DNA can be combined. In theory, any colonies after antibiotic selection will be positive transformants. If all goes well, putative constructs can be obtained in a few days starting from restriction enzyme digestion to verification of positive colonies. (2) Choice B: if two unique restriction enzyme sites cannot be found at the right place, one unique site will be acceptable. For instance, if each of the vectors and inserts has a unique EcoR I site at the right site for cloning, both vector and insert can then be digested with EcoR I. The disadvantage is that, compared with choice A, choice B will have less efficiency of ligation between vector and insert and lower efficiency of transformation. Because the same single sticky end is present in both vector and DNA insert, the vector or insert will undergo a high percentage of self-ligation even though the vectors are treated with alkaline phosphatase. As a result, most of the colonies selected by antibiotics may be from self-ligation vectors. This will be time consuming and costly to verify the selected colonies. Nonetheless, single, sticky-end ligation will allow simultaneously making sense and antisense orientations of the DNA insert. In other words, the DNA of interest will be inserted into the vector in both directions, with a probability of 1:1. This is particularly useful if it is desirable to make both sense and antisense constructs of cDNA to be expressed. (3) Choice C: f a unique enzyme site available at the right place is a blunt-end enzyme, it may be necessary to use a blunt-end ligation approach. Compared with choices A and B, choice C has the lowest efficiency of ligation and transformation. It is known that blunt-end ligation is tough. A very small percentage of digested vectors and inserts can be ligated together at 4 or 16°C. Even worse, a significant portion of the ligated DNA molecules is self-ligated DNA. It usually takes weeks or months to obtain putative DNA constructs. Obviously, this is not effective way to clone and may cause frustration.*

Selection of Cloning Vectors

Commercial plasmids such as pcDNA3, pcDNA 3.1, pMAMneo, pGEM series and pBluescript-SK II are available for subcloning. Selection of particular vectors depends on individual investigators. In general, a standard plasmid for subcloning of a DNA fragment should contain the following necessary fragments: (1) a poly-cloning site for the insertion of the DNA of interest, (2) SP6, T7, or T3, or equivalent promoters upstream from the multiple cloning site (MCS) in opposite directions in order to express the DNA insert for sense RNA, or antisense RNA, or for protein analysis, (3) the origin of replication for the duplication of the recombinant plasmid in the host cells, and (4) a selectable marker gene such as *Amp*^{r} for antibiotic selection of transformants.

Protocols for Restriction Enzyme Digestion

1. For single-restriction enzyme digestion, set up the following reaction on ice:
   DNA (vector or insert DNA), 10 to 12 μg in 6 μl
   10X Appropriate restriction enzyme buffer, 2.5 μl
   Appropriate restriction enzyme, 3.3 units/μg DNA
   Add dd.H_2O to a final volume of 25 μl.

   The following reaction is designed for double-restriction enzyme digestion:
   DNA (vector or insert DNA), 10 to 12 μg 6 μl
   10X Appropriate restriction enzyme buffer, 3 μl
   Appropriate restriction enzyme A, 3 units/μg DNA
   Appropriate restriction enzyme B, 3 units/μg DNA
   Add dd.H_2O to a final volume of 30 μl.

   Notes: (1) The restriction enzymes used for vector and insert DNA digestions should be the same in order to ensure optimal ligation. The digestions should be carried out in separate tubes for the vector and insert DNA. (2) For double-restriction enzyme digestions, the appropriate 10X buffer containing a higher NaCl concentration than the other buffer may be used for the double-enzyme digestion. (3) If two restriction enzymes require different reaction buffers, it is better to set up two consecutive single-restriction digestions to ensure the DNA will be completely cut. Specifically, perform one enzyme digestion of the DNA and precipitate the DNA afterwards. The DNA can be dissolved in dd.H_2O and carry out the second enzyme digestion.

2. Incubate the reaction at an appropriate temperature (e.g., 37°C) for 2 h. For single-enzyme digested DNA, proceed to step 3. For double-enzyme digested DNA, proceed to step 5.

Tips: (1) To obtain an optimal ligation, the vector and insert DNA should be completely digested. The digestion efficiency may be checked by loading 1 μg of the digested DNA with loading buffer in a 1% agarose minigel. Meanwhile, the undigested vector and insert DNA (1 μg) and standard DNA markers should be loaded into the adjacent wells. After electrophoresis, the undigested plasmid DNA may display multiple bands because of different levels of supercoiled plasmids. However, one band should be visible for a complete, single-enzyme digestion; one major band and one tiny band (<70 bp) may be visible after digestion with two different restriction enzyme digestions of the vectors. (2) After completion of restriction enzyme digestion, calf intestinal alkaline phosphatase (CIAP) treatment should be carried out for single-restriction enzyme digestion of vectors. This treatment serves to remove 5′-phosphate groups, thus preventing recircularization of the vector during ligation. Otherwise, the efficiency of ligation between the vector and insert DNA will be very low. For vectors digested with double-restriction enzymes, the CIAP treatment is not necessary.

3. Carry out the CIAP treatment as follows.
   10X CIAP buffer, 5 μl
   Single-enzyme digested DNA, 25 μl
   CIAP diluted in 10X CIAP buffer, 0.01 unit/pmol ends
   Add dd.H_2O to a final volume of 50 μl.

   Notes: (1) CIAP and 10X CIAP should be kept at 4°C. (2) Calculate pmol ends; for example, 9 μg digested DNA is left after using 1 μg of 10 μg digested DNA for analysis in an agarose gel. If the vector size is 3.2 kb, the pmol ends can be calculated by the formula below:

$$\textit{pmol ends} = \frac{\textit{amount of DNA}}{\textit{base pairs} \times 660/\textit{pair}} \times 2$$

$$= \frac{9}{3.2 \times 1000 \times 660} \times 2$$

$$= 4.2 \times 10^{-6} \times 2$$

$$= 8.4 \times 10^{-6}\ \mu M$$

$$8.4 \times 10^{-6} \times 10^{6} = 8.4 \text{ pmol ends}$$

4. Incubate the reaction at 37°C for 1 h and add 2 μl of 0.5 *M* EDTA buffer (pH 8.0) to terminate the reaction. At this point, one of two options can be chosen. If the vector or insert DNA will be a single fragment following restriction enzyme, the digested DNA can be purified by protein extraction

and ethanol precipitation as described next. However, if restriction enzyme digestion results in more than one fragment or band, the fragment of interest should be purified by agarose gel electrophoresis as described in the next section.

5. Add one volume of TE-saturated phenol/chloroform to the reaction. Mix well by vortexing the tube for 1 min and centrifuge at 11,000 × *g* for 5 min at room temperature.
6. Carefully transfer the top, aqueous phase to a fresh tube and add one volume of chloroform:isoamyl alcohol (24:1) to the supernatant to further extract proteins. Mix well and centrifuge as in step 5.
7. Carefully transfer the upper, aqueous phase to a fresh tube and add 0.1 volume of 3 *M* sodium acetate buffer (pH 5.2) or 0.5 volume of 7.5 *M* ammonium acetate to the supernatant for efficient precipitation. Briefly mix and add 2 to 2.5 volumes of chilled 100% ethanol to the supernatant and precipitate the DNA at –80°C for 1 h or at –20°C for 2 h.
8. Centrifuge at 12,000 × *g* for 10 min and carefully decant the supernatant. Add 1 ml of 70% ethanol to the tube and centrifuge at 12,000 × *g* for 5 min. Dry the DNA pellet under vacuum for 20 min. Dissolve the DNA in 10 μl dd.H_2O. Take 1 μl of the sample to measure the concentration of the DNA at A_{260} nm. Store the sample at –20°C prior to use. Proceed to DNA ligation.

Reagents Needed

Appropriate enzymes
10X Appropriate restriction enzyme buffer
1% Agarose minigel
TE-saturated phenol/chloroform
Chloroform:isoamyl alcohol (24:1)
7.5 *M* Ammonium acetate
3 *M* Sodium acetate buffer, pH 5.2
Ethanol (100%, 70%)
0.5 *M* EDTA, pH 8.0
Calf intestinal alkaline phosphates (CIAP)
TE buffer

10X CIAP Buffer

0.5 *M* Tris-HCl, pH 9.0
1 m*M* $ZnCl_2$
10 m*M* $MgCl_2$
10 m*M* Spermidine

PURIFICATION OF DNA FRAGMENTS FROM AGAROSE GELS

ELUTION OF DNA BANDS BY HIGH-SPEED CENTRIFUGATION OF AGAROSE GEL SLICES

We have developed a rapid, simple method for eluting DNA bands from regular agarose gels by high-speed centrifugation. The agarose matrix containing DNA of interest will be compressed by the force of centrifugation, releasing DNA into the supernatant fluid. The method allows recovery of large as well as small DNA molecules. In addition to being rapid and simple, this method is of low cost and it does not require the use of phenol and chloroform as compared with traditional methods. The eluted DNA is of high quality and is suitable for restriction enzyme digestion, ligation, cloning or labeling without further purification.[1] The protocol is described next:

1. Carry out electrophoresis as described in Chapter 7.
2. When electrophoresis is complete, quickly locate the DNA band of interest by illuminating the gel on a long wavelength (>300 nm) UV transilluminator. Quickly slice out the band of interest using a sharp, clean razor blade.

Note: To avoid potential damage to the DNA molecule, the UV light should be turned on as briefly as possible.

3. To enhance the yield of DNA, trim away extra agarose gel outside the band and cut the gel slice into tiny pieces with the razor blade on a clean glass plate.
4. Transfer the fine slices into a 1.5-ml microcentrifuge tube (Eppendorf).

Notes: (1) If it is not necessary to maximally elute DNA out of the gel slices, the slices do not need to be further sliced into tiny pieces. They can be directly placed in a tube. (2) At this point, there are two options for eluting DNA from the agarose gel pieces. The first is to carry out high-speed centrifugation (step 5) immediately. The second option is to elute the DNA as high yield as possible (see below).

5. Centrifuge at 12,000 to 14,000 × *g* or at the highest speed using an Eppendorf centrifuge 5415C (Brinkmann Instruments, Inc.) for 15 min at room temperature.

Principles: With high-speed centrifugation, the agarose matrix is compressed or even partially destroyed by the strong force of centrifugation. The DNA molecules contained in the matrix are released into the supernatant fluid.

6. Following centrifugation, carefully transfer the supernatant fluid containing DNA into a clean microcentrifuge tube. The DNA can be used directly for ligation, cloning, and labeling as well as restriction enzyme digestion without ethanol precipitation. Store the DNA solution at 4 or –20°C until use.

Tips: (1) In order to confirm that the DNA is released from the gel pieces, the tube can be briefly illuminated with long wavelength UV light after centrifugation. An orange-red fluid color indicates the presence of DNA in the fluid. (2) The supernatant fluid should be immediately transferred from the agarose pellet; within minutes, the temporarily compressed agarose pellet may swell, absorbing the supernatant fluid.

High Yield and Cleaner Elution of DNA

1. Add 100 to 300 µl of dd.H_2O or TE buffer (10 m*M* Tris-HCl, 1 m*M* EDTA, pH 8.0) to the gel pieces at the preceding step 4.

Function: TE buffer or dd.H_2O is added to the gel pieces to help elute the DNA and simultaneously to wash EtBr from the DNA during high-speed centrifugation.

2. Centrifuge at 12,000 to 14,000 × *g* at room temperature for 20 min. When the centrifugation is complete, briefly visualize the eluted DNA by illumination using a long-wavelength UV light.
3. Immediately transfer the supernatant that contains the eluted DNA into a fresh tube.
4. To increase the recovery of DNA, extract the agarose pellet by resuspending the pellet in 100 µl of dd.H_2O or TE buffer followed by another cycle of centrifugation.
5. Pool the DNA supernatant and precipitate the DNA with ethanol. Dissolve DNA 10 µl of dd.H_2O or TE buffer. Store the DNA solution at 4 or –20°C until use.

Elution of DNA Fragment by Melting and Thawing of Agarose Gel Slices[2,3]

1. After restriction enzyme digestion, carry out electrophoresis as described in Southern blotting in Chapter 7, except use 0.5 to 1% low melting-temperature agarose instead of normal agarose.
2. Transfer the gel to a long-wavelength UV transilluminator (305 to 327 nm) to visualize the DNA bands in the gel. Excise the band area of interest with a clean razor blade and transfer the slices into a microcentrifuge tube.

Notes: (a) The gel should not be placed on a short-wavelength (e.g., <270 nm) UV transilluminator because this type of UV light may cause breakage inside the DNA fragment. (b) The gel slices containing the DNA band of interest should be trimmed as much as possible in order to remove excess unstained gel areas.

Caution: UV light is harmful to the human body. Protective eye glasses, gloves, and a lab coat should be worn.

3. Add two volumes of TE buffer to the gel slices and completely melt the gel in a 60 to 70°C water bath.

Note: The gel slices can be directly melted without adding TE buffer. The DNA concentration is usually higher, but the total yield of the DNA fragments is much lower than that if TE buffer is added.

4. Immediately chill the melted gel solution on dry ice or its equivalent and leave the tube at –70°C for at least 20 min.
5. Thaw the gel mixture by vigorously tapping the tube. It takes 5 to 10 min to thaw the gel into a resuspension state.
6. Centrifuge at 12,000 × *g* for 5 min at room temperature.
7. Carefully transfer the liquid phase containing the eluted DNA fragments into a fresh tube, which can be used directly for labeling, although the concentration is low.
8. Extract EtBr with three volumes of water-saturated n-butanol.
9. Precipitate the DNA by adding 0.1 volume of 3 *M* sodium acetate buffer (pH 5.2) and 2.5 volumes of chilled 100% ethanol to the DNA solution. Precipitate DNA at –70°C for 1 h and centrifuge at 12,000 × *g* for 5 min at room temperature.
10. Discard the supernatant and briefly rinse the DNA pellet with 1 ml of 70% ethanol. Dry the DNA under vacuum for 15 min. The DNA can be dissolved in an appropriate volume of dd.H_2O or TE buffer. Store at –20°C before use.

Elution of DNA Fragment Using NA45 DEAE Membrane

1. Perform step 1 in the previous section, but use 1% normal agarose.
2. Soak a piece of NA45 DEAE membrane (Scheicher and Schuell) in 10 m*M* EDTA buffer (pH 7.6) for 10 to 20 min at room temperature and in 0.5 *N* NaOH solution for 5 min. Wash the membrane with dd.H_2O three to four times and store at 4°C prior to use.
3. During the electrophoresis, monitor the migration of the DNA bands stained with EtBr in the gel using a long-wavelength UV lamp. Turn off the power, make an incision in front of the DNA band of interest and insert a piece of the prepared membrane into the incision in the gel.
4. Continue electrophoresis until the DNA fragments migrate onto the membrane by monitoring the migration of the stained band.
5. Remove the membrane strip from the gel and transfer it in a solution containing 20 m*M* Tris-HCl (pH 8.0), 0.15 *M* NaCl, 0.1 m*M* EDTA. Rinse the membrane briefly to remove any agarose.
6. Transfer the strip into 0.2 to 0.3 ml of elution buffer containing 20 m*M* Tris-HCl (pH 8.0), 1*M* NaCl, and 0.1 m*M* EDTA.
7. Incubate the strip at 55 to 68°C for 30 to 60 min with occasional agitation.
8. Rinse the strip with 0.1 ml of elution buffer and extract EtBr with three volumes of water-saturated n-butanol.
9. Precipitate and dissolve the eluted DNA as described in the preceding section.

Elution of DNA Fragments in Agarose Gel Well

1. Carry out DNA electrophoresis as described in the section on elution by melting and thawing agarose gel slices, except add the running buffer up to the upper edges of the gel instead of covering the gel.
2. During electrophoresis, check the separation of DNA bands stained by EtBr in the gel using a long-wavelength UV lamp. Stop the electrophoresis and use a razor blade or a spatula to make a well in front of a specific band. Add 10 to 30 μl of running buffer into the well.
3. Continue electrophoresis until the band migrates into the well by monitoring the band.
4. Stop electrophoresis and transfer the solution containing DNA of interest from the well to a fresh tube.
5. Extract EtBr with three volumes of water-saturated n-butanol, precipitate and dissolve the DNA as described in the section on elution by melting and thawing agarose gel slices.

LIGATION OF DNA FRAGMENTS

Once the digested vectors and insert DNA are ready, ligation can be carried out. To achieve an optimal ligation, the ratio of vector to insert DNA (1:1, 1:2, 1:3 and 3:1 molar ratios) should be optimized using a small-scale ligation. The following reaction is standard for sticky end ligation of a 3.2-kb plasmid vector and a 3.0-Kb insert DNA.

1. Calculate the molar weights of the vector and insert DNA:
   1 M plasmid vector = $3.2 \times 1000 \times 660 = 2.112 \times 10^6$
   1 M insert DNA = $3 \times 1000 \times 660 = 1.98 \times 10^6$
2. Calculate the molar ratio of the vector to insert DNA:

Vector DNA:insert DNA	Amount of DNA: Vector	Amount of DNA: Insert
1:1	1 μg	0.792 μg
1:2	1 μg	1.584 μg
1:3	1 μg	2.376 μg
3:1	1 μg	0.264 μg

3. Set up the following ligation on ice:

Components	Ligation reactions: 1 (1:1)	2 (1:2)	3 (1:3)	4 (3:1)
Vector DNA	1 μg	1 μg	1 μg	1 μg
Insert DNA	0.792 μg	1.584 μg	2.376 μg	0.244 μg
10X Ligase buffer	1 μl	1 μl	1 μl	1 μl
T4 DNA ligase (Weiss units)	5 units	5 units	5 units	5 units
Add dd.H_2O to	10 μl	10 μl	10 μl	10 μl

4. Incubate the reactions at 25°C or room temperature for 3 h for sticky-end ligation; incubate at 16°C for 6 h to overnight or 4°C 12 to 20 h for blunt-end ligation.
5. Determine the efficiency of the ligations by 1% agarose electrophoresis. When the electrophoresis is complete, take a picture of the gel stained with EtBr under UV light. As compared with unligated vector or insert DNA, high efficiency of ligation should visualize less than approximately 20% of unligated vector and insert DNA by estimation of the intensity of fluorescence. Approximately 80% of the vector and insert DNA are ligated to each other and show strong bands with molecular weight shifts. By comparing the efficiency of ligations using different molar ratios, the optimal concentration of DNA can be used as a guide for large-scale ligation.
6. Carry out large-scale ligation of the vector and insert DNA based on the optimal conditions determined by small-scale ligations. For example, if a 1:2 molar ratio of vector DNA:insert DNA is used as the optimal ligation condition, a large-scale ligation can be carried out as follows:
   Vector DNA, 3 μg
   Insert DNA, 4.75 μg
   10X Ligase buffer, 3 μl
   T4 DNA ligase (Weiss units), 15 to 50 units
   Add dd.H_2O to a final volume of 30 μl.

Incubate the ligation as described in small-scale ligation. Proceed to transformation.

SINGLE-STEP CLONING BY PCR

This new strategy is particularly useful if no restriction enzyme site can match both the vector and insert DNA. In spite of the fact that the nonmatch sticky ends can be blunt-ended followed by blunt-end ligation, the ligation efficiency will be extremely low, as will be the efficiency of transformation. This is because blunt-end ligation is usually difficult and many of the ligated molecules are self-ligated. Therefore, a single-step cloning by PCR can eliminate such a problem. No ligase and ligation will be involved. The principles and procedures are outlined in Figure 4.1.

1. Design and synthesize a pair of forward (sense or upstream) and reverse (antisense or downstream) primers based on the known sequences at the cloning sites. Specifically, the forward primer is designed from the 5′ → 3′ direction and has approximately 36 to 43 bases that the first part of the primer (about 18- to 20-mer) matches the sequence at 3′ end of the first or (–)strand of vector DNA. The rest of the sequence is the same as that of 5′ end of the first or (–)strand of insert DNA. On the other hand, a reverse primer is also designed from the 5′ → 3′ direction but its first part is the same as the sequence at the 3′-end sequence of the second or (+)strand of the vector DNA. The rest of the base sequence is the same as that of the 5′ end of the second or (+)strand of the insert DNA.
2. Perform simultaneous amplification and ligation by PCR.

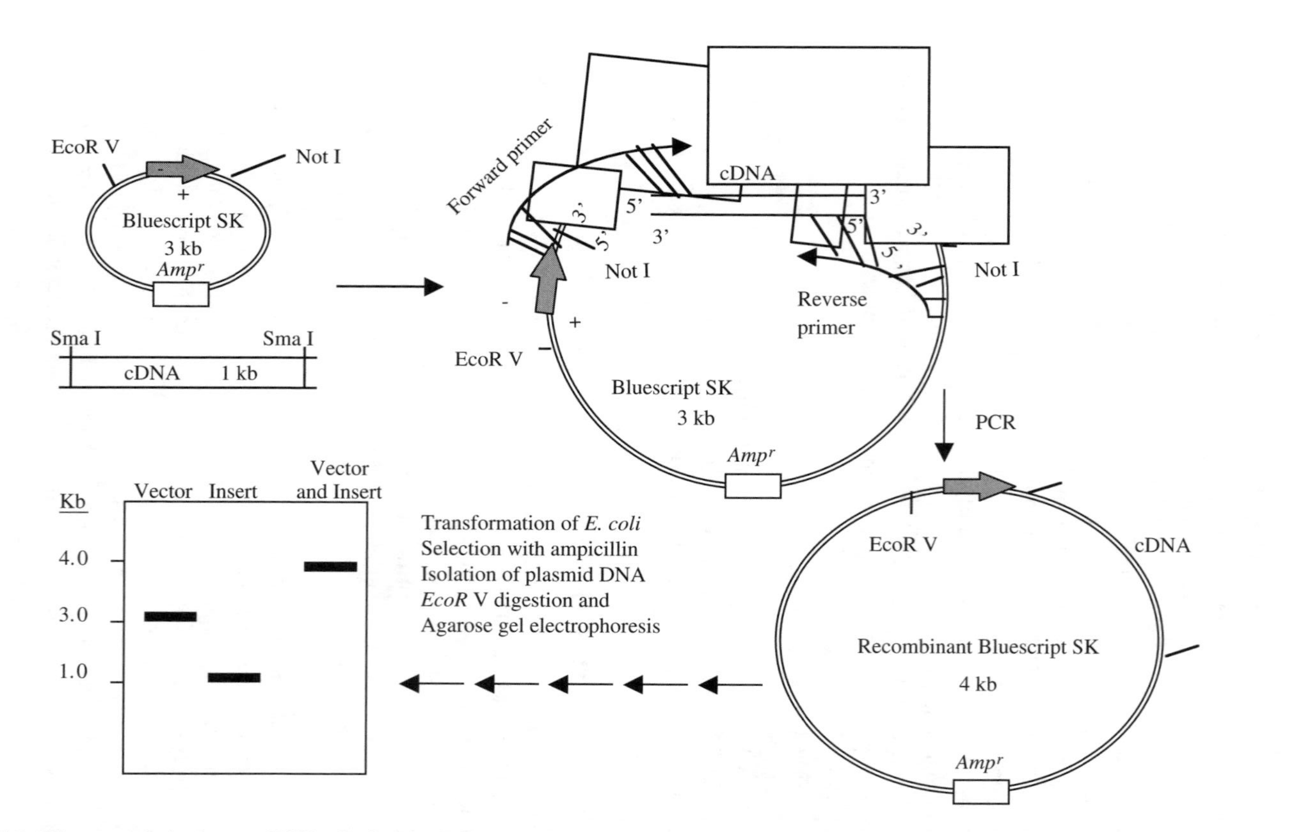

FIGURE 4.1 Diagram of single-step DNA cloning by PCR, transformation, selection and verification of recombinant plasmid DNA by agarose gel electrophoresis.

a. Prepare the following mixture in a 0.5-ml microcentrifuge tube on ice.
   Forward primer (36- to 43-mer, 10 to 30 ng/μl), 5 to 8 pmol (30 to 40 ng) depending on the size of the primer
   Reverse primer (36- to 43-mer, 10 to 30 ng/μl), 5 to 8 pmol (30 to 40 ng) depending on the size of the primer
   DNA template (vector or insert; 0.4 to 7 kb, 10 to 100 ng/μl), 100 to 1000 ng depending on the size of the template
   10X Amplification buffer, 4 μl
   dNTPs (2.5 m*M* each), 4 μl
   Taq or Tth DNA polymerase, 5 to 10 units
   Add dd.H_2O to a final volume of 40 μl.
   Note: The DNA polymerase should be the high-fidelity and long-expand variety, which is commercially available.
b. Overlay the mixture in each tube with approximately 30 μl of mineral oil to prevent evaporation of the samples during the PCR amplification. Place the tubes in a thermal cycler and perform PCR.

	Cycling (4 to 12 cycles)				
Profile	**Predenaturation**	**Denaturation**	**Annealing**	**Extension**	**Last**
Template (<4 kb) Primer is <24 bases or with <40% G–C content	94°C, 3 min	94°C, 1 min	60°C, 1 min	70°C, 1.5 min	4°C
Template is >4 kb Primer is >24-mer or <24 bases with >50% G–C content	95°C, 3 min	5°C, 1 min	60°C, 1 min	72°C, 2 min	4°C

3. When the PCR is complete, carry out transformation of the appropriate bacterial strain using 2 to 10 μl of the amplified mixture as described previously.
4. Select and verify recombinant plasmid DNA as described previously.

TRANSFORMATION OF LIGATED DNA INTO BACTERIA

Protocol 1. Preparation of Competent Cells for Transformation

Competent Cells for $CaCl_2$ Transformation prior to Ligation

1. Prepare LB medium and LB plates. This should be done 2 to 3 days prior to ligation.
2. Streak an appropriate *E. coli* strain (DH5αF′, Top 10 or JM109) directly from a small amount of frozen stock stored at –80°C onto the surface of

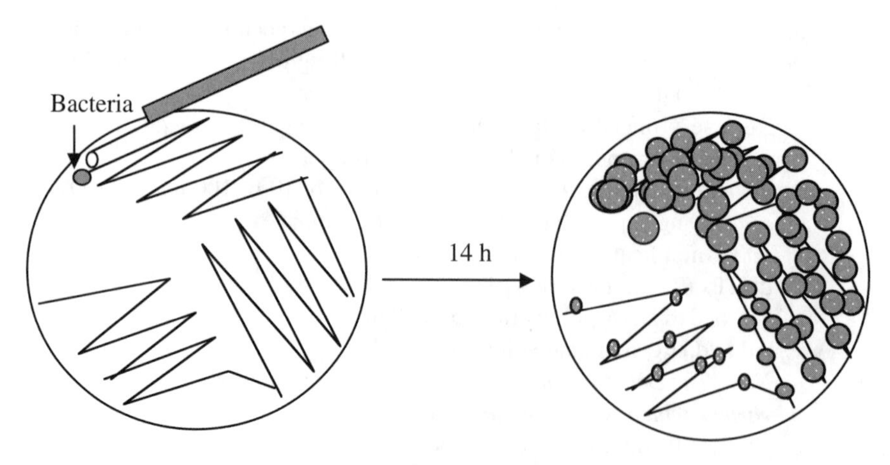

FIGURE 4.2 Streaking of bacterial stock in a zig-zag pattern on an LB plate. Well-isolated bacterial colonies will be generated 14 h later in the last streaking area in the plate.

an LB plate in a zig-zag pattern (Figure 4.2) using a sterile platinum wire loop. Invert the plate and incubate at 37°C for 12 to 16 h. Bacterial colonies will be formed.

Notes: It is not necessary to thaw frozen bacteria completely at room temperature or 0°C. A tiny bit of bacteria adhering to the wire loop is sufficient for inoculation.

3. Inoculate a well-isolated colony from the plate at step 2 into 5 ml of LB medium. Incubate at 37°C overnight with shaking at 160 rpm.
4. Add 0.5 ml of cells from step 3 to 100 ml of LB medium. Incubate at 37°C with shaking at 160 rpm. Four hours later, take 1 ml of the culture and measure the OD_{600} or A_{600} every 20 to 30 min until the A_{600} reaches 0.5 to 0.6. This usually takes 5 to 8 h.

Tip: Never allow bacteria to overgrow. If the density is higher than 0.65, the culture is too dense and transformation efficiency will be low.

5. Chill the cells on ice for 20 min and centrifuge at 5000 × *g* for 5 min at 4°C. Aspirate the medium.
6. Resuspend the cells in half of the original culture volume of sterile-filtered 0.1 *M* $CaCl_2$ (50 ml) that has been prechilled on ice.
7. Incubate the cells on ice for 45 to 60 min.
8. Centrifuge at 5000 × *g* for 10 min at 4°C and gently resuspend the cells in 20 ml (1/20 original volume) of ice-cold 0.1 *M* $CaCl_2$ solution.
9. Add sterile glycerol dropwise to the cell solution with gentle swirling to a final concentration of 15% (v/v). Aliquot the cells at 0.2 ml/tube, freeze and store at –80°C until use.
10. Take one aliquot to test efficiency of transformation.

PREPARATION OF COMPETENT CELLS FOR ELECTROPORATION

1. Carry out step 1 through step 4 in the preceding section.
2. Extensively wash the cells with 100 ml distilled water or low-salt buffer in order to reduce the ionic strength of the cell suspension.
3. Centrifuge at 5000 × *g* for 10 min at 4°C and carefully decant the supernatant.
4. Repeat step 2 and step 3 twice.
5. Resuspend the cells in 20 ml of low-salt buffer or distilled water. Add glycerol dropwise with gentle swirling to 10% (v/v). Aliquot the cells at 0.2 ml/tube, freeze and store at –80°C until use.

Reagents Needed

LB (Luria-Bertaini) Medium (per Liter)

10 g Bacto-tryptone
5 g Bacto-yeast extract
5 g NaCl
Adjust to pH 7.5 with 0.2 *N* NaOH and autoclave.

LB Plates

Add 15 g of Bacto-agar to 1 l of freshly prepared LB medium and autoclave. Allow the mixture to cool to about 55°C and add the appropriate amount of antibiotic stock. Pour 25 to 30 ml of the mixture into each 100-mm Petri dish in a sterile laminar flow hood with filtered air flowing. Remove any bubbles with a pipette tip and let the plates cool for 5 min prior to covering them. Allow the agar to harden for 2 h and partially dry at room temperature for a couple of days. Store the plates at room temperature for up to 10 days or at 4°C in a bag for 1 month. Allow the cold plates to warm up at room temperature for 1 to 2 h prior to use.

PROTOCOL 2. TRANSFORMATION OF CELLS BY $CACL_2$ METHOD

1. Thaw an aliquot of 0.2 ml of frozen $CaCl_2$-treated competent cells on ice.
2. Assembly transformation as follows:

	Tube		
Components	**1**	**2**	**3**
Cells	50 μl	50 μl	50 μl
DNA	50 ng in 5 μl	200 ng in 5 μl	500 ng in 5 μl

Tip: 15 to 20 μl of ligated mixture can be directly added into 50 to 60 μl of competent cells.

3. Incubate on ice for 50 to 60 min.

4. Heat shock at 42°C for 1 to 1.5 min and quickly place the tubes on ice for 1 min.

Tip: Heat-shock and chilling on ice should not extend past the time indicated. Otherwise, the efficiency of transformation will decrease 10- to 20-fold.

5. Transfer the cell suspension to sterile culture tubes containing 1 ml of LB medium without antibiotics. Incubate at 37°C for 1 to 2 h with shaking at 140 rpm to recover the cells.
6. Transfer 50 to 150 μl of the cultured cells per plate onto the center of the LB plates containing 50 μg/ml ampicillin (or 0.5 m*M* IPTG and 40 μg/ml X-Gal for color selection). Immediately spread the cells over the entire surface of the LB plates using a sterile, bent glass rod.
7. Invert the plates and incubate them at 37°C overnight. Selected colonies should be visible 14 h later.

Protocol 3. Transformation by Electroporation

1. Thaw an aliquot of 200 μl of frozen, non-$CaCl_2$-treated competent cells on ice.
2. Chill three disposable microelectroporation chambers (BRL) on ice.
3. Connect the power cable to a BRL-Porator pulse control + power supply apparatus, to BRL-Porator voltage booster, and between these two units.
4. Set a pulse control as follows:

Power	Charge
Capacitance (μF)	330
High ohm/low ohm	Low ohm
Charge rate	Fast
Set up the voltage booster at 4 kV for *E. coli*	

5. Add ice water to the chamber safe up to 4/5 of the volume, and place the chamber rack in the chamber safe.
6. Thaw three aliquots (A, B, C) of 20 μl of competent cells on ice. To aliquot A, add 3 μl of DNA solution containing 1 μg DNA. To aliquot B, add 3 μl of DNA solution containing 2 μg DNA. To aliquot C, add 3 μl of DNA solution containing 4 μg DNA. Gently mix and place on ice until use.

Note: Try to avoid any air bubbles during mixing. The total volume of the transformation mixture should be ≤25 μl.

7. Use a pipette tip to transfer one aliquot and carefully place the mixture drop between the electrode poles in the microelectroporation chamber. Carefully cover the chamber.

Note: The liquid should not drop to the bottom of the chamber.

8. Gently place the chamber into the cell in the safe rack, cover the chamber and turn the electric-shock pointer toward the cell containing the transformation chamber.

9. Connect the power from the voltage booster to the chamber safe, and turn on power for the pulse control and voltage booster.
10. Press charge button on the pulse control up to 365. When the DC voltage drops down to 345 to 350 V, turn the button from charge to arm. Quickly push the trigger button for 1 to 2 s. The voltage goes down to <10 and the kV number in the voltage booster should be 1.9 to 2.0.
11. Turn off the power and gently remove the chamber from the cell.

Note: The transformed cell suspension should still be between the positive and negative electrode poles. It is not a good sign if the liquid drops to the bottom.

12. Quickly transfer the transformed cell suspension into a test tube containing 0.5 ml LB medium without ampicillin because the cells are quite weak. Mix well and place at room temperature for not longer than 15 min.
13. Repeat the preceding steps until all the samples are transformed.
14. Incubate at 37°C for 1 to 2 h with shaking at 150 rpm to recover the cells.
15. Use a sterile, bent glass rod to spread 20, 50, 100, and 200 µl of each of the three transformant cells over the entire surface of LB plates containing 50 µg/ml ampicillin.
16. Invert all the plates prepared and incubate at 37°C for 12 to 16 h until colonies are visible.

ISOLATION AND PURIFICATION OF PLASMID DNA BY ALKALINE METHOD

PROTOCOL 1. MINIPREPARATION OF PLASMID DNA

Plasmids can be purified using the CsCl gradient method or minipreparation method. Although we have found that the CsCl gradient method yields excellent results, it is time consuming and relatively expensive. The minipreparation method, on the other hand, is simple, less expensive and gives excellent sequencing results in our hands. The isolated DNA is suitable for restriction enzyme digestion, *in vitro* transcription, DNA subcloning and DNA sequencing.

1. Inoculate a single colony or 10 µl of previously frozen bacteria containing the plasmid DNA of interest in 5 ml of LB medium and 50 µg/ml of appropriate antibiotics (e.g., ampicillin), depending on the specific antibiotic-resistant gene carried by the specific plasmid. Culture the bacteria at 37°C overnight with shaking at 200 rpm.
2. Transfer the overnight culture into microcentrifuge tubes (1.5 ml/tube) and centrifuge at 6000 rpm for 4 min. Carefully aspirate the supernatant.

Note: If the plasmid carried by bacteria is a high-copy number plasmid, 1.5 ml of cell culture usually yields approximately 40 µg plasmid DNA using the present method. In this way, 10 tubes of preparation can generate about 400 µg plasmid DNA, which is equivalent to a large-scale preparation. Therefore, the volume of bacterial culture used for purification of plasmid DNA depends on the particular experiment.

3. Add 0.1 ml of ice-cold lysis buffer to each tube and gently vortex for 1 min. Incubate at room temperature for 5 min.

Note: The function of this step is to lyse the bacteria by hyperlytic osmosis, releasing DNA and other contents.

4. Add 0.2 ml of a freshly prepared alkaline solution to each tube and mix by inversion. **Never VORTEX.** Place the tubes on ice for 5 min.

Note: This step denatures chromosomal DNA and proteins.

5. Add 0.15 ml of ice-cold potassium acetate solution to each tube. Mix by inversion for 10 s and incubate on ice for 5 min.

Note: The principle of this step is to precipitate proteins, polysaccharides and genomic DNA selectively. Under this particular buffer condition, plasmid DNA remains in the supernatant without being precipitated.

6. Centrifuge at 12,000 × *g* for 5 min and **carefully** transfer the supernatant to fresh tubes.
7. To degrade total RNAs from the sample, add RNase (DNase free, Boehringer Mannheim) into the supernatant at 1.5 μl/10 μl supernatant. Incubate the tubes at 37°C for 30 min.

Tips: At this point, RNase and other remaining proteins can be removed by either of the following two options. Option A is faster, cheaper and more efficient than option B.

Option A: Filtration Using Probind Filters (Millipore)

Transfer the RNase-treated supernatant into a minifilter cup (0.4 ml/cup) placed in an Eppendorf tube. Cap the tube and centrifuge at 12,000 × *g* for 30 s. Remove the filter cup and the collected supernatant in the Eppendorf tube is ready for DNA precipitation. Proceed to step 8.

Note: The membrane filter at the bottom of the cup is specially designed to retain proteins and allow nucleic acids to pass through. One cup can be reused four to five times if it is used for the same supernatant pool. However, it is not wise to reuse it for different plasmid samples due to potential contamination. The author routinely utilizes this type of filter for plasmid DNA preparation. DNA obtained is pure and high yield with a ratio (A_{260}/A_{280}) of approximately 2.0.

Option B: Phenol/Chloroform Extraction

a. Add an equal volume of TE-saturated phenol/chloroform to each tube and mix by vortexing for 1 min.
b. Centrifuge at 12,000 × *g* for 5 min and transfer the upper, aqueous phase into fresh tubes.
c. Add an equal volume of chloroform:isoamyl alcohol (24:1) and mix by vortexing for 1 min and centrifuge as in step (b).
d. Centrifuge at 12,000 × *g* for 5 min and transfer the supernatant to fresh tubes. Proceed to step 8.

8. Add 2 to 2.5 volumes of 100% ethanol to each tube. Mix by inversion and allow the plasmid DNA to precipitate for 10 min at –80°C or 30 min at –20°C.
9. Centrifuge at 12,000 × *g* for 10 min. Carefully decant the supernatant, add 1 ml of prechilled 70% ethanol and centrifuge at 12,000 × *g* for 5 min. Carefully aspirate the ethanol and dry the plasmid DNA under vacuum for 15 min.
10. Dissolve the plasmid DNA in 15 µl of TE buffer or sterile deionized water. Take 1 µl to measure the concentration and store the sample at –20°C until use.

Reagents Needed

LB (Luria-Bertaini) Medium (per Liter)

10 g Bacto-tryptone
5 g Bacto-yeast extract
5 g NaCl
Adjust pH to 7.5 with 2 *N* NaOH and autoclave. If needed, add antibiotics after the autoclaved solution has cooled to less than 50°C.

TE Buffer

10 m*M* Tris-HCl, pH 8.0
1 m*M* EDTA

Lysis Buffer

25 m*M* Tris-HCl, pH 8.0
10 m*M* EDTA
50 m*M* Glucose

Alkaline Solution

0.2 *N* NaOH
1% SDS

Potassium Acetate Solution (pH 4.8)

Prepare 60 ml of 5 *M* potassium acetate. Add 11.5 ml of glacial acetic acid and 28.5 ml of H_2O to total volume of 100 ml. This solution is 3 *M* with respect to potassium and 5 *M* with respect to acetate. Store on ice prior to use.

RNase (DNase-Free)

Boehringer Mannheim, Cat. No: 119915, 500 µg/ml.

TE-Saturated Phenol/Chloroform

Thaw crystals of phenol in 65°C water bath with occasional shaking. Mix equal parts of TE buffer and thawed phenol. Let it stand until the phases separate at room temperature. Mix an equal part of the lower, phenol phase with an equal part of chloroform:isoamyl alcohol (24:1).

Chloroform:Isoamyl Alcohol (24:1)

Ethanol: 100% and 70%

Protocol 2. Large-Scale Preparation of Plasmid DNA

These principles are the same as those for the minipreparation method.

1. Inoculate a single colony or 50 μl of previously frozen cells in 100 ml of LB medium containing 50 μg/ml of appropriate antibiotics (e.g., ampicillin). Grow the bacteria at 37°C overnight with shaking at 160 rpm.

Note: The volume of culture can be scaled up, depending on particular experiments, to a few liters and cultured for 16 h with shaking at 240 rpm.

2. Centrifuge at 6000 × *g* for 5 min at 4°C and completely aspirate the liquid.
3. Resuspend the pellet in an appropriate volume (1 ml/15 ml cell culture) of ice-cold lysis buffer and vortex for 2 min. Incubate the tubes for 5 to 10 min at room temperature.
4. Add an appropriate volume (2 ml/15 ml cell culture) of a freshly prepared alkaline denaturation solution and mix by inversion. Incubate the tubes on ice for 5 to 10 min.
5. Add an appropriate volume (1.5 ml/15 ml cell culture) of ice-cold potassium acetate solution, mix by inversion for 20 s and incubate on ice for 5 to 10 min.
6. Centrifuge at 12,000 × *g* for 8 min at 4°C and gently transfer the supernatant to a fresh tube.
7. To degrade total RNAs from the sample, add RNase (DNase free, Boehringer Mannheim) into the supernatant at 10 μl/ml supernatant. Incubate the tubes at 37°C for 30 to 50 min.

Tips: At this point, RNase and other remaining proteins can be removed using either of the following two options.

Option A: Filtration Using Probind Filters (Millipore)

1. Add 2 to 2.5 volumes of 100% ethanol to the supernatant. Mix by inversion and allow the plasmid DNA to precipitate for 10 min at –80°C or 20 min at –20°C.
2. Centrifuge at 12,000 × *g* for 10 min. Carefully decant the supernatant and dry the plasmid DNA under vacuum for 10 min.
3. Dissolve the plasmid DNA in 0.8 ml of TE buffer or sterile deionized water.

Tip: The DNA sample contains RNase and other proteins, which can be readily removed by filtration. The volume (0.8 ml) is suitable for minifiltration using a Probind-filter from Millipore.

4. Transfer the DNA solution into a minifilter cup (0.4 ml/cup) placed in an Eppendorf tube. Cap the tube and centrifuge at 12,000 × *g* for 40 s. Remove the filter cup and the supernatant collected in the Eppendorf tube is ready for DNA precipitation. Proceed to step 7 in Protocol 3.

Note: The membrane filter fixed at the bottom of the cup is specially designed to retain proteins and allow nucleic acids to pass through. One cup can be reused four to five times if it is used for the same supernatant pool. However, it is not wise to reuse it for different plasmid samples due to potential contamination. The author routinely utilizes this type of filter for plasmid DNA preparation. DNA obtained is pure and of high yield with a ratio (A_{260}/A_{280}) of approximately 2.0.

Option B: Phenol/Chloroform Extraction

This detailed protocol has been described earlier in this chapter.

Protocol 3. Purification of Plasmid DNA by CsCl Gradient

Compared with the protocol described previously, this method is relatively complicated, time consuming and costly. Although it usually needs a large-scale culture of bacteria, many investigators comment that the plasmid DNA purified by CsCl is pure and may be more suitable for DNA sequencing.

1. Inoculate a single colony or 50 µl of previously frozen cells in 100 ml of LB medium containing 50 µg/ml of appropriate antibiotics (e.g., ampicillin). Grow the bacteria at 37°C overnight with shaking at 160 rpm.

Note: The volume of culture can be scaled up, depending on particular experiments, to a few liters and cultured for 16 h with shaking at 240 rpm.

2. Centrifuge at 6000 × *g* for 5 min at 4°C and completely aspirate the liquid.
3. Resuspend the pellet in an appropriate volume (1 ml/15 ml cell culture) of ice-cold lysis buffer and vortex for 2 min. Incubate the tubes for 5 to 10 min at room temperature.
4. Add an appropriate volume (2 ml/15 ml cell culture) of a freshly prepared alkaline denaturation solution and mix by inversion. Incubate the tubes on ice for 5 to 10 min.
5. Add an appropriate volume (1.5 ml/15 ml cell culture) of ice-cold potassium acetate solution, mix by inversion for 20 s and incubate on ice for 5 to 10 min.
6. Centrifuge at 12,000 × *g* for 8 min at 4°C and gently transfer the supernatant to a fresh tube.
7. Add 0.8 volumes of isopropanol to the supernatant, mix well, and allow the sample to precipitate for 15 min at room temperature.
8. Centrifuge at 12,000 × *g* for 10 min at 4°C, aspirate the supernatant, and gently rinse the pellet with 6 ml of prechilled 70% ethanol. Dry the DNA under vacuum for 20 min.

9. Dissolve the DNA in 4 ml of TE buffer or sterile deionized water. Store at 4°C until use for CsCl gradient centrifugation.
10. Accurately measure the volume of the DNA sample and add 1 g/ml of solid CsCl. Warm the mixture at 30°C to dissolve the CsCl and gently mix well.
11. Add 0.35 ml of 10 mg/ml of ethidium bromide (EB) per 5 ml of the DNA–CsCl solution. Quickly and gently mix the solution.

Caution: Ethidium bromide is a powerful mutagen, so gloves should always be worn when working with the solutions. The used EtBr solution must by collected in a special container to inactivate the chemical gradually before it is disposed under specific regulation. Stock solutions of EtBr should be stored in light-tight bottle at 4°C.

12. Centrifuge at 8000 rpm for 6 min at room temperature with a Sorvall SS34 rotor (or its equivalent).
13. Gently transfer the clear, reddish DNA–CsCl solution under the surface scum into Quick-seal ultracentrifuge tubes (Beckman) using a Pasteur pipette or a disposable syringe fitted with a large-gauge needle.

Tip: Fill the tubes up to the neck of the tube, which is about 5.2 ml/tube. If the tube is overfilled, it will cause a problem when sealing the tube. On the other hand, if the tube is underfilled, a large air bubble may present after sealing the tube and cause the tube to collapse. Carefully balance the tubes using CsCl balance solution.

14. Turn on and warm the sealing machine for 5 min and place the tubes on a metal sealing plate. Cover each tube with an appropriate metal cap and place, one at a time, under the sealing rack. Push and hold the hot rack against the metal cap of the tube for 30 to 60 s. The neck of the tube will gradually melt until a sound is heard from the sealing machine. Immediately remove the tube from the melt rack and place the cap of the tube against a cold metal strip. Press and hold the strip against the cap for about 1 to 2 min to seal the tube completely and then take out the cap.
15. Attach a cap and a metal adapter to each sealed tube for ultracentrifugation and carefully balance again with a piece of tape if necessary. Place the tubes symmetrically into the Beckman vertical rotor and make sure to cover them with their own caps and adapters.
16. Carry out centrifugation of the CsCl density gradients at 45,000 rpm for 16 h (VTi65), 45,000 rpm for 48 h (Ti50), 60,000 rpm for 24 h (Ti65), or 60,000 rpm for 24 h (Ti70.1) at 20°C.

Note: Never centrifuge at 4°C or it will cause the CsCl to precipitate out of solution, reducing the density of CsCl in the gradient.

17. Stop centrifugation and slowly take the tubes out of the rotor and place them in a plastic rack. Use long-wavelength UV light (360 nm) to visualize the relaxed and supercoiled DNA bands.

Tips: Two bands of DNA are visible in the center of the gradient, which usually contains linear bacterial chromosomal DNA and nicked circular plasmid DNA. The lower band consists of circular plasmid DNA. The deep-red pellet on the bottom of the tube is ethidium bromide/RNA. The materials on the top surface are proteins.

18. To remove the DNA, place the tube in a ringstand rack. Locate the DNA bands using 360 nm UV light. Wipe dry one side of the DNA band area using ethanol and place a piece of tape on the area. Carefully puncture through the tube with a needle and syringe (16 to 20 gauge) until the needle has penetrated into the closed circular plasmid DNA band area. Slowly withdraw a minimal volume of DNA sample and then remove the needle. Transfer the sample into a fresh tube. Discard the tubes containing the rest of the solution in a special container for disposal.
19. Extract ethidium bromide from the DNA sample by adding one volume of isopropanol. Gently invert the tube and allow the phases to separate by gravity. The EtBr usually partitions into the upper propanol phase and can be removed with a pipette. The lower aqueous phase contains DNA–CsCl with some EtBr, which should be reextracted four to five times until no pink color is visible.
20. Transfer the lower phase into a fresh Corex tube and dilute the CsCl with two volumes of TE buffer. Add two volumes of chilled 100% ethanol into the diluted sample and precipitate the DNA at –20°C for 2 h to overnight.

Note: Never place the tubes at –80°C because the CsCl will precipitate.

21. Centrifuge at 12,000 × *g* for 15 min at 4°C and dry the DNA pellet under vacuum for 15 min. Resuspend the DNA in 100 to 200 μl of TE buffer.
22. Repeat step 19 to further extract CsCl.
23. Centrifuge at 12,000 × *g* for 15 min at 4°C and dry the DNA pellet under vacuum for 15 min. Resuspend the DNA in 100 to 200 μl of TE buffer. Store the sample at –20°C until use.

VERIFICATION OF DNA INSERTION BY RESTRICTION ENZYME DIGESTION AND AGAROSE GEL ELECTROPHORESIS

This is a mandatory step to provide evidence whether insert DNA is cloned at the right site in the vector. Detailed protocols for restriction enzyme digestion and agarose gel electrophoresis were given earlier. The present section mainly focuses on analysis of the restriction enzyme digestion pattern — for example, if a cDNA of interest is cloned between *Not* I and *Xho* I sites at the MCS of vector Bluescript SK(+), assuming that *Not* I and *Xho* I are unique in vector and insert. As shown in Figure 4.3, after ligation and transformation, plasmid DNA is isolated from selected colonies. After digestion of the plasmid DNA with the same restriction enzymes, *Not* I and *Xho* I, electrophoresis will be carried out. If all goes well, the pattern of DNA bands should be as expected. That is, two bands are the same sizes as the insert and vector used for ligation. If the insert size or vector size does not match the original size, it is likely that some mutation may occur during ligation or transformation. In this case, the best way to solve the problem is to forget about the colony and start again.

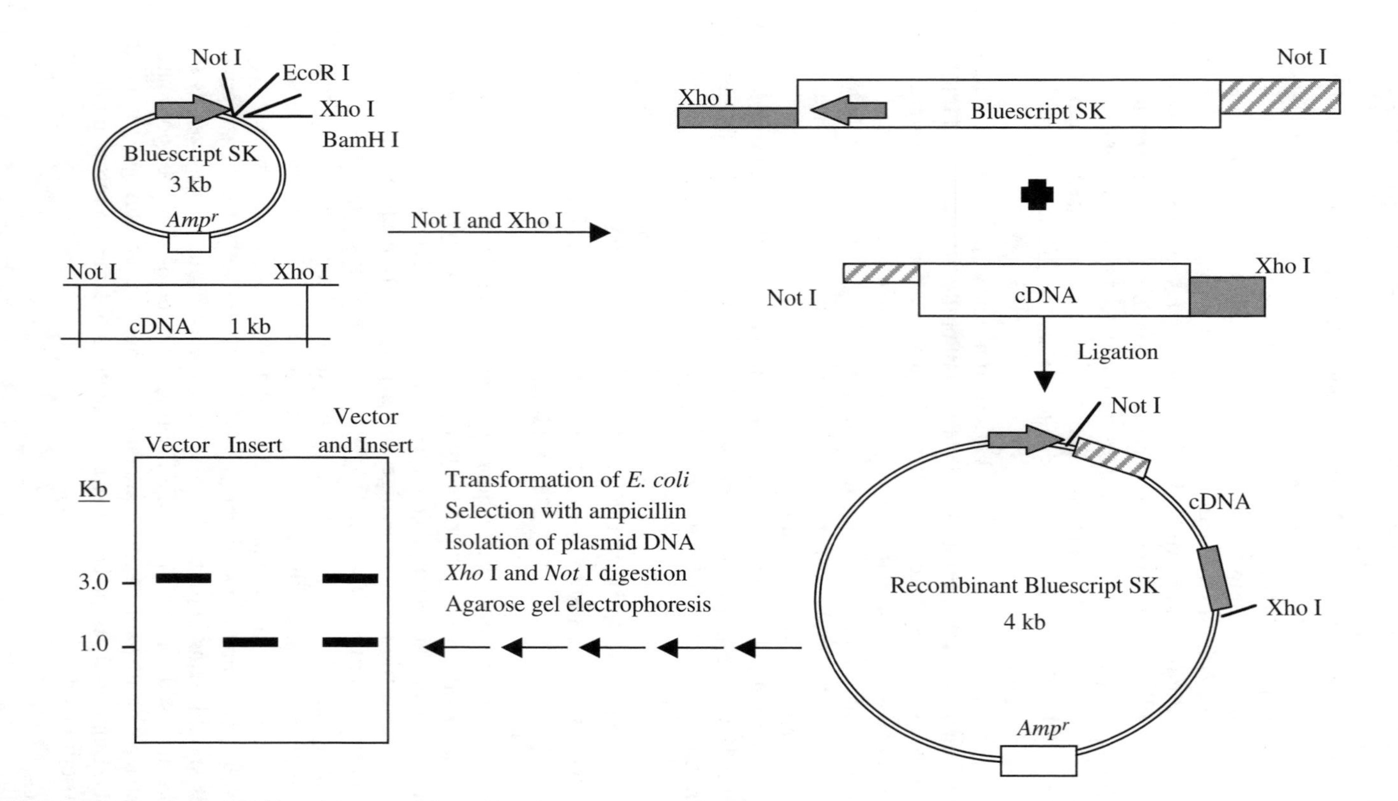

FIGURE 4.3 Diagram of DNA ligation, transformation, selection and verification of recombinant plasmid DNA by agarose gel electrophoresis.

VERIFICATION OF INSERTION SITE BY DNA SEQUENCING

The detailed protocols for DNA sequencing are described in Chapter 5. The major purpose of doing this is to design primers that can be used to specifically sequence the ligation sites to ensure that the sequence at the insertion site is not altered, including potentially missing, adding, or substituting a base.

TROUBLESHOOTING GUIDE

1. **Following restriction enzyme digestion and electrophoresis, the number of bands is as expected but the sizes of one or more bands are not.** This is most likely due to inaccurate information of the restriction enzyme site in the vector or insert DNA. Try to be careful when analyzing the restriction enzyme map, cutting site and sizes of DNA fragments.
2. **Many unexpected bands show up after a single restriction enzyme digestion of the vector or insert DNA.** It seems that the conditions set up for restriction enzyme digestion are not appropriate. As a result, the enzyme activity is not specific, but has cut DNA into pieces rather randomly. Make sure that the reagents and temperature are optimal for the enzyme.
3. **Compared with undigested DNA, incomplete restriction enzyme digestion occurs.** The solution is to increase slightly units of the restriction enzymes or increase digestion time for one more hour. Additionally, try to use fresh enzymes.
4. **No colonies are observed following antibiotic selection.** This is a complicated situation. Most likely, no ligation occurred between the vector and insert DNA. All the DNA molecules transformed into bacteria are linear DNA degraded rapidly in bacteria. Or, extremely high concentration of antibiotics such as ampicillin may be used in LB plates, which will kill all the bacteria, including transformants. Try to increase the efficiency of ligation and make sure that an appropriate antibiotic is utilized in the LB plates.
5. **A large percentage of colonies is due to self-ligation.** This is a common problem when ligation occurs between the single-restriction, enzyme-digested vector and the insert DNA. The effective solution is to increase the units of alkaline phosphatase to remove the 5′ phosphate group from the vector prior to its ligation with insert DNA. Whenever possible, try to choose two unique enzymes that can generate two incompatible sticky ends to eliminate self-ligation.
6. **The size or pattern of digested DNA does not match those of the vector and insert used for ligation.** This may be due to mutation occurring during transformation or replication of plasmid DNA in bacteria. If this happens, there is no effective way to solve the problem except for eliminating the bacterial colonies.

REFERENCES

1. Wu, W. and Welsh, M.J., A method for purification of DNA species from agarose gels by high speed centrifugation, *Anal. Biochem.*, 229, 350–352, 1995.
2. Sambrook, J., Fritsch, E.F., and Maniatis, T., *Molecular Cloning: A Laboratory Manual*, Cold Spring Harbor University Press, Cold Spring Harbor, NY, 1889.
3. Wu, W., Gene transfer and expression in animals, in *Handbook of Molecular and Cellular Methods in Biology and Medicine,* Kaufman, P.B., Wu, W., Kim, D., and Cseke, L., CRC Press, Boca Raton, FL, 1995.

5 Nonisotopic and Isotopic DNA or RNA Sequencing

CONTENTS

INTRODUCTION

DNA and RNA sequencing is an essential technique that is mandated for DNA cloning, characterization, mutagenesis, DNA recombination and gene expression.[1–3] Very recently, nucleic acid sequencing has become a rapid and powerful tool to identify novel genes targeted by a novel drug or protein. To discover new target molecules, one may not need to utilize traditional gene or cDNA cloning methodologies nor does the isolated gene or cDNA need to be completely sequenced. Instead, as long as a portion of exon sequences is sequenced, a potentially novel DNA molecule can be identified by computer database searching. Nevertheless, nucleic acid sequencing is a crucial tool for one to fulfill such a task. How does one sequence the DNA of interest? How does one identify the existence of mRNA expression in a cell or tissue type of interest? Which methodology is the best selection for one's purposes? The present chapter summarizes the author's extensive experience and will help one achieve one's goals.

There are several well-established methods for nucleic acid sequencing. Based on the author's years of hands-on sequencing experience, this chapter describes in detail the protocols for DNA or RNA sequencing by nonisotopic or isotopic sequencing approaches. Specific methodologies include nonisotopic and isotopic DNA sequencing by dideoxynucleotides chain termination,[2,3] formamide gel sequencing,[3,4] primer walking,[5] unidirectional deletions,[6] direct sequencing by PCR and RNA sequencing[2,7–9] with modifications. These methods have been successfully employed in our laboratories. Although automatic fluorescent sequencing is a very attractive methodology, because of the high cost of the DNA sequencer, we do not consider this method to be within the scope of the present chapter, which is designed for regular nucleic acid sequencing in modestly equipped laboratories.

Since Sanger et al.[2] developed the dideoxynucleotides chain termination method of DNA sequencing, it has been extensively modified by the use of some superior enzymes and DNA cloning vectors. The general principles of the method include: (1) a synthesized oligonucleotide primer anneals to the 3′ end of the DNA template to be sequenced; (2) a DNA polymerase catalyzes the *in vitro* synthesis (5′ Ø 3′) of a new DNA strand complementary to the template, starting from the primer site, using deoxynucleoside 5′-triphosphates (dNTPs). One of the dNTPs is biotinylated dATP or dCTP used for nonisotopic sequencing, or α-^{35}SdATP or α-^{35}SdCTP utilized for isotopic sequencing; and (3) following the synthesis of the new DNA strand, it is terminated by the incorporation of a nucleotide analog that is an appropriate 2′,3′-dideoxynucleotide 5′-triphosphate (ddNTP). All four ddNTPs lack the 3′-OH group required for DNA chain elongation. Based on the nucleotide bases of the DNA template, one of the four ddNTPs in each reaction is incorporated. The enzyme-catalyzed polymerization will then be stopped at each site where a ddNTP is incorporated, generating a population of chains of different sizes. Therefore, by setting up four separate reactions, each with a different ddNTP, complete nucleotide sequence information of the DNA strand will be revealed.

In DNA or RNA sequencing, different factors and strategies should be considered. The following factors greatly influence nucleic acid sequencing.

1. **Quality of template.** The purity and integrity of DNA or RNA to be sequenced are essential for obtaining an accurate and complete nucleotide sequence. Make sure that the DNA or RNA template is not degraded prior to sequencing. The DNA template can be single-stranded DNA from M13 cloning vectors, double-stranded plasmid DNA or double-stranded bacteriophage DNA with appropriate pretreatment. mRNA template is usually transcribed into cDNA followed by standard DNA sequencing.
2. **DNA polymerase.** United States Biochemical (USB) and Amersham Life Science produce excellent DNA polymerase for nucleic acid sequencing. These include Sequenase version 2.0 DNA polymerase and AMV reverse transcriptase, which are routinely used in our laboratory with successful results. Sequenase version 2.0 DNA polymerase is a superior enzyme that has no $3' \rightarrow 5'$ exonuclease activity. These enzymes should be kept at –20°C prior to use.
3. **Primers.** For sequencing the region of the DNA template close to the cloning site of the vector, primers are complementary to the vector DNA strands and are commercially available. For the sequence beyond 250 to 500 nucleotides from the cloning site, new primers should be designed based on the nucleotide sequence information obtained from the very last sequencing. The quality of primers is critical for the success of DNA or RNA sequencing. Primers can be designed by computer programs such as Oligo version 4.0 or CPrimer f. We routinely check with the Tm value, potential intrahairpin structures, potential interloop formation and GC contents of the designed primers before employing them in sequencing.
4. **Nonisotopic or isotopic dNTP.** Biotinylated dATP or dCTP is commonly used for nonisotopic DNA or RNA sequencing. The drawback is potential background due to the high sensitivity. Nonetheless, the nonisotopic approach is faster and safer compared with radioactive sequencing. In the near future, traditional isotopic methods may be completely replaced by nonisotopic methodology. ^{32}P-labeled dNTP (dATP or dCTP) has a high energy level and a short half-life (usually 14 days). A major disadvantage of using ^{32}P is that it gives diffuse bands on autoradiographic x-ray film, limiting readable information of DNA sequences. However, [^{32}S]dATP has a lower energy level and a longer half-life (usually 84 to 90 days) and greatly improves autoradiographic resolution. Therefore, [^{32}S]dATP is recommended for isotopic DNA or RNA sequencing. The activity and amount of labeled nucleotide play a significant role in nucleic acid sequencing.
5. **Compressions.** Compressions represent a common problem in DNA sequencing, which is primarily due to dG-dC rich regions that may not be fully denatured during electrophoresis. As a result, interruption of the normal pattern of migration of DNA fragments becomes obvious. The bands are usually spaced closer than usual (compressed together) or occur further apart than usual, resulting in a significant loss of sequence information. In order to solve this problem, we use dITP or 7-deaza-dGTP (USB) to replace the nucleotide dGTP, which forms a weaker secondary structure that can

be readily denatured during electrophoresis. In addition, a formamide gel is helpful in this compression situation. With these modifications, we have found that bands are sharper and compressions eliminated by the use of dITP and formamide. In some cases, some bands appear weak using both dGTP and dITP. This limitation can be eliminated using pyrophosphatase (USB) in the presence of Mg^{2+}, Mn^{2+}, or both. The manganese can allow one to obtain sequence information close to the primer site.

NONISOTOPIC DNA SEQUENCING METHOD

The following protocols are adapted from Sanger et al.:[2] Sequenase Images™ nonisotopic sequencing (USB) and Sequenase version 2.0 (USB) with some modifications.

PROTOCOL 1. PREPARATION OF DNA TEMPLATES FOR SEQUENCING

Purification of Double-Stranded Plasmid DNA by the Alkaline Method

Following denaturation, double-stranded plasmids containing DNA inserts of interest can be directly sequenced. One can simultaneously sequence both strands of the DNA insert at opposite directions using forward or reverse primers. The primers are annealed to the appropriate sites at their 3′ ends of the DNA insert, in two separate sequencing reactions. This simultaneous sequencing approach can speed up the sequencing procedure and is especially useful for sequencing large DNA fragments. Double-stranded plasmids can be purified using the CsCl gradient method or the minipreparation method. Although we have found that the CsCl gradient method produces excellent results, it is time consuming and relatively expensive. The minipreparation method, on the other hand, is simple, less expensive and gives excellent sequencing results. The detailed protocol for isolation and purification of plasmid DNA is described next.

1. Inoculate a single colony or 10 μl of previously frozen bacteria containing the plasmid DNA of interest in 5 ml of LB medium and 50 μg/ml of appropriate antibiotics (e.g., ampicillin), depending on the specific antibiotic-resistant gene carried by the specific plasmid. Culture the bacteria at 37°C overnight with shaking at 200 rpm.
2. Transfer the overnight culture into a microcentrifuge tubes (1.5 ml/tube) and centrifuge at 6000 rpm for 4 min. Carefully aspirate the supernatant.

Note: If the plasmid carried by bacteria is a high-copy number plasmid, 1.5 ml of cell culture usually gives a yield of approximately 40 μg plasmid DNA using the present method. In this way, 10 tubes of preparation can generate about 400 μg plasmid DNA, which is equivalent to a large-scale preparation. Therefore, the volume of bacterial culture used for purification of plasmid DNA depends on the particular experiment.

3. Add 0.1 ml of ice-cold lysis buffer to each tube and gently vortex for 1 min. Incubate at room temperature for 5 min.

Principle: The function of this step is to lyse the bacteria by hyperlytic osmosis, releasing DNA and other contents.

4. Add 0.2 ml of a freshly prepared alkaline solution to each tube and mix by inversion. **NEVER VORTEX.** Place the tubes on ice for 5 min.

Principle: This step denatures chromosomal DNA and proteins.

5. Add 0.15 ml of ice-cold potassium acetate solution to each tube. Mix by inversion for 10 s and incubate on ice for 5 min.

Principle: This step is to precipitate proteins, polysaccharides, and genomic DNA selectively. Under these particular buffer conditions, plasmid DNA still remains in the supernatant without being precipitated.

6. Centrifuge at 12,000 × *g* for 5 min and carefully transfer the supernatant to fresh tubes.
7. Add RNase (DNase-free, Boehringer Mannheim) to the supernatant at 1.5 µl/0.1 ml supernatant. Incubate the tubes at 37°C for 30 min.

Principle: RNase degrades total RNAs from the sample.

8. Remove RNase and other remaining proteins by using one of the following two options. Option A is faster, less expensive and more efficient than Option B.

Option A: Filtration Using a Probind Filter (Millipore)

Transfer the RNase-treated supernatant into a minifilter cup (0.4 ml/cup) placed in an Eppendorf tube. Cap the tube and centrifuge at 12,000 × *g* for 30 s. Remove the filter cup and the collected supernatant in the Eppendorf tube is ready for DNA precipitation. Proceed to step 9.

Note: The membrane filter attached at the bottom of the cup is specially designed to retain proteins and to allow nucleic acids to pass through. One cup can be reused for four to five times if one uses it for the same supernatant pool. However, it is not wise to reuse it for different plasmid samples due to potential contamination. The author routinely utilizes this type of filter for plasmid DNA preparations. DNA obtained is pure and of high yield with a ratio (A_{260}/A_{280}) of approximately 2.0.

Option B: Phenol/Chloroform Extraction

a. Add an equal volume of TE-saturated phenol/chloroform to each tube and mix by vortexing for 1 min.
b. Centrifuge at 12,000 × *g* for 4 min and transfer the upper, aqueous phase into fresh tubes.
c. Add an equal volume of chloroform:isoamyl alcohol (24:1), mix by vortexing for 1 min and centrifuge as in the previous step.
d. Centrifuge at 12,000 × *g* for 4 min and transfer the supernatant to fresh tubes. Proceed to step 9.

9. Add 2 to 2.5 volumes of 100% ethanol to each tube. Mix by repeat inversions and allow the plasmid DNA to precipitate for 10 min at –80°C or 30 min at –20°C.
10. Centrifuge at 12,000 × *g* for 10 min. Carefully decant the supernatant, add 1 ml of prechilled 70% ethanol and centrifuge at 12,000 × *g* for 5 min. Carefully aspirate the ethanol and dry the plasmid DNA under vacuum for 15 min.
11. Dissolve the plasmid DNA in 15 μl of TE buffer or sterile deionized water. Take 1 μl to measure the concentration and store the sample at –20°C until use.

Reagents Needed

LB (Luria-Bertaini) Medium (per liter)

10 g Bacto-tryptone
5 g Bacto-yeast extract
5 g NaCl
Adjust pH to 7.5 with 2 *N* NaOH and autoclave. If needed, add antibiotics after the autoclaved solution has cooled to less than 50°C.

TE Buffer

10 m*M* Tris-HCl, pH 8.0
1 m*M* EDTA

Lysis Buffer

25 m*M* Tris-HCl, pH 8.0
10 m*M* EDTA
50 m*M* Glucose

Alkaline Solution

0.2 *N* NaOH
1% SDS

Potassium Acetate Solution (pH 4.8)

Prepare 60 ml of 5 *M* potassium acetate. Add 11.5 ml of glacial acetic acid and 28.5 ml of H_2O. Total volume is 100 ml. This solution is 3 *M* with respect to potassium and 5 *M* with respect to acetate. Store on ice prior to use.

RNase (DNase-Free)

Boehringer Mannheim, Cat. No: 119915, 500 μg/ml.

TE-Saturated Phenol/Chloroform

Thaw crystals of phenol in 65°C water bath with occasional shaking. Mix equal parts of TE buffer and thawed phenol. Let it stand until the phases separate at room temperature. Mix an equal part of the lower phenol phase with an equal part of chloroform:isoamyl alcohol (24:1).

Chloroform:Isoamyl Alcohol (24:1)

Ethanol: 100% and 70%

Purification of Single-Stranded DNA

Single-stranded DNA (ssDNA) utilized as a template usually gives excellent sequencing results. The following protocol works quite well for purification of ssDNA from plasmid-phage (phagemid) vectors, which include M13mp9, M13mp12, M13mp13, M13mp18 and M13mp19.

1. Streak an appropriate *E. coli* strain on an LB plate, which carries the putative phagemids containing the DNA insert of interest. Invert the plate and incubate at 37°C overnight in order to obtain individual colonies.
2. Inoculate a single colony in 3 ml of LB medium containing 50 μg/ml ampicillin and 12 μg/ml tetracycline or appropriate antibiotic. Incubate at 37°C overnight with shaking at 200 rpm.
3. Add 0.3 ml of the cell culture to 3 ml of superbroth in a 50-ml conical tube. Incubate at 37°C with shaking at 200 rpm for 2 to 3 h.
4. Add 8 ml of helper phage R408 (pfu = 1×10^{11}, available from Stratagene) to the culture at step 3 and continue to culture for 8 to 10 h.
5. Centrifuge at 6000 rpm for 4 min and carefully transfer the supernatant to a fresh tube.
6. Based on the volume of the supernatant, add 1/4 volume of polyethylene glycol (PEG) precipitation buffer containing 3.5 *M* ammonium acetate buffer (pH 7.5) and 20% (w/v) PEG to the tube. Vortex for 1 min and leave it at room temperature for 20 min.
7. Centrifuge at 12,000 × *g* for 15 min and aspirate the supernatant completely.
8. Resuspend the PEG pellet in 1/4 volume (vs. the volume of the supernatant at step 5) of TE buffer (pH 8.0) and extract it with 1 volume of TE-saturated phenol:chloroform:isoamyl alcohol (25:24:1). Mix by vortexing and centrifuge at 11,000 × *g* for 5 min.
9. Carefully transfer the top, aqueous phase to a fresh tube and repeat extraction twice as in the previous step.
10. Measure the volume of the supernatant and precipitate the single-stranded DNA by adding 0.5 volume of 7.5 *M* ammonium acetate (pH 7.5) and 2.5 volumes of chilled 100% ethanol to the supernatant. Place at –70°C for 30 min.
11. Centrifuge at 12,000 × *g* for 20 min at 4°C and aspirate the supernatant. Wash the DNA pellet by adding 5 to 10 ml of 70% ethanol and spin down at 12,000 × *g* for 5 min. Carefully aspirate the ethanol and dry the pellet under vacuum for 20 min. Dissolve the DNA in 40 μl of TE buffer (pH 7.6). Take 2 μl to measure the concentration of single-stranded DNA at A_{260} and A_{280} nm. Store the sample at –20°C until use.

Reagents Needed

Superbroth Medium

Bacto-tryptone, 12 g
Bacto-yeast extract, 24 g
0.4% (v/v) Glycerol
Dissolve in a total of 900 ml in dd.H_2O and autoclave. Cool to about 50°C and add 100 ml of phosphate buffer containing 170 m*M* KH_2PO_4 and 720 m*M* K_2HPO_4. Autoclave again.

PEG Precipitation Buffer

3.5 *M* Ammonium acetate, pH 7.5
20% (w/v) Polyethylene glycol (PEG) 8000

PROTOCOL 2. SEQUENCING REACTIONS

METHOD A. SEQUENCING REACTIONS FOR DOUBLE-STRANDED PLASMID DNA

1. Denature double-stranded DNA as follows:
   a. To denature plasmid DNA, the alkaline denaturation method is recommended. Transfer an appropriate volume of purified plasmid DNA (approximate 1 μg/μl) to a microcentrifuge tube and add one volume of freshly prepared alkaline solution containing 0.4 *M* NaOH and 0.4 m*M* EDTA (pH 8.0). Incubate at 37°C for 35 min.
   Tip: The amount of DNA to be denatured should be in excess of the amount of DNA to be sequenced. For example, if one uses 3 to 5 μg DNA for one primer sequencing reaction, the amount of DNA for two primer sequencing reactions for both strands at opposite directions will be 6 to 7 μg. If the yield of DNA following denaturation, neutralization and precipitation is 80%, the amount of DNA used for denaturation should be 7.2 to 8.4 μg. We routinely double the amount of DNA for denaturation and measure the DNA concentration after it is precipitated. This ensures a sufficient amount of DNA for sequencing.
   b. After denaturation is complete, add 0.1 volume of 3 *M* sodium acetate buffer (pH 5.2) to the denatured sample to neutralize the mixture.
   c. Add 2 to 5 volumes of 100% ethanol to the mixture and precipitate DNA at –70°C for 15 min.
   d. Centrifuge at 12,000 × *g* for 10 min and decant the supernatant. Wash any remaining salt from the DNA pellet with 1 ml of chilled 70% ethanol and spin down at 12,000 × *g* for 5 min. Dry the DNA under vacuum for 15 min.
   e. Dissolve the DNA in 10 μl dd.H_2O and place the tube on ice. Quickly measure the concentration of DNA using 1 to 2 μl of the sample at A_{260} and A_{280} nm. Immediately proceed to the primer annealing reaction.
2. Carry out template–primer annealing as follows:

a. Transfer 1 to 2 μg or 1 pmol of freshly denatured plasmid DNA to a microcentrifuge tube on ice and add dd.H_2O to a total volume of 7 μl. Add 2 μl of reaction buffer and 1 μl of 5′-biotinylated sequencing primer (2 to 3 pmol). This gives approximately 1:1 (template:primer) molar stoichiometry with a total volume of 10 μl.

   DNA + dd.H_2O, 7 μl
   5X Reaction buffer, 2 μl
   5′-Biotinylated primer, 1 μl
   Total = 10 μl

b. Place the tube in a plastic rack or its equivalent and heat at 65°C for 2 to 3 min. Quickly transfer the rack together with the tube to a beaker or tray containing an appropriate volume of 60 to 63°C water and then let cool slowly to <30°C or room temperature over 20 to 30 min. If one uses a heating block, the tubes can be heated at 65°C for 2 min followed by slow cooling to room temperature by turning off the heat so that the heating block will slowly cool down. When the temperature drops to <30°C, the annealing is complete. The real annealing temperature is 50 to 52°C. Some laboratories prefer to anneal at 50 to 55°C for 15 to 30 min followed by slow cooling.

3. Prepare labeling and termination.
   a. While the annealing mixture is cooling, remove the necessary materials from a commercial sequencing kit that is usually stored at –20°C. Thaw the materials on ice.
   b. For each template–primer mixture, label four microcentrifuge tubes A, G, T and C, representing ddATP, ddGTP, ddTTP and ddCTP, respectively.
   c. Transfer 2.5 μl of the termination mixture of ddATP, ddGTP, ddTTP and ddCTP to the labeled tubes A, G, T, and C, respectively. Cap the tubes and keep at room temperature until use.
   d. Dilute the labeling mixture fivefold as a working concentration and store on ice prior to use. For example, 2 μl of labeling mixture is diluted to total 10 μl with dd.H_2O.

      Tips: (1) Mix the mixture in each stock tube well by pipetting it up and down prior to removal of an appropriate amount of mixture from each tube. (2) Two sets of labeling mixtures and termination mixtures are in the sequencing kit (USB). For regular noncompression sequencing, the one labeled dGTP should be used. However, if compression appears due to G–C-rich sequences, the one labeled dITP is strongly recommended. dITP replaces the nucleotide analog dGTP. We have found that it significantly eliminates the compressed bands.
   e. Dilute Sequenase version 2.0 T7 DNA polymerase (1:8 dilution) as follows:

      Ice-cold enzyme dilution buffer, 6.5 μl
      Pyrophosphatase, 0.5 μl
      DNA polymerase stored at –20°C, 1 μl

Tips: The enzyme should not be diluted in glycerol enzyme dilution buffer if one uses TBE in the gel and in the running buffer. The diluted enzyme should be stored on ice and be used within 1 h.

4. Perform a labeling reaction when the template–primer annealing is complete.
   a. Briefly spin down the annealed mixture and store on ice until use.
   b. Add the following to the tube containing 10 µl of annealed mixture in the order shown:
      For use of a biotinylated primer:
         DTT (0.1 M), 1 µl
         dd.H_2O, 2.5 µl
         Diluted Sequenase DNA polymerase, 2 µl
      For use of a biotinylated nucleotide
         DTT (0.1 M), 1 µl
         Diluted labeling mixture, 2.5 µl
         Diluted Sequenase DNA polymerase, 2 µl
      Tip: Add 1 µl of Mn buffer to enhance the bands close to the primer.
   c. Gently mix well to avoid any air bubbles. For use of a biotinylated nucleotide in a labeling reaction, incubate at room temperature for 3 to 5 min. Proceed to the next step.
5. Carry out the termination reactions as follows:
   a. Carefully and quickly transfer 3.5 µl of the mixture to each of the termination tubes (A, G, T, C) prewarmed at 37°C for 1 min. Mix and continue to incubate the reaction at 37°C in a heating block for 4 to 5 min.
   b. Add 4 µl of stopping solution to each tube, mix and cap the tubes. Store at –20°C until use.
      Tip: The samples should be electrophoresed within 4 days even when stored at –20°C. Denature the sample at 75 to 80°C for 2 to 3 min prior to loading into a sequencing gel.

Reagents Needed

5X Reaction Buffer

0.2 *M* Tris-HCl, pH 7.5
0.1 *M* $MgCl_2$
0.25 *M* NaCl

DTT Solution

100 m*M* Dithiothreitol

5X Labeling Solution for dGTP

7.5 µ*M* dGTP
7.5 µ*M* dTTP
7.5 µ*M* dCTP or dATP
Biotinylated dATP or dCTP

5X Labeling Mixture for dITP

7.5 μ*M* dITP
7.5 μ*M* dTTP
7.5 μ*M* dCTP or dATP
Biotinylated dATP or dCTP

ddATP Termination Mixture for dGTP

80 μ*M* dATP
80 μ*M* dGTP
80 μ*M* dCTP
80 μ*M* dTTP
8 μ*M* ddATP

ddGTP Termination Mixture for dGTP

80 μ*M* dATP
80 μ*M* dGTP
80 μ*M* dCTP
80 μ*M* dTTP
8 μ*M* ddGTP

ddCTP Termination Mixture for dGTP

80 μ*M* dATP
80 μ*M* dGTP
80 μ*M* dCTP
80 μ*M* dTTP
8 μ*M* ddCTP

ddTTP Termination Mixture for dGTP

80 μ*M* dATP
80 μ*M* dGTP
80 μ*M* dCTP
80 μ*M* dTTP
8 μ*M* ddTTP

ddATP Termination Mixture for dITP

80 μ*M* dATP
80 μ*M* dITP
80 μ*M* dCTP
80 μ*M* dTTP
8 μ*M* ddATP

ddGTP Termination Mixture for dITP

80 μ*M* dATP
160 μ*M* dITP
80 μ*M* dCTP
80 μ*M* dTTP
8 μ*M* ddGTP

ddCTP Termination Mixture for dITP

80 μ*M* dATP
80 μ*M* dITP
80 μ*M* dCTP
80 μ*M* dTTP
8 μ*M* ddCTP

ddTTP Termination Mixture for dITP

80 μ*M* dATP
80 μ*M* dITP
80 μ*M* dCTP
80 μ*M* dTTP
8 μ*M* ddTTP

Sequence Extending Mixture for dGTP

180 μ*M* dATP
180 μ*M* dGTP
180 μ*M* dCTP
180 μ*M* dTTP

Sequence Extending Mixture for dITP

180 μ*M* dATP
360 μ*M* dITP
180 μ*M* dCTP
180 μ*M* dTTP

Mn Buffer (only for dGTP)

150 m*M* Sodium isocitrate
100 m*M* $MnCl_2$

Enzyme Dilution Buffer

10 m*M* Tris-HCl, pH 7.5
5 m*M* DTT
0.5 mg/ml BSA

Glycerol Enzyme Dilution Buffer

20 m*M* Tris-HCl, pH 7.5
2 m*M* DTT
0.1 m*M* EDTA
50% Glycerol

Sequenase Version 2.0 T7 DNA Polymerase

13 units/μl in
20 m*M* KPO_4, pH 7.4
1 m*M* DTT

0.1 m*M* EDTA, pH 7.4
50% Glycerol

Pyrophosphatase
5 units/ml in
10 m*M* Tris-HCl, pH 7.5
0.1 m*M* EDTA, pH 7.5
50% Glycerol

Stop Solution
20 m*M* EDTA, pH 8.0
95% (v/v) Formamide
0.05% Bromophenol blue
0.05% Xylene cyanol FF

Method B. Sequencing Reactions for Single-Stranded DNA

The procedures in Method B are very similar to those described in Method A except for the following:

- No double-stranded DNA denaturation is mandated. Instead, the appropriate primer and template DNA can be directly annealed in an annealing reaction.
- The temperature for primer–template annealing can be 50 to 55°C in a heating block for 15 to 25 min followed by slow cooling by turning off the heat.

PROTOCOL 3. PREPARATION OF SEQUENCING GELS

Gel preparation is one of the most important steps for success in DNA sequencing. The DNA sequencing gel is very thin (0.2 to 0.4 mm thick) and its quality and integrity directly influence the resolution and maximum reading information of DNA sequences. Two common problems may be encountered when preparing a sequencing gel. One is the generation of air bubbles in the gel, while the other is potential leaking of gel mixture from the bottom or the side edges of the gel apparatus. Either problem can cause failure or poor quality of DNA sequencing. In our experience, the following detailed protocol, which is routinely and successfully used in our laboratory, can help solve these problems.

1. Prepare a gel mixture as follows:
   a. Dissolve an appropriate amount of urea in dd.H_2O in a clean beaker using a stir bar.

For Use of Long Ranger Gel Mixture from AT Biochemicals

Components	Small size gel (50 ml)	Large size gel (100 ml)
Ultrapure urea (7 *M*)	21 g	42 g
dd.H_2O	12 ml	24 ml

For Use of Regular Acrylamide Gel

Components	Small size gel (50 ml)	Large size gel (100 ml)
Ultrapure urea (7 to 8.3 *M*)	21 to 25 g	42 to 50g
dd.H_2O	12 ml	24 ml

b. Warm at approximately 40 to 50°C to dissolve the urea with stirring. Once the urea is dissolved, add the following to the mixture with stirring at room temperature.

Recipe for Using Long Ranger Gel Mixture from AT Biochemicals

Components	Small size gel (50 ml)	Large size gel (100 ml)
5X TBE	12 ml	24 ml
Long Ranger mixture	5 ml	10 ml
Add dd.H_2O up to	50 ml	100 ml

Recipe for Regular 5 or 8% Acrylamide Gel Using TBE Buffer

Components	Small size gel (50 ml)	Large size gel (100 ml)
5X TBE	10 ml	20 ml
5% Acrylamide/*bis*-acrylamide	2.37 g/0.13 g	4.75 g/0.25 g
8% Acrylamide/*bis*-acrylamide	3.8 g/0.2 g	7.6 g/0.4 g
Add dd.H_2O up to	50 ml	100 ml

Recipe for Regular 5% or 8% Acrylamide Gel when Using Glycerol Gel Buffer (for Sequenase Verison 2.0 DNA Polymerase Diluted with Glycerol Enzyme Dilution Buffer)

Components	Small size gel (50 ml)	Large size gel (100 ml)
10X TBE	5 ml	10 ml
5% Acrylamide/*bis*-acrylamide	2.37 g/0.13 g	4.75 g/0.25 g
8% Acrylamide/*bis*-acrylamide	3.8 g/0.2 g	7.6 g/0.4 g
Add dd.H_2O up to	50 ml	100 ml

*Caution: The Long Ranger mixture and the acrylamide/*bis*-acrylamide are very toxic. Care should be taken when using these chemicals. Gloves should be worn when handling the gel mixture.*

c. Filter and degas the mixture under vacuum for 3 to 5 min (optional) and transfer the gel mixture to a plastic squeeze bottle or leave it in the beaker if a syringe is used to fill a gel apparatus with the liquid.

2. While the gel mixture is cooling to 25 to 30°C with a stirring bar, one can start to clean glass plates. Gently place glass plates, one by one, in a sink and thoroughly clean them with detergent using a sponge. Wash away the soap residue with running tap water followed by distilled water. Spray (95 or 100%) ethanol onto the plates to remove the soap and oily residues thoroughly. Finally, wipe the plate dry using clean paper towels (e.g., bleached singlefold towels, James River Corporation, Norwalk, CT).
3. Carefully lay the glass plate down on the lab bench and coat its surface with 0.5 to 1.0 ml of Sigmacoat (Sigma Chemicals) to form a thin film. This helps prevent the gel from sticking onto the glass plate when the glass plates are separated from each other after electrophoresis. Mark the coated plate on the back with a marker pen for identification. The other plate should not be coated and serves as the bottom plate when the plates are separated from each other.
4. Assemble a glass plate sandwich by placing a spacer (0.2 or 0.4 mm thick) on each of two side edges of the bottom glass plate and cover the plate with the top plate. There are different ways to assemble glass plates (depending on the manufacturer's instructions) and to pour the gel mixture into the sandwich.

METHOD A. POURING THE GEL MIXTURE HORIZONTALLY INTO THE SANDWICH

1. Place two styrofoam supporters (about 5 to 10 cm thick, one for the top and the other for the bottom of the sandwich) or two plastic pipette-tip boxes with an even surface on a flat bench. Lay the glass sandwich horizontally on the two supports and check the flatness with a level. It is not necessary to seal the edges of the glass sandwich with tape. Clamp the side edges of the plates with the manufacturer's clamps (optional).
2. Immediately prior to pouring the gel, add *N,N,N′N′*-tetramethylethylenediamine (TEMED) and freshly prepared 10% ammonium persulfate solution (APS) to the gel mixture (for 100 ml of gel mixture, the volume for TEMED and APS should be 48 μl and 0.49 ml, respectively.). Quickly and gently mix the mixture by swirling the squeeze bottle to avoid generating air bubbles, and immediately begin to pour the gel mixture into the sandwich. Starting at the middle region between the plates, slowly and continuously squeeze the bottle with one hand to cause outward flow of the mixture. Simultaneously use a roll of masking tape (1 to 2 cm wide and 10 to 12 cm in diameter) to tap the top glass plate along the front of the flowing solution with the other hand until the sandwich is completely filled. The gel mixture is distributed into the sandwich by capillary suction and the tapping helps cause its even flow. If squeezing stops at any time,

make sure to squeeze the bottle to remove any air bubbles prior to refilling the sandwich with gel mixture. It is strongly recommended that two people work together while pouring the gel in such a way that one person can focus on squeezing the bottle to pour the gel, while the other is responsible for tapping the top glass plate to help flow and prevent bubble formation. Alternatively, the gel mixture can also be loaded into the sandwich by using a syringe or a pipette, depending on personal preference.

3. Immediately and carefully insert the comb (0.2 or 0.4 mm thick depending on the thickness of the spacer), upside down, into the gel mixture to make a straight, even front edge of the gel. Avoid producing any bubbles underneath the comb. To prevent the gel from polymerizing between the comb and the glass plates, clamp the comb in place together with the glass plates with two to three appropriately sized clamps. It is recommended that a 24-well comb be used for a small size gel and a 48- or 60-well comb for a large size of gel. The color of the comb should be off-white or white so that each well can be clearly visible when loading the DNA samples.
4. Allow the gel to polymerize for 0.5 to 1.5 h at room temperature. The gel can be subjected to electrophoresis directly by removing the comb and mounting the cassette onto the sequencing apparatus according to the manufacturer's instructions. For multiple loadings at approximately 2-h intervals, we recommend that the gel be poured late in the day or at night, and that electrophoresis be performed early the next morning. If this is the option, the polymerized gel should be covered at the top and bottom edges of the cassette with clean paper towels wetted with distilled water. Wrap the paper towels with SaranWrap to keep the gel from drying. Leave it overnight so that the electrophoresis can be carried out the next day.

Method B. Pouring the Gel at an Angle

1. Carefully seal the two side edges of the sandwich using appropriate tape. Clamping both sides of the sandwich is optional.
2. Seal the bottom of the sandwich in two ways: (1) tightly seal the bottom edge with an appropriate tape; and (2) leave the bottom unsealed and attach the special bottom tray (provided by the manufacturer) to the bottom of the sandwich. Some types of commercial apparatus have disposable bottom spacers available that can be inserted into the sandwich at the bottom and may then be sealed by tape.
3. Pour the gel mixture into the sandwich. One option is to preseal the bottom first with some gel mixture to ensure against leaking. This can be done by transferring approximately 3 ml of the gel mixture to a tray and adding 5 μl of TEMED and 45 μl of freshly prepared 10% APS. Mix well, hold the sandwich vertically and immediately immerse the bottom of the sandwich in the gel mixture. The mixture containing a high concentration of APS can quickly polymerize and seal the bottom; after that, the rest of the gel mixture can be poured into the sandwich.

4. Immediately prior to pouring the gel, add TEMED and freshly prepared 10% APS to the gel mixture contained in a clean squeeze bottle (for 100 ml of gel mixture, the volume for TEMED and APS should be 48 μl and 0.49 ml, respectively). Quickly and carefully mix by swirling the squeeze bottle to avoid bubbles, then immediately start to pour the gel into the sandwich. Raise the top of the sandwich to about a 45° angle with one hand. Using the other hand, slowly squeeze the bottle to cause flow of the gel mixture at the top corner along one side of the sandwich. The angle of the plate and the rate of flow should be adjusted to avoid forming air bubbles. When the sandwich is half full, place the top end of the plate on a small box or support on the lab bench and the bottom end of the plate on the bench, forming an angle of 25 to 30°. Continuously fill the gel mixture into the sandwich and simultaneously use a roll of masking tape (1 to 2 cm thick and 10 to 12 cm diameter) or its equivalent to tap the top glass plate along the front of flowing gel with the other hand in order to prevent any air bubbles until the sandwich is completely filled. On the other hand, the gel mixture can also be loaded into the sandwich with a syringe or pipette, depending on personal preference.
5. Immediately and carefully insert a clean comb (0.2 or 0.4 mm thick depending on the thickness of the spacer), upside down, into the gel mixture to make a straight, even front edge of the gel. Avoid generating any air bubbles underneath the comb. Clamp the comb in place between the glass plates with two to three appropriately sized clamps.
6. Carefully lay the plates on a flat surface and allow the gel to polymerize for 0.5 to 1.5 h at room temperature. At this point, the gel can be directly subjected to electrophoresis by removing the comb and mounting the gel cassette onto the sequencing apparatus according to the manufacturer's instructions. For multiple loadings (e.g., four loadings at 2-h intervals), we recommend that the gel be poured late in the day or at night, and that the electrophoresis be carried out early the next morning. To leave the gel overnight, the polymerized gel should be covered at the top and bottom edges of the cassette with clean paper towels wetted with distilled water. Wrap the paper towels with SaranWrap to keep the gel from drying.

Materials Needed

5X TBE Buffer

Boric acid, 27.5 g
Tris base, 54 g
0.5 *M* EDTA buffer (pH 8.0), 20 ml
Dissolve well after each addition in 600 ml dd.H_2O. Add dd.H_2O to a final volume of 1000 ml. Autoclave and then store at room temperature.

10X Glycerol-Tolerant Gel Buffer

Tris base, 108 g

Taurine, 36 g

$Na_2EDTA \cdot 2H_2O$ buffer (pH 8.0), 4.65 g

Dissolve well after each addition in 600 ml dd.H_2O. Add dd.H_2O to a final volume of 1000 ml. Autoclave and then store at room temperature.

Urea

Ultrapure grade

Long Ranger Gel Mixture™

AT Biochem (Cat. # 211ISI).

10% Ammonium Persulfate Solution (APS)

0.1 g in 1.0 ml dd.H_2O (fresh)

High-Power Supply

It should have an upper limit of 2500 to 3000 V.

PROTOCOL 4. ELECTROPHORESIS

1. Carefully remove the comb and insert the cassette into the gel apparatus according to the manufacturer's instructions.
2. Add sufficient volume of 0.6X TBE running buffer (diluted from 5X TBE stock buffer) to top (cathode) and bottom (anode) chambers. Gently wash the top surface of the gel three times by pipetting the running buffer up and down.
3. Vertically hold the left and right sides of the comb using two hands and carefully insert the comb into the top of the sandwich to form the wells. The insertion of the comb should be done slowly and evenly until all the teeth just touch the surface of the gel. Remove any air bubbles trapped in the wells using a pipette.

Tips: It is acceptable if the teeth of the comb penetrate a little bit into the gel (<0.5 mm from the surface of the gel). However, if the teeth are inserted too deeply into the gel or the comb repeatedly pulled and inserted, the surface of the gel will be badly damaged and cause leaking when loading samples. This results in contamination of the wells and inaccurate sequence data.

4. Calculate the volume of the gel (the length × the width × the thickness of the spacer) and set the power supply. We strongly recommend setting it at a constant power (watts) using 1.0 to 1.1 W/cm^3. If constant power is not available from the power supply, 0.5 to 0.8 mA/cm^3 constant current should be set up. For example:

Gel volume	Constant watts	Electrophoresis temperature
35 cm^3	35 to 38.5	45 to 55°C
70 cm^3	70 to 77	45 to 55°C

Caution: If the power supply is set at a nonconstant level (e.g., high voltage or high current), the gel may burn or melt during the electrophoresis due to the amount of heat generated.

5. Connect the power supply unit to the gel apparatus with the cathode at the top of the gel and anode at the bottom of the gel so that the negatively charged DNA will migrate downwards.
6. Turn on the power and prerun the gel for 10 to 12 min.
7. After prerunning for 7 to 9 min, denature the labeled DNA samples at 75°C for 2.5 to 3 min prior to loading in wells. The tubes containing the samples should be capped tightly while heating in order to prevent evaporation of the samples.

Tip: For heating, placing the tubes in a heating block in the order of A, T, G, C or A, G, T, C is recommended to in order to prevent a potential mix-up when loading the samples into the gel.

8. When pre-electrophoresis is complete, turn off the power supply unit, wash the wells briefly to remove urea, and then quickly and carefully load the samples into appropriate wells (3.5 to 4 μl/well).

Tips: (a) No leaking is allowed between wells. Otherwise, the DNA samples will be contaminated and result in a mix-up of the nucleotide sequences. (b) Special pipette tips are available for DNA sample loading. In order to prevent any contamination, using one tip for loading multiple samples is not recommended . If a 0.4-mm spacer is used, then using normal 100- to 200-μl pipette tips works fine for loading. This is faster and no air bubbles develop, which is quite common when using the commercial sequencing loading tips with a long and flat tip. (c) After taking a sample from each tube, the tubes should be capped immediately. All tubes should be briefly spun down and stored at 4 or –20°C until the next loading. (d) It is very important that the samples loaded into the wells be recorded in a notebook in the order of the DNA samples loaded (e.g., A, G, T, C or A, T, G, C depending on particular loading sequence). This will help in reading the DNA sequence following autoradiography.

9. After all the samples are loaded, turn on the power supply and allow electrophoresis to occur at an appropriate constant voltage. The running time depends on the size of the DNA. Generally, the run times for 250 bp, 400 to 450 bp and > 500 bp are 2, 4, and 6 to 8 h, respectively. For multiple-loading (e.g., two to five times loading depending on the size of the gel and volume of the DNA sample), monitor the migration of the first blue dye (Bromophenol blue or BPB) that migrates at 40 bp. When the BPB reaches approximately 2 to 3 cm from the bottom, turn off the power supply and carry out the next loading so that there are overlapping sequences between the adjacent loads. This is very useful for correctly reading DNA sequences after autoradiography.

Tips: The power supply must be turned off and the wells for sample loading should be washed to remove air bubbles with the running buffer using a pipette. DNA samples may be denatured again at 75°C for 2 min prior to loading. The loading order of the samples should be the same as for the previous loading to avoid misreading DNA sequences.

10. When the electrophoresis is complete, disconnect the power supply and decant the upper chamber buffer. Lay the cassette containing the gel on a table or bench and allow it to cool for 20 to 25 min prior to separating the glass plates. Mark and measure the area of the gel to be transferred.

Note: If the glass plates are separated immediately following electrophoresis, the gel may not be flat, may become broken or cause difficulty in separating the glass plates.

11. While the gel is cooling, wear clean gloves and cut one piece of nylon membrane and three pieces of 3MM Whatman paper. The size of membrane or filter paper should be slightly larger than the area of the gel to be transferred. Using a pencil, label one corner of the membrane for orientation and prewet the membrane in dd.H_2O for 5 to 10 min.
12. Remove the clamps from the glass plates. Starting at one corner of the spacer site, carefully insert a spatula into the sandwich and slowly lift the top glass plate (the Sigma-coated plate) with one hand. Meanwhile, hold the bottom plate with the other hand until the top glass plate is completely separated from the gel. The gel should remain stuck to the bottom plate.

Tip: Separation of glass plates is critical for the success of DNA sequencing. As long as the top plate is loosened, continue to lift it until it is completely separated from the other plate. Do not allow the top plate to come back in contact with the gel at any time during separation or it may cause trouble for the gel.

13. Perform fixation. If Long Ranger Gel mixture is used, the electrophoresed gel does not need to be fixed. However, if a regular acrylamide gel mixture is used, the gel should be fixed. Immerse the gel together with the bottom plate in a relatively large tray filled with a sufficient volume of fixing solution containing 15 to 20% of ethanol or methanol and 5 to 10% acetic acid for 15 to 20 min. Remove the gel together with the plate from the tray, drain off excess fixing solution and gently place bleached, clean paper towels on the gel to remove excess solution.

PROTOCOL 5. TRANSFERRING DNA FROM GEL ONTO A NYLON MEMBRANE

1. Transfer the sequenced DNA from the gel onto the prewetted membrane in the order shown: wear clean gloves and lay the membrane on the area of the gel to be transferred with the marked corner facing the gel. Remove any air bubbles underneath the membrane by gently rolling a clean 25-ml pipette on the membrane. Carefully place three layers, one by one, of 3MM Whatman dry filter papers on the membrane. Place the other glass

plate or its equivalent on top of the filter papers and then a weight on the top (e.g., a bottle or beaker containing 1000 ml water). Leave undisturbed for 1 to 1.5 h.

Note: Any bubbles generated will block local transfer of DNA bands from the gel to the membrane. Once this occurs, reading of DNA sequences will not be accurate. One virtually must start all over again.

2. Remove the weight, top glass plate and 3MM filters. Wear gloves and gently peel the membrane off the gel and place the membrane with the DNA side facing up on a sheet of clean 3MM Whatman paper. Allow the membrane to dry for 30 to 45 min at room temperature. The dried membrane may be stored at 4°C until use.

PROTOCOL 6. DETECTION

1. Block nonspecific binding sites on the membrane with 0.4 ml/cm^2 of blocking solution using an appropriate size of tray. Place the tray on an orbital shaker with shaking at 60 rpm for 40 min.
2. Add streptavidin–alkaline phosphatase conjugate (SAAP) to the blocking solution (1:2000 to 5000 dilution) in the tray. Allow SAAP to hybridize with biotinylated primer or nucleotides on the membrane for 40 min with shaking at 60 rpm.
3. Briefly rinse the membrane with dd.H_2O and wash it with 1 ml/ cm^2 of 0.5 to 1.0X SSC containing 0.1% SDS by agitating for 10 min on a shaker at 60 rpm. Repeat this step twice.
4. Briefly rinse the membrane with dd.H_2O and wash it once with 1 ml/ cm^2 of 0.5 to 1.0X SSC without SDS for 5 min on a shaker at 60 rpm. The membrane can be subjected to detection by the following methods.

METHOD A. CHEMILUMINESCENT DETECTION

1. Place a piece of SaranWrap film on a bench. According to the size of the membrane, add 10 μl/cm^2 of Lumi-Phos 530 (Boehringer Mannheim Corporation) to the center of the SaranWrap.
2. Wear gloves and briefly dampen the membrane to remove excess washing solution and thoroughly wet the DNA-binding side of the membrane by lifting and overlaying the membrane several times.
3. Wrap the membrane with SaranWrap film, leaving two ends of the film unfolded. Place the wrapped membrane on a paper towel and carefully press the wrapped membrane, using another piece of paper towel, to wipe off excess detection solution through the unfolded ends of the SaranWrap. Excess detection solution will most likely cause a high background.
4. Completely wrap the membrane and place it in an exposure cassette with the DNA side facing up. Tape the four corners of the wrapped membrane.
5. In a darkroom with the safe light on, overlay the membrane with a sheet of Kodak® XAR or XRP x-ray film and close the cassette. Allow exposure

to proceed at room temperature for 20 s to 12 h, depending on the intensity of the detected signal.

6. In a darkroom, develop and fix the film in an appropriate developer and fixer, respectively. If an x-ray processor is available, development, fixation, washing and drying of the film can be completed in 2 min.

Tips: Multiple films may be needed to be exposed and processed until the desired signals are obtained. Exposure for more than 4 h may generate a high black background. In our experience, a good hybridization and detection should display sharp bands within 1.5 h (Figure 5.1).

7. Proceed to DNA sequence reading (Figure 5.2).

Method B. Colorimetric Detection

1. Following washing, place the membrane in color developing solution (20 μl/cm^2) containing 0.08 μl of NBT stock solution and 0.06 μl of BCIP stock solution. NBT and BCIP stock solutions are commercially available.
2. Allow color to develop in a dark place for 15 to 120 min at room temperature or until a desired level of detection is obtained. Detected bands should appear as a blue–purple color on the membrane.

Reagents Needed

5X TBE Buffer

600 ml dd.H_2O
0.45 *M* Boric acid (27.5 g)
0.45 *M* Tris base (54 g)
10 m*M* EDTA (20 ml 0.5 *M* EDTA, pH 8.0)
Dissolve well after each addition. Add dd.H_2O to 1 l. Autoclave.

20X SSC Solution (1 l)

175.3 g NaCl
88.4 Sodium citrate
Adjust the pH to 7.5 with HCl.

Blocking Buffer

3% (w/v) Nucleic acid blocking reagents or 5% (w/v) nonfat dry milk
150 m*M* NaCl
100 m*M* Tris-HCl, pH 7.5
Dissolve well with stirring.

Color Developing Buffer

0.1 *M* Tris-HCl, pH 9.5
0.1 *M* NaCl
50 m*M* $MgCl_2$

NBT Stock Solution

75 mg/ml Nitroblue tetrazolium (NBT) salt in 70% (v/v) dimethylformamide

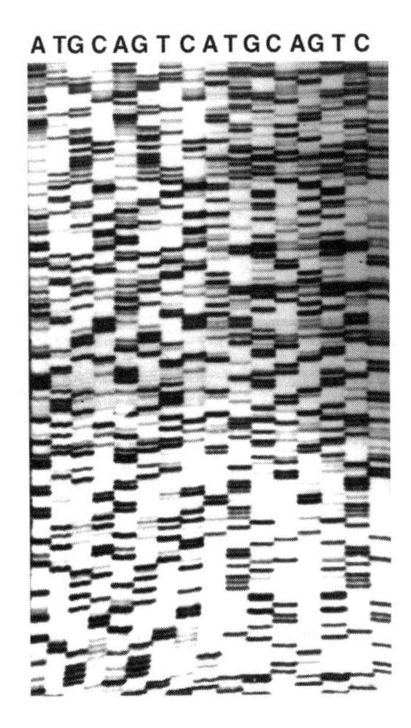

FIGURE 5.1 Chemiluminescent detection of DNA sequences.

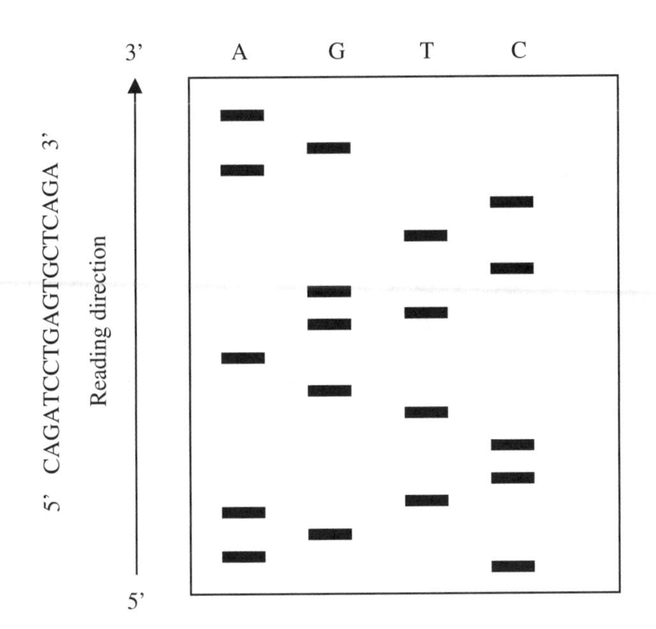

FIGURE 5.2 Reading of DNA sequences.

BCIP Stock Solution

50 mg/ml 5-Bromo-4-chloro-3-indolyl phosphate (BCIP or X-phosphate), in 100% dimethylformamide

Streptavidin-Alkaline Phosphatase Conjugate (SAAP)

Lumi-Phos 530

ISOTOPIC DNA SEQUENCING METHOD

Procedures for isotopic DNA sequencing are essentially the same as those described for the nonisotopic sequencing method except for the following differences.

The first difference is the labeling reaction in which [^{35}S]dATP or [^{35}S]dCTP (not a biotinylated nucleotide) is employed. Add the following components to the annealed primer–template mixture (10 μl).

DTT, 1 μl
Diluted labeling mixture, 2 μl
[^{35}S]dATP or [^{35}S]dCTP (1000 to 1500 Ci/mmol), 1 μl
Diluted Sequenase DNA polymerase, 2 μl

The second difference is in the drying of the gel and autoradiography. No DNA transfer is involved. Specific steps are:

1. Once the top glass plate is separated from the gel on the bottom plate, gently and slowly lay the precut 3MM Whatman paper on the gel from the bottom side or the upper side of the gel until it lies on the entire gel. Gently and firmly press the paper thoroughly onto the gel surface with a styrofoam block (approximately 50 cm long, 5 cm wide and 5 cm thick). Then, starting from one corner, slowly peel the 3MM Whatman paper with the gel on it from the bottom plate. Place it on a flat surface with the gel side up and carefully cover the gel with SaranWrap without any air bubbles or wrinkles generated between the SaranWrap and the gel.
2. Assemble the gel on the drying apparatus according to the manufacturer's instructions and dry the gel at 70 to 80°C under vacuum for 30 to 55 min, depending on the strength of the vacuum pump. Turn off the heat and allow cooling for another 20 to 30 min under vacuum. Proceed to autoradiography.

Tips: (a) Place a water trap at a position lower than the gel dryer for effective suction of the water. If no freeze or refrigerated trap is available, place a water-trap flask or bottle in a bucket with some dry ice and ethanol. Connect the trap to a vacuum pump and to the gel dryer. This works fine in our laboratory. (b) If the gel is removed immediately after the heat is turned off, it may be cracked or have an uneven surface. Cooling the gel for a while under vacuum before taking it out of the gel dryer is recommended.

3. Peel the SaranWrap from the gel and put the gel on an appropriate exposure cassette. In a darkroom under a safe light, place an appropriately sized piece of x-ray film (Kodak XAR-film or Amersham x-ray film) on

the surface of the gel and close the cassette. Allow exposure to take place at room temperature (for ^{35}S) by placing the cassette in a dark place for 1 to 4 days, depending on the intensity of the signal. It is recommended that the film be developed after exposure for 24 h and then, if necessary, the length of further exposure decided.

Caution: [^{35}S] dATP or dCTP is dangerous. Care should be taken when using the nucleotides. Gloves should be worn and an appropriate Plexiglas protector should be used. Any used isotopic buffer should be collected in a special container for disposal.

THE USE OF FORMAMIDE GELS

Compression is a common phenomenon that usually causes unreadability on a developed sequencing film. The problem is primarily due to intramolecular base pairing or G–C rich in a primer extension. The local folded structures or hairpin loops migrate faster through the gel matrix than unfolded structures, and persist during the electrophoresis. As a result, bands run very close together with a gap or increased band spacing in the region. One way to solve this problem is to increase the denaturing conditions in the gel matrix using urea and formamide. The procedures for preparing a formamide gel, electrophoresis and autoradiography are very similar to those described in earlier protocols except for the following steps.

1. The gel mixture:

Use of Long Ranger Gel Mixture from AT Biochemicals

Components	Small size gel (50 ml)	Large size gel (100 ml)
Ultrapure urea	21 g	42 g
Long Ranger mixture	8 ml	16 ml
Ultrapure formamide	20 ml	40 ml
5X TBE	10 ml	20 ml
Add dd.H_2O up to	50 ml	100 ml

Use of Regular Acrylamide Gel Mixture if Sequenase Version 2.0 DNA Polymerase is Diluted with a Buffer without Glycerol

Components	Small size gel (50 ml)	Large size gel (100 ml)
Ultrapure urea (7 *M*)	21 g	42 g
Ultrapure formamide	15 to 20 ml	30 to 40 ml
6% Acrylamide/*bis*-acrylamide	2.8 g/0.15 g	5.7 g/0.3 g
8% Acrylamide/*bis*-acrylamide	3.8 g/0.2 g	7.6 g/0.4 g
5X TBE	5 ml	10 ml
Add dd.H_2O up to	50 ml	100 ml

Recipe for a Regular Acrylamide Gel Mixture if Sequenase Version 2.0 DNA Polymerase is Diluted in Glycerol Enzyme Dilution Buffer

Components	Small size gel (50 ml)	Large size gel (100 ml)
Ultrapure urea (7 *M*)	21 g	42 g
Ultrapure formamide	15 to 20 ml	30 to 40 ml
6% Acrylamide/*bis*-acrylamide	2.8 g/0.15 g	5.7 g/0.3 g
8% Acrylamide/bis-acrylamide	3.8 g/0.2 g	7.6 g/0.4 g
10X Glycerol tolerant gel buffer	5 ml	10 ml
Add dd.H_2O up to	50 ml	100 ml

2. Electrophoresis: double the electrophoresis time because formamide slows the migration of DNA by about 50%.

EXTENDING SEQUENCING FAR FROM THE PRIMERS

We routinely use the sequence-extending mixture (USB) to obtain sequences beyond 600 to 800 bases from the primer with high resolution. Using multiple loadings, more than 1500 to 2000 bases from the primer can be extended with ease. The protocols of extending sequencing are quite similar to those described in the previous protocols except for the following steps.

1. The labeling mixture should not be undiluted or it may be diluted twofold, instead of using fivefold dilution.
2. Increase the amount of biotinylated nucleotide or [^{35}S]dATP (1000 to 1500 Ci/mmol) from the regular volume of 2 to 4 μl (biotinylated nucleotide) or 1 to 2 μl (isotopic nucleotide).
3. Extend the labeling reaction from the regular time of 2 to 5 min to 4 to 8 min at room temperature.
4. In the termination reaction, instead of adding 2.5 μl of ddNTPs to the appropriate tubes (A,G,T,C), add 2.5 μl of a mixture of the appropriate sequence extending mixture (USB) and the appropriate termination mixture (USB) at ratios of 2:0.5, 2:1.0, and 1.5:1 (v/v), depending on the particular DNA sequence, to the appropriate tubes.
5. Utilize a relatively long gel (80 to 100 cm) and extend electrophoresis for 12 to 16 h using a high quality sequencing apparatus.

DNA SEQUENCING BY PRIMER WALKING

Primer walking refers to stepwise sequencing starting from the 5′ end of the target DNA using a series of primers. The first primer is usually a universal primer that is built into the plasmid vector, such as T3, T7 or SP6 primers. Following each DNA sequencing, a new primer will be designed based on the very last sequence. For instance, if the first sequencing has revealed 450 bases with 3′-end sequences of 5′GCATGTAGACGATAGGATACAG-3′, to continue sequencing of the DNA, the

GCATGTAGACGATAGGATACAG can be chosen as the walking primer for the next round of sequencing. In the same way, new walking primers can be selected for subsequent sequencing until the entire target DNA has been completely sequenced.

DNA SEQUENCING BY UNIDIRECTIONAL DELETIONS

It takes a longer time to sequence large DNA molecules compared with regular stepwise or primer-walking sequencing methods. However, using exonuclease III, the gene or large DNA fragment can be unidirectionally deleted in order to generate a series of shorter fragments with overlapping ends. In this way, these progressively deleted fragments can then be simultaneously sequenced in a short time. The general principles and procedures are outlined in Figure 5.3.

1. Recombinant plasmid, phagemid, or bacteriophage M13 replicate form of DNA that contains the cloned DNA of interest is first linearized with two appropriate restriction enzymes. Both enzymes should cut DNA between one end of the target DNA and the binding site for the universal sequencing primer. One enzyme should cleave near the target DNA and generate either a recessed 3′ end (or 5′ protruding end) or a blunt end. The other enzyme will cleave the DNA near the binding site for the universal sequencing primer and must produce a 4-base protruding 3′ end or be filled in with α-phosphorothioate dNTPs.
2. The linearized DNA is then progressively deleted with exonuclease III that only digests the DNA from the blunt or 5′protruding terminus, leaving the 3′ protruding (overhang) or α-phosphorothioate-filled end intact. The digestion proceeds unidirectionally from the site of cleavage to the target DNA sequence. The digestion is terminated by taking an appropriate volume of the samples at appropriate time intervals, thus generating a series and progressive deletions of shorter DNA fragments.
3. The exposed single-stranded DNAs are then cleaved by nuclease S1 or mung-bean nuclease. This produces blunt termini at both ends of the DNA fragments to be sequenced.
4. The shortened DNA is then recircularized using T4 DNA ligase and transformed into an appropriate bacterial host. Transformants can be selected with appropriate antibiotics in the culture medium and the recombinant plasmids will be purified and subjected to DNA sequencing.

Plasmid-based vectors for subcloning the DNA to be sequenced are commercially available. These include pGEM-5Zf or pGEM-7Zf (Promega Incorporation) whose multiple cloning sites contain two unique restriction sites lying between the end of the DNA insert to be deleted and the binding site for the universal sequencing primer. One of the two unique restriction sites should be near the end of the insert and must generate a blunt or 5′-overhang end that is necessary for exonuclease

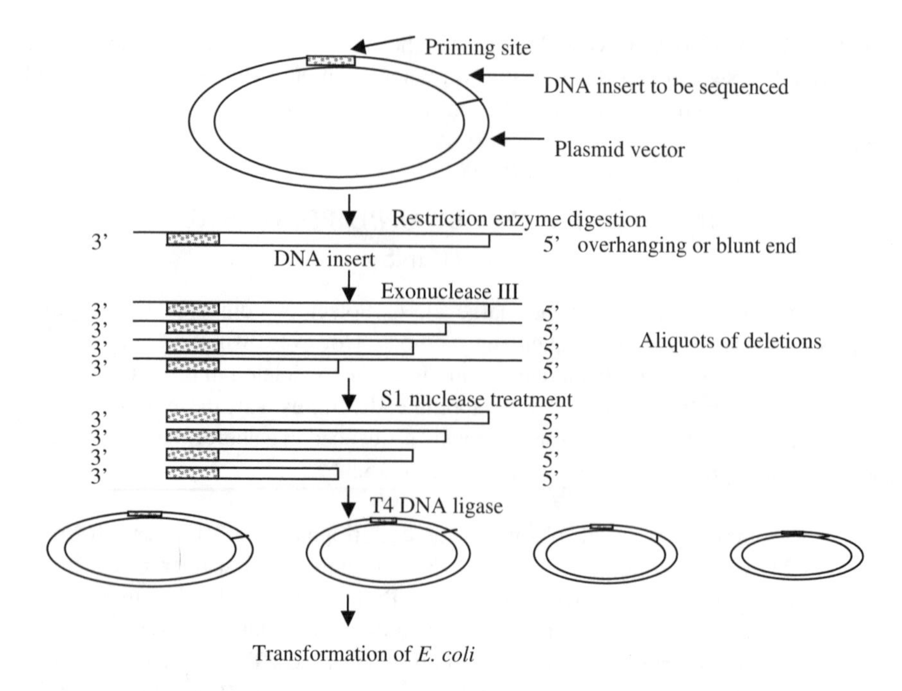

FIGURE 5.3 Diagram of progressive deletions of DNA insert for sequencing.

III deletion. The other enzyme should cut near the sequencing primer binding site and must produce a 3′-overhang end to protect the end from exonuclease III deletion. Thus, the exonuclease digestion will be unidirectional and proceed to the insert DNA sequence. The unique restriction enzymes that can be used are listed below.

Unique Enzymes for Generating 5′-Protruding or Blunt Ends

Restriction enzyme	Recognition sequence	Enzyme	Recognition sequence
Not I	5′..GC*GGCC GC..3′	*Sma I*	5′..CCC*GGG..3′
	3′..CG CCGG*CG..5′		3′..GGG*CCC..5′
Xba I	5′..T*CTAG A..3′	*Xho I*	5′..CT*CGA G..3′
	3′..A GATC*T..5′		3′..GA GCT*C..5′
Sal I	5′..G*TCGA C..3′		
	3′..C AGCT*G..5′		

Note: * indicates the cutting site. In spite of the fact that Hind III, EcoR I and BamH I can also generate 5′ overhangs, they are not usually unique because most of the DNA inserts contain the recognition sites for these enzymes.

Unique Enzymes That Produce Exonuclease III-Resistant 3′-Protruding Termini

Enzyme	Recognition sequence	Enzyme	Recognition sequence
Sph I	5′..G CATG*C.. 3′ 3′..C*GTAC G.. 5′	*Pvu* I	5′..CG AT*CG.. 3′ 3′..GC*TA GC.. 5′
Sac I	5′..G AGCT*C.. 3′ 3′..C*TCGA G.. 5′	*Kpn* I	5′..G GTAC*C.. 3′ 3′..C*CATG G.. 5′
Aat II	5′..G ACGT*C.. 3′ 3′..C*TGCA G.. 5′	*Bgl* I	5′..GCCN NNN*NGGC.. 3′ 3′..CGGN*NNN NCCG.. 5′

Note: *indicates the restriction enzyme cutting site. Plasmid DNA purification and restriction enzyme digestion are described in appropriate chapters of this book.

PROTOCOL 7. PERFORMING A SERIES OF DELETIONS OF THE LINEARIZED DNA WITH EXONUCLEASE III AND RECIRCULARIZATION OF DNA WITH T4 DNA LIGASE (FIGURE 5.3)

1. Label 15 to 30 microcentrifuge tubes (0.5 ml) depending on the size of insert to be deleted, and add 7.5 to 8.0 µl of nuclease S1 mixture to each tube. Store on ice prior to use.
2. Dissolve or dilute 6 to 8 µg of linearized DNA in a total 60 µl of 1X exonuclease III buffer.
3. Warm the sample to an appropriate temperature and start digestion by adding 250 to 550 units of exonuclease III to the sample. Mix as quickly as possible. Transfer 2.5 to 3.0 µl of the reaction mixture at 0.5-min intervals to the labeled tubes containing nuclease S1 prepared at step 1. Quickly mix by pipetting up and down several times and place on ice before using.

Notes: (1) The digestion rate depends on the reaction temperature.

Temperature (°C)	Digestion rate (bp/min)
4	25 to 30
22	80 to 85
25	90 to 100
30	200 to 220
37	455 to 465
45	600 to 630

(2) It is not necessary to change the buffer between exonuclease III and nuclease S1 because the S1 buffer contains zinc cations and has a low pH, which can inactivate the exonuclease III. However, some exonuclease III buffer will not inhibit the activity of nuclease S1.

4. After all of the samples have been taken, place all the tubes at room temperature and carry out the nuclease S1 digestion of single strands by incubating at room temperature for 30 min.
5. Terminate the reaction by adding 1 µl of S1 stopping buffer and heat at 70°C for 10 min.
6. Check the efficiency of digestion by preparing a 1% agarose gel using 2 µl of sample from each time point.
7. Blunt termini by adding 1 µl of Klenow mixture to each tube and incubating at 37°C for 4 min followed by adding 1 µl of dNTP mixture to each tube and incubate at 37°C for 5 to 6 min.
8. Recircularize the DNA by adding 40 µl of ligase mixture to each tube, mixing and incubating at 16°C overnight. Carry out transformation of the plasmid DNA into an appropriate bacterial host as described in the appropriate chapter in this book. The series of plasmids with progressively reduced sizes can now be simultaneously sequenced.

Reagents Needed

10X Exonuclease III Buffer

0.66 *M* Tris-HCl, pH 8.0
6.6 m*M* $MgCl_2$

S1 Buffer

2.5 *M* NaCl
300 m*M* Potassium acetate, pH 4.6
10 m*M* $ZnSO_4$
50% Glycerol

Nuclease S1 Mixture (Fresh)

54 µl S1 buffer
0.344 ml dd. H_2O
120 units Nuclease S1

S1 Stopping Buffer

300 m*M* Tris base
50 m*M* EDTA, pH 8.0

Klenow Buffer

20 m*M* Tris-HCl, pH 8.0
100 m*M* $MgCl_2$

Klenow Mixture

6 to 12 units Klenow DNA polymerase
60 µl Klenow buffer

dNTP Mixture

0.13 m*M* dATP
0.13 m*M* dGTP
0.13 m*M* dCTP
0.13 m*M* dTTP

10X T4 DNA Ligase Buffer

0.5 *M* Tris-HCl, pH 7.6
0.1 *M* $MgCl_2$
10 m*M* ATP

T4 DNA Ligase Mixture

0.1 ml 10X Ligase buffer
0.79 ml dd.H_2O
0.1 ml 50% PEG
10 µl of 0.1 *M* DTT
5 units of T4 DNA ligase

DIRECT DNA SEQUENCING BY PCR

DNA molecules or fragments or RT-PCR products (single-stranded, double-stranded and plasmid DNA) can be directly sequenced by coupling the polymerase chain reaction (PCR) technology and the dideoxynucleotide chain termination method. This is a particularly powerful technique when one wants to simultaneously amplify and sequence a DNA fragment of interest.

PREPARATION OF PCR AND SEQUENCING REACTIONS

1. Label four 0.5-ml microcentrifuge tubes as A, G, T, or C for each set of sequencing reactions for each PCR primer. A, G, T, and C represent ddATP, ddGTP, ddTTP and ddCTP, respectively.
2. Transfer 0.5 µl of 2X stock mixture of dNTPs/ddATP, dNTPs/ddGTP, dNTPs/ddTTP and dNTPs/ddCTP to the labeled tubes A, G, T and C, respectively. To each tube, add 0.5 µl dd.H_2O, generating 1X working mixture solution. Cap the tubes and store on ice prior to use.
3. Prepare the following mixture for each set of four sequencing reactions for each primer (forward or reverse primer) in a 0.5-ml microcentrifuge tube on ice.

   Forward or reverse primer, 2 to 5 pmol (15 to 27 ng) depending on size of the primer (15- to 27-mer, 10 to 30 ng/µl)
   DNA template, 100 to 1000 ng depending on the size of the template (0.4 to 7 kb, 10 to 100 ng/µl)
   [α-^{35}S]dATP, 1 to 1.2 µl (>1000 Ci/mmol)
   or [α-^{32}P]dATP, 0.5 µl (800 Ci/mmol)
   5X Sequencing buffer 4.0 µl
   Add dd.H_2O to a final volume of 17 µl.

4. Add 1 μ (5 units/μl) of *Taq* DNA polymerase to the mixture at step (c). Gently mix and transfer 4 μl of the primer–template-enzyme mixture to each tube containing 1 μl of dNTPs and appropriate ddNTP prepared at step (b).
5. Overlay the mixture in each tube with approximately 20 μl of mineral oil to prevent evaporation of the samples during the PCR amplification. Place the tubes in a thermal cycler and perform PCR as follows:

	Cycling				
Profile	**Predenaturation**	**Denaturation**	**Annealing**	**Extension**	**Last**
Template (<4 kb) Primer is <24 bases or with <40% G–C content	94°C, 3 min	94°C, 1 min	52°C, 1 min	70°C, 1.5 min	4°C
Template is >4 kb Primer is >24-mer or <24 bases with >50% G–C content	95°C, 3 min	95°C, 1 min	60°C, 1 min	72°C, 2 min	4°C

6. When the PCR cycling is complete, carefully remove the mineral oil from each tube using pipette tips (optional) and add 3.5 μl of stop solution to inactivate the enzyme activity. Proceed to DNA gel sequencing as described earlier in this chapter.

Reagents Needed

2X dNTPs/ddATP Mixture

dATP, 80 μ*M*
dCTP, 80 μ*M*
dTTP, 80 μ*M*
7-Deaza dGTP, 80 μ*M*
ddATP, 1.4 m*M*

2X dNTPs/ddTTP Mixture

dATP, 80 μ*M*
dCTP, 80 μ*M*
dTTP, 80 μ*M*
7-Deaza dGTP, 80 μ*M*
ddTTP, 2.4 m*M*

2X dNTPs/ddGTP Mixture

dTTP, 80 μ*M*
dATP, 80 μ*M*
dCTP, 80 μ*M*
7-Deaza dGTP, 80 μ*M*
ddGTP, 120 μ*M*

2X dNTPs/ddCTP Mixture

dTTP, 80 μ*M*
dCTP, 80 μ*M*
dATP, 80 μ*M*
7-Deaza dGTP, 80 μ*M*
ddCTP, 800 μ*M*

Amount of Primer per pmol

15-mer or 15 bases, 5 ng
16-mer, 5.3 ng
17-mer, 5.7 ng
18-mer, 6.0 ng
19-mer, 6.3 ng
20-mer, 6.7 ng
21-mer, 7.0 ng
22-mer, 7.3 ng
23-mer, 7.6 ng
24-mer, 8.0 ng
25-mer, 8.3 ng
26-mer, 8.6 ng
27-mer, 9.0 ng
28-mer, 9.3 ng
29-mer, 9.6 ng
30-mer, 10.0 ng

5X Sequencing Buffer

0.25 *M* Tris-HCl, pH 9.0 at room temperature
10 m*M* $MgCl_2$

Stop Solution

10 m*M* NaOH
95% Formamide
0.05% Bromophenol blue
0.05% Xylene cyanole

RNA SEQUENCING

Like DNA, RNA can also be sequenced by using nonisotopic or isotopic sequencing methods. A reverse transcriptase, avian myeloblastosis virus (AMV), is a DNA polymerase that catalyzes the polymerization of nucleotides using RNA or DNA template. In the labeling reaction, a biotinylated or isotopic nucleotide is incorporated into the cDNA strands transcribed from specific mRNA. The power of this technique is that it allows searching for the potential existence of the mRNA of interest or the mRNA expression in a specific cell or tissue type. Specific primers can be designed based on DNA sequences or amino acid sequences of interest. Total RNA can be directly utilized in specific primer–mRNA template annealing reactions.

PROTOCOL 1. ANNEALING OF PRIMER AND RNA TEMPLATE

1. For each sequencing reaction (one set of four lanes), add the following components in a 0.5-ml microcentrifuge tube.
   Primer, 12 pmol
   Total RNA template, 5 μg
   or mRNA, 100 ng
   Add DEPC-dd.H_2O to final volume of 10 μl
2. Cap the tube and denature RNA secondary structures at 90°C for 3 min and allow slow cooling to 30°C over a period of 30 min. Briefly spin down and place the tube on ice prior to use.

PROTOCOL 2. LABELING REACTIONS

1. While the annealing reaction is cooling, label four microcentrifuge tubes for each template–primer mixture, A, G, T and C, that respectively represent ddATP, ddGTP, ddTTP and ddCTP.
2. Transfer 4 μl of the termination mixture of ddATP, ddGTP, ddTTP and ddCTP to the labeled tubes A, G, T, and C, respectively. Cap the tubes and keep at room temperature until use.
3. Dilute labeling mixture to fivefold as a working concentration and store on ice prior to use. For example, 2 μl of labeling mixture is diluted to total 10 μl with dd.H_2O.
4. Add the following components in a 0.5-ml microcentrifuge tube for each annealing reaction.
   5X AMV RT buffer, 5 μl
   Diluted (1:5) labeling mixture, 2.5 μl
   Biotinylated dCTP or dATP, 1 μl or [α-^{32}P] dATP or dCTP, 1 μl
   Add DEPC-dd.H_2O to final volume of 10 μl.
5. When the annealing is complete, combine the annealing reaction mixture with the labeling reaction. Then, add 1 μl of AMV reverse transcriptase to the combined mixtures.
6. Carry out reverse transcription by incubating the tube at 42°C for 8 min.

PROTOCOL 3. TERMINATION REACTION

1. Carefully and quickly transfer 4 μl of the labeling mixture to each of the termination tubes (A, G, T, C) prewarmed at 42°C for 1 min. Mix and continue to incubate the reaction at 42°C in a heating block for 10 min.

2. Add 4 µl of stopping solution to each tube, mix and cap the tubes. Store at –20°C until electrophoresis is carried out. Proceed to gel electrophoresis and detection or autoradiography as described earlier in this chapter.

Reagents Needed

AMV Reverse Transcriptase

5X AMV Reverse Transcription Buffer
0.25 *M* Tris-HCl, pH 8.3
40 m*M* $MgCl_2$
0.25 *M* NaCl
5 m*M* DTT

DTT Stock Solution
100 m*M* Dithiothreitol

5X Labeling Solution for dGTP (Nonisotopic)
7.5 µ*M* dGTP
7.5 µ*M* dTTP
7.5 µ*M* dCTP or dATP
Biotinylated dATP or dCTP

5X Labeling Mixture for dITP (Nonisotopic)
7.5 µ*M* dITP
7.5 µ*M* dTTP
7.5 µ*M* dCTP or dATP
Biotinylated dATP or dCTP

ddATP Termination Mixture for dGTP
0.16 *M* dATP
0.2 *M* dGTP
0.2 *M* dCTP
0.2 *M* dTTP
0.04 *M* ddATP

ddGTP Termination Mixture for dGTP
0.2 *M* dATP
0.16 *M* dGTP
0.2 *M* dCTP
0.2 *M* dTTP
0.04 *M* ddGTP

ddCTP Termination Mixture for dGTP
0.2 *M* dATP
0.2 *M* dGTP
0.16 *M* dCTP
0.2 *M* dTTP
0.04 *M* ddCTP

ddTTP Termination Mixture for dGTP

0.2 *M* dATP
0.2 *M* dGTP
0.2 *M* dCTP
0.16 *M* dTTP
0.04 *M* ddTTP

Sequence Extending Mixture for dGTP

0.4 *M* dATP
0.4 *M* dGTP
0.4 *M* dCTP
0.4*M* dTTP

Enzyme Dilution Buffer

10 m*M* Tris-HCl, pH 7.5
5 m*M* DTT
0.5 mg/ml BSA

Stop Solution

20 m*M* EDTA, pH 8.0
95% (v/v) Formamide
0.05% Bromophenol blue
0.05% Xylene cyanol FF

TROUBLESHOOTING GUIDE

1. **Gel melts away from the comb during electrophoresis.**
   Cause: Too much heat is built up.
   Solution: Be sure to set the power supply at a constant power or a constant current. Do not set the voltage at some level; otherwise, the current undergoes changes during electrophoresis, producing high temperatures that melt the surface of the gel from the top toward the bottom. When that occurs, multiple loadings of the samples are out of the question.
2. **No bands appear at all on the developed x-ray film.**
   Possible causes: (a) The quality of primer is poor and cannot be annealed with DNA or RNA template. (b) Double-stranded DNA is not well denatured, so that the primer fails to anneal to the template. (c) Some components are missed during the labeling reaction. (d) Sequenase version 2.0 T7 DNA polymerase has lost its activity.
   Solution: Make sure to denature the DNA template completely and add all components mandated for the labeling reaction. Try to use control template and primer provided.
3. **Bands are fuzzy.**
   Possible causes: (a) Urea is not washed away from the wells prior to loading the samples. (b) Labeled samples are overheated during denatur-

ation. (c) It takes too long to finish loading all the samples, generating some reannealing of DNA.

Solutions: (a) Make sure to rinse the surface of the gel prior to trying the prerun and repeat washing after the prerun prior to loading the samples into the wells using a pipette. (b) Control the time for denaturing the labeled samples between 2 to 3 min and immediately load the sample into the wells. In the case of many samples, the loading should be carried out quickly so that all the samples are loaded within 2 min.

4. **No clear bands appear except for a smear in each lane.**

   Possible causes: (a) Preparation of DNA template is poor. (b) Labeled DNA samples are not well denatured at 75°C prior to being loaded into the gel. (c) Gel polymerizes too rapidly (10 to 15 min) due to excess 10% APS added. (d) The gel is electrophoresed at too cold or too hot a temperature.

   Solutions: (a) Make sure the DNA template is very pure without any nicks. (b) Use 0.5 ml of freshly prepared APS per 100 ml of gel mixture and make sure the gel mixture is cooled to room temperatue prior to pouring into the glass sandwich. (c) Keep the labeling reaction time to 2 to 5 min for regular sequencing and 4 to 7 min for extending sequencing. (d) Make sure to denature the labeled samples at 75 to 80°C for 2 to 3 min before loading into the gel. (e) Dry the gel at 75 to 80°C under vacuum but not above 80°C.

5. **All the bands are weak.**

   Possible causes: (a) Primer concentration is too low or the annealing of primer and template does not work well. (b) Double-stranded linear DNA and double-stranded plasmid DNA are too large due to the presence of a large DNA insert, resulting in difficulty in denaturation. (c) Biotinylated or isotopic nucleotide has lost its activity. (d) Labeled DNA samples are not completely denatured before loading into the gel.

   Solutions: (a) Heat the primer and double-stranded DNA template at 65°C for 3 to 4 min and slowly cool to room temperature over 20 to 35 min. (b) Use the alkaline-denaturing method to denature large-size DNA template. If this still does not work well, try to fragment the DNA insert to be sequenced and subclone for further sequencing. (c) Make sure the labeling reaction is carried out properly and denature the labeled sample at 75 to 80°C for 3 to 4 min prior to loading into the gel. (d) Try to use fresh biotinylated or isotopic nucleotide with high activity.

6. **Bands occur across all four lanes in some areas that are called compressions.**

   Possible cause: Target DNA sequences with strong secondary structure or G–C rich.

   Solution: Use an appropriate amount of dITP to replace dGTP and an appropriate amount of pyrophosphatase in the labeling reaction, or try formamide gel sequencing.

7. **Bands are faint near the primer.**

   Possible cause: Insufficient DNA template or insufficient primer.

Solution: (a) Use 1 to 1.5 μg single-stranded M13 DNA or 3 to 5 μg of plasmid DNA per reaction. (b) Increase the molar ratio of primer:DNA template from 1:1 to 1:4 or 1.5. (c) Use 1 μl of Mn buffer per regular labeling reaction.

8. **Bands are faint or blank in one or two lanes.**
   Possible cause: Some components may have been improperly added or missed in the samples loaded in the appropriate lanes.
   Solution: Be sure that all the components are added properly.
9. **No bands are observed in PCR-directed DNA sequencing, including the positive control.**
   Possible cause: *Taq* DNA polymerase has lost polymerization activity or primer is missing.
   Solution: Try fresh *Taq* DNA polymerase and ensure that the primer is included in the annealing reactions.
10. **No bands in RNA sequencing, including the positive RNA control lane.**
    Cause: AMV reverse transcriptase is missing or has lost its activity in the labeling reaction.
    Solution: Make sure that the reverse transcriptase is functional.
11. **No bands occur in sample RNA sequencing but visible bands are shown in the positive RNA control lane.**
    Cause: The RNA template is degraded or the primer is missing.
    Solution: Make sure that the RNA is of good quality and that the primer is added in the annealing reaction.
12. **High background occurs or no detection at all is obtained in nonisotopic sequencing.**
    Cause: (a) Blocking is not efficient. (b) Detection reagents have lost their activity.
    Solution: Make sure that the nonspecific binding sites are efficiently blocked and try fresh detection reagents.

REFERENCES

1. Cullmann, G., Hubscher, U., and Berchtold, M.W., A reliable protocol for dsDNA and PCR product sequencing, *BioTechniques*, 15, 578, 1993.
2. Sanger, F., Nicken, S., and Coulson, A.R., DNA sequencing with chain termination inhibitors, *Proc. Natl. Acad. Sci., USA,* 74, 5463, 1977.
3. Church, G.M. and Gilbert, W., Genomic sequencing, *Proc. Natl. Acad. Sci., USA,* 81, 1991, 1984.
4. Bishop, M.J. and Rawlings, C.J., *Nucleic Acid and Protein Sequence Analysis: A Practical Approach,* IRC Press, Oxford, 1987.
5. Wiemann, H.V., Grothues, D., Sensen, C., Zimmermann, C.S., Stegemann, H.E., Rupp, T., and Ansorge, W., Automated low-redundancy large-scale DNA sequencing by primer walking, *BioTechniques,* 15, 714, 1993.

6. Reynolds, T.R., Uliana, S.R.B., Floeter-Winter, L.M., and Buck, G.A., Optimization of coupled PCR amplification and cycle sequencing of cloned and genomic DNA, *BioTechniques,* 15, 462, 1993.
7. Wu, (W.)L., Song, I., Karuppiah, R., and Kaufman, P.B., Kinetic induction of oat shoot *Pulvinus* invertase mRNA by gravistimulation and partial cDNA cloning by the polymerase chain reaction, *Plant Molecular Biol.*, 21(6), 1175, 1993.
8. Wu, W., DNA sequencing, in *Handbook of Molecular and Cellular Methods in Biology and Medicine,* Kaufman, P.B., Wu, W., Kim, D., and Cseke, L., pp. 211–242, CRC Press, Boca Raton, FL, 1995.
9. Wu, (W.)L., Mitchell, J.P., Cohn, N.S., and Kaufman, P.B., Gibberellin (GA_3) enhances cell wall invertase activity and mRNA levels in elongating dwarf pea (*Pisum sativum*) shoots, *Int. J. Plant Sci.*, 154(2), 278, 1993.

6 Information Superhighway and Computer Databases of Nucleic Acids and Proteins

CONTENTS

INTRODUCTION

We are living in an era of information exposure. To many molecular biologists, the flood of information is particularly overwhelming because gene cloning, gene mapping, human genome projects, DNA, RNA and protein sequences are growing so rapidly. The fundamental questions concern how such an enormous volume of information is systematically organized and how to find the information desired from a sea of information. Obviously, development of superpower computers and smart computer programs plays a major role in the informatics. The birth of the Internet, particularly, has brought together scientists worldwide, creating the "information superhighway."

The present chapter provides investigators with an introduction to GenBank, the "headquarters of sequence information."[1–5] With the help of the Internet, it is easy to send or receive sequences and look for information inside headquarters without the office or laboratory. The topics covered in this chapter include the use of NCBI programs to deposit a sequence to the database, retrieve sequences from the database and carry out database searching from GenBank.[3] The second part of the chapter provides an introduction to computer analysis of DNA or protein sequences via a widely used program: the genetics computer group, or the GCG package, originally developed by the Department of Genetics at the University of Wisconsin.[4]

PART A. COMMUNICATION WITH GENBANK VIA THE INTERNET

SUBMISSION OF A SEQUENCE TO GENBANK

Nucleic acid and protein sequences can be submitted to GenBank, the EMBL Data Library, or the DNA Database of Japan (DDBJ), regardless of whether they have been published. In fact, most journals require that the sequence to be published first be submitted to whichever of the three libraries is the most convenient. GenBank, the EMBL and DDBJ are international database partners. Data submitted to one site will be exchanged with another site on a daily basis. Data to be submitted can be saved on a floppy disk or printed out and sent to:

National Center for Biotechnology Information (NCBI)
National Library of Medicine
National Institutes of Health
Building 38A, Room 8N-803
8600 Rockville Pike

The fastest and easiest way of submitting DNA, RNA and protein sequences to the three database sites is via e-mail using the following addresses.

GenBank: gb-sub@ncbi.nlm.nih.gov
EMBL: datasubs@ebi.ac.uk
DDBJ: ddbjsub@ddbj.nig.ac.jp

Special forms for sequence submission are available from these sites by mail or e-mail or from appropriate journals. In general, the questions on the forms are straightforward, including the name and features of the sequence. The common questions are:

1. What is the name of the sequence? (The name is given by the authors, e.g., Rat HSP27 cDNA.
2. What organism is the source of the DNA?
3. Is it an mRNA (cDNA) or genomic DNA sequence?
4. If it is a genomic DNA, are the intron sequences determined?
5. Has the sequence been published? What is the title of the publication? What is the journal? etc.

A sequence to be submitted should be prepared according to the detailed instructions on the forms. The newest way to prepare and submit a sequence is to obtain access to the World Wide Web. Here is how it is done:

1. Access the World Wide Web via Netscape.
2. Type NCBI World Wide Web home page at http://**www.ncbi.nlm.nih.gov**. Press the **Enter** key. You are in the NCBI home page. Several programs are exhibited on the screen such as **Entrez, BLAST, BankIt, OMIM, Taxonomy** and **Structure.**
3. Click BankIt. The following information will be displayed on the screen, including "To prepare a New GenBank submission, enter the size in nucleotides of your DNA sequence here [ ] and click **New**," and "To update an existing GenBank submission, press **Update**," depending on a specific sequence.
4. Click **Continue**. The electronic submission forms will show up on the screen.
5. Follow the instructions and fill in the appropriate answers in the blanks. Submit the sequence.

For each sequence submitted, a unique accession number will be given, e.g., M86389. The accession number is very important because it is considered to be the name of the sequence in the database. After submission, authors are encouraged to update the sequence, including corrections and publication.

SEQUENCE SIMILARITY SEARCHING USING THE BLAST PROGRAMS

Searches for sequence similarity or homology is performed hundreds and thousands of times every day worldwide. It provides a powerful tool for scientists to determine whether a newly isolated gene, DNA or protein is novel. If it is a known DNA or protein sequence, what is the percentage of similarity or homology compared to other species? How large is the gene family? These interesting questions can be answered by searching the GenBank database. Fortunately, the search does not cost anything as long as a computer and access to the World Wide Web are available. Based on our experience, we will introduce BLAST, which is by no means the most powerful or fastest program. BLAST stands for basic local alignment search tool and represents a family of programs for database searching. This section primarily focuses on three programs: BLASTN, BLASTX and BLASTP.

BLASTN

In this program, the sequence submitted for search is called the query sequence; it can be submitted in a single strand of nucleotides or in both strands. The database will search for any similarity among nucleotide sequences and display similar alignments on the computer screen. If there is not too much of an "information traffic jam," the searching speed is unbelievably fast. The entire search and exhibition take a few seconds or less than 1 min.

Let us take an example of how to carry out a database search. Assume that we have a nucleotide sequence named DNA X and that we wish to know whether any nucleotide sequences in the database are similar to DNA X. The search can be performed using the following procedures.

1. Access the **World Wide Web** system via **Netscape**. In the location box, type **http://www.ncbi.nlm.nih.gov/** and press the **Enter** key. Several programs are displayed on the computer screen, including **BLAST** and **NCBI Services**.
2. Find **NCBI Services** and click **•Blast Sequence Similarity Searching.**
3. Click **•Basic Blast search** or **•Advanced Blast search** (advanced is recommended). We are now in the **NCBI BLAST program**. From the top to the bottom, there are a number of blank boxes to be filled in or prechosen as default.
4. Choose **blastn** for the program and **nr** or **GenBank** for the database. A relatively large box in the middle of the screen is for entering the sequence to be submitted. The nucleotide sequence should be in the **FASTA format** or written in a **Courier font**.
5. In the sequence box, type the name of the sequence on the first line. The > must be included immediately before the first letter. Otherwise, the database will not recognize the name and treat it as an unknown sequence. For example, **>DNA X** is the name here. Starting from the second line,

type the sequence or paste a sequence cut from another file. An example is given below (only part of the sequence is shown here for brevity):

```
>DNA X
  1 CAGTGCTTCT  AGATCCTGAG  CCCTGACCAG  CTCAGCCAAG  ACCATGACCG
 51 AGCGCCGCGT  GCCCTTCTCG  CTACTGCGGA  GCCCCAGCTG  GGAGCCGTTC
101 CGGGACTGGT  ACCCTGCCCA  CAGCCGCCTC  TTCGATCAAG  CTTTCGGGGT
151 GCCTCGGTTT  CCCGATGAGT  GGTCTCAGTG  GTTCAGCTCC  GCTGGTTGGC
201 CCGGCTATGT  GCGCCCTCTG  CCCGCCGCGA  CCGCCGAGGG  CCCCGCAGCA
```

6. Edit the sequence as desired. Choose an appropriate option as illustrated below (bold letters):
   Advanced options for the Blast service:
   Expect **default**Cut off **default**Matrix **default**
   Strand **both**Filter **default**
   Description **default**Alignments **default**
7. Check **In HTML format** to display the search results on the screen or check **send reply** to send to an e-mail address. However, the format of the results on the screen in an e-mail system may be strange. We recommend that checking **In HTML format** for a perfect display.
8. Click **Submit Query**. A warning sign of a possible review of the sequence by a third party will show up and ask whether to continue or to cancel.
9. Click **Continue** and the search is carried out. Once it is complete, the sequence alignments will be exhibited, two by two, on the screen.
10. The results can be copied and then pasted into a regular Microsoft Word™ file. The font of Courier should be used to be compatible to the searching format. The sequences shown next are examples. The submitted sequence is called **Query** and the aligned sequence from the database is **Sbjct**. For brevity, only a part of the sequences is shown here. The first part of the results is a summary of different sequences in the database that show similarity to the **Query** sequence. The second part is the similarity alignments of individual sequences. If the name at the beginning of each sequence is clicked, detailed features of the searched sequences appear, including name, features, journal and title and full length of DNA, cDNA/mRNA and protein sequences. One can readily copy the sequences into an MS Word file or other files for further analysis. The last part of the results is mainly statistical analysis.

```
                DNA X
                (787 letters)
   Database:    Non-redundant GenBank+EMBL+DDBJ+PDB sequences
                243,087 sequences; 337,947,647 total letters.
                                                               Smallest
                                                                 Sum
                                                      High    Probability
Sequences producing High-scoring Segment Pairs        Score      P(N)     N
gb|M86389|RATHSP27A  Rat heat shock protein (Hsp27) mRNA,...3935 0.0        1
gb|S67755|S67755     hsp 27=heat shock protein 27 [rats,... 1440 4.9e-301 4
```

```
gb|L11610|MUSHSP25PS Mus musculus heat shock 25 (HSP25)p... 2665 1.3e-266 4
emb|X51747|CLSHSP    Cricetulus longicaudatus mRNA for sm...1711 9.3e-264 4
emb|X14686|MURSPH    Murine mRNA(pP25a) for 25-kDa mammal...1361 5.0e-250 4
```

```
gb|M86389|RATHSP27A Rat heat shock protein (Hsp27) mRNA, complete cds.
         Length = 787

  Plus Strand HSPs:

Score = 3935 (1087.3 bits), Expect = 0.0, P = 0.0
Identities = 787/787 (100%), Positives = 787/787 (100%), Strand = Plus /
Plus

Query:1 CAGTGCTTCTAGATCCTGAGCCCTGACCAGCTCAGCCAAGACCATGACCGAGCGCCGCGT    60
        ||||||||||||||||||||||||||||||||||||||||||||||||||||||||||||
Sbjct:1 CAGTGCTTCTAGATCCTGAGCCCTGACCAGCTCAGCCAAGACCATGACCGAGCGCCGCGT    60

Query:61 GCCCTTCTCGCTACTGCGGAGCCCCAGCTGGGAGCCGTTCCGGGACTGGTACCCTGCCCA  120
         ||||||||||||||||||||||||||||||||||||||||||||||||||||||||||||
Sbjct:61 GCCCTTCTCGCTACTGCGGAGCCCCAGCTGGGAGCCGTTCCGGGACTGGTACCCTGCCCA  120

Query:121 CAGCCGCCTCTTCGATCAAGCTTTCGGGGTGCCTCGGTTTCCCGATGAGTGGTCTCAGTG 180
          ||||||||||||||||||||||||||||||||||||||||||||||||||||||||||||
Sbjct:121 CAGCCGCCTCTTCGATCAAGCTTTCGGGGTGCCTCGGTTTCCCGATGAGTGGTCTCAGTG  18

gb|S67755|S67755 hsp 27=heat shock protein 27 [rats, Sprague-Dawley,
         Genomic, 1891 nt]
         Length = 1891
   Plus Strand HSPs:
Score = 1440 (397.9 bits), Expect = 4.9e-301, Sum P(4) = 4.9e-301
Identities = 296/306 (96%), Positives = 296/306 (96%), Strand = Plus / Plus

Query:482 ACGCTCCCTCCAGGTGTGGACCCCACCCTGGTGTCCTCTTCCCTGTCCCCTGAGGGCACA 541
           | |||||||||||||||||||||||| ||||||||||||||||||||||||||||||||
Sbjct:1570 AGGCTCCCTCCAGGTGTGGACCCCACCTTGGTGTCCTCTTCCCTGTCCCCTGAGGGCACA 1629

Query:542 CTCACCGTGGAGGCTCCGCTGCCCAAAGCAGTCACACAATCAGCGGAGATCACCATTCCG 601
           ||||| ||||||||||||||||||||||||||||||||||||||||||||||||||||||
Sbjct:1630 CTCACGGTGGAGGCTCCGCTGCCCAAAGCAGTCACACAATCAGCGGAGATCACCATTCCG 1689

gb|L11608|MUSHSP25A1 Mus musculus heat shock protein 25 (HSP25) gene, exon 1.
         Length = 760

  Plus Strand HSPs:

Score = 1690 (467.0 bits), Expect = 6.9e-146, Sum P(2) = 6.9e-146
Identities = 362/392 (92%), Positives = 362/392 (92%), Strand = Plus / Plus

Query:33 CAGCCAAGACCATGACCGAGCGCCGCGTGCCCTTCTCGCTACTGCGGAGCCCCAGCTGGG 92
          ||||||||| ||||||||||||||||||||||||||||||| ||||||||||| |||||||
Sbjct:272 CAGCCAAGAACATGACCGAGCGCCGCGTGCCCTTCTCGCTGCTGCGGAGCCCGAGCTGGG 331
```

```
Query:93 AGCCGTTCCGGGACTGGTACCCTGCCCACAGCCGCCTCTTCGATCAAGCTTTCGGGGTGC  152
          | || |||||||||||||||||||| |||||||||||||||||||||||||||||||||
Sbjct:332 AACCATTCCGGGACTGGTACCCTGCACACAGCCGCCTCTTCGATCAAGCTTTCGGGGTGC 391
```

BLASTX

This program allows submitting a query sequence in nucleotides; the database will translate the nucleotide sequence into an amino acid sequence. After that, the database will search for similarity between the translated protein sequence and other protein sequences in the GenBank. The results will be displayed on the screen. The protocol for protein searching is similar to **blastn** except for use of **blastx** for the program at step 4 described previously. The example is as follows:

```
Query= cDNA X
         (100 letters)


Translating both strands of query sequence in all 6 reading frames
Database:Non-redundant GenBank CDS
translations+PDB+SwissProt+SPupdate+PIR
202,545 sequences; 57,487,450 total letters.
                                                                 Smallest
                                                                    Sum
                                                  Reading High  Probability
Sequences producing High-scoring Segment Pairs:         Frame  Score P(N)
N
gi|55514             (X14686) p25 growth-related prote...         +2   103
1.8e-071
pir||JN0924         heat shock 27 protein - rat   +2103 1.9e-071
gi|544758            (S67755) heat shock protein 27, H...         +2   103
1.9e-071
sp|P42930|HS27_RAT HEAT SHOCK 27 KD PROTEIN (HSP 27)              +2   103
1.9e-071
gi|55516(X14687) p25 growth-related prote...                      +2   103
1.9e-071

gi|55514 (X14686) p25 growth-related protein (pP25a) (AA 1-208) [Mus sp.]
         Length = 199

Plus Strand HSPs:

Score = 103 (47.4 bits), Expect = 1.8e-07, P = 1.8e-07
Identities = 19/19 (100%), Positives = 19/19 (100%), Frame = +2
Query:    44     MTERRVPFSLLRSPSWEPF 100
                 MTERRVPFSLLRSPSWEPF
```

```
Sbjct:   1      MTERRVPFSLLRSPSWEPF 19

sp|P80492|BRA2_BRAFL BRACHYURY PROTEIN HOMOLOG 2 (T PROTEIN).
         Length = 440

Plus Strand HSPs:

Score = 39 (17.9 bits), Expect = 7.5, Sum P(2) = 1.0
Identities = 8/18 (44%), Positives = 10/18 (55%), Frame = +2
Query:   47     TERRVPFSLLRSPSWEPF 100
                TER + +L   P WE  F
Sbjct:   31     TERDLKVTLGEKPLWEKF  48

Score = 29 (13.3 bits), Expect = 7.5, Sum P(2) = 1.0
Identities = 5/9 (55%), Positives = 6/9 (66%), Frame = +
```

BLASTP

This program is used for similarity searches between query protein and any proteins in the database. The procedures are the same as those described earlier except that **a peptide or protein sequence** needs to be typed or pasted in the sequence box and use **blastp** for program. A part of the results is shown below.

```
Query=Protein X
(28 letters)
Database:Non-redundant GenBank CDS
translations+PDB+SwissProt+SPupdate+PIR
202,545 sequences; 57,487,450 total letters.
                                                               Smallest
                                                                 Sum
                                                       High   Probability
Sequences producing High-scoring Segment Pairs:        Score   P(N)     N

gi|55514           (X14686) p25 growth-related protein ...   160  21e-16  1
pir||JN0924        heat shock 27 protein - rat               160  2.2e-16 1
gi|544758          (S67755) heat shock protein 27, HSP ...   160  2.2e-16 1
sp|P42930|HS27_RAT HEAT SHOCK 27 KD PROTEIN (HSP 27).  ...   160  2.2e-16 1
gi|55514 (X14686) p25 growth-related protein (pP25a) (AA 1-208) [Mus sp.]
         Length = 199

Score = 160 (74.9 bits), Expect = 2.1e-16, P = 2.1e-16
Identities = 28/28 (100%), Positives = 28/28 (100%)
```

```
Query:   1      MTERRVPFSLLRSPSWEPFRDWYPAHSR 28
                MTERRVPFSLLRSPSWEPFRDWYPAHSR
Sbjct:   1      MTERRVPFSLLRSPSWEPFRDWYPAHSR 28

gnl|PID|e213334 (Z68160) D1046.3 [Caenorhabditis elegans]
          Length = 269

Score = 50 (23.4 bits), Expect = 1.2, P = 0.71
Identities = 7/17 (41%), Positives = 12/17 (70%)
Query:   4      RRVPFSLLRSPSWEPFR 20
                R +PFS+++ P WE +
Sbjct:  151     REIPFSIIQFPIWEALK 167
```

PART B. COMPUTER ANALYSIS OF DNA SEQUENCES BY THE GCG PROGRAM

In order to obtain access to the GCG program, a username and a password must be issued by appropriate persons in charge of the program. Once the username and password are typed, the message "Welcome to the......" will be displayed on the screen followed by a prompt or dollar sign ($). We routinely use the GCG program to analyze nucleic acid and amino acid sequences. This section will give a tour of how to use the GCG program for research.

ENTRY AND EDITING OF A SEQUENCE USING GCG

It is necessary to use the program **SeqEd,** which is designed especially for sequence entry and editing or modification.

SEQUENCE ENTRY

A sequence can be entered by typing on the keyboard or by cutting and pasting existing sequences from other files or sources.

1. At the prompt or after the **$** sign, type **SeqEd** and press **Enter**. What appears on the screen follows:

```
                         ***** K E Y B O A R D *****

:Type heading of text as you wish.
:Make comments as you wish.
:
:
|.........|.........|.........|.........|.........|.........|
0         10        20        30        40        50        60
|......|......|......|......|......|......|......|......|......|......|
0      10     20     30     40     50     60     70     80     90
```

SeqEd of what sequence?

2. Type a name after the question. Make sure to type **.seq** after the name, for example, SeqEd of what sequence? **Test.seq** and **enter**. The cursor will move to the first base position and the sequence can be entered.
3. Type a sequence with the keyboard or paste sequences from other files or GenBank. The length of the entered sequence will be denoted by numbers as follows and the cursor will appear at the end of the sequence. By keyboard typing:

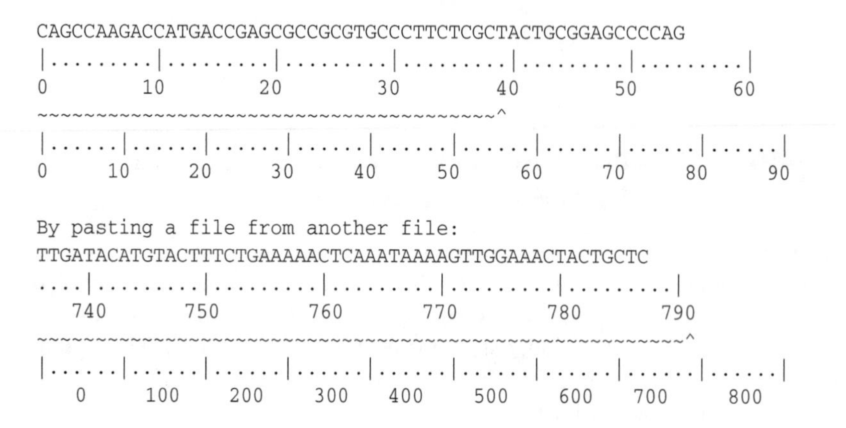

4. Once the nucleotide sequence entry is complete, if editing or modification is not necessary, hold **Ctrl** and press the **Z** key. The cursor will move to the bottom left-hand corner of the screen.
5. At the cursor, type **Exit** and then **Enter**; the sequence is saved. The name and length of the sequence are shown at the bottom of the screen. Two examples are given below:

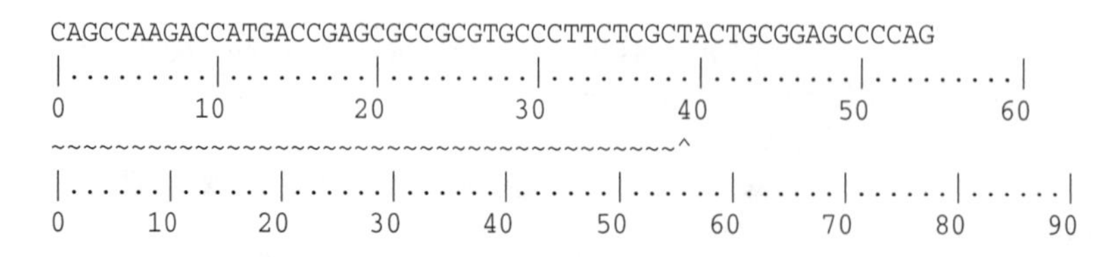


```
"Test.seq" 55 nucleotides

TTGATACATGTACTTTCTGAAAAACTCAAATAAAAGTTGGAAACTACTGCTC
....|.........|.........|.........|.........|.........|
   740       750       760       770       780       790
~~~~~~~~~~~~~~~~~~~~~~~~~~~~~~~~~~~~~~~~~~~~~~~~~~~~~~^
|......|......|......|......|......|......|......|......|......|
   0     100    200    300   400    500    600   700     800

"HSP27.seq" 787 nucleotides
```

If **Quit** and **Enter** are typed, the sequence has not been saved; it must be started all over again.

Sequence Editing or Modification

1. To edit a sequence, following the **$** sign, type **SeqEd** and press **Enter**. At the end of "**SeqEd of what sequence?**" type the sequence name to be edited, for instance, **Test.seq,** and press **Enter**. The sequence with position numbers will show up on the screen as shown earlier.
2. To edit, move the cursor with the left or right arrow key to an appropriate position in the sequence. Hold **Ctrl** and press the **Backspace** or **Delete** key to delete bases or insert more bases by keyboard or by cutting and pasting a sequence from another file.
3. When editing is complete, hold **Ctrl** and press **Z** to quit. Then, type **Exit** to save the edited sequence; however, if **Quit** and **Enter** are typed, the sequence is not saved.

Review of Sequence Output

To see the entire sequence, at the prompt, type "Type," then space and type the name of the sequence to be reviewed. For example, $**Type HSP27.seq** and press **Enter**. The entire sequence will be displayed on the screen as follows:

```
HSP27.seq  Length:787  June 17, 1996  21:00  Type:  N  Check:  740
        1  CAGTGCTTCT  AGATCCTGAG  CCCTGACCAG  CTCAGCCAAG  ACCATGACCG
       51  AGCGCCGCGT  GCCCTTCTCG  CTACTGCGGA  GCCCCAGCTG  GGAGCCGTTC
      101  CGGGACTGGT  ACCCTGCCCA  CAGCCGCCTC  TTCGATCAAG  //
```

COMBINATION OR ASSEMBLY OF MULTIPLE FRAGMENTS INTO A SINGLE SEQUENCE

When sequencing of large DNA molecules by primer walking is performed, it is quite often the case that multiple short DNA sequences or fragments will be generated. Because these fragments contain overlapping ends, they can be assembled into a single sequence using the program of Fragment Assembly, which includes four procedures: GelStart, GenEnter, GelMerge or GelOverlap, and GelAssemble. Let us take the following fragments 1 to 3 as an example, assuming that they have 10-base overlapping ends and that their sequences are as follows:

```
Fragment 1 (30 bases): GACTAGACGATGCTGCTAGGCCTGACGTCG
Fragment 2 (30 bases): CCTGACGTCGGACTAGACGGCTGCGCGCTG
Fragment 3 (30 bases): CTGCGCGCTGGGGGACACACCATAGCTCGC
```

Generating a New Project File Using the GelStart Program

Following the $ sign, type **GelStart** (that is, $GelStart) and press the **Enter** key. The following information will show up on the screen. After each question, type the appropriate answer (see the words in bold typeface).

What is the name of your fragment assembly project? **Fragment**
GELSTART cannot find this project. Is it a new one (* No *)? **yes**
You have a new project named "FRAGMENT."
Which vector sequence(s) would you like to be highlighted? **GACACACCA** [You can type any short sequence.]
STOREVECTORS could not open the file GACACACCA.
This vector will not be highlighted. Check the file specification and run GELStart/VECtors to reenter this vector. [You can ignore it.]
Which restriction site(s) would you like highlighted? **CTCAGA,GAATTC** [You can enter any two restriction enzyme sequences, with a comma between them.]
Project FRAGMENT has 0 fragments in 0 contigs.
You are now ready to run the other fragment assembly programs.
$

Enter Sequences to be Assembled into the Project File Generated in A (e.g., FRAGMENT) Using the GelEnter Program

1. Following the **$** sign, type **GelEnter** and press **Enter**. At the end of "**GelEnter of what sequence?**" type the sequence name to be entered, for instance, **Fragment1** and press **Enter**. The sequence with position numbers will show up on the screen.
2. Press **Ctrl** and **Z** keys, and type a heading or comments as desired. If a heading and comments are not wanted, press **Ctrl** and **Z** keys and begin to type or paste a sequence of interest cut from another file at the cursor position. If the sequence to be entered is already in the GCG and saved as xx.seq (e.g, Fragment1.seq), simply type Fragment1.seq (e.g., "**GelEnter of what sequence?**" **Fragment1.seq**). In this case, the sequence will be automatically pulled out, so the sequence does not need to be retyped.
3. To edit, move the cursor with the left or right arrow keys to an appropriate position in the sequence. Hold **Ctrl** and press the **Backspace** or **Delete** key to delete bases or, at the cursor, insert more bases by keyboard or by cutting and pasting a sequence from another file.
4. When editing is complete, hold **Ctrl** and press **Z** key to quit editing. Then, type **Exit** to save the edited sequence; however, if **Quit** and **Enter** are typed, the sequence is not saved in the file. The following information will be displayed on your screen.

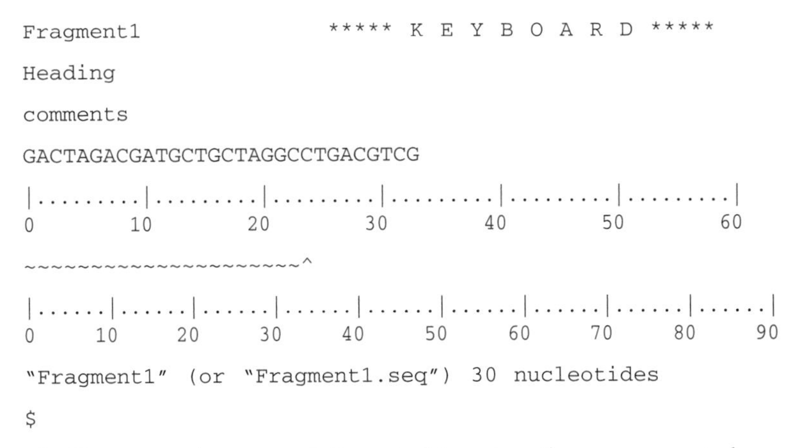

5. Repeat previous steps 1 through 4 to enter other sequences to be assembled. For example, fragments 2 and 3 are displayed as follows:

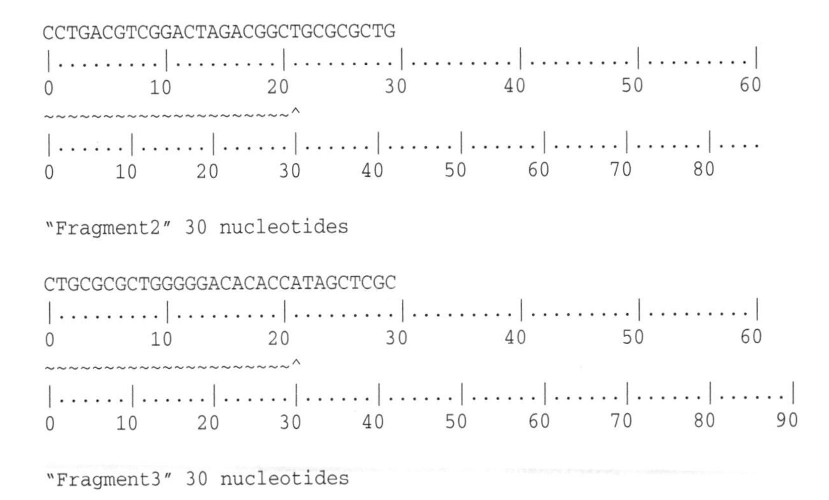

Compare and Identify Overlap Points of Entered Fragments Using the GelMerge or GelOverlap Program

Once the fragment sequences have been entered and saved, their sequences can be analyzed to find the overlapping regions with the **GelMerge** program. This program will align the fragments into assemblies called contigs; however, the alignment cannot be seen. To do that, it is necessary to proceed to **D. GelAssemble**.

Following the $ sign, type **GelMerge** ($GelMerge) and press **Enter**. The information will be shown on the screen as follows:

What word size (* 7 *)? **7**
What fraction of the words in an overlap must match (*0.80*)? **0.8**
What is the minimum overlap length (* 14 *)? **10**
Reading...

Comparing…
Aligning…
Writing…
  Input Contigs:3
  Output Contigs:1
  CPU time:01.87
$

Assemble and Review the Combined Sequence by Using the GelAssemble Program

1. Following the $ sign, type **GelAssemble** ($GelAssemble) and press **Enter**. The information will be shown on the screen as follows:

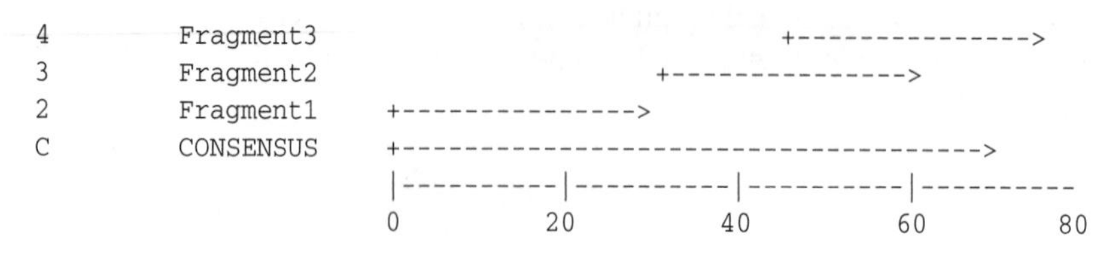


```
Contig   1 of   1

<Right-arrow> for next contig, <Left-arrow> for previous contig
<Ctrl>Z for Screen Mode, <Ctrl>K to load a contig:
```

2. Press the **Ctrl** and **K** keys to load a contig. The assembled sequences will be displayed on the screen.

```
GelAssemble                      Fragment                    GCG
                     Absolute:          1    Relative:  1
                                     CTGCGCGCTGGGGGACACACCATAGCTCGC
                      CCTGACGTCGGACTAGACGGCTGCGCGCTG
```

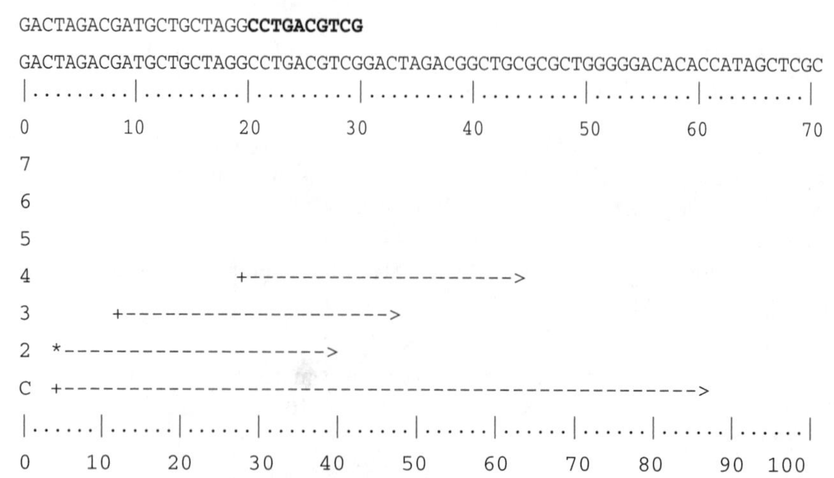

3. Press the **Ctrl** and **Z** keys to move the cursor to the sequence. Review the entire sequence using the right or left arrow keys. The overlapping sequences of the present three fragments are indicated in bold letters.

IDENTIFICATION OF RESTRICTION ENZYME DIGESTION SITES, FRAGMENT SIZES, AND POTENTIAL PROTEIN TRANSLATIONS OF A DNA SEQUENCE

Cloning, digestion, manipulation and characterization of DNA molecules mandate detailed analysis of restriction enzyme digests. This section will serve as a simplified guide for investigators who want to find out information about restriction enzyme digests of a specific DNA via using the mapping program in GCG.

EXHIBITION OF RESTRICTION ENZYMES ABOVE BOTH STRANDS OF A DNA SEQUENCE AND POSSIBLE PROTEIN TRANSLATION BELOW THE SEQUENCE USING THE MAP PROGRAM

1. Enter and save the DNA sequence of interest using the SeqEd program described earlier. For example, a short sequence is saved as DNAX.seq.
2. Following the $ sign, type **Map** ($Map) and press **Enter**. Information will be displayed on the screen. Type appropriate answers to the questions (e.g., bold letters below) and press the **Enter** key:
   (Linear) MAP of what sequence ? **DNAX.seq**
   Begin (* 1 *) ? **1**
   End (* 60 *) ? **60**
   Select the enzymes: Type nothing or "*" to get all enzymes. Type "?"
   Enzyme(* * *): *****
   What protein translations do you want:
   a) frame 1, b) frame 2, c) frame 3, d) frame 4, e) frame 5
   f) frame 6, t)hree forward frames, s)ix frames, o)pen frames only, n)o protein translation, q)uit
   Please select (capitalize for 3-letter) (* t *): **o**
   What should I call the output file (* Dnax.map *)? **DNAX.map**
   $
3. At the $, type **DNAX.map** and press the **Enter** key. The output will be displayed as follows. Therefore, one can choose appropriate enzymes of interest to cut the DNA.

```
(Linear) MAP of:  DNAX.seq check: 6764 from:  1  to:  60
With 215 enzymes: *

                              B                          B
                              s                          s
                        N   F p                          p    M
                    C   l   nH C T1               NC     B1   CsP
                D   v   a   HAuaMaTBa2T  M    A   lv     a2   Avpv
                d   I   I   hc4ewcamq8h  w    c   ai     n8   liAu
                e   J   I   aiHIo8ugI6a  o    I   IJ     I6   uJ1I
                I   I   I   IIIIIIIIIII  I    I   VI     II   IIII
                              //  /   /                  /    ///

           CTCAGCCAAGACCATGACCGAGCGCCGCGTGCCCTTCTCGCTACTGCGGAGCCCCAGCTG

1 ---------+---------+---------+---------+---------+---------+ 60

GAGTCGGTTCTGGTACTGGCTCGCGGCGCACGGGAAGAGCGATGACGCCTCGGGGTCGAC
a    L  S  Q  D  H  D  R  A  P  R  A  L  L  A  T  A  E  P  Q  L            -
b     S  A  K  T  M  T  E  R  R  V  P  F  S  L  L  R  S  P  S             -
c    Q  P  R  P  *                                                        -
  1 ---------+---------+---------+---------+---------+---------+ 60
d                                                 *  Q  P  A  G  A         -
e             L  W  S  W  S  R  A  G  R  A  R  R  A  V  A  S  G  W  S      -
f            E  A  L  V  M  V  S  R  R  T  G  K  E  S  S  R  L  G  L  Q    -

Enzymes that do cut:

   AciI    AluI   BanII    BmgI  Bsp1286I   Cac8I   CviJI    DdeI
 Fnu4HI   HaeII    HhaI  MspA1I      MwoI  NlaIII   NlaIV   PvuII

Enzymes that do not cut: (Here is part of it.)

  AatII    AccI  AceIII   AflII    AflIII    AhdI    AlwI  Alw26I
  AlwNI    ApaI   ApaBI   ApaLI      ApoI    AscI    AvaI   AvaII
$
```

Identification of Specific Restriction Enzyme Cutting Sites and Sizes of Fragments by Using the MapSort Program

1. Enter and save the DNA sequence of interest using the SeqEd program as described earlier. For example, a short sequence is saved as DNAX.seq.
2. Following the $ sign, type **MapSort** ($MapSort) and press **Enter**. Information will be displayed on the screen. Type appropriate answers to questions (e.g., bold-face letters below) and press the **Enter** key.

```
(Linear) MAPSORT of what sequence? DNAX.seq

                Begin (* 1 *)? 1
                End (* 60 *)? 60
Is this sequence circular (* No *)? no
Select the enzymes:  Type nothing or "*" to get all enzymes. Type "?"
Enzyme(* * *):  *

What should I call the output file (* Dnax.mapsort *)? Enter
Mapping ....
$
```

3. At the $, type **Dnax.mapsort** and press the **Enter** key. The output will be displayed on the screen. Here is part of the output.

```
  (Linear) MAPSORT of: DNAX.seq  check: 6764  from: 1  to: 60
  With 215 enzymes: *
  June 18, 1996 16:33

  AciI C'CG_C

Cuts at:        0      25      46      60
  Size:        25      21      14
  Fragments arranged by size:
               25      21      14

  Bsp1286I G_dGCh'C

Cuts at:        0      33      53      60
  Size:        33      20       7
    Fragments arranged by size:
               33      20       7

  HaeII r_GCGC'y
```

```
Cuts at:          0        25       60
   Size:         25        35

  PvuII  CAG'CTG

Cuts at:          0        57       60
   Size:         57         3

  Enzymes  that  do  cut:

    AciI     AluI     BanII     BmgI Bsp1286I     Cac8I     CviJI     DdeI
  Fnu4HI    HaeII      HhaI   MspA1I      MwoI    NlaIII     NlaIV    PvuII

  Enzymes  that  do  not  cut:

   AatII     AccI   AceIII    AflII    AflIII     AhdI      AlwI   Alw26I
   AlwNI     ApaI    ApaBI    ApaLI      ApoI     AscI      AvaI    AvaII
   AvrII     BaeI     BaeI    BamHI      BanI     BbsI      BbvI     BccI
  $
```

COMPARISON OF SIMILARITY BETWEEN TWO SEQUENCES

To obtain this information, it may be necessary to use the comparison program of the GCG. Specifically, programs such as BestFit, Gap or Gapshow will do the job. Gapshow requires a plotter to be attached to a LaserWriter. Here is the introduction of an optimal alignment of the best segment of similarity between two sequences by the BestFit program.

1. Enter and save the two DNA sequences to be analyzed using the SeqEd program as described previously. For example, two short DNA sequences are saved as sequence1.seq and sequence2.seq.
2. Following the $ sign, type **BestFit** ($BestFit) and press **Enter**. Information will be displayed on the screen. Type appropriate answers to questions (e.g., bold-face letters below) and press the **Enter** key each time.

```
$ BestFit

BESTFIT of what sequence 1? sequence1.seq

                    Begin (* 1 *)? 1

                    End (* 200 *)? 200

                    Reverse (* No *)? no

to what sequence 2 (* sequence1.seq *)? sequence2.seq

                    Begin (* 1 *)? 1
```

```
                    End (* 150 *)? 150
                    Reverse (* No *)?  nmo

What is the gap creation penalty (* 5.00 *)? 3
What is the gap extension penalty (* 0.30 *)? 0.3
What should I call the paired output display file (*
Sequence1.pair *)? Enter to accept the name

Aligning ........
            Gaps:     0
         Quality:   150.0
Quality Ratio:        1.000
  % Similarity:     100.000
          Length:   150

$
```

3. At the $, type **type Sequence1.pair** and press the **Enter** key. The output will be displayed on the screen.

```
$ type Sequence1.pair
BESTFIT of: Sequence1.Seq  check: 7766  from: 1  to: 200
to: Sequence2.Seq  check: 4913  from: 1  to: 150
        Gap Weight:  3.000      Average Match:  1.000
     Length Weight:  0.300   Average Mismatch: -0.900

           Quality:  150.0            Length:    150
             Ratio:  1.000              Gaps:      0
 Percent Similarity: 100.000   Percent Identity: 100.000

Sequence1.Seq x Sequence2.Seq June 18, 1996 17:15  ..

                       .         .         .         .         .
      51 CGGGACTGGTACCCTGCCCACAGCCGCCTCTTCGATCAAGCTTTCGGGGT 100
         ||||||||||||||||||||||||||||||||||||||||||||||||||
       1 CGGGACTGGTACCCTGCCCACAGCCGCCTCTTCGATCAAGCTTTCGGGGT 50
                       .         .         .         .         .
     101 GCCTCGGTTTCCCGATGAGTGGTCTCAGTGGTTCAGCTCCGCTGGTTGGC 150
         ||||||||||||||||||||||||||||||||||||||||||||||||||
      51 GCCTCGGTTTCCCGATGAGTGGTCTCAGTGGTTCAGCTCCGCTGGTTGGC 100
                       .         .         .         .         .
```

```
    151 CCGGCTATGTGCGCCCTCTGCCCGCCGCGACCGCCGAGGGCCCCGCAGCA 200
        ||||||||||||||||||||||||||||||||||||||||||||||||||
    101 CCGGCTATGTGCGCCCTCTGCCCGCCGCGACCGCCGAGGGCCCCGCAGCA 150
$
```

TRANSLATION OF NUCLEIC ACID SEQUENCES INTO AMINO ACID SEQUENCES OR AN AMINO ACID SEQUENCE INTO A NUCLEIC ACID SEQUENCE

The translate program can be used for translation of nucleic acid sequences into amino acid sequences. On the other hand, the BackTranslate program is used to reversely translate amino acid sequences into nucleic acid sequences.

TRANSLATE

1. Enter and save the DNA sequences to be translated using the SeqEd program as described earlier. For example, the short DNA sequences are saved as sequence1.seq.
2. Following the $ sign, type **Translate** ($translate) and press **Enter**. Information will be displayed on the screen. Type appropriate answers to questions (e.g., bold-face letters below) and press the **Enter** key each time.

```
$ Translate
TRANSLATE from what sequence(s)? sequence1.seq

                  Begin (* 1 *)? 1
                  End (* 200 *)? 200
              Reverse (* No *)? yes
Range begins TGCTG and ends GCGCT. Is this correct (* Yes *)? y

That is done, now would you like to:

   A) Add another exon from this sequence
   B) Add another exon from a new sequence
   C) Translate and then add more genes from this sequence
   D) Translate and then add more genes from a new sequence
   W) Translate assembly and write everything into a file

Please choose one (* W *): w
What should I call the output file (* Sequence1.pep *)? Enter
```

3. At the $, type **sequence1.pep** and press **Enter** key. The output will be displayed on the screen.

```
$ type sequence1.pep
TRANSLATE of: sequence1.seq /rev check: 7766 from: 1 to: 200
generated symbols 1 to: 66.
Sequence1.pep  Length: 66  June 18, 1996 18:16  Type: P Check: 7296

      1  CCGALGGRGG QRAHIAGPTS GAEPLRPLIG KPRHPESLIE EAAVGRVPVP

     51  ERLPAGAPQ* REGHAA
$
```

BackTranslate (Using the Sequence1.pep as an Example)

Here is a part of the result that can help in recognizing some ambiguous regions that may be useful for synthesis of probes:

```
$ Backtranslate
BACKTRANSLATE what sequence? sequence1.pep
              Begin (* 1 *)? 1
               End (* 66 *)? 66
Would you like to see:
     a) table of back-translations and most probable sequence
     b) table of back-translations and most ambiguous sequence
     c) most probable sequence only
     d) most ambiguous sequence only

Please choose one (* b *): b

Use what codon frequency file (*GenRunData:ecohigh.cod*)?Enter
 What should I call the output file (*Sequence1.seq*)? Enter

$ type sequence1.seq

BACKTRANSLATE of: : Sequence1.Pep  check: 7314  from: 1  to: 66
TRANSLATE of: sequence1.seq check: 7766 from: 1 to: 200
generated symbols 1 to: 66.

Codon usage for enteric bacterial (highly expressed) genes 7/19/83
  Ser        Ala        Ala        Cys        Pro        Ser        Arg
```

```
 UCC 0.37   GCU 0.35   GCU 0.35   UGC 0.51   CCG 0.77   UCC 0.37   CGU 0.74
 UCU 0.34   GCA 0.28   GCA 0.28   UGU 0.49   CCA 0.15   UCU 0.34   CGC 0.25
 AGC 0.20   GCG 0.26   GCG 0.26              CCU 0.08   AGC 0.20   CGA 0.01
 UCG 0.04   GCC 0.10   GCC 0.10              CCC 0.00   UCG 0.04   AGG 0.00
 AGU 0.03                                               AGU 0.03   AGA
Sequence1.Seq  Length: 198  June 18, 1996 18:31  Type: N  Check: 7709
       1  WSNGCNGCNT GYCCNWSNMG NTAYTGYGGN GCNCCNGCNG GNWSNMGNWS
      51  NGGNACNGGN ACNYTNCCNA CNGCNGCNWS NWSNATHAAR YTNWSNGGNT
     101  GYYTNGGNTT YCCNATGWSN GGNYTNWSNG GNWSNGCNCC NYTNGTNGGN
     151  CCNGCNATGT GYGCNYTNTG YCCNCCNMGN CCNCCNMGNG CNCCNCAR
$
```

IDENTIFICATION OF ENZYME DIGESTION SITES WITHIN A PEPTIDE OR PROTEIN

To obtain this information, it may be necessary to use the PeptideMap program. For example, if a nucleic acid sequence is saved as a **Nuc.seq** file and its amino acid file is **Nuc.pep**, the PeptideMap program can be used to find the enzyme cutting site in the **Nuc.pep**.

1. Enter and save a new DNA sequence to be translated using the **SeqEd** program as described previously. For example, a short DNA sequence is saved as **Nuc.seq**.
2. Translate **Nuc.seq** into protein and save it as **Nuc.pep**. If protein files are already in the GCG, the files can be directly used for mapping.
3. Following the $ sign, type **PeptideMap** ($PeptideMap) and press **Enter**. Information will be displayed on the screen. Type appropriate answers to questions (e.g., bold letters below) and press the **Enter** key each time.

```
$ Peptidemap
(Linear) (Peptide) MAP of what sequence ? Nuc.pep
               Begin (* 1 *)? 1
                End (* 20 *)? 20
Select the enzymes: Type nothing or "*" to get all enzymes. Type "?"
Enzyme(* * *):*

What should I call the output file (*Nuc.map*)? Enter

$ type nuc.map
(Linear) (Peptide) MAP of: Nuc.Pep  check: 6438  from: 1  to: 20
TRANSLATE of: nuc.seq check: 6445 from: 1 to: 60
generated symbols 1 to: 20.
With 11 enzymes: *
```

```
                  S   PC
      T        T C tTT rh
      r        r n arr oy
      y        y B pyy Em
      p        p r hpp no
      RS*ALTSSAKTMTERRVPFS
    1 ---------+---------+ 20

Enzymes that do cut:
Chymo     CnBr    ProEn    Staph     Tryp

Enzymes that do not cut:
NH2OH     NTCB    pH2.5
$
```

OBTAINING NUCLEOTIDE AND AMINO ACID SEQUENCES FROM GENBANK

As long as the accession number of the sequence of interest is known, the sequences can be easily accessed and pulled out using the **Fetch** program. If the accession number is not known, the common name for the sequences may be typed in just like a key word search. All the sequences under that name will be pulled out and copied into the directory.

1. At the $ sign, type **fetch**, a space and **accession number** or **name** of a sequence such as M86389 or hsp27 (**$fetch M86389** or **$fetch hsp27**) and press **Enter**. The program will search for the sequences and make copies from the GCG database into a directory with a specific name such as **M86389.Gb_Ro** or **Mhsp27.Gb_Pr**. Information will displayed on the screen. Type appropriate answers to questions (e.g., bold letters below) and press the **Enter** key each time.

   $ fetch M86389
   M86389.Gb_Ro

   $fetch hsp27
   Mhsp27.Gb_Pr

2. At the prompt, type **type,** space, **M86389.Gb_Ro** (**$type M86389.Gb_Ro**) or **type**, space, **Mhsp27.Gb_Pr**, (**$type Mhsp27.Gb_Pr**) and press the **Enter** key. The sequences from the directory will be displayed on the computer screen. Detailed information of each sequence includes three parts: (1) the name and features of the

sequence; (2) the translated amino acid sequence; and (3) the entire length of DNA sequences. Examples are given next. Only part of the information is shown here for brevity.

```
$ Type M86389.Gb_Ro
LOCUS        RATHSP27A     787 bp ss-RNA            ROD       03-FEB-1992
DEFINITION   Rat heat shock protein (Hsp27) mRNA, complete cds.
ACCESSION    M86389
NID          g204664
KEYWORDS     heat shock protein 27.
SOURCE       Rattus norvegicus (strain Fisher) (library: Lambda
translation="MTERRVPFSLLRSPSWEPFRDWYPAHSRLFDQAFGVPRFPDEWS
                   QWFSSAGWPGYVRPLPAATAEGPAAVTLARPAFSRALNRQLSSGVSEIRQTADRWRVS
                   LDVNHFAPEELTVKTKEGVVEITGKHEERQDEHGYISRCFTRKYTLPPGVDPTLVSSS

M86389  Length: 787  May 24, 1996 15:15  Type: N  Check: 740  ..

       1  CAGTGCTTCT AGATCCTGAG CCCTGACCAG CTCAGCCAAG ACCATGACCG
      51  AGCGCCGCGT GCCCTTCTCG CTACTGCGGA GCCCCAGCTG GGAGCCGTTC
     101  CGGGACTGGT ACCCTGCCCA CAGCCGCCTC TTCGATCAAG CTTTCGGGGT
     151  GCCTCGGTTT CCCGATGAGT GGTCTCAGTG GTTCAGCTCC GCTGGTTGGC
```

$ fetch hsp27
MHsp27.Gb_Pr

$ type hsp27.Gb_Pr

All the hsp27 sequences will be displayed in a similar pattern to the preceding one (omitting the output for brevity).

REFERENCES

1. Altschul, S.F., Boguski, M.S., Gish, W., and Wootton, J.C., Issues in searching molecular sequence databases, *Nature Genet.*, 6, 119–129, 1994.
2. Altschul, S.F., Gish, W., Miller, W., Myers, E.W., and Lipman, D.J., Basic local alignment search tools, *J. Mol. Biol.*, 215, 403–410, 1990.
3. Recipon, H.E., Schuler, G.D., and Boguski, M.S., Informatics for molecular biologists, in *Current Protocols in Molecular Biology,* Susubel, F.M., Brent, R., Kingston, R.E., Moore, D.D., Seidman, J.G., Smith, J.A., and Struhl, K., Eds., John Wiley & Sons, New York, 19.0.1–19.3.38, 1996.
4. Berg, L., Butler, B., Devereux, J., Edelman, I., Katz, D., Kiefer, A., King, J., Murphy, J., Rose, S., Smith, M., and Schultz, M., *The Sequence Analysis Software Package by Genetics Computer, Inc.* (the GCG Package), Version 7.0, 1991.

5. Wootton, J.C., Nonglobular domains in protein sequences: automated segmentation using complexity measures, *Comput. Chem.*, 18, 269–285, 1994.

7 Characterization of DNA or Genes by Southern Blot Hybridization

CONTENTS

INTRODUCTION

Characterization of DNA or the gene plays a crucial role in current molecular biology. A powerful, widely used tool is Southern blot hybridization, which refers to a procedure in which different sizes of DNA molecules are immobilized from agarose gels onto a solid support such as nylon or nitrocellulose membranes and then hybridized with a labeled DNA probe. In other words, Southern blot refers to DNA–DNA hybridization. Developed by Southern in 1975, this blotting technique was later subjected to modification[1,2] and has now become one of the most fundamental and powerful tools used in molecular biology studies. It is a simple, sensitive and reliable method. Its applications are incredibly broad, including general DNA analysis, gene cloning, screening and isolation, DNA mapping, mutant detection, identification of genetic diseases and DNA amplification.[2–5]

Based on our extensive experience, detailed protocols for successful Southern blot hybridization using nonradioactive or radioactive probes are described in this chapter.[3–9]

PRINCIPLES AND GENERAL CONSIDERATIONS

The general principles and procedures of Southern blot hybridization are outlined in Figure 7.1. Based on their molecular weights, different DNA molecules are first separated by standard agarose gel electrophoresis in which agarose gel serves as a molecular sieve and electrophoresis is the force for migration of negatively charged DNA in the electrical field. The separated double-stranded DNA species are denatured into single-stranded DNAs and then blotted or immobilized onto a nylon or a nitrocellulose membrane. Usually, different positions of DNA molecules in an agarose gel are exactly blotted on their related positions on the membrane. Following covalent cross-linking of DNA molecules on the membrane, specific bands of interest can then be hybridized and detected with a specific DNA probe. The nucleotide sequence of one strand of the DNA molecule is complementary to the nucleotide sequence of the other strand. It is exactly the base pairing or complementary rule that makes it possible for a single-stranded DNA probe to hybridize with its target sequences of single-stranded DNA on the membrane.

Agarose gel electrophoresis is commonly used for separation of DNA fragments. A linear polymer that basically consists of D-galactose and L-galactose, agarose is extracted from seaweed algae. Once agarose is melted and then allowed to harden, it forms a matrix that serves as a molecular sieve by which different sizes of DNA molecules can be separated from each other. Agarose is commercially available with varying degrees of purity. An ultrapure-grade agarose that is DNase-free is recommended for Southern blotting.

General considerations should be kept in mind when one carries out agarose gel electrophoresis of DNA. For an optimal separation of DNA species, agarose concentration should be determined based on the following principle. The larger the size of DNA is, the lower the percentage of agarose that should be used.

In general, linear duplex molecules move through the gel matrix at a rate that is inversely proportional to the log of their molecular weight. Small DNA molecules move faster than large ones. Also, DNA conformation influences the migration rate. The speed of DNA migration is: supercoiled DNA > slightly coiled DNA > linear DNA > open circular DNA.

Transfer of DNA from agarose onto a solid nitrocellulose or nylon membrane can be carried out by capillary blot, electrophoretic blot or vacuum blot methods.

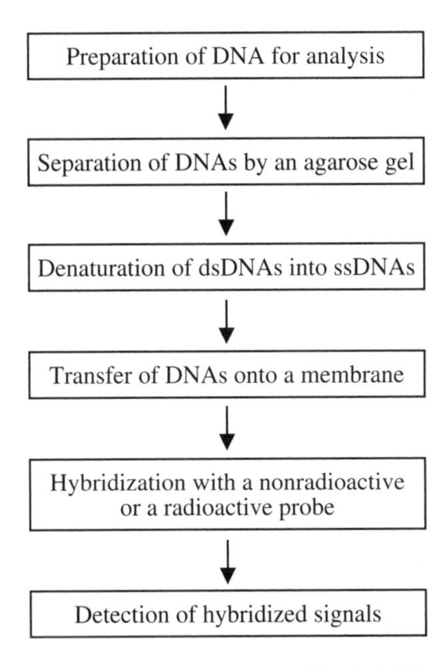

FIGURE 7.1 Scheme for DNA analysis by Southern blot hybridization.

Agarose % (w/v)	DNA size (kb)
0.6	1.0 to 20
0.07	0.8 to 10
1.0	0.5 to 8.0
1.2	0.4 to 6.0
1.4	0.2 to 4.0

Each method has its merits and drawbacks. Although electrophoretic and vacuum blotting can offer faster transfer of DNA than capillary blot does, special instruments or systems are required and are not suitable for simultaneous blotting of multiple filters. The present chapter describes the capillary transfer method that is the most convenient and least expensive method commonly used in most laboratories. Nylon membranes are selected as transfer solid supports because they are more flexible and tear resistant compared with traditional nitrocellulose membranes. Tear-resistant nylon membranes are certainly more desirable for inexperienced students to handle.

ISOLATION OF DNA FOR ANALYSIS

Detailed protocols for isolation of high quality of DNA are described in Chapter 15.

Restriction Enzyme Digestion of DNAs

Prior to agarose gel electrophoresis, DNA molecules may need to be digested into fragments using appropriate restriction enzymes. Detailed instructions for standard restriction enzyme digestion are given in Chapter 4.

AGAROSE GEL ELECTROPHORESIS OF DNAs

1. Prepare agarose gel mixture in a clean bottle or a beaker as follows:

Component	Minigel	Medium gel	Large gel
1X TBE buffer	30 ml	75 ml	120 ml
Agarose (1% w/v)	0.3 g	0.75 g	1.2 g

Note: 1X TBE buffer may be diluted from 5X TBE stock solution in sterile dd.H_2O. Because the agarose is only 1%, its volume can be ignored when calculating the total volume.

2. Place the bottle (with the cap loose) or beaker in a microwave oven and slightly heat the agarose mixture for 1 min. Briefly shake the bottle to rinse off any agarose powder that is stuck onto the glass walls. To melt the agarose completely, carry out gentle boiling for 1 to 3 min, depending on the volume of the mixture. Gently mix, open the cap and allow the mixture to cool to 50 to 60°C at room temperature.
3. While the gel mixture is cooling, seal a clean gel tray at the two open ends with tape or a gasket and insert the comb in place. Add 1 μl of ethidium bromide (EtBr, 10 mg/ml in dd.H_2O) per 10 ml of agarose gel solution at 50 to 60°C. Gently mix and slowly pour the mixture into the assembled gel tray and allow the gel to harden for 20 to 30 min at room temperature.

Caution: EtBr is a mutagen and a potential carcinogen and should be handled carefully. Gloves should be worn when working with this material. The gel running buffer containing EtBr should be collected in a special container for toxic waste disposal.

Notes: The function of EtBr is to stain DNA molecules by interlacing between the complementary strands of double-stranded DNA or in the regions of secondary structure in the case of ssDNA. It will fluoresce orange when illuminated with UV light. EtBr can be added to DNA samples using 1 μl of EtBr (1 mg/ml) per 10 μl of DNA solution, instead of using EtBr in agarose gel. However, the advantage of using EtBr in gel is that DNA bands can be stained and monitored with a UV lamp during electrophoresis. During the electrophoresis, the positively charged EtBr moves towards the negative pole but negatively charged DNA molecules migrate towards the positive pole. The drawback, however, is that running buffer and gel apparatus are likely contaminated with EtBr. Thus, an alternative is to stain the gel after electrophoresis is completed. The gel can be stained with EtBr solution for 10 to 30 min followed by washing in distilled water for 3 to 5 min.

4. Slowly and vertically pull the comb out of the gel and remove the sealing tape or gasket from the gel tray. Place the gel tray in an electrophoresis apparatus and add a sufficient volume of 0.5X TBE buffer or 1X TAE buffer to the apparatus until the gel is covered to a depth of 1.5 to 2 mm above the gel.

The comb should be pulled out from the gel slowly and vertically because any cracks that occur inside the wells of the gel will cause sample leaking when the sample is loaded. The top side of the gel where wells are located must be placed at the negative pole in the apparatus because the negatively charged DNA molecules will migrate toward the positive pole. If any wells contain visible bubbles, they should be flushed out of the wells using a small pipette tip to flush the buffer up and down the well several times. Bubbles may adversely influence the loading of the samples and the electrophoresis. The concentration of TBE buffer should never be lower than 0.2X. Otherwise, the gel may melt during electrophoresis. Prerunning the gel at a constant voltage for 10 min is optional.

5. Add 5X loading buffer to the digested DNA sample and DNA standard marker to a final concentration of 1X. Mix and carefully load DNA standard markers (0.2 to 2 μg/well) into the very left-hand or the right-hand well, or into both left- and right-hand wells of the gel. Carefully load the samples, one by one, into appropriate wells in the submerged gel.
6. After all the samples are loaded, estimate the length of the gel between the two electrodes and set the power at 5 to 15 V/cm gel. Allow electrophoresis to proceed for 2 h or until the first blue dye reaches a distance of 1 cm from the end of the gel. If overnight running of the gel is desired, the total voltage should be determined empirically.
7. Stop electrophoresis of the gel and visualize DNA bands in the gel under UV light. Photograph the gel with a Polaroid camera.

Caution: Wear safety glasses and gloves for protection against UV light.

Notes: The purpose of taking a picture is to record different DNA molecules at different positions following electrophoresis, which is useful once a hybridized signal is detected later. It is recommended that pictures be taken with a fluorescent ruler placed beside the gel. Whereas a short exposure time is needed to obtain relatively sharp bands and a clear background, a longer exposure time is needed to visualize very weak bands. For genomic DNA, smears and some visible bands should appear in each of the lanes.

BLOTTING OF DNAs ONTO NYLON MEMBRANES

Method A. Upward Capillary Transfer (6 to 12 h)

1. Following photography, soak the gel in two volumes of 0.25 *N* HCl for 5 to 8 min with gentle agitating to depurinate DNA. Acid treatment can increase the efficiency of blotting of DNA species larger than 8 kb in length.
2. Replace the HCl solution with four volumes of denaturing solution (approximate 200 ml/gel) to denature dsDNA molecules into ssDNA for later hybridization with a probe. Allow the denaturation to proceed for 30 min at room temperature with shaking at 60 rpm.

3. While the gel is being denatured, cut a piece of nylon membrane the same size as the gel, two pieces of 3MM Whatman filter paper the same size as the membrane, and a piece of a relatively larger and longer size of 3MM Whatman filter paper compared with the size of the gel. The membrane filter should be marked at the upper left or right corner with a pencil as a positional marker.

Notes: (1) Gloves should be worn while cutting the membrane. (2) Positively charged nylon membranes are better than neutral nylon membranes to bind to negatively charged DNA. However, neutral nylon membranes usually have a lower background than positively charged membranes following detection.

4. Briefly rinse the denatured gel with 500 ml/gel of distilled water twice and immerse the gel in four volumes of neutralization buffer to partially neutralize the gel conditions but avoid renaturation of the ssDNA molecules. Allow the neutralization to proceed for 30 min at room temperature with gentle shaking.
5. Quickly rinse the gel once in distilled water and soak it and the cut membrane filter in 10X SSC solution for 20 min.
6. Place a clean glass plate or an equivalent Plexiglas blotting plate on top of a tray or container containing 500 to 1000 ml of 10X SSC buffer. Alternatively, a gel casting tray can be turned upside down to serve as a platform and then placed in a clean container filled with 10X SSC up to the edges of the platform.

Do not overfill the container to immerse the plate.

7. Assemble blotting apparatus in the order shown (Figure 7.2).

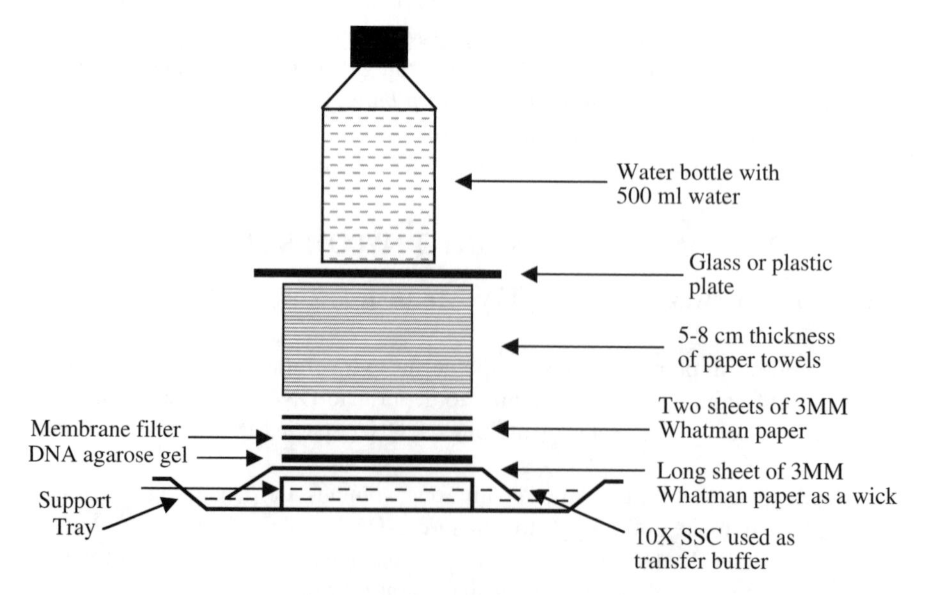

FIGURE 7.2 Standard assembly of upward capillary transfer of DNA from an agarose gel onto a nylon membrane filter.

Note: Air bubbles should be removed from each layer of the blot by carefully rolling a pipette over the surface. It is important not to move the gel, membrane, or blotting paper until transfer is complete. Otherwise, bubbles will be generated and band positions are likely to be changed.

a. Place the presoaked large piece of 3MM Whatman filter paper on the flat plate with its two ends dipping into the 10X SSC buffer reservoir, which serves as a wick. Remove any bubbles underneath it by pressing or lifting one end of the filter and slowly laying it down on the plate.
b. Turn the gel over with the well side facing down and carefully place the gel, starting from one side of the gel, onto the 3MM Whatman filter paper. Gently press the gel to remove any bubbles underneath.
c. Surround the top edges of the gel with Parafilm™ strips to prevent any of the top layers of the gel blot paper from coming into contact with the 10X SSC buffer reservoir, thus eliminating a potential short circuit.
d. Carefully overlay the gel with the soaked membrane, starting at one side of the gel, then slowly proceed to the other end. The marked side of the membrane should face the gel. Wet the filter with some 10X SSC buffer if necessary and carefully remove any bubbles between the membrane and the gel; this can be done by lifting up and laying down the membrane.
   It is recommended not to press the membrane. Otherwise, bubbles that are not visible may be produced underneath the gel.
e. Gently overlay the membrane with two pieces of the same size of 3MM Whatman paper that have been prewetted in 10X SSC buffer.
f. Gently place four pieces of dry, precut 3MM Whatman paper and a stack of regular paper towels (5 to 10 cm thick) the same size as the membrane on top of the 3MM Whatman filter paper.
   Do not distribute the filters lying underneath to prevent from producing bubbles.
g. Put a glass plate, or an equivalent plate, on top of the paper towel stack. Place a bottle or beaker containing approximate 500 ml water on top of the plate to serve as a weight.
   Do not place too heavy a weight on the plate; otherwise, the gel may be crushed.

8. Allow DNA to be transferred onto the membrane filter by capillary action for 6 h or overnight.
9. Remove the paper towel stack and the Whatman filter papers. Mark each of the well positions, if possible, on the membrane filter. The filter is then subjected to cross-linking using one of the methods below.
   a. Use UV-induced cross-linking for 30 to 60 s at the optimal setting in a UV cross-linker (Stratagen, CA) according to the manufacturer's instructions.
   b. Alternatively, place the membrane on a piece of SaranWrap™, which is then placed on a UV light box. Allow exposure to proceed at 254 to 312 nm for 5 min to induce cross-linking between the DNA and the membrane.

c. Another option is to place the membrane between two pieces of 3MM Whatman filter paper to form a sandwich that is then placed in a vacuum oven. Place a glass plate on top of the sandwich to flatten the membrane during the baking process. Close the door of the oven and turn on the vacuum, and allow the membrane to bake for 1 to 2 h at 70 to 80°C. After baking is complete, allow the oven to cool for a while. Open the door and allow the sandwich to cool for 5 min prior to removing the membrane; otherwise, the membrane may curl up.

Note: After transfer, the gel should be rather thin; it may be checked under UV light. An efficient transfer should have very little or no visible DNA staining left in the gel.

10. Immediately proceed to prehybridization or, if the schedule does not permit, air-dry the membrane filter at room temperature. Wrap the membrane with aluminum foil or with SaranWrap and store it at 4°C until prehybridization. It can be stored for up to 6 months.

Gloves should be worn and changed when handling the membrane.

Method B. Downward Capillary Transfer (1 to 1.5 h Using Alkaline Buffer or 3 h Using Neutral Buffer)

1. Carry out steps 1 to 5 in Method A if a neutral transfer buffer is used, or perform step 1 in Method A if an alkaline transfer buffer is to be utilized. For alkaline transfer, the HCl-treated gel is subjected to denaturation in alkaline denaturing solution for 2 × 30 min. The gel is then soaked in alkaline transfer buffer for 10 min. Proceed to the next step.
2. Carry out downward capillary transfer as shown in Figure 7.3.

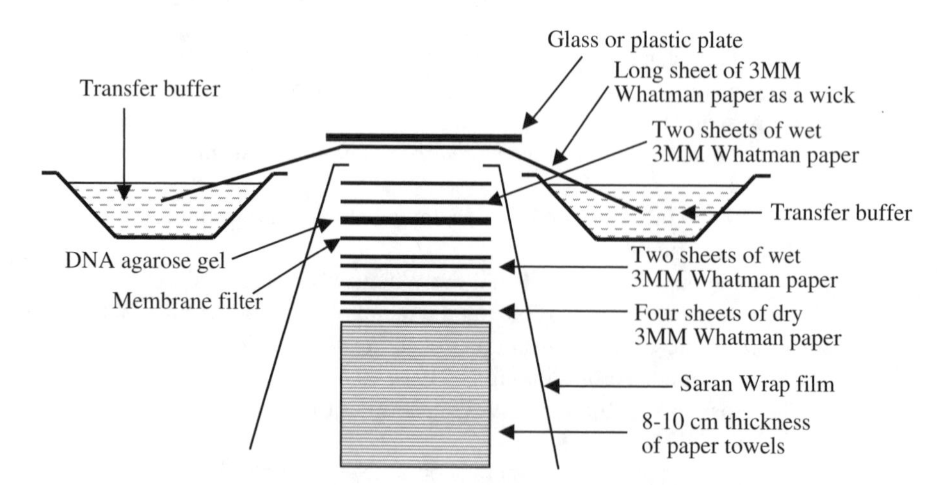

FIGURE 7.3 Standard assembly of downward capillary transfer of DNA from an agarose gel onto a nylon membrane filter.

a. Place a stack of precut paper towels (8 to 10 cm thick) on a level bench.
b. Place four pieces of dry 3MM Whatman filter paper on the paper towel stack.
c. Put two pieces of 3MM Whatman paper, presoaked in transfer buffer, on the dry 3MM Whatman filter papers.
d. Carefully overlay the wet 3MM Whatman filters with a presoaked nylon membrane and remove any bubbles underneath.
e. Place the soaked gel on the membrane. Make sure that no bubbles are evident between the gel and the membrane.
f. Carefully overlay the gel with two pieces of prewet 3MM Whatman filter paper.
g. Gently cover the entire stack with two relatively large pieces of SaranWrap film that has a window precut the same size as the 3MM Whatman filter papers on top of the gel.
h. Fill two containers with approximately 200 ml of transfer buffer and place them near the two sides of the stack unit.
i. Carefully place a relatively long and large piece of 3MM Whatman filter paper on top of the 3MM Whatman paper on the gel, which serves as a wick. Lay the two sides of the wick over the SaranWrap and place the ends into the transfer buffer reservoir in the containers on both sides.
j. Place a light glass plate, or its equivalent, on top of the stack and allow transfer to proceed for 1 to 1.5 h in alkaline transfer buffer or 3 h in neutral transfer buffer.

   Air bubbles should be removed from each layer of the blot by carefully rolling a pipette over the surface. It is important not to move the gel, membrane and blotting paper until transfer is complete. Otherwise, bubbles will be generated and band positions are likely to be changed.

3. After transfer is complete, remove the Whatman filter papers. Following alkaline transfer, neutralize the membrane in 50 ml per membrane in 0.2 *M* sodium phosphate buffer (pH 6.8) for 4 min. Following neutral transfer, soak the membrane in 5X SSC for 4 min to remove bits of gel from the membrane.

   Note: After transfer, the gel should be rather thin; it may be checked under UV light. An efficient transfer should have very little or no visible DNA staining left in the gel.

4. Mark each of the well positions, if possible, on the membrane filter. The filter is then subjected to cross-linking using one of the methods below.
   a. Perform UV-induced cross-linking for 30 to 60 s at an optimal setting in a UV cross-linker (Stratagene, CA) according to the manufacturer's instructions.
   b. Alternatively, place the membrane on a piece of SaranWrap that is then placed on a UV light box. Allow exposure to proceed at 254 to 312 nm for 5 min to induce cross-linking between the DNA and the membrane.
   c. Another option is to place the membrane between two pieces of 3MM Whatman filter paper to form a sandwich, which is then placed in a vacuum oven. Place a glass plate on top of the sandwich to flatten the

membrane during baking. Close the door of the oven, turn on the vacuum, and allow the membrane to bake for 1 to 2 h at 70 to 80°C. After baking is complete, allow the oven to cool for a while. Open the door and allow the sandwich to cool for 5 min prior to removing the membrane; otherwise, the membrane may roll up.

5. Immediately proceed to prehybridization or, if the schedule does not permit, air-dry the membrane filter at room temperature. Wrap the membrane with aluminum foil or SaranWrap and store at 4°C until prehybridization or store for up to 6 months.

Gloves should be worn and changed when handling the membrane.

PREPARATION OF PROBES

Preparation of Nonisotopic DNA Probes

There are three well-tested and commonly used labeling methods for preparation of nonisotopic probes. Each methodology works equally well but the time and costs needed are different. Protocol 1 is a modified direct labeling using the ECL kit from Amersham Life Science Corporation. One of the major advantages of this method is that it only takes 20 to 30 min to complete DNA labeling without using expensive dNTPs. The probes do not need to be denatured; instead, they can be directly used for hybridization. Following hybridization and washing, the membrane filter is ready for detection of hybridized signals, thus, eliminating the antibody incubation required in Protocols 2 and 3. The reagents needed are commercially available (e.g., Boehringer Mannheim Corporation, Indianapolis, IN; or Amersham Life Science Corporation, Arlington Heights, IL).

Protocol 1. Direct Labeling of ssDNA Using the ECL Kit

The primary principles of direct labeling begin with denaturation of dsDNA into ssDNA by heat. Negatively charged ssDNA species then interact with positively charged polymers conjugated to horseradish peroxidase. This enzyme can hydrolyze substrates such as luminol and generate photons or light that will be recorded on an x-ray film as a result (Figure 7.4).

1. Transfer 100 to 200 ng of dsDNA of interest into a microcentrifuge tube and bring the total volume to 8 μl with dd.H_2O. Tightly cap the tube and denature the DNA by boiling for 10 min and then quickly chill the tube on ice for 3 min. Briefly spin down and place it on ice until use.
2. Add 5 μl of DNA labeling mixture and 8 μl of fixer from the ECL labeling kit to the tube containing the denatured DNA.
3. Incubate the tube at 37°C for 20 min or at room temperature for 30 h.
4. Purify labeled DNA from free positively charged polymer molecules using an appropriate Sephadex column (e.g., G-10 or G-50) (Figure 7.5).

Tip: Purification is very important to reduce hybridization background. Otherwise, the free positively charged polymers will interact with nontargeting

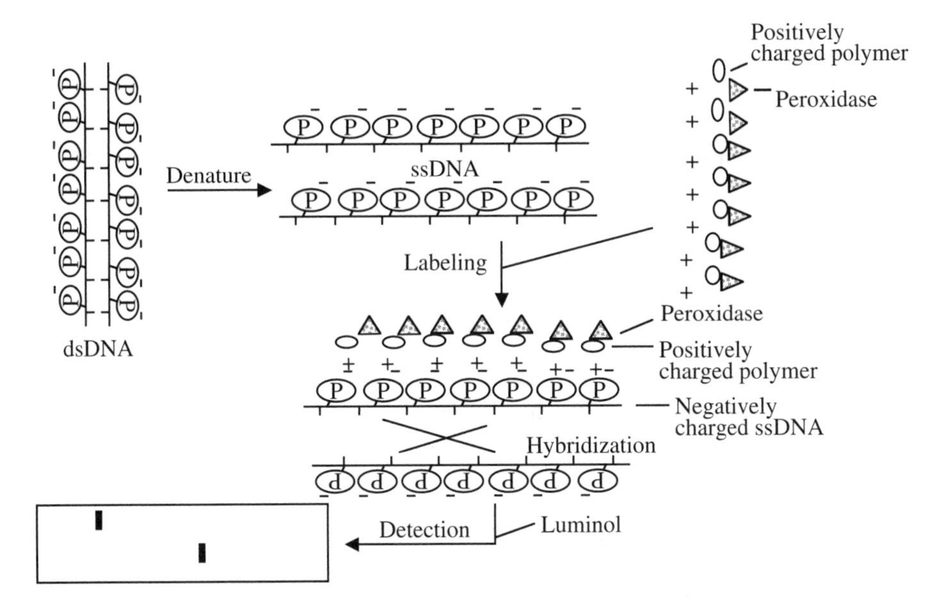

FIGURE 7.4 Nonradioactive labeling and detection of DNA in Southern blotting.

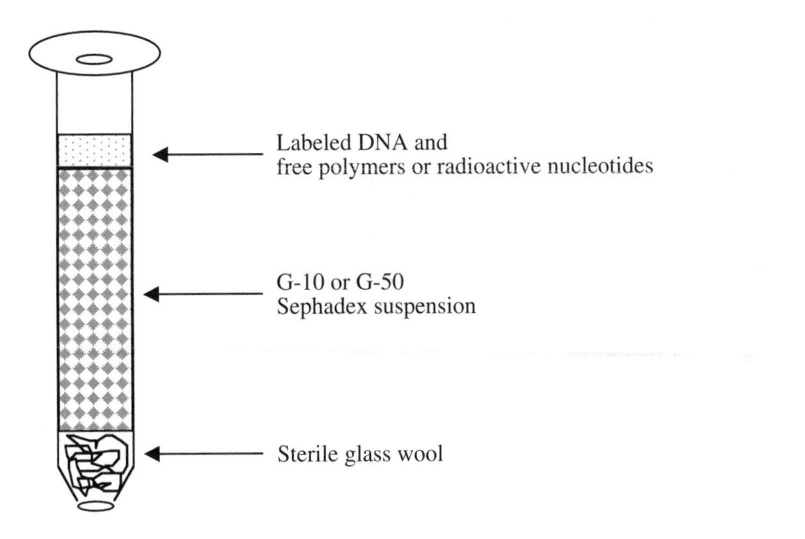

FIGURE 7.5 Purification of labeled DNA using a G-10 or G-50 Sephadex chromatography column.

DNA molecules on the membrane. As a result, all DNA or RNA bands will be detected. Sephadex G-10 or G-50 spin columns or Bio-Gel P-60 spin columns are very effective chromatography for separating labeled DNA from free polymers or unincorporated radioactive precursors such as [α-^{32}P]dCTP or [α-^{32}P]dATP and oligomers, which will be retained in the *column. This is very useful when an optimal signal-to-noise ratio with 150 to 1500 bases in length of probe is generated for an optimal hybridization.*

a. Resuspend 2 to 5 g Sephadex G-10, G-50 or Bio-Gel P-60 in 50 to 100 ml of TEN buffer and equilibrate for at least 1 h at 4°C.
b. Insert a small amount of sterile glass wool into the bottom of a 1 ml disposable syringe or an equivalent using the barrel of the syringe to tamp the glass wool in place.
c. Fill the syringe with the Sephadex suspension until it is completely full.
d. Insert the column containing the suspension into a 15-ml disposable plastic tube and place the tube in a swinging-bucket rotor in a benchtop centrifuge. Centrifuge at 1600 × *g* for 4 min at room temperature.
e. Repeat, adding the suspended resin to the column and centrifuging at 1600 × *g* for 4 min until the packed volume reaches 0.9 ml in the syringe and remains unchanged after centrifugation.
f. Add 0.1 ml of 1X TEN buffer (10 m*M* Tris-HCl, pH 8.0, 1 m*M* EDTA, pH 8.0, 100 m*M* NaCl) to the top of the column and recentrifuge as conducted earlier. Repeat this step twice.
g. Transfer the spin column to a fresh 15-ml disposable tube. Add the labeled DNA sample onto the top of the resin dropwise, using a pipette. Note: If the labeled DNA sample is less than 0.1 ml, dilute it to 0.1 ml in 1X TEN buffer.
h. Centrifuge at 1600 × *g* for 4 min at room temperature. Collect the void volume fluid containing labeled DNA. Store at –20°C until use. Proceed to hybridization.

Protocol 2. Random Primer Digoxigenin Labeling of dsDNA

The basic principle of biotin or digoxigenin labeling is that, in a random labeling reaction, random hexanucleotide primers anneal to denatured DNA template. A new strand of DNA complementary to template DNA is catalyzed by the Klenow enzyme. During the incorporation of four nucleotides, one of them is prelabeled by digoxigenin such as biotin-dUTP or DIG-dUTP. As a result, dsDNA probes are generated. After hybridization of the probe with the target DNA, an antidigoxigenin antibody conjugated with an alkaline phosphatase will bind to the bound biotin-dUTP or digoxigenin-dUTP. The hybridized signal can then be detected using a chemiluminescent or colorimetric substrate. If chemiluinescent substrates are used, the detected signals can be exposed onto an x-ray film, whereas colorimetric substrates, NBT and BCIP, or X-Phosphate, can produce a purple/blue color (Figure 7.6).

1. Add an appropriate amount of dsDNA template of interest to a microcentrifuge tube, denature the DNA by boiling for 10 min. and then quickly chill the tube on ice for 4 min. Briefly spin down and place it on ice until use.
2. Set up a standard labeling reaction as follows:
   Denatured DNA template (50 ng to 2 μg), 5 μl
   10X Hexanucleotide primers mixture, 2.5 μl

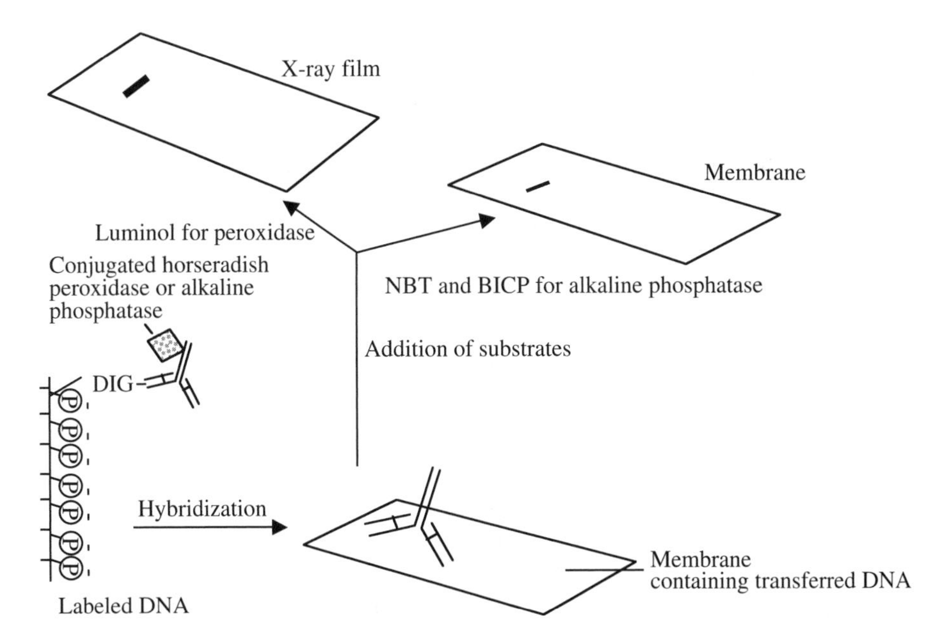

FIGURE 7.6 Random labeling and detection of DNA using biotin-dUTP or DIG-dUTP.

10X dNTPs mixture containing biotin-dUTP or DIG-dUTP, 2.5 µl
dd$\cdot H_2O$, 13.5 µl
Klenow enzyme (2 units/µl), 1.5 µl
Total volume = 25 µl

3. Incubate at 37°C for 1 to 2 h.

Note: The amount of labeled DNA depends on the amount of DNA template and on the length of incubation at 37°C. In our experience, the longer the incubation within a 12-h time period, the more DNAs are labeled.

4. Purify the labeled DNA using G-10 or G-50 Sephadex column chromatography as described in Protocol 1.
5. After the labeling is complete, the mixture can be boiled for 10 min and chilled on ice for 3 min. The probes can be immediately used for hybridization or stored at –20 or –80°C until use. However, if precipitation is desired, add 2.5 µl of 0.5 M EDTA solution to stop the labeling reaction and carry out the following steps.
   a. Add 0.15 volumes of 3 *M* sodium acetate buffer (pH 5.2) and 2.5 volumes of chilled 100% ethanol to the mixture. Allow DNA to precipitate at –80°C for 0.5 hr.
   b. Centrifuge at 12,000 × *g* for 6 min at room temperature, and carefully aspirate the supernatant. Briefly rinse the DNA pellet with 1 ml of 70% ethanol and dry the pellet under vacuum for 10 min.
   c. Dissolve the labeled DNA in 50 µl dd.H_20 or TE buffer. Store at –20°C prior to use.

Notes: (1) Labeled dsDNA must be denatured prior to use for hybridization. This can be done by boiling the DNA for 10 min then quickly chilling it on ice for 3 min. (2) Following hybridization, the probe contained in the hybridization buffer can be stored at –20°C and reused up to four times. (3) If necessary, the yield of the labeled DNA can be estimated by dot blotting serial dilutions of commercially labeled control DNA and labeled sample DNA on a nylon membrane filter. After hybridization and detection, compare the spot intensities of the control and of the labeled DNA. However, this is optional.

Protocol 3. Nick Translation Labeling of dsDNA with Biotin-11-dUTP or Digoxigenin-11-dUTP

Nick translation labeling is also a widely used method for labeling DNA probes. In this method, dsDNA templates do not need to be denatured prior to labeling. The primary principle of nick translation labeling is that single-stranded nicks in dsDNA are first created by DNase I. *E. coli* DNA polymerase I with 5′ to 3′ exonuclease activity removes stretches of ssDNA in the direction of 5′ to 3′. Almost simultaneously, the DNA polymerase I catalyzes the synthesis of a new strand of DNA from 5′ to 3′, in which a prelabeled deoxyribonucleotide such as DIG-11-dUTP or biotin-11-dUTP is incorporated into the new strand DNA, producing dsDNA probes.

1. Set up, on ice, a labeling reaction as follows:
   dsDNA template (1 to 3 μg), 2 to 5 μl
   10X Biotin or DIG DNA labeling mixture, 2.5 μl
   10X Reaction buffer, 2.5 μl
   DNase I/DNA polymerase I mixture, 2.5 μl
   Add dd.H_2O to a final volume of 25 μl.
2. Incubate at 15°C for 70 min. Process the probes as described in Protocol 2.

Preparation of Isotopic DNA Probes

Protocol 1. Nick Translation Labeling of dsDNA

These procedures are very similar to those described for nonradioactive labeling except that one of four deoxynucleotides that is radioactively labeled (e.g., α-^{32}PdATP or α-^{32}PdCTP, commercially available) is used and incorporated into the new DNA strand by a base complementary to those of the template. A high specific activity (10^8cpm/μg) of labeled DNA can be obtained using ^{32}PdATP or ^{32}PdCTP.

1. Set up a reaction in a tube on ice:
   10X Nick translation buffer, 5 μl
   DNA sample, 0.4 to 1 μg in <2 μl
   Mixture of three unlabeled dNTPs, 10 μl
   [α-^{32}P]dATP or [α-^{32}P]dCTP(>3000 Ci/mmol), 4 to 7 μl
   Diluted DNase I (10 ng/ml), 5 μl
   E. coli DNA polymerase I, 2.5 to 5 units

Add dd.$H_2$0 to a final volume of 50 μl.

2. Incubate the reaction for 60 min at 15°C.

Notes: (1) The temperature should not be higher than 18°C, which can generate "snapback" DNA by E. coli *DNA polymerase I. Snapback DNA can lower the efficiency of hybridization. (2) Relatively longer incubation time (1 to 2 h) is acceptable. However, incubation for too long may reduce the overall length of the labeled DNA.*

3. Stop the reaction by adding 2.5 μl of 0.2 *M* EDTA (pH 8.0) solution. Place the tube on ice.
4. Determine the percentage of [α-^{32}P]dCTP incorporated into the DNA using one of the following protocols. The labeled DNA should be purified using Sephadex column chromatography as described previously.

Protocol 2. DE-81 Filter-Binding Assay

1. Dilute 1 μl of the labeled DNA in 99 μl (1:100) of 0.2 *M* EDTA solution. Spot 3 μl of the diluted sample, in duplicate, onto Whatman DE-81 circular filters (2.3 cm in diameter). Air-dry the filters.
2. Wash one filter in 50 ml of 0.5 *M* sodium phosphate buffer (pH 6.8) for 5 min to remove unincorporated cpm. Repeat washing once. The other filter will be utilized directly for determination of total cpm in the sample.
3. Add an appropriate volume of scintillation fluid (about 10 ml) to each tube containing one of the filters. Determine the cpm in a scintillation counter according to the manufacturer's instructions.

Protocol 3. TCA Precipitation

1. Dilute 1 μl of the labeled mixture in 99 μl (1:100) of 0.2 *M* EDTA solution. Spot 3 μl of the diluted sample onto a glass fiber filter or a nitrocellulose filter for determining the total cpm in the sample. Air-dry the filter.
2. Add 3 μl of the same diluted sample to a tube containing 100 μl of 0.1 mg/ml carrier DNA or acetylated BSA and 20 m*M* EDTA.
3. Add 1.3 ml of ice-cold 10% trichloroacetic acid (TCA) and 1% sodium pyrophosphate to the mixture at step 2. Mix well and incubate the tube on ice for 20 to 25 min to precipitate the DNA.
4. Filter the precipitated DNA on a glass fiber filter or a nitrocellulose filter using a vacuum. Wash the filter with 5 ml of ice-cold 10% TCA four times using a vacuum. Rinse the filter with 5 ml acetone (for glass fiber filters only) or 5 ml of 95% ethanol. Air-dry the filter
5. Add 10 to 15 ml of scintillation fluid to each tube. Count the total cpm and incorporated cpm in a scintillation counter according to the manufacturer's instructions.
6. Calculate the specific activity of the probe as follows:
   a. Calculate the theoretical yield:

$$\text{ng theoretical yield} = \frac{\text{mCidNTP added } 4 \quad 330 \text{ ng / nmol}}{\text{specific activity of the labeled dNTP (mCi / nmol)}}$$

b. Calculate the percentage of incorporation:

$$\%\ \text{incorporation} = \frac{\text{cpm incorporated}}{\text{total cpm}} \times 100$$

c. Calculate the amount of DNA synthesized:

$$\text{ng DNA synthesized} = \%\ \text{incorporation} \times 0.01 \times \text{theoretical yield}$$

d. Calculate the specific activity of the prepared probe:

$$\text{cpm / } \mu\text{g} = \frac{\text{total cpm incorporated (cpm incorporated} \times 33.3 \times 50)}{\text{(ng DNA synthesized + ng input DNA)} \quad 0.001 \text{ mg / ng}} \times 100$$

Note: The factor 33.3 comes from using 3 µl of 1:100 dilution for the filter-binding or TCA precipitation assay. The factor 50 is derived from using 1 µl of the total 50 µl reaction for 1:100 dilution. For example: Given that 25 ng DNA is to be labeled and that 50 µCi[α-^{32}P]dCTP (3,000 Ci/mmol) is used in 50 µl of a standard reaction, and assuming that 4.92 × 10^4 cpm is precipitated by TCA and that 5.28 × 10^4 cpm is the total cpm in the sample, the calculations are performed as follows:

$$\%\ \text{incorporation} = \frac{4.92 \times 10^4}{5.28 \times 10^4} \times 100 = 93\%$$

$$\text{DNA synthesized} = 0.93 \times 0.01 \times 22 = 20.5 \text{ ng}$$

$$\text{Specific activity} = \frac{4.92 \times 104 \times 33.3 \times 50}{20.5 \text{ ng} + 25 \text{ ng}) \times 0.001\ \mu\text{g / ng}} \times 100 = 1.8 \quad 10^9 \text{ cpm / } \mu\text{g DNA}$$

Note: The specific activity of 2 × 10^{8-9} cpm/µg is considered to be a high specific activity that should be used in Southern, northern, or equivalent hybridizations.

Protocol 4. Random Primer Labeling of dsDNA

These general procedures are similar to those described for nonradioactive labeling of DNA except that a radioactively labeled nucleotide is used.

1. Set up a reaction mixture by adding the following components, in the order shown below, into a fresh microcentrifuge tube placed on ice.
   5X Labeling buffer, 10 μl
   Denatured DNA template (25 to 50 ng), 1 μl
   Three unlabeled dNTP mixture (500 μ*M* each), 2 μl
   [α-^{32}P]dCTP or [α-^{32}P]dATP(>3000 Ci/mmol), 2.5 to 4 μl
   Klenow enzyme, 5 units
   Add dd.H_20 to a final volume of 50 μl.
   Note: [α-^{32}P]dCTP or [α-^{32}P]dATP (3000 Ci/mmol) should not be >5 μl used in the reaction; otherwise, the background will be high.
2. Mix well and incubate the reaction at room temperature for 1 to 1.5 h.
3. Stop the reaction mixture by adding 2 μl of 0.2 *M* EDTA and place on ice. Purify the labeled DNA as described previously. Store at –20°C until use.

Protocol 5. 3′-end Labeling of ssDNA (Oligonucleotides) with a Terminal Transferase

1. 3′ tailing of ssDNA primers
   a. Set up a reaction as follows:
      5X Terminal transferase buffer, 5 μl
      ssDNA primers, 2 pmol
      [α-^{32}P]dATP (800 Ci/mmol), 1.6 μl
      Terminal transferase (10 to 20 units/μl), 1 μl
      Add dd.H_2O to a final volume of 25 μl.
   b. Incubate at 37°C for 1 h and stop the reaction by heating at 70°C for 10 min to inactivate the enzyme.
      Note: Multiple [α-^{32}P]dATPs were added to the 3′ end of the DNA.
   c. Calculate the percentage of incorporation and the specific activity of the probe.
      Note: The procedures for the TCA precipitation assay and the DE81 filter binding assay are described in the section on nick translation labeling.

$$\%\ \text{incorporation} = \frac{\text{cpm incorporated}}{\text{total cpm}} \times 100$$

$$\text{Specific activity} = \frac{\%\ \text{incorporation}\ \ \text{total cpm added to the reaction}}{\text{mg of DNA in the reaction mixture}} \times 100$$

Example: $49{,}000/51{,}000 \times 100 = 96\%$

$$\frac{0.96\ \ 3.5\ \ 106}{0.09\ \text{mg}}\ \ 100 = 3.7 \times 107\ \text{cpm} / \mu\text{g DNA}$$

2. 3′-end labeling of ssDNA with a single [α-32P]cordycepin-5′-triphosphate lacking the 3′-OH.
   a. Set up a reaction as follows:
      5X Terminal transferase buffer, 10 μl
      ssDNA primers, 10 pmol
      [α-^{32}P]cordycepin-5′-phosphate (3000 Ci/mmol), 7.5 μl
      Terminal transferase (10 to 20 u/μl), 2 μl
      Add dd.H_2O to a final volume of 50 μl.
   b. Incubate the reaction at 37°C for 1 h.
   c. Stop the reaction by incubating at 70°C for 10 min. The labeled reaction can be directly used in hybridization or purified as described for 3′-end labeling in order to fill the recessed 3' ends of dsDNA.

PREHYBRIDIZATION AND HYBRIDIZATION

1. Prepare hybridization buffer based on nonradioactively or radioactively labeled DNA. One filter (10 × 12 cm^2) needs 10 to 15 ml buffer, or 30 to 40 ml of buffer is sufficient for four to six filters in a single hybridization container.
2. If the blotted membrane is predried, immerse it in 5X SSC for 5 min at room temperature to equilibrate the filter.

Note: Do not allow the filters to dry out during subsequent steps. Otherwise, a high background or anomalous results will occur.

3. Carefully roll the filter and insert it into a hybridization bottle or an equivalent container containing 10 to 15 ml/filter of appropriate prehybridization solution, depending on the isotopic or nonisotopic DNA probe used. Cap the bottle or container and place it in a hybridization oven or equivalent shaker with controlled speed and temperature. Allow prehybridization to proceed for 1.5 h with gentle agitation.

Notes: (a) It is strongly recommended not to use traditional plastic hybridization bags because it is usually not easy to get rid of air bubbles and they cannot be well sealed. This will result in leaking and contamination. An appropriate size of plastic beaker or tray or hybridization bottle is the best type of hybridization container to use for this purpose. (b) The temperature depends on the prehybridization buffer; it should be 42 to 55°C if the buffer contains 30 to 40% (for low-stringency conditions) or 50% (for high-stringency conditions) formamide. If the buffer contains no formamide, the temperature can be 65°C. High-stringency conditions can prevent nonspecific cross-binding during hybridization. Assuming that the buffer used for this experiment contains 50% formamide, in our experience, the temperature for this experiment can be set up at 42°C regardless of which DNA labeling method in step 3 is employed. (c) It is highly recommended that a regular culture shaker with a cover, controlled speed, and temperature be used as a hybridization chamber. Such a chamber is easy to handle by simply placing the hybridization beaker or tray containing the filters and buffer on the shaker in the chamber. Multiple filters can be placed in one container containing an appropriate volume of hybridization buffer. This is particularly useful for biotech lab classes. In contrast, a commercial hybridization oven may be difficult to operate. Rolling up the filters with a matrix screen and placing the roll inside the hybridization bottle takes time and air bubbles are easily generated during that period.

4. If the probes are dsDNA, denature the probe to ssDNA in boiling water for 10 min and immediately chill on ice for 4 min. Briefly spin down.

Note: This is a critical step. If the probes are not completely denatured, a weak, or no hybridization signal will occur. However, the probes prepared by Method A do not need denaturation.

5. Dilute the probes with 0.8 ml of hybridization solution. Replace the prehybridization buffer with fresh prehybridization buffer and add the diluted probes to the freshly replaced prehybridization buffer. Alternatively, diluted probes may be directly added to the old prehybridization buffer without replacing it. Place the container back in the hybridization oven or shaker and allow hybridization to proceed for 6 h to overnight under the same conditions as prehybridization.

WASHING OR INCUBATION OF ANTIBODIES

Protocol A. Washing of Filters Hybridized with ECL-Labeled Probes

1. Wash the filters in 200 ml of 2X SSC solution containing 0.1% SDS (w/v) for 10 min at 42°C with gentle agitation.
2. Wash the filters in 200 ml of 1X SSC solution containing 0.1% SDS (w/v) for 10 min at room temperature with gentle agitation.

Protocol B. Washing and Antibody Incubation of Filters Hybridized with Biotin-dUTP- or DIG-UTP-Labeled Probes

1. Wash the filters in 200 ml of 2X SSC solution containing 0.1% SDS (w/v) for 15 min at 55 to 60°C with slow shaking. Repeat once.
2. Wash the filters in 200 ml of 1X SSC solution containing 0.1% SDS (w/v) for 15 min at room temperature with slow shaking.
3. Block nonspecific binding sites on the membrane in 200 ml of blocking buffer for 1 h at room temperature.
4. Incubate the filters in a solution (10 to 15 ml/filter) containing anti-DIG-alkaline phosphatase antibodies (Boehringer Mannheim Corporation) diluted at 1:5000 to 10,000 in blocking buffer with 1% nonfat dry milk. Allow incubation to proceed for 60 min at room temperature with gentle agitation.
5. Transfer the filter to a clean tray containing 200 ml of washing buffer for 2×15 min at room temperature.

Note: The used antibody solution may be stored at 4°C for up to 1 month and reused four to five times.

6. Equilibrate the filter in 100 ml of predetection buffer for 2 to 3 min. Proceed to detection.

Protocol C. Washing of Filters Hybridized with Isotopic Probes

1. Wash the filters in 200 ml of 2X SSC solution containing 0.1% SDS (w/v) for 15 min at 65°C with slow shaking. Repeat once.
2. Wash the filters in 200 ml of 0.5X SSC solution containing 0.1% SDS (w/v) for 15 min at room temperature with slow shaking. Proceed to detection.

DETECTION OF HYBRIDIZED SIGNALS

Method A. Chemiluminescent Detection

1. Place a piece of SaranWrap film on a bench and add 0.4 ml/filter (10 × 12 cm^2) of detection reagents 1 and 2 from the ECL kit (Amersham Life Science Corporation) or 0.8 ml of the Lumi-Phos 530 (Boehringer Mannheim Corporation) to the center of the SaranWrap film.
2. Wear gloves and briefly damp the filter to remove excess washing solution and thoroughly wet the DNA binding side by lifting and overlaying the filter with the solution several times.
3. Wrap the filter with SaranWrap film, leaving two ends of the film unfolded. Place the wrapped filter on a paper towel and, using another piece of paper towel, carefully press the wrapped filter to remove excess detection solution from the unfolded ends of the SaranWrap film. Excess detection solution will most likely cause a high background.
4. Completely wrap the filter and place it in an exposure cassette with DNA binding side facing up. Tape the four corners of the filter.
5. In a darkroom with the safe light on, overlay the filter with a piece of x-ray film and close the cassette. Allow exposure to proceed at room temperature for 10 s to 15 h, depending on the intensity of the detected signal.
6. In a darkroom, develop and fix the film in an appropriate developer and fixer, respectively. If an x-ray processor is available, developing, fixing, washing and drying the film can be completed in 2 min. If a hybridized signal is detected, it appears as black band on the film.

Note: It may be necessary to expose and process multiple films until the appropriate intensity of the signals is obtained. Exposure for more than 4 h may generate a high black background. In our experience, good hybridization and detection should display sharp bands within 1.5 h. In addition, the film should be slightly overexposed in order to obtain a relatively black background that will help identify the sizes of the bands compared with the marks made previously.

METHOD B. COLORIMETRIC DETECTION OF FILTERS HYBRIDIZED WITH ANTIBODY-CONJUGATED PROBES

1. Following hybridization and washing, place the filter in color developing solution (10 ml/filter, 10×12 cm^2) containing 40 µl of NBT stock solution and 30 µl of BCIP stock solution. These stock solutions are commercially available.
2. Allow color to develop in the dark at room temperature for 15 to 120 min or until the desired level of detection is obtained. Positive signals should appear as a blue/purple color.

METHOD C. DETECTION OF SIGNALS BY AUTORADIOGRAPHY

If the isotopic probes are used for hybridization, autoradiography should be employed for the detection of signals.

1. Air-dry the washed filters and individually wrap the filters with a piece of SaranWrap film and place the wrapped filter in an exposure cassette with the DNA binding side facing up. Tape the four corners of the filter.
2. In a darkroom with the safe light on, overlay the filter with an x-ray film and close the cassette. Allow exposure to proceed at –80°C for 2 h to 5 days, depending on the intensity of the detected signals.
3. In a darkroom, develop and fix the film in an appropriate developer and fixer, respectively. If an x-ray processor is available, developing, fixing, washing and drying the film can be completed in 2 min. If a hybridized signal is detected, it appears as a black band on the film.

Materials Needed

5X TBE Buffer

600 ml dd.H_2O
0.45 *M* Boric acid (27.5 g)
0.45 *M* Tris base (54 g)
10 m*M* EDTA (20 ml 0.5 *M* EDTA,_pH 8.0)
Dissolve well after each addition. Add dd.H_2O to 1 l. Autoclave.

Ethidium Bromide (EtBr)

10 mg/ml in dd.H_2O, dissolve well and keep in a dark or brown bottle at 4°C.
Caution: EtBr is extremely toxic and should be handled carefully.

5X Loading Buffer

50% Glycerol
2 m*M* EDTA
0.25% Bromphenol blue
0.25% Xylene cyanol
Dissolve well and store at 4°C.

10X Nick Translation Buffer

0.5 *M* Tris-HCl, pH 7.5
0.1 *M* $MgS0_4$
1 m *M* DTT
500 μg/ml BSA (Fraction V, Sigma Chemical Corporation) (optional)
Aliquot the stock solution and store at –20°C until use.

Unlabeled dNTP Stock Solutions

1.5m*M* each dNTP

Radioactive Labeled dNTP (Commercially Available)

[α-^{32}P]dATP or [α-^{32}P]dCTP (3000 Ci/mmol)

Pancreatic DNase I Solution

DNase I (1 mg/ml) in a solution containing 0.15 *M* NaCl and 50% glycerol.
Aliquot and store at –20°C.

E. coli DNA Polymerase I Solution

Commercial suppliers

Stop Solution

0.2 *M* EDTA, pH 8.0

1X TEN Buffer

10 m*M* Tris-HCl, pH 8.0
1 m*M* EDTA, pH 8.0
0.1 *M* NaCl

Sephadex G-10, G-50 or Bio-Gel P-60 Powder

Commercial suppliers

5X Labeling Buffer

0.25 *M* Tris-HCl, pH 8.0 (from stock solution)
25 m*M* $MgCl_2$
10 m*M* DTT (dithiothreitol)
1 m*M* HEPES buffer, pH 6.6 (from stock solution)
26 A_{260} units/ml random hexadeoxyribonucleotides

Three Unlabeled dNTPs Solution

1.5 m*M* of each dNTP

Klenow enzyme

5 unit/μl, labeling grade

[α-^{32}P]dCTP or [α-^{32}P]dATP

Specific activity 3000 Ci/mmol

TE Buffer

10 m*M* Tris-HCl, pH 8.0
1 m*M* EDTA

5X Terminal Transferase Buffer

0.5 *M* Cacodylate, pH 6.8
1 m*M* $CoCl_2$
0.5 m*M* DTT
500 μg/ml BSA

ssDNA Primers

Prepared from bacteriophage M13 or phagemid, sscDNA or ssDNA isolated from DNA

Terminal Transferase

From available commercial source (Promega Inc.)

[α-^{32}P]cordycepin-5′-Phosphate Analog

3000 Ci/mmol

Depurination Solution

0.25 *M* HCl

Denaturing Solution

0.5 *M* NaOH
1.5 *M* NaCl

Neutralization Solution

1.5 *M* NaCl
1*M* Tris-HCl, pH 7.5

Alkaline Denaturing Solution

3 *M* NaCl
0.5 *N* NaOH

Alkaline Transfer Solution

3*M* NaCl
8 m*M* NaOH

20X SSC Solution (1 l)

175.3 g NaCl
88.4 g Sodium citrate
Adjust the pH to 7.5 with HCl.

Prehybridization Buffer for Probes Prepared with the ECL Kit

Add 0.3 g blocking reagent per 10 ml of hybridization buffer. Allow the blocking powder to dissolve in the buffer for 60 min at room temperature or 30 min at 42°C with vigorous agitation.

Hybridization Buffer for Probes Prepared Using the ECL Kit

Add probe to an appropriate volume of freshly prepared prehybridization buffer.

Prehybridization Buffer for Probes Prepared with Biotin-dUTP or DIG-dUTP

5X SSC
0.1% (w/v) *N*-lauroylsarcosine
0.02% (w/v) Sodium dodecyl sulfate (SDS)
1% (w/v) Blocking reagent
50% (v/v) Formamide
Dissolve well on a heating plate with stirring at 65°C after each addition. Add sterile water to final volume.

Hybridization Buffer for Probes Prepared with Biotin-dUTP or DIG-dUTP

Add denatured, biotin-dUTP or DIG-dUTP-labeled probe to an appropriate volume of fresh prehybridization buffer.

50X Denhardt's Solution

1% (w/v) BSA (bovine serum albumin)
1% (w/v) Ficoll (type 400, Pharmacia)
1% (w/v) PVP (polyvinylpyrrolidone)
Dissolve well after each addition; adjust to the final volume into 500 ml aliquot with distilled water and sterile filter. Divide the solution to 50 ml each and store at –20°C. Dilute 10-fold into prehybridization and hybridization buffers.

Prehybridization Buffer for Isotopic Probes

5X SSC
0.5% SDS
5X Denhardt's reagent
0.2% Denatured and sheared salmon sperm DNA

Hybridization Buffer for Isotopic Probes

5X SSC
0.5% SDS
5X Denhardt's reagents
0.2% Denatured salmon sperm DNA
[α-^{32}P]-labeled DNA probe

Washing Buffer

150 m*M* NaCl
100 m*M* Tris-HCl, pH 7.5

Blocking Buffer

5% (w/v) Nonfat dry milk
150 m*M* NaCl
100 m*M* Tris-HCl, pH 7.5
Dissolve well with stirring at room temperature.

Predetection/Color Developing Buffer

0.1 *M* Tris-HCl, pH 9.5
0.1 *M* NaCl
50 m*M* $MgCl_2$

NBT Stock Solution

75 mg/ml Nitroblue tetrazolium (NBT) salt in 70% (v/v) dimethylformamide

BCIP Stock Solution

50 mg/ml 5-Bromo-4-chloro-3-indolyl phosphate (BCIP or X-phosphate) in 100% dimethylformamide

Other Equipment and Supplies

Microcentrifuge
Sterile microcentrifuge tubes (0.5 ml)
Pipettes or pipetman (0 to 200 µl, 0 to 1000 µl)
Sterile pipette tips (0 to 200 µl, 0 to 1000 µl)
Gel casting tray
Gel combs
Electrophoresis apparatus
DC power supply
Small or medium size DNA electrophoresis apparatus, depending on samples to be run
Ultrapure agarose powder
Nylon membranes
3MM Whatman filters
Blotting paper towels
Anti-DIG-alkaline phosphatase antibodies
Chemiluminescent substrates

TROUBLESHOOTING GUIDE

1. **Following electrophoresis, distribution of DNA species stained by EtBr is near the very top or the very bottom in some lanes instead of a long smear ranging from the top to the bottom of the lanes.** These problems are most likely caused by two factors. If genomic DNA is not completely digested, DNA molecules with high molecular weights remain near the loading wells during electrophoresis. On the other hand, if DNA is somehow degraded by Dnases, they become small fragments and are distributed at the bottom of a lane. To prevent these problems, genomic DNA should be properly digested. Ensure that DNA is handled without DNase contamination.
2. **After transfer is complete, obvious DNA staining remains in the agarose gel.** This indicates that transfer is not efficient. Try to set up DNA transfer carefully and allow blotting to proceed for a longer time.
3. **The blotted membrane shows some trace of bubbles.** This is due to bubbles formed between gel and membrane. Carefully follow instructions when assembling blotting apparatus.
4. **No signal is detected at all following hybridization.** This is the worst and most disappointing problem in Southern blot hybridization. Genomic DNA may not be denatured before being transferred onto the membrane, or dsDNA probes may not be denatured prior to hybridization. When this occurs, it is not surprising to see a zero hybridization signal. Keep in mind that ssDNA species are the basis for hybridization.
5. **Hybridized signals are quite weak on film or filter.** The activities of DNA probes may be low or hybridization efficiency is not so good. The solution to this problem is to allow hybridization to proceed for a longer time and increase exposure time for x-ray film or develop it for a longer time on filter.
6. **A black background occurs on x-ray film using the chemiluminescent detection method.** Multiple factors may be responsible for such a common problem. The membrane filter may have dried out during hybridization or the washing process. Excess detection solution may not have been removed prior to exposure. Exposure time may have been too long. To solve this problem, make sure that the filter is kept wet and that excess detection reagents are completely removed. Try to reduce exposure time for the x-ray film.
7. **A strong purple/blue background occurs on filter using the colorimetric detection method.** Excess detection reagents may have been used or color allowed to develop for too long. Try to apply appropriate amounts of detection reagents and closely monitor the color development. Once major bands become visible, stop development immediately by rinsing the filter with distilled water several times.
8. **Unexpected bands show up on the x-ray film or membrane filter.** This problem is most likely caused by nonspecific binding between the probes and different DNA species. To solve or prevent such a problem

from occurring, increase the blocking time for the blotted membrane and elevate the stringency conditions for the hybridization and washing processes.

REFERENCES

1. Southern, E.M., Detection of specific sequences among DNA fragments separated by gel electrophoresis, *J. Mol. Biol.,* 98, 503, 1975.
2. Wu, W., Electrophoresis, blotting, and hybridization, in *Handbook of Molecular and Cellular Methods in Biology and Medicine,* Kaufman, P.B., Wu, W., Kim, D., and Leland, C., pp. 87–122, CRC Press, Boca Raton, FL, 1995.
3. Ausubel, F.M., Brent, R., Kingston, R.E., Moore, D.D., Seidman, J.G., Smith, J.A., and Struhl, K., *Current Protocols in Molecular Biology,* Greene Publishing Associates and Wiley-Interscience, John Wiley & Sons, New York, 1995.
4. Fourney, R.M., Aubin, R., Dietrich, K.D., and Paterson, M.C., Determination of foreign gene copy number in stably transfected cell lines by Southern transfer analysis, in *Gene Transfer and Expression Protocols,* Murray, E.J., Ed., The Humana Press Inc., Clifton, NJ, 1991.
5. Sambrook, J., Fritsch, E.F., and Maniatis, T., *Molecular Cloning: A Laboratory Manual,* 2nd ed., Cold Spring Harbor Press, Cold Spring Harbor, NY, 1989.
6. Reed, K.C. and Mann, D.A., Rapid transfer of DNA from agarose gels to nylon membranes, *Nucl. Acids Res.,* 13, 7207, 1985.
7. Khandjian, E.W., Optimized hybridization of DNA blotted and fixed to nitrocellulose and nylon membranes, *Biotechnology,* 5, 165, 1987.
8. Amasino, R.M., Acceleration of nucleic acid hybridization rate by polyethylene glycol, *Anal. Biochem.,* 152, 304, 1986.
9. Thomas, P.S., Hybridization of denatured RNA and small DNA fragments transferred to nitrocellulose, *Proc. Natl. Acad. Sci. USA,* 77, 5201, 1980.

8 Gene Overexpression by Sense RNA in Mammalian Systems

CONTENTS

INTRODUCTION

Well-established DNA recombination and gene transfer technologies are widely used in mammalian cells and in animals to address a vast spectrum of biological questions. In order to gain an insight into the function of the gene of interest, we and others use the gene overexpression approach to up-regulate the expression of a specific gene followed by analysis of the potential roles of the gene in cell or animal systems.[1–5] Gene overexpression refers to an increase in the amount of a given protein or the product level of a gene, which may result in an alteration in the function of the gene. We have successfully transfected mammalian cells with DNA constructs containing sense HSP27 cDNA and developed stably transfected cell lines in which the level of HSP27 increased four- to eightfold compared with nontransfected cells. As a result, cells containing higher levels of HSP27 are protected against heat and toxicants such as heavy metals, demonstrating that one of the functions of the HSP27 protein is to enable cells to survive and recover from stress conditions.[1] Additionally, gene overexpression has a broad range of applications in animal systems.[2–6] It is also applied to treat some human diseases resulting from defects in single genes by transferring the cDNA of a normal gene into patients to compensate for the lack of a given protein; this application is termed gene therapy. Therefore, gene overexpression technology constitutes one of the major advances in current medicine and molecular genetics.[3,6–11]

How is a gene of interest overexpressed? The general procedures and principles are outlined in Figure 8.1. The primary principles are that the sense cDNA constructs of interest are first introduced into cells and integrated into the chromosomal DNA of the host, and that the expression of exogenous cDNA is driven by an appropriate promoter to produce sense RNA that is translated into protein. In this way, the endogenous protein plus the exogenous protein greatly elevate the level of the protein in the cell or animal. There are two types of transfection and expression in

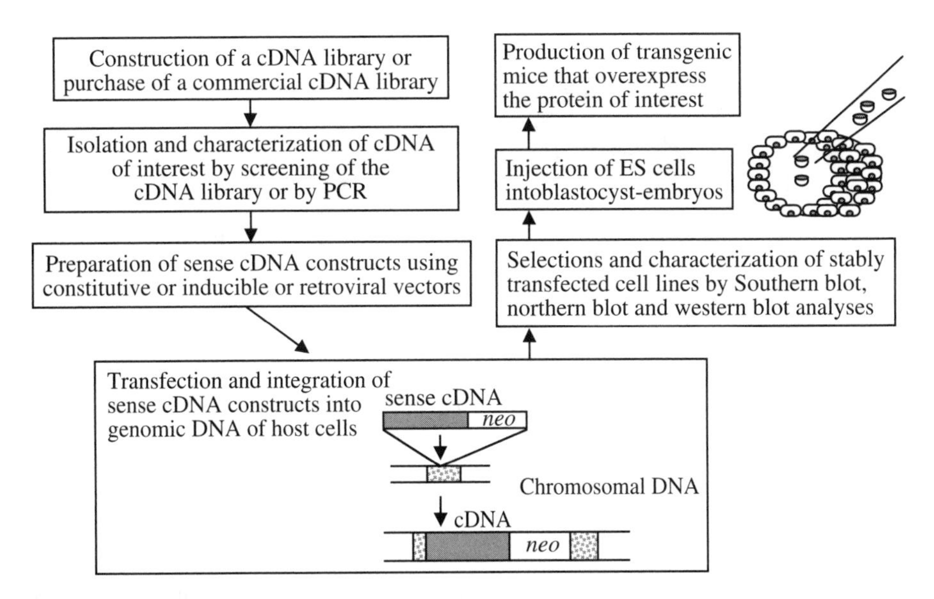

FIGURE 8.1 Flowchart of gene overexpression in cells and animals.

mammalian cells. One is transient transfection in which exogenous cDNA is introduced into cells and allowed to be expressed for 1 to 3 days. The transfected cells are lysed and the proteins are analyzed. The purpose of transient transfection is usually to obtain a burst in the expression of the transferred cDNA; however, such transfection is not suitable for the selection of stably transfected cell lines. The other type is stable transfection in which foreign genes are introduced into cells and stably integrated into the chromosomes or genomes of the host cells. The integrated DNA can replicate efficiently and is maintained during cell division. The products expressed can be analyzed from successive generations of divided cells, establishing genetically altered cell lines.

Overall, gene overexpression by sense cDNA or RNA approaches consists of highly involved techniques that include isolation and characterization of a specific cDNA, DNA recombination, gene transfer and analysis of the gene expression in transfected cells or transgenic animals. The present chapter offers detailed protocols that have been successfully utilized in our laboratories.

DESIGN AND SELECTION OF PLASMID-BASED EXPRESSION VECTORS

Selection of an appropriate vector, including promoters, enhancers, selectable markers, reporter genes and poly(A) signals, is very important for the success of gene transfer and analysis of gene expression. A number of vectors are commercially available. Each has its own strengths and weaknesses with regard to gene transfection and expression. Based on our experience, we recommend and describe several plasmid-based vectors widely used in mammalian systems.

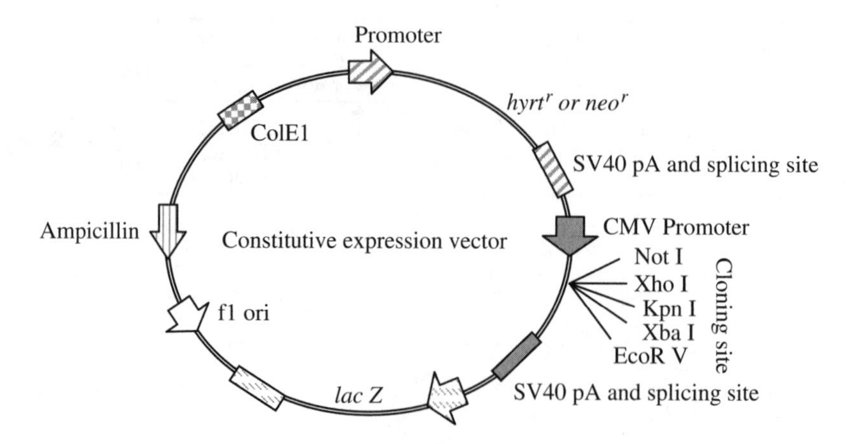

FIGURE 8.2 Anatomy of a constitutive expression vector.

Constitutive Promoter Vectors

The anatomy of a constitutive vector is shown in Figure 8.2. The function of each component is described next.

Constitutive Promoters

The function of this type of promoter is to drive the expression of the cDNA cloned downstream from the promoter. Its driving activity is constitutive. A typical promoter is human cytomegalovirus (CMV), a type of vector that is commercially available. For example, pcDNA3 and pcDNA3.1 vectors from Invitrogen contain PCMV consisting of major intermediate early promoter and enhancer regions.

The SV40 system from Promega has also been demonstrated to be a strong expression vector. The simian virus 40 (SV40) is a small double-stranded circular DNA tumor virus with a 5.2-kb genome. This genome contains an early region encoding the tumor (T) antigen, a late region encoding the viral coat proteins, the origin of replication (ori), and enhancer elements near the ori. The ori and early region play essential roles for the expression of genes. Two divergent transcription units are produced from a single complex promoter or replication region. These viral transcripts are named "early" and "late" because of the time of maximal expression during infection. Both transcripts have introns and are polyadenylated. Using DNA manipulation technology, the SV40 genome has been mutated and fused with a plasmid such as pBR322 and generated a series of very valuable vectors used for gene transfer and expression.

Selectable Marker Genes

Stable transfection mandates a selectable marker gene in a vector so that stably transfected cells will confer resistance to drug selection. The commonly used marker genes include *neo*^{r}, *hyg*^{r}, and *hyrt*^{r}. Detailed selection principles are described under the section on drug selection.

Reporter Genes

Even though a reporter gene is not required in an expression vector, it is recommended that an appropriate reporter gene be utilized in the expression constructs. This will allow use of the activity assay of the reporter gene for selection and characterization of stably transfected cells or transgenic animals. The reporter genes described next are widely utilized.

β-Galactosidase gene: *Lac Z* gene encoding for β-galactosidase is widely used as a reporter gene. The cell extracts of transfected cells can be directly assayed for β-galactosidase activity with spectrophotometric methods. The cells and embryos can be directly stained blue using X-gal. Tissues from adult animals can also be stained for tissue-specific expression. Therefore, we strongly recommend that this gene be constructed in transfection vectors.

CAT gene: A bacterial gene encoding for chloramphenicol acetyl transferase (CAT) has proven to be quite useful as a reporter gene for monitoring the expression of transferred genes in transfected cells or transgenic animals because eukaryotic cells contain no endogenous CAT activity. The CAT gene is isolated from the *E. coli* transposon, Tn9, and its coding region is fused to an appropriate promoter. CAT enzyme activity can be readily assayed by incubating the cell extracts with acetyl Co-A and ^{14}C-chloramphenicol. This enzyme acylates the chloramphenicol, and products can be separated by thin-layer chromatography (TLC) on silica gel plates followed by autoradiography.

Luciferase gene: The luciferase gene that encodes for firefly luciferase has been isolated and widely utilized as a highly effective reporter gene. Compared with the CAT assay, the assay for luciferase activity is sensitive by more than 100-fold. It is simple, rapid and relatively inexpensive. Luciferase is a small, single polypeptide with a molecular weight of 62 kDa. Furthermore, it does not require any posttranslational modification for the activity. Other advantages of using the luciferase gene are that mammalian cells do not have endogenous luciferase activity, and that luciferase can produce, with very high efficiency, chemiluminescent light that can be easily detected.

β-glucuronidase (GUS) gene: Some investigators have employed *uidA* or the gusA gene from *E. coli*, which encodes for β-glucuronidase (GUS), as a reporter gene used for gene transfer. However, the disadvantage is that mammalian tissues contain endogenous GUS activity, thus making the enzyme assay more difficult. Nonetheless, the GUS reporter gene is an excellent reporter gene widely used in higher plant systems because plants do not contain detectable endogenous GUS activity.

Splicing Regions

A sequence of an appropriate polyadenylation site and a splicing region should be incorporated downstream from each individual gene in an expression vector so that individual mRNA species to be transcribed in the host cell can be spliced out. Thus, fusion proteins that are functionally different from the nonfusion or native protein of interest can be avoided.

Kozak Sequence and Enhancer Element

In order to enhance the expression of the cloned cDNA further, it is a good idea to incorporate an enhancer element upstream from the promoter. It has been reported that a Kozak sequence, GCC(A/G)CCAUGG, that includes the first codon, AUG, can increase the efficiency of translation up to 10-fold.

Inducible Promoter Vectors

Unlike constitutive promoter vectors, the expression of a cloned cDNA in an inducible vector requires an appropriate inducer. The general structure is shown in Figure 8.3. These vectors are commercially available. For instance, the pMAM*neo* (Clontech) contains the dexamethasone-inducible MMTV-LTR promoter linked to the RSV-LTR enhancer.

Retrovirus Vectors

It has been demonstrated that retroviruses can be used as effective vectors for gene transfer in mammalian system, especially for transient transfection.[1,5] There are several advantages over other vector systems, which include (1) the retroviral genome can stably integrate into the host chromosome of the host cell and can be passed from generation to generation, thus providing an excellent vector for stable transfection; (2) retroviruses have a great range of infectivity and expression host for any animal cells via viral particles; (3) integration is site-specific with respect to the viral genome at long terminal repeats (LTRs); in other words, any DNA cloned within two LTRs will be expected to be integrated into chromosomal DNA of the host cell, which can preserve the structure of the gene intact with ease after integration; and (4) viral genomes are very plastic and manifest a high degree of natural size manipulation for DNA recombination. The drawback, however, is that the techniques are relatively sophisticated. Therefore, for successful transfection, it is

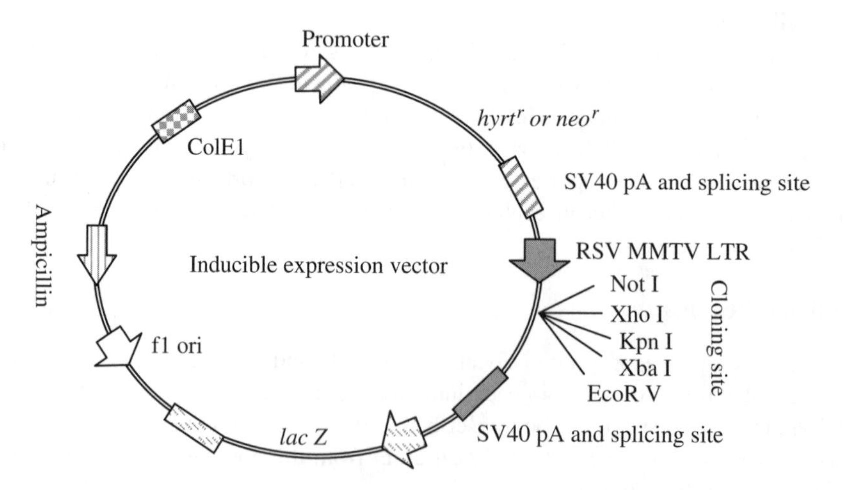

FIGURE 8.3 Diagram of an inducible expression vector.

necessary to elaborate briefly on the life cycle of retroviruses prior to describing the construction of retroviruse-based vectors for mediating gene transfer.

Retroviruses, such as Rous sarcoma virus (RSV) and Moloney murine leukemia virus (MoMLV), are RNA viruses that cause a variety of diseases (e.g., tumors) in humans. The virus has two tRNA primer molecules, two copies of genomic RNA (38S), reverse transcriptase, RNase H and integrase, which are packaged with an envelope. The viral envelope contains glycoproteins that can determine the host range of infection. As shown in Figure 8.4, when the virus or virion attaches to a cell, the viral glycoproteins in the envelope bind to specific receptors in the plasma membrane of the host cell. The bound complex facilitates the internalization of the virus that is now uncoated as it passes through the cytoplasm of the host cell. In the cytoplasm, reverse transcriptase in the viral genome catalyzes the formation of a double-stranded DNA molecule from the single-stranded virion RNA. The DNA molecule undergoes circularizing, enters the nucleus, and becomes integrated into the chromosome of the host cell, forming a provirus. Subsequently, the integrated provirus serves as the transcriptional template for mRNAs and virion genomic RNA. Interestingly, such transcription is catalyzed by the host RNA polymerase II. The mRNAs then undergo translation to produce viral proteins and enzymes using the host machinery. These components are packaged into viral core particles that move through the cytoplasm, attach to the inner side of the plasma membrane and then bud off. This cycle of infection, reverse transcription, transcription, translation, virion assembly and budding is repeated again and again, infecting new host cells.

A well-understood retrovirus is RSV. The mechanism of synthesis of double-stranded DNA intermediates (provirus) from viral RNA is unique and quite complex. The nucleotide sequence of the DNA molecule is different from that of the viral RNA. The sequence U3-R-U5, which is a combination of the 5′ r-u5 segment and

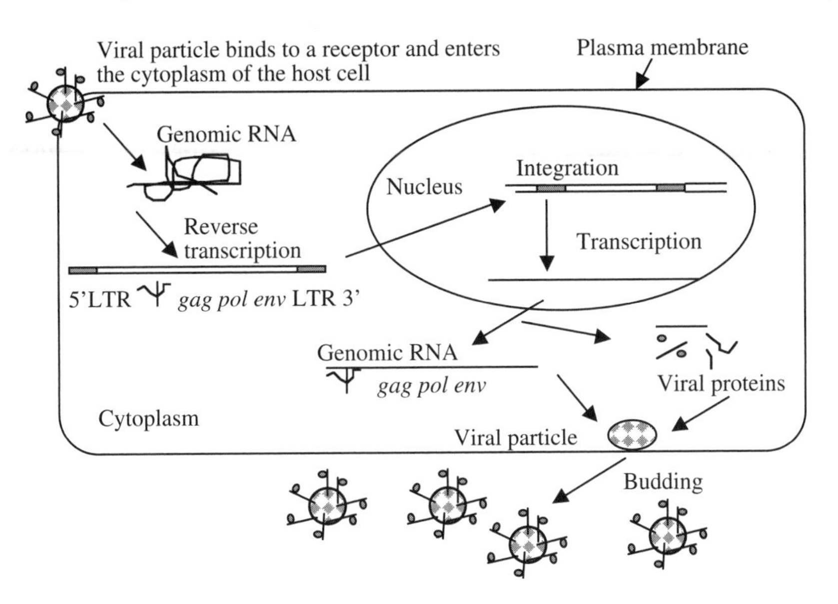

FIGURE 8.4 Life cycle of a replication-competent retrovirus.

the 3′ u3-r segment of the RNA, is present at 5′ and 3′ ends of the double-stranded DNA molecule. The U3-R-U5 is designated the long terminal repeat (LTR). This complete scheme can be divided into the eight steps shown in Figure 8.5:

Step 1. One of the proline tRNA primers first anneals to the pbs region in the 5′ r-u5 of the viral genome RNA. Reverse transcriptase catalyzes the extension of the tRNA primer from its 3′-OH end to the 5′ end, producing a DNA fragment called 3′ R′-(U5)′-tRNA.

Step 2. RNase H removes the cap and poly(A) tail from the viral RNA as well as the viral r-u5 segment in the double-stranded region.

Step 3. The 3′ R′-(U5)′-tRNA separates from the pbs region, jumps to the 3′ end of the viral RNA and forms an R′/r duplex.

Step 4. The 3′ R′-(U5)′-tRNA undergoes elongation up to the pbs region of the viral RNA by reverse transcriptase, producing the minus (–)strand of DNA.

Step 5. RNase H removes the u3-r from the 3′ end of the viral RNA, followed by synthesis of DNA from the 3′ end of the RNA via reverse transcriptase, generating the first LTR (U3-R-U5) that contains the promoter sequence. This serves as a part of the plus (+)strand of DNA.

Step 6. All RNA, including the tRNA primer, is removed by RNase H.

Step 7. The U3-R-U5 (PBS) separates from (U3)′R′(U5)′, jumps to the 3′ end of the complementary strand of DNA and forms a (PBS)/(PBS)′ complex. This is an essential process for the virus because the jumping action brings the promoter in the U3 region from the 3′ end to the 5′ end of the plus

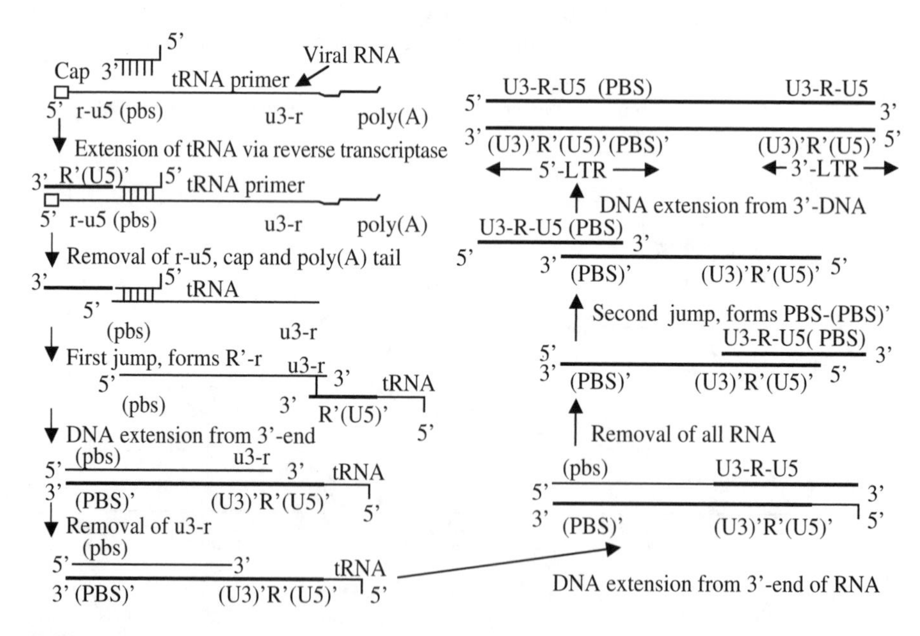

FIGURE 8.5 Scheme for synthesis of double-stranded DNA containing two long terminal repeats (LTR) from Rouse sarcoma viral RNA.

strand of DNA. Thus, the promoter is now upstream from the coding region for gag-pol-env-src in the 38S RNA genome.

Step 8. Reverse transcriptase catalyzes the extension of DNA from the 3′-termini of both strands, producing a double-stranded DNA (provirus). The LTRs at both ends of the DNA molecule contain promoter and enhance elements for transcription of the virus genome. The double-stranded DNA molecule can now become integrated into the chromosomes of the infected cell.

How can the harmful infectious retrovirus be used for gene transfer and expression? Because retroviruses can cause tumors in animals and human beings, we obviously do not want the whole virus genome to be used for gene transfer or as an expression vector. Recent DNA recombination technology makes it possible to manipulate the retrovirus genome so that it is a powerful tool for gene transfer. The simplest type of gene transfer system is one in which all or most of the *gag, pol,* and *env* genes in the provirus are deleted. Nonetheless, all of the *cis*-active elements such as the 5′ and 3′ LTRs, PBS(+), PBS(–) and psi (Ψ) packaging sequence are left intact. A foreign cDNA of interest or a selectable gene such as *neo, gpt, dhfr,* or *hprt* can be inserted at the initiating ATG site for *gag*. The expression of the inserted gene is driven by the 5′ LTR. This manipulated vector is then fused with a plasmid fragment such as pBR322 containing the origin of replication (ori) and an antibiotic-resistant gene. The recombinant plasmids are propagated in a bacterial strain of *E. coli*. The vectors are then utilized for standard DNA-mediated transfection of suitable recipient cells and stable transformants can be selected by using an appropriate antibiotic chemical such as G418. The partial viral RNA transcribed in the host cell can be further packaged into retroviral particles that become an infectious recombinant retrovirus. This can be done by cotransfecting the host cells with a helper virus that produces *gag* and *env* proteins, which can recognize the psi (Ψ) sequence on the recombinant transcript and become packaged to form viral particles.

The disadvantage of this strategy is that the culture supernatant contains recombinant and wild-type viruses. To overcome this problem, some vectors are constructed by deleting the Ψ sequence, replacing the 3′LTR with an SV40 terminator, the poly(A) signal, and fusing to the backbone of pBR322. Another type of expression vector may be constructed by deleting *gag* and *env* genes and by inserting a selectable marker gene (e.g., *neo*) or a reporter gene followed by polylinker sites for the insertion of the foreign gene or cDNA of interest at the ATG site for *gag* downstream from the 5′ LTR. The 3′ LTR can be replaced with the SV40 poly(A) signal downstream from the introduced cDNA or gene.

This chapter focuses on a replication-incompetent viral vector (Figure 8.6) This vector lacks *gag, pol* and *env* genes encoding virus structural proteins. These deleted viral genes are replaced by a marker gene (e.g., *neo*) and the cDNA of interest. cDNA to be expressed can be cloned at the multiple cloning site and be driven by the CMV promoter or other promoters such as the herpes simplex thymidine kinase promoter (TK). The major advantage of this vector is that any genes cloned between the 5′-LTR and 3′-LTR can be efficiently integrated into the chromosomes of the host cell because the LTRs contain a short sequence that facilitates the integration. The recombinant vector is then fused with the pBR322 backbone containing the

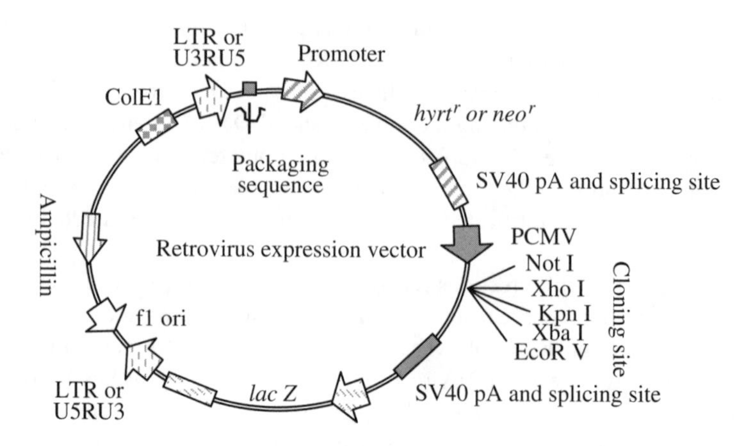

FIGURE 8.6 Diagram of a retrovirus expression vector.

origin of replication and the Amp^r gene. The recombinant plasmids are then amplified in *E. coli*. The foreign gene or cDNA of interest can be cloned at the polylinker site downstream from the 5′ LTR of recombinant vectors, and gene transfer can be carried out by a standard plasmid DNA-mediated procedure.

PREPARATION OF PLASMID SENSE cDNA CONSTRUCTS

cDNA isolation and characterization are described in Chapter 3. Restriction enzyme digestion of DNA, ligation, bacterial transformation, and purification of plasmid DNA are given in Chapter 4. The subject of this section is to emphasize the following special points with respect to the preparation of sense cDNA constructs. Any mistakes will cause failure in the expression of the cloned cDNA.

1. Length of cDNA to be cloned: For the overexpression of a functional protein from a cDNA, the cDNA must include the entire coding region or ORF. In our experience, 10 to 20 base pairs in 5′UTR or 3′UTR are recommended for subcloning. Long 5′UTR or 3′UTR sequences may not be good for overexpression of the cloned cDNA.
2. Orientation of the cDNA: It is absolutely essential that the cDNA be placed in the sense orientation downstream from the promoter. In other words, the poly(T) strand or (–)strand of the cDNA should be 3′ → 5′ downstream from the driving promoter in order to make 5′ → 3′ mRNA.
3. Start or stop codon: Make sure that no start codon is between the driving promoter and the first start codon ATG of the cDNA. Otherwise, the protein to be expressed will be a fusion protein other than the protein of interest. For this reason, we strongly recommend that the plasmid construct DNA be partially sequenced at the multiple cloning site (MCS) to make sure that the cDNA is cloned at the right place without any ATG codon upstream from the first ATG codon in the cDNA. However, it is

acceptable if a stop codon is present upstream from the first codon ATG because, during translation, the small subunit of ribosome will locate the ATG codon instead of the stop codon.

TRANSIENT TRANSFECTION OF MAMMALIAN CELLS WITH SENSE CONSTRUCTS

METHOD A. TRANSFECTION BY CALCIUM PHOSPHATE PRECIPITATION

The principle of calcium phosphate transfection is that the DNA to be transferred is mixed with $CaCl_2$ and phosphate buffer to form a fine calcium phosphate precipitate. The precipitated complexes or particles are then placed on a cell monolayer. They bind or attach to the plasma membrane and are taken into the cell by endocytosis. This is one of the most widely used methods for both transient and stable transfection.

1. Perform trypsinizing to remove adherent cells for subculture or cell counting as well as for monolayer cell preparation.
   a. Aspirate the medium from the cell culture or the tissue culture dish and wash the cells twice with 10 to 15 ml of PBS buffer or other calcium- and magnesium-free salt solutions.
   b. Remove the washing solution and add 1 ml of trypsin solution per 100-mm dish. Quickly overlay the cells with the solution and incubate at 37°C for 5 to 10 min, depending on the cell type (e.g., NIH 3T3, 2 min; CHO, 2 min; HeLa, 2.5 min; COS, 3 min).
   c. Closely monitor the cells. Once they begin to detach, immediately remove the trypsin solution and tap the bottom and sides of the culture flask or dish to loosen the remaining adherent cells.
   d. Add culture medium containing 10% serum (e.g., FBS or FCS) to the cells to inactivate the trypsin. Gently shake the culture dish. The cells can be used for cell subculturing or counting.
2. Prepare cell monolayers the day prior to the transfection experiment. The cell monolayers should be at 60 to 70% confluence. Aspirate the old culture medium and feed the cells with 5 to 10 ml of fresh culture medium.

Notes: (1) It is recommended that this procedure be carried out in a sterile laminar flow hood. Gloves should be worn to prevent contamination of the cells. (2) The common cell lines used for transfection are NIH 3T3, CHO, HeLa, COS and mouse embryonic stem (ES) cells. Other mammalian cells of interest can also be transfected in this way. (3) The density of the cells to be seeded depends on the particular cell line. A general guideline is 1 to 5 × 10^6 cells per 100-mm culture dish. (4) All of the solutions and buffers should be sterilized by autoclaving or by sterile-filtration.

3. Prepare a transfection mixture for each 100-mm dish as follows:
   2.5 M $CaCl_2$ solution, 50 µl
   DNA construct, 15 to 30 µg
   Add dd.H_2O to a final volume of 0.5 ml.

Mix well, and slowly add the DNA mixture dropwise to 0.5 ml of well-suspended 2X HBS buffer with continuous mixing by vortexing. Incubate the mixture at room temperature for 30 to 45 min to allow coprecipitation to occur.

Notes: (1) The DNA to be transferred can be in a circular form (e.g., recombinant plasmids) or be linear chimeric genes. (2) After the DNA mixture has been added to the 2X HBS buffer, the mixture should be slightly opaque due to the formation of fine coprecipitate particles of DNA with calcium phosphate.

4. Slowly add the precipitated mixture dropwise to the dish containing the cells while continuously swirling the plate. For each 100-m dish, add 1 ml of the DNA mixture, or add 0.5 ml of the DNA mixture to each 60-mm dish.
5. Place the dishes in a culture incubator overnight at 37°C and provide 5% CO_2.
6. Aspirate the medium and wash the monolayer once with 10 ml of serum-free DMEM medium and once with 10 ml of PBS. Add fresh culture medium and return the dishes to the incubator. In general, cells can be harvested 48 to 72 h after transfection for transient expression analysis.

Reagents Needed

10X HBS Buffer

5.94% (w/v) HEPES [4-(2-hydroxyethyl)-1-piperazine ethanesulfonic acid], pH 7.2
8.18% (w/v) NaCl
0.2% (w/v) Na_2HPO_4

2X HBS Buffer

Dilute the 10X HBS in dd.H_2O.
Adjust the pH to 7.2 with 1 *N* NaOH solution.

DNA to be Transferred

Circular or linear recombinant DNA constructs in 50 μl of TE buffer, pH 7.5

2.5 M $CaCl_2$

DMEM Culture Medium

Dissolve one bottle of Dulbecco's modified eagle's medium (DMEM) powder in 800 ml dd.H_2O; add 3.7 g $NaHCO_3$; adjust the pH to 7.2 with 1 *N* HCl. Add dd.H_2O to 1 l. Sterile-filter the medium (do not autoclave) in a laminar flow hood. Aliquot the medium into 200- to 500-ml portions in sterile bottles. Add 10 to 20% FBS, depending on the particular cell type, to the culture medium. Store at 4°C until use.

PBS Buffer

0.137 *M* NaCl

4.3 m*M* Na_2HPO_4
2.7 m*M* KCl
1.47 m*M* KH_2PO_4
The final pH is 7.2 to 7.4.

1X Trypsin-EDTA Solution
0.05% (w/v) Trypsin
0.53 m*M* EDTA, pH 7.6
Dissolve in PBS buffer, pH 7.4 to 7.6.

Method B. Transfection by Retrovirus Vectors

This section focuses on a replication-incompetent viral vector (Figure 8.6) and describes in detail the protocols for retroviral transfection. As shown in Figure 8.6, a replication-incompetent vector lacks *gag, pol,* and *env* genes that encode virus structural proteins, which are replaced by marker *neo* gene and cDNA to be expressed. To generate a retrovirus producer cell line, replication-incompetent viral vectors are first introduced into a murine packaging cell line able to produce the proteins (e.g., virus structural proteins) necessary for production of retroviral particles. The cloned cDNA and *neo* gene between the two LTRs will integrate into the genome of the packaging cells. After drug selection (e.g., use of G418), stably transfected packaging cells will survive and produce retroviral particles whose genomic RNA contains the packaging sequence, the *neo* gene and the cDNA of interest, generating retrovirus producer cells. The retrovirus particles will leave the host cells by budding and are able to reinfect these host cells. After serial drug selections, a high titer of specific retrovirus producer cell lines can be established. These cells are then utilized to prepare a large-scale, high concentration of retrovirus stocks that can be used to infect other animal cells of interest for high efficiency of transfection. These retrovirus particles cannot replicate themselves because they lack *gag, pol,* and *env* genes that encode virus structural proteins. Thus, retrovirus particles are solely generated from the producer cell lines.

Caution: The authors wish to remind investigators about the biological safety requirements involved in the use of retroviruses. Occasionally, a replication-incompetent retrovirus stock may contain a helper virus that is a replication-competent virus. The presence of a helper-virus could be problematic if such a virus infects animals because leukemia and other diseases may be generated. Therefore, extreme precautions should be taken when using retroviruses for research. A proven proposal may be needed from an appropriate committee. Any solutions and media containing potential retrovirus particles should be collected in special containers and be autoclaved before being disposed of or being given to authorized people to handle.

Protocol 1. Preparation of Viral Supernatant by Transient Transfection of a Packaging Cell Line with Retrovirus Vector Constructs

As shown in Figure 8.6, replication-incompetent retrovirus DNA constructs lack the structural genes (*gag, pol, env*) that encode the retrovirus structural proteins.

Therefore, production of infectious viral particles mandates transfection of a packaging cell line with the plasmid-based retrovirus constructs. Packaging cell lines are derived from mouse-fibroblast cells or Avian cells. These cell lines have been stably transfected with retroviruses whose genomic RNA does not contain the packaging sequence Ψ. In this way, even though the cells contain *gag, pol* and *env* genes encoding viral structural proteins, the cells cannot package the viral RNA. However, these cells are very valuable packaging cell lines that can be used for transfection with a plasmid-based, replication-incompetent retrovirus vector (Figure 8.6). Because the packaging sequence Ψ is incorporated into the construct, viral RNA molecules can be readily packaged. As a result, infectious viral particles will be produced.

Commonly used packaging cell lines include murine-ecotropic GP+E-86, ΩE and murine-amphotropic ψCRIP. These cell lines cannot produce the replication-competent helper virus and they provide safe cell lines for packaging. A number of packaging cell lines are commercially available or can be obtained from other investigators with permission from appropriate authorities. The general procedure is outlined in Figure 8.7.

1. Grow packaging cells to 40 to 50% confluence prior to transfection.
2. Using the calcium phosphate precipitation method, carry out transfection of cells with replication-incompetent retrovirus constructs (Figure 8.6). The detailed protocol is described in Method A.

Notes: (1) Electroporation can also be applied for transfection except that approximately 50% of cells may be killed by a high-voltage pulse. (2) The DNA constructs can be linear. However, because the vector contains 5′-LTR and 3′-LTR, circular plasmid DNA works better.

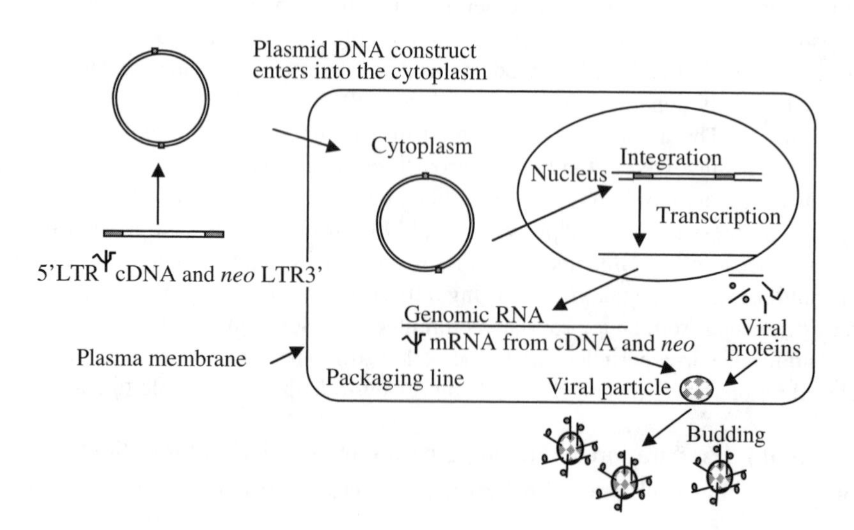

FIGURE 8.7 Transfection of a packaging cell line with a replication-incompetent retrovirus vector and production of infectious retroviral particles.

3. For transient transfection, harvest the medium 24 to 48 h following transfection and filter through a 0.45-µm filter. Now, the supernatant should contain infectious viruses and can be used immediately for titer or stored at –80°C until use. The cells can be utilized for drug selection to generate stable producer cells (see the next protocol).

Protocol 2. Production of Stably Transfected Producer Cell Lines

1. Trypsinize the cells (see Chapter 16 for detailed protocols) remaining at step 3 in the preceding protocol and plate them at a 1:10 to 50 dilution. Add an appropriate volume of medium (10 ml/100 mm-dish or 15 ml/T75 flask) and allow the cells to grow for 24 to 48 h.
2. Replace the medium with selection medium containing an appropriate amount of drug (e.g., G418). The cells are cultured in the selection medium for 10 to 14 days with a change of medium once every 2 to 7 days, depending on the specific cell lines, to remove dead cells and cellular debris. In theory, only those cells that bear and express the drug-selectable marker gene can survive in the selection medium.

Notes: For selection, a killing curve should first be established from non-transfected cells cultured in the medium containing serial concentrations of an appropriate drug. Choose the concentration that kills 100% of the cells in 5 days as the concentration to be used for selection of stably transfected cells. On the other hand, the concentration used for selection can be found from previous studies using the same cell line.

3. Pick up and transfer well-isolated individual colonies to separate dishes or flasks and expand these clones in the selection medium until 70 to 90% confluence is obtained.
4. Harvest the medium and filter it with a 0.45-µm filter. The virus stock can be immediately used for titer or stored at –80°C until use.
5. Freeze back as many vials of the stable producer cell lines as possible. The detailed protocols for cell freezing are described in Chapter 16.

Protocol 3. Determination of Viral Titer

After individual clones have been isolated, it is necessary to produce virus stocks from each of the clones and titer the virus in the stocks. To do so, a targeting or titering cell line is needed. The most widely used cell lines are NIH 3T3 (mouse fibroblast), Avian QT6 and chicken embryo fibroblast CEF.

1. Grow stable producer clones to 80 to 90% confluence in 100-mm dishes or in T25 or T75 flasks.
2. Individually harvest the medium and filter it through a 0.45-µm filter. Store the medium at –80°C until use.
3. Plate 1×10^4 NIH 3T3 into each 6-cm dish 1 day before infection. Allow the cells to grow for 24 h in an incubator at 37°C, 5% CO_2.

4. Collect the medium and filter it through a 0.45-µm filter. Store the supernatant containing viruses at –80°C until use. This supernatant will serve as the virus stock.
5. Make serial 10-fold dilutions of the virus stock into the culture medium for the target cells. It is recommended to start from 1000 µl/ml → 100 µl/ml → 10 µl/ml → 1 µl/ml → 0.1 µl/ml → 0.01 µl/ml → 0.001 µl/ml. For each 6-cm or 100-mm dish, make 5 or 10 ml of dilution viral medium. Add polybrene (Sigma Chemical Co., St. Louis, MO) to final concentration of 2 to 4 µg/ml using 100X stock in dd.H_2O.

Note: The diluted viral medium should be used within 1 h.

6. Aspirate the medium from the target cells at step 3 and add 5 ml of the diluted medium prepared at step 5 to each 6-cm or 10 ml- to 100-mm dish. Incubate the cells at 37°C for 24 h.
7. Replace the medium with fresh culture medium without virus and allow the cells to grow for 24 h. At this point, there are two options. If the retroviral particles contain a reporter gene, such as *lac Z,* the infected cells can be stained with X-gal (see reporter gene assay in this chapter). No drug selection is necessary. However, if the viruses carry a selectable marker gene such as *neo,* the infected cells can be identified by drug (G418) selection, which is described in the next steps.
8. Replace the medium at step 7 with drug selection medium containing an appropriate concentration of G418, which should be based on the killing curve. The concentration of G418 that results in 100% killing of the target cells will be chosen as the selection medium.
9. Allow the cells to be selected and cultured for 10 to 12 days with changes of the culture medium once every 3 days. Stably infected cells should survive and expand to form colonies within 2 weeks.
10. Count the colonies and calculate the titer as CFU/ml. Titer is expressed as the number of colonies per ml of virus stock. It is easy to use the highest dilution for calculation. For example, if a dish has five stably infected colonies in 10^6 dilution medium, the titer will be 5×10^6 $G418^r$ CFU/ml of virus stock. In the same way, if the cells are stained with X-gal, then X-gal CFU/ml = number of blue colonies/virus volume (ml) × dilution factor. 10^{6-8} CFU/ml is considered a good titer of virus stock.

Protocol 4. Amplification of Virus Stock by Serial Reinfection of Fresh Target Cells

A good titer of virus stock can be utilized to reinfect fresh target cells to obtain a large scale of a high-titer virus stock (Figure 8.8).

1. Plate 1×10^4 NIH 3T3 in each 6-cm dish and allow the cells to grow to about 20 to 30% confluence in an incubator at 37°C with 5% CO_2.

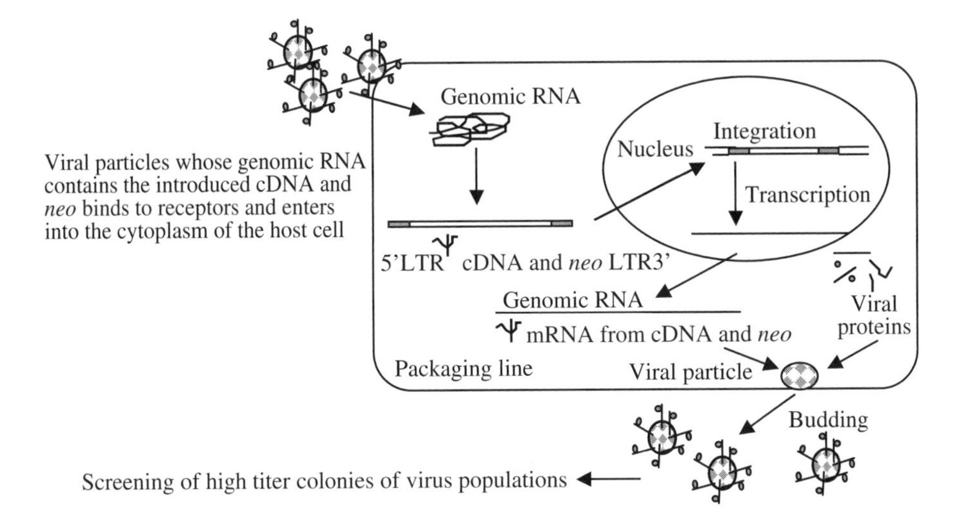

FIGURE 8.8 Transfection of a packaging cell line with a replication-incompetent retrovirus vector and production of infectious retroviral particles.

2. Collect a good titer of virus stock and filter through a 0.45-μm filter. Store the supernatant containing viruses at –80°C until use.
3. Make a dilution of the culture medium using 10^5 CFU/ml. For each 6-cm or 100-mm dish, make 5 or 10 ml of diluted viral medium. Add polybrene to a final concentration of 2 to 4 μg/ml using 100X stock in dd.H_2O.
4. Aspirate the medium from the target cells and add 5 ml of the diluted medium to each 6-cm dish or 10 ml to each 100-mm dish. Incubate the cells at 37°C for 24 h.
5. Replace the medium with fresh culture medium without virus and allow the cells to grow for 24 h.
6. Replace the medium with drug selection medium containing an appropriate concentration of G418, which should be based on the killing curve. The concentration of G418 that results in 100% killing of the target cells will be chosen as the selection medium.
7. Allow the cells to be selected and cultured for 15 to 21 days with changes of the medium once every 3 days. Stably infected cells should be expanded to 90% confluence.
8. Pick up well-isolated colonies and expand them in individual flasks or dishes. Grow the cells to confluence.
9. Harvest the virus supernatant and filter it. Freeze back as many vials of stably infected target cells as possible. The filtered virus stock with high titer can be used to infect other animal cells of interest (see Protocol 5).

Protocol 5. Transfection of Cells of Interest with High-Titer Stock of Replication-Incompetent Retroviruses (Figure 8.9)

1. Plate 1×10^4 cells of interest in each 6-cm dish and allow the cells to grow to about 20 to 30% confluence in an incubator at 37°C with 5% CO_2.
2. Filter the high-titer stock of the virus prepared in Protocol 4 using a 0.45-μm filter. Store the supernatant containing viruses at –80°C until use.
3. Make a dilution of culture medium using 10^{6-7} CFU/ml. For each 6-cm or 100-mm dish, make 5 or 10 ml of diluted viral medium. Add polybrene to a final concentration of 2 to 4 μg/ml using 100X stock in dd.H_2O.
4. Aspirate the medium from the cultured cells and add 5 ml of the diluted medium to each 6-cm dish or 10 ml to each 100-mm dish. Incubate the cells at 37°C for 24 h.
5. Replace the medium with fresh culture medium without virus and allow the cells to grow for 24 h.
6. Replace the medium with drug selection medium containing an appropriate concentration of G418, which should be based on the killing curve.
7. Allow the cells to be selected and cultured for 15 to 21 days with changes of the medium once every 3 days. Stably infected cells should be expanded to 90% confluence.
8. Pick up well-isolated colonies and expand them in individual flasks or dishes. Allow the cells to grow to confluence.
9. Freeze back the stably infected cells.

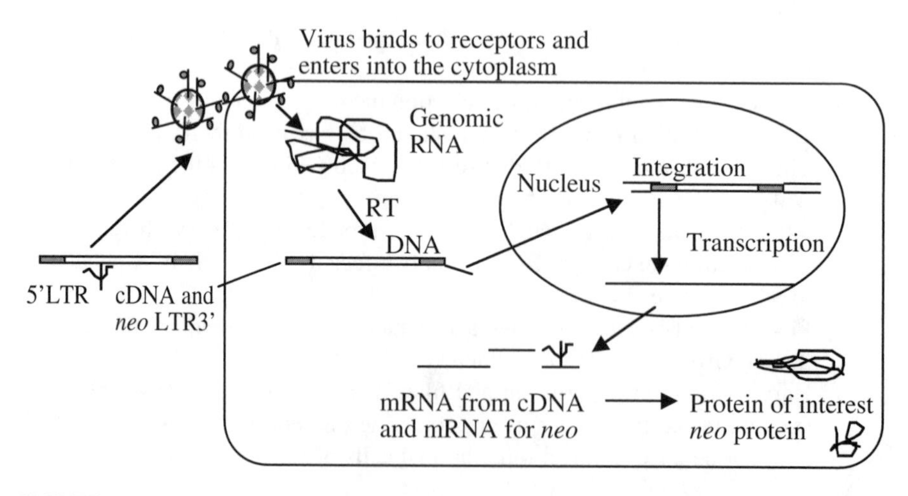

FIGURE 8.9 Infection of an animal cell of interest by a replication-incompetent retrovirus containing the introduced cDNA and *neo* gene.

STABLE TRANSFECTION OF MAMMALIAN CELLS WITH SENSE DNA CONSTRUCTS

METHOD A. TRANSFECTION BY LIPOSOMES

Detailed protocols are described under Transient Transfection in Chapter 9.

METHOD B. TRANSFECTION BY ELECTROPORATION

The principle underlying this simple, fast and effective method is that cells and DNA constructs to be transferred are mixed and subjected to a high-voltage electrical pulse that creates pores in the plasma membranes of the cells. This allows the exogenous DNA to enter the cytoplasm of the cell and the nucleus through the nuclear pores and become integrated into the genome. This method is suitable for transient and stable transformation experiments; however, the pulse conditions should be adjusted empirically for the particular cell type. In spite of the fact that this method has a high efficiency of transfection, its drawback is that approximately 50% of the cells are killed by the high-voltage electrical pulse. These dead cells should be removed on a daily basis in order to reduce toxicity.

1. Linearize the DNA constructs to be transferred by a unique restriction enzyme. Protocols for appropriate restriction enzyme digestion are described under Subcloning of Gene and DNA Fragments in Chapter 4.

Tips: (1) The unique restriction enzyme cut should be in the plasmid backbone area such as ColE1, fi ori or ampicillin gene. It is extremely important that this cut not be in any region that will affect gene targeting by homologous recombination. The activities of Lac Z *and selectable marker genes such as* neo, hyg, hyrt *and* HSV-tk *should not be affected. In the case of double-knockout rescue DNA, ensure that the cut is not in any area in the promoter, inserted cDNA, or the splicing site. (2) In our experience, plasmid DNA constructs do not need to be linearized for homologous recombination. In fact, plasmid DNA enhances the transfection efficiency up to 30%.*

2. Harvest the cells by gentle trypsinization as described in Chapter 16. Centrifuge the cells at 150 × *g* for 5 min and resuspend the cells well in normal culture medium at about 10^6 cells/ml.

Note: Make sure that the cells are gently but well suspended prior to transfection because aggregation of cells may cause difficulty in isolation of stably transfected cell clones.

3. In a laminar flow hood, gently mix 0.6 ml of the cell suspension with 15 to 30 μl of DNA constructs (25 to 50 μg) in a disposable electroporation cuvette (Bio-Rad, Hercules, CA).
4. According to the manufacturer's instructions, carry out electroporation using a Gene Pulser (Bio-Rad) with a pulse controller and capacitance extender at 300 V and 250 μF.
5. The electroporated cells are immediately plated in culture dishes at a density of about 3000 cells/100-mm Petri dish and cultured at 37°C with 5% CO_2. Proceed to drug selection.

METHOD C. TRANSFECTION BY RETROVIRUS VECTORS

Detailed protocols are described in Method B under Transient Transfection.

SELECTION OF STABLY TRANSFECTED CELL LINES WITH APPROPRIATE DRUGS

As shown in Figure 8.1, foreign DNA enters the cells after transfection, but only a small portion of the DNA gets transferred to the nucleus in which some of the DNA might be transiently expressed for a few days. A smaller portion of the DNA introduced into the nuclei of some of the transfected cells undergoes integration into the chromosomes of the host and will be expressed in future generations of the cell population, generating stably transfected cells. The selection of the transfected cells depends on the drug-selectable marker gene fused in the recombinant plasmid, linear DNA construct or a separate plasmid that can be cotransfected into the cells. Generally, the transfected cells are cultured for 30 to 60 h in the medium lacking specific antibiotics in order to allow the cells to express the selectable-marker gene. The cells are then subcultured in a 1:10 to 100 dilution in a selective medium containing an appropriate drug. The cells are cultured in the selective medium for 10 to 14 days with changes of medium once every 2 to 7 days, depending on specific cell lines, to remove dead cells and cellular debris. In theory, only those cells that bear and express the drug-selectable marker gene can survive in the selective medium. The common marker genes that have been used for transfection and the selection can be carried out as follows.

USE OF AMINOGLYCOSIDE ANTIBIOTICS

If cells are transfected with recombinant gene constructs that contain a bacterial gene *neomycin* or Tn60 for aminoglycoside 3′-phosphotransferase (*aph*), the stably transfected cells can confer resistance to aminoglycoside antibiotics such as kanamycin (*kan*) and geneticin (G418). For selection, a killing curve should first be established from nontransfected cells cultured in the medium containing different concentrations of an appropriate drug. Choose the concentration that kills 100% of the cells in 5 days as the concentration for selection of stably transfected cell. On the other hand, the concentration used for selection can be found from previous studies using the same cell line. For instance, we use 250 μg/ml and 1 mg/ml G418 to select mouse ES cells and fibroblast STO cells, respectively.

HYGROMYCIN-β-PHOSPHOTRANSFERASE

Stably transfected cells containing the hygromycin-β-phosphotransferase gene can be selected with hygromycin B, which is a protein synthesis inhibitor. Hygromycin B stocks are available from Sigma Chemical Co. 10 to 500 μg/ml is used in the culture medium, depending on the prekilling curve of a specific cell line.

Thymidine Kinase

If cells lacking the thymidine kinase gene (*tk*⁻) are transfected by recombinant gene constructs containing the *tk* gene, the stable transformants can be selected by culturing the transfected cells in a medium containing hypoxanthine, aminopterin, and thymidine (HAT). Nonstable transfected cells cannot metabolize this lethal analog and will be killed. The HAT medium is prepared as follows:

Dissolve 15 g hypoxanthine and 1 mg aminopterin in 8 ml of 0.1 *N* NaOH solution. Adjust the pH to 7.0 with 1 *N* HCl and then add 5 mg thymidine. Add water to 10 ml and sterile-filter it. Dilute the stock HAT to 100-fold in the culture medium used for selection. Again, the dilution or concentration of the HAT should be based on the prekilling curve. Different cell lines confer different degrees of resistance.

Tryptophan Synthetase

If cells are transfected with DNA constructs containing the tryptophan synthetase gene (the *trp B*), the stably transfected cells can be selected by culturing the cells in a medium lacking tryptophan, an essential amino acid necessary for the growth of cells.

CHARACTERIZATION OF STABLY TRANSFECTED CELL CLONES

Characterization of stably transfected cell lines will provide information about whether the developed clones contain and express the genes introduced into the cells. This verification step is required because drug-resistant cells may not contain or express the gene of interest. Traditionally, characterization begins with DNA analysis by Southern blotting, proceeds to the mRNA level by northern blotting, followed by protein analysis by western blotting. Nonetheless, in practice, the putative clones that express the cDNA introduced are only approximately 10% of the stably selected cell clones. Obviously, if one starts from the Southern blot analysis, there will be a tremendous amount of work, assuming that 100 clones have been obtained after drug selection and only 10 or fewer clones really express the cDNA or gene of interest. In our experience, Southern blotting as the first step of analysis may not be a good idea because it is time consuming and costly.

The same situation prevails, however, if one begins with northern blotting, which is a relatively complicated procedure. We strongly recommend starting with protein analysis by dot blotting or western blot analysis using antibodies that are specific against the protein expressed from the introduced cDNA or gene. Keep in mind that the bottom line of gene expression is the protein or the product of interest. We first choose western blot analysis[12] because it is faster, less expensive, and a more reliable procedure compared with Southern and northern blot analyses. We have found that a majority of nonputative clones can be ignored after western blot analysis. Once the putative clones have been identified to express the protein of interest, one can go back to Southern or northern blot analysis. Therefore, this can be considered the reverse sequence to traditional characterization.

Analysis of Gene Overexpression at the Protein Level by Western Blotting

The detailed protocols for protein extraction and western blot hybridization are described in Chapter 11. This section focuses primarily on how to analyze western blot data. Assuming that the product or protein of the cDNA is 50 kDa and that monoclonal antibodies have been raised to be against the 50-kDa protein (Figure 8.10) in control cells (lane 1), only a weak 50-kDa band is detected. This is the endogenous level of the protein in the cells. In contrast, putative clones show overexpression of the 50-kDa protein (lanes 2 to 4) compared with the control cells.

Examination of the Expression of Sense RNA by Northern Blotting

The detailed protocols for RNA isolation and northern blot hybridization are described in Chapter 10. The major focus of this section is on how to analyze northern blot data. Assuming that the mRNA of the gene is 1 kb and that specific oligonucleotides or the cDNA is labeled as the probe (Figure 8.11), in control cells (lane 1), only a weak 1-kb band is detected. This is the endogenous mRNA level in the cells. In contrast, putative clones show overexpression of the mRNA (lanes 2 to 4) compared with the control cells.

Determination of Integration Copy Number by Southern Blot Analysis

Isolation of Genomic DNA from Cultured Cells

1. Aspirate the culture medium from cultured cells and rinse the cells twice in PBS.
2. Add 10 ml/60-mm culture plate (with approximately 10^5 cells) of DNA isolation buffer to the cells and allow the cells to be lysed for 5 min at room temperature with shaking at 60 rpm.

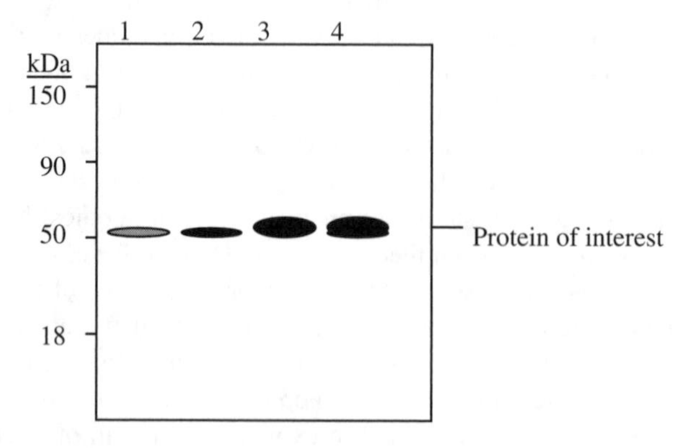

FIGURE 8.10 Protein analysis by western blotting. Lane 1: nontransfected cells. Lane 2 to lane 4: overexpression of the 50-kDa protein in different cell clones.

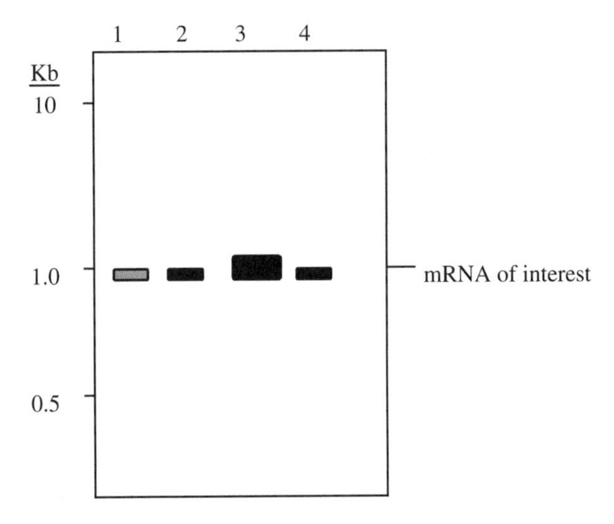

FIGURE 8.11 mRNA analysis by northern blotting. Lane 1: nontransfected cells. Lane 2 to lane 4: overexpression of the 1-kb mRNA in different transfected cell clones.

3. Incubate the plate at 37°C for 5 h or overnight to degrade proteins and extract the genomic DNA.
4. Allow the lysate to cool to room temperature and add three volumes of 100% ethanol (precold at –20°C) to the lysate. Gently mix it and allow the DNA to be precipitated at room temperature with slow shaking at 60 rpm. Precipitated DNA, like white fibers, should be visible in 20 min.
5. "Fish" out the DNA using a sterile glass hook or spin it down. Rinse the DNA twice in 70% ethanol and then partially dry the it for 40 to 60 min at room temperature to evaporate the ethanol. The DNA can be directly subjected to restriction enzyme digestion without being dissolved in TE buffer or dd.H_2O. We usually overlay the precipitated DNA with an appropriate volume of restriction enzyme digestion cocktail and incubate the mixture at the appropriate temperature for 12 to 24 h. The digested DNA is mixed with an appropriate volume of DNA loading buffer and is now ready for agarose gel electrophoresis.

DNA Isolation Buffer

75 m*M* Tris-HCl, pH 6.8
100 m*M* NaCl
1 mg/ml Protease K
0.1% (w/v) *N*-Lauroylsarcosine

Analysis of Southern Blot Data

Chapter 7 describes the detailed protocols for Southern blot hybridization. Assuming that genomic DNA is digested with a unique restriction enzyme that cuts outside the endogenous gene or cDNA introduced, and that the cDNA is labeled as the probe for hybridization, as shown in Figure 8.8 to Figure 8.12, lane 1 in the control cells contains one band (5 kb) of the gene. In lane 2, a putative clone carries one copy insertion of the cDNA introduced. In lane 3, a putative clone shows two copies of integration of the cDNA.

EXPRESSION ASSAY OF A REPORTER GENE

Activity Assay of Chloramphenical Acetyl Transferase (CAT)

Protocol 1. Preparation of a Cytoplasmic Extract from Transfected Cells for the CAT Assay

1. For transient transfection, approximately 48 to 72 h after transfection, remove the culture medium from the dish and wash the cells twice with 10 ml of PBS. For stable transfection, proteins can be extracted from stably transfected cells.
2. Aspirate the PBS buffer and add 1 ml of fresh PBS. Loosen and resuspend the cells into the buffer using a sterile rubber policeman.
3. Transfer the cell suspension into a fresh microcentrifuge tube, centrifuge at 1200 × *g* for 5 min and decant the supernatant.

Note: Do not centrifuge at top speed. Otherwise, the cells may rupture.

4. Resuspend the cells in 0.3 ml of 250 m*M* Tris-HCl, pH 8.0.
5. Lyse the cells by three to four cycles of freezing (dry ice or dry ice–methanol bath, 5 min) and thawing (37°C, 5 to 10 min).
6. Centrifuge at 10,000 × *g* for 10 min in a microcentrifuge and transfer the supernatant (extract) to a fresh tube. Store the extract at –70°C prior to use.

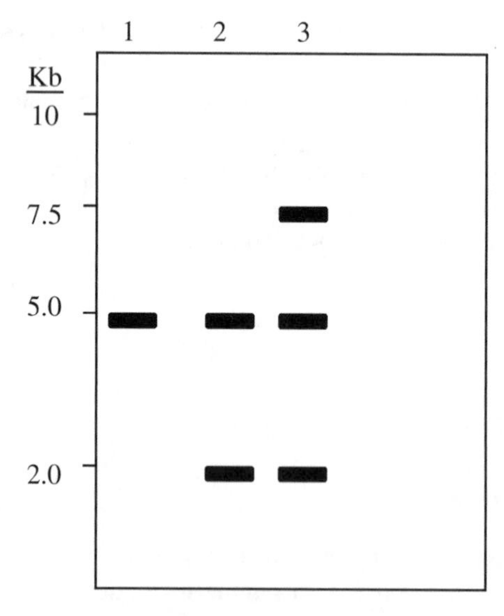

FIGURE 8.12 DNA analysis by Southern blotting. Lane 1: nontransfected cells. Lane 2 and lane 3: putatively transfected cell lines.

Protocol 2. Chloramphenicol Acetyl Transferase (CAT) Enzyme Assay

The CAT enzyme, expressed by the CAT gene, originally comes from bacteria. It catalyzes the conversion of [^{14}C]chloramphenicol to its 1-acetyl and 3-acetyl derivatives. The derived products are separated from the unconverted compound by thin-layer chromatography (TLC) on plates precoated with silica gel, which are then exposed to x-ray film for autoradiography.

1. Add the following components to a microcentrifuge tube:
   Prepared cell extract, 0.1 ml
   500 m*M* Tris-HCl, pH 8.0, 60 µl
   [^{14}C]chloramphenicol (0.025 mCi/ml), 10 µl
   n-Butyryl coenzyme A (5 mg/ml), 10 µl
   Total volume of 180 µl

   Note: A standard curve for CAT activity includes one blank with the CAT enzyme.
2. Incubate reaction at 37°C for the optimum time period (0.5 to 20 h) as determined from prior experiments.
3. Terminate the reaction by adding 0.5 ml of ethyl acetate to the tube for the TLC assay and vortex for 1 min.
4. Centrifuge at 10,000 × *g* for 4 min and transfer the upper, organic phase to a fresh tube and dry down the ethyl acetate in a vacuum evaporator.
5. Resuspend the residue in 20 µl of ethyl acetate and spot 10 µl of each sample onto a silica gel TLC plate and air-dry.
6. Place the TLC plate in a tank containing 100 ml of solvent (chloroform:methanol, 95:5). Once the solvents migrate up to about 1 cm from the top of the plate, remove the plate and air-dry.
7. Expose the plate to x-ray film for 10 h or overnight at room temperature. In order to carry out a quantitative assay, for each sample, slice a square corresponding to the monoacetylated form from the silica plate and place in a vial. Add 5 to 10 ml of scintillation fluid to the vial and count in a scintillation spectrometer.

Luciferase Assay

Luciferase is encoded by the luciferase gene. Its assay is based on an oxidation reaction mediated by luciferyl-CoA and light is produced as a result of this reaction.

1. For each sample, add 30 µl of the cell extract as described previously to a vial containing 150 µl of luciferase assay reagent at room temperature.
2. Quickly place the reaction mixture in a luminometer and measure light produced for a period of 10 s over a 2- to 3-min period, depending on the sensitivity.

Note: The luciferase activity decreases very rapidly with time.

Reagents Needed

Luciferase Assay Reagents

20 m*M* Tricine
1.07 m*M* $(MgCO_3)_4$ $Mg(OH)_2$ $5H_2O$
0.1 m*M* EDTA, pH 7.8
2.67 m*M* $MgSO_4$
33.3 m*M* DTT
0.27 m*M* Coenzyme A
0.47 m*M* Luciferin
0.53 m*M* ATP
Adjust the pH to 7.8.

β-Galactosidase Assay

1. For each sample, add the following components to a microcentrifuge tube:
   Magnesium solution, 3 µl
   CPRC (4 mg/ml), 66 µl
   0.1 *M* Sodium phosphate buffer, 200 µl
   Cell extract, 31 µl
   Total volume of 300 µl

   For blank:
   Magnesium solution, 3 µl
   CPRC (4 mg/ml), 66 µl
   0.1 *M* Sodium phosphate buffer, 200 µl
   0.25 *M* Tris-HCl, pH 8.0, 31 µl
   Total volume of 300 µl
2. Incubate at 37°C for 30 to 60 min and add 0.7 ml of 1 *M* of Na_2CO_3 to terminate the reaction.
3. Measure the absorbency at $A_{574\ nm}$ against blank reference.

Reagents Needed

Magnesium Solution

0.1 *M* $MgCl_2$
5 *M* 2-Mercaptoethanol

0.1 M Sodium Phosphate Buffer, pH 7.3

CPRG Solution

4 mg/ml Chlorophenol red-β-D-galactopyranoside in 0.1 *M* sodium phosphate buffer

β-Galactosidase Staining of Cells

This is a fast procedure to assay the activity of the *lac Z* gene if it is incorporated into the DNA construct. β-galactosidase can hydrolyze X-gal (5-bromo-4-chloro-3-

indolyl-β-D-galactoside) and yield a blue precipitate, staining the cells blue. Protocol is as follows:

1. Culture the stably transfected cells in a Petri dish for 1 to 2 days.
2. Aspirate the medium and rinse the cells with PBS.
3. Fix the cells in 0.2% glutaraldehyde or 4% paraformaldehyde (PFA) for 5 to 10 min at room temperature.
4. Wash the cells for 3 to 5 min.
5. Stain the cells with X-gal solution for 1 to 24 h at 37°C.
6. Wash the cells for 3 to 5 min in PBS. Any blue cells or colonies indicate that the cells express β-galactosidase and that they most likely contain the targeted gene.

Reagents Needed

Potassium Phosphate Buffer (PBS)

2.7 m*M* KCl
1.5 m*M* KH_2PO_4
135 m*M* NaCl
15 m*M* Na_2HPO_4
Adjust pH to 7.2, then autoclave.

Preparation of 4% (w/v) Paraformaldehyde (PFA) Fixative (1 l)

a. Carefully weigh out 40 g paraformaldehyde powder and add it to 600 ml preheated dd.H_2O (55 to 60°C) and allow the paraformaldehyde to dissolve at the same temperature with stirring.

Caution: *Paraformaldehyde is toxic and should be handled carefully. The preparation should be carried out in a fume hood on a heating plate. The temperature should not exceed 65°C and the fixative should not be overheated.*

b. Add 1 drop of 2 *N* NaOH solution to clear the fixative.
c. Remove from heat source and add 333.3 ml (1/3 total volume) of 3X PBS (8.1 m*M* KCl, 4.5 m*M* KH_2PO_4, 411 m*M* NaCl, 45 m*M* Na_2HPO_4, pH 7.2, autoclaved).
d. Adjust the pH to 7.2 with 4 *N* HCl and bring the final volume to 1 l with dd.H_2O.
e. Filter the solution to remove undissolved fine particles, cool to room temperature and store at 4°C until use.

X-Gal Solution

PBS solution contains 0.5 mg/ml X-gal and 10 m*M* $K_3[Fe(CN)_6]$. X-gal stock solution contains 50 mg/ml in dimethylsulfoxide or dimethylformamide. The stock solution should be kept in a brown bottle at –20°C prior to use.

GENERATION OF TRANSGENIC MICE FROM SENSE ES CLONES

Detailed protocols are described in Chapter 9.

CHARACTERIZATION OF TRANSGENIC MICE

Detailed protocols are given in Chapter 9.

TROUBLESHOOTING GUIDE

1. **Low efficiency of transfection of cells by liposomes, electroporation and calcium phosphate methods.** If cells are normal and the cell density is appropriate, it is likely that the amount of DNA constructs used and parameters applied in transfection are not optimal. Make sure to optimize the conditions for transfection of the cells, including DNA concentration, lipids ratio, and voltage pulse. In case of stable transfection, ensure that the concentration of the drug used in the selection is not too high. A killing curve should be established.
2. **Low titer of retroviruses.** The virus stock is overdiluted or there is a low density of retrovirus particles due to low efficiency of stable transfection of packaging cells with modified virus DNA constructs. Try to use a good packaging cell line and optimize the parameters for transfection. Reinfection may be performed to amplify the virus titer.
3. **Low efficiency of transfection with frozen high-titer viruses.** Assuming transfection conditions are optimal, it is very likely that the viruses have lost significant activity. It is recommended that high-titer viruses be used in a couple of hours. Frozen aliquots should not be frozen and thawed.
4. **Southern blot analysis shows that sense cDNA is correctly integrated into the genome of the host cell, but no exogenous mRNA and protein are expressed.** Something is wrong with the inserted cDNA. This could be due to a minor mutation that occurs during integration. Once this occurs, the best way to deal with it is to forget it. However, if verifying the cell clone is desirable, a PCR approach may be used to sequence the insertion site, especially the 5′ end of the cDNA introduced. If some bases are missing or ORF has been shifted, there is no way to rescue the cell clone.
5. **Low-level expression of mRNA and protein of interest.** This is obviously related to the activity of the promoter driving the expression of the introduced cDNA. If it is a constitutive promoter, its activity is low. Try to switch to other, stronger promoters. If the driving promoter is inducible, try to optimize the concentration of the appropriate inducer.
6. **Some of drug selected stably transfected cell clones show 100% death during subsequent cloning or culture in the selection medium.** If these clones are not frozen, it is certain that these isolated clones are false stably

transfected cells. It is likely that during trypsinization and resuspension, the cells were not well suspended; instead, they aggregated. After transfection, some of the cells within the aggregated colonies receive fewer drugs and still undergo proliferation, forming false colonies that are picked up in the process. During the trypsinizing and picking up, these cells become loosened by physical force. Because they are not particularly stable colonies, they are killed by drug selection medium during the subsequent expanding or culture.

7. **Very few frozen, stably transfected cells survive during subsequent culture.** Certainly, this is due to inappropriate freezing. Make sure to use 5 to 10% (v/v) DMSO in the freezing medium.

REFERENCES

1. Wu, W. and Welsh, M.J., Expression of HSP27 correlates with resistance to metal toxicity in mouse embryonic stem cells transfected with sense or antisense HSP27 cDNA, *Appl. Toxicol. Pharmacol.,* 141, 330, 1996.
2. Mehlen, P., Preville, X., Chareyron, P., Briolay, J., Klemenz, R., and Arrigo, A-P., Constitutive expression of human hsp27, *Drosophila* hsp27, or human αB-crystallin confers resistance to TNF- and oxidative stress-induced cytotoxicity in stably transfected murine L929 fibroblasts, *J. Immunol.*, 363–374, 1995.
3. Iwaki, T., Iwaki, A., Tateishi, J., and Goldman, J.E., Sense and antisense modification of glial αB-crystallin production results in alterations of stress fiber formation and thermoresistance, *J. Cell Biol.*, 15, 1385–1393, 1994.
4. Yang, N.S., Burkholder, J., Roberts, B., Martinell, B., and McCabe, D., *In vitro* and *in vivo* gene transfer to mammalian somatic cells by particle bombardment, *Proc. Natl. Acad. Sci., USA,* 87, 9568, 1990.
5. Carmeliet, P., Schoonjans, L., Kiechens, L., Ream, B., Degen, J., Bronson, R., Vos, R.D., Oord, J.J., Collen, D., and Mulligan, R.C., Physiological consequences of loss of plasminogen activator gene function in mice, *Nature,* 368, 419, 1994.
6. Zelenin, A.V., Titomirov, A.V., and Kolesnikov, V.A., Genetic transformation of mouse cultured cells with the help of high-velocity mechanical DNA injection, *FEBS Lett.,* 244, 65, 1989.
7. Johnston, S.A., Biolistic transformation: microbes to mice, *Nature,* 346, 776, 1990.
8. Moore, K.A. and Belmont, J.W., Analysis of gene transfer in bone marrow stem cells, in *Gene Targeting: A Practical Approach,* Joyner, A.L., Ed., IRL Press, Oxford University Press, Inc., New York, 1993.
9. Hasty, P. and Bradley, A., Gene targeting vectors for mammalian cells, in *Gene Targeting: A Practical Approach,* Joyner, A.L., Ed., IRL Press, Oxford University Press, Inc., New York, 1993.
10. Papaioannou, V. and Johnson, R., Production of chimeras and genetically defined offspring from targeted ES cells, in *Gene Targeting: A Practical Approach,* Joyner, A.L., Ed., IRL Press, Oxford University Press, Inc., New York, 1993.
11. Gossler, A. and Zachgo, J., Gene and enhancer trap screens in ES cell chimeras, in *Gene Targeting: A Practical Approach,* Joyner, A.L., Ed., IRL Press, Oxford University Press, Inc., New York, 1993.

12. Wu, W., Gene transfer and expression in animals, in *Handbook of Molecular and Cellular Methods in Biology and Medicine,* Kaufman, P.B., Wu, W., Kim, D., and Cseke, L.J., CRC Press, Boca Raton, FL, 1995.

9 Gene Underexpression in Cultured Cells and Animals by Antisense DNA and RNA Strategies

CONTENTS

INTRODUCTION

In order to gain insight into the function of the protein product of a gene of interest, antisense DNA and RNA approaches have been widely applied in eukaryotes in which the antisense oligonucleotides or the antisense RNA transcripts expressed successfully inhibit the expression of specifically targeted sense RNA or mRNA.[1–6] In recent years, the antisense approach has been used to address fundamental questions in molecular and cellular biology in animals, human diseases and plants.[1,3,6–10] Gene underexpression refers to a decrease in the protein level of a gene, which in turn may alter the function of the gene.

Although the precise mechanism by which antisense DNA–RNA inhibits gene expression is not well known, it is generally assumed that antisense oligonucleotides or antisense RNA can interact with complementary sequence or sense RNA transcripts by base pairing, thus blocking the processing or translation of the sense RNA or mRNA. The duplex of antisense RNA and sense RNA may be rapidly degraded, resulting in a decrease in specific mRNA levels. There is no evidence, however, that the antisense RNA can directly affect the target gene or DNA sequence. Therefore, the primary principle of using antisense DNA and antisense RNA is to inhibit the expression of the target gene or mRNA level partially or completely.

How is a gene of interest underexpressed? In general, gene underexpression by antisense DNA or RNA approaches consists of highly involved techniques that include design of antisense oligonucleotides, isolation and characterization of a specific cDNA, gene transfer and analysis of gene expression in transfected cells or transgenic animals. The present chapter will focus on techniques using exogenous antisense oligonucleotides and antisense cDNA constructs to down-regulate the expression of a specific gene. These protocols have been successfully utilized in our laboratories.

ANTISENSE OLIGONUCLEOTIDE APPROACHES

DESIGN AND SYNTHESIS OF ANTISENSE OLIGONUCLEOTIDES

Antisense oligomers can be chemically synthesized based on any known sequence of DNA or RNA of interest. If the DNA sequence is not known, a specific peptide sequence is available. Antisense oligonucleotides can be designed based on the amino acid sequence of the peptide.

1. Design appropriate antisense oligomers based on the target, known sequence of the mRNA (sense RNA) or its first-strand cDNA, which is the poly(T) or (–)strand. For example, if the target sequence is 5′-ACAUGCCCCUCAACGUUAGC-3′ (mRNA) the antisense oligonucleotide should be designed as 3′-TGTACGGGGAGTTGCAATCG-5′. The oligo is complementary to the target sequence and both can form a hybrid as follows:

5′-ACAUGCCCCUCAACGUUAGC-3′ mRNA

3′-TGTACGGGGAGTTGCAATCG-5′ antisense oligomer

If, on the other hand, the amino acid sequence is known, the nucleotide sequence can be derived:

N----Pro-Ser-His-Gly-Arg-Ser-Pro----C amino acid sequence

↓

5′-CUUUCGCACGGUAGAUCGCCA-3′ mRNA sequence

↓

3′-GAAAGCGTGCCATCTAGCGGT-5′ antisense oligonucleotide

Tips: Target regions: *(a) The upstream nucleotide sequence from the initiation codon AUG and the first codon AUG should be selected as the most effective target region when designing an antisense oligomer. The hybridization of the antisense oligomers may prevent the small subunit of the ribosome from interacting with the upstream region of the initiating AUG codon, thus blocking the initiation of translation. (b) Antisense oligonucleotides, such as conserved motifs to block the chain elongation of the specific polypeptide, can also be designed to be complementary to the coding regions. However, this will be less effective than (a). (c) The terminal region of translation is a good target to block the termination of translation. The partial-length polypeptide may not be folded properly and will be rapidly degraded. (d) The most effective strategy for blocking translation of an mRNA is to use antisense oligonucleotides complementary to the upstream region, including the first codon AUG, and to the terminal site of the coding region. The efficiency of inhibition of specific gene expression can be up to 95%. (e) Unlike the design of primers for PCR that involves the Tm value, the oligomers for inhibition of gene expression do not need computer analysis of Tm.* ***Oligomer modifications:*** *Conventional or unmodified oligonucleotides may be substrates for nucleases (DNases) such as RNase H and cannot effectively block the expression of a specific gene. To prevent oligonucleotides from being degraded by nucleases, the antisense oligomers may be chemically modified prior to use. Active compounds such as cross-linking and cleaving reagents can be coupled with the oligomers so that they will be much more stable and cause irreversible damage to the target sequence. Subsequently, this will result in very effective blocking of translation. For example, oligomers may be conjugated to photosensitizers. Antisense oligomer linked to nuclease or to chemical reagents will produce sequence-dependent cleavage of the target RNA following the binding of the oligomer to its complementary sequence. The cleavage site in the target RNA is usually in the area surrounding the modified antisense oligomer. This will prevent the translation of the target RNA. Chemical groups linked to the end of the oligomers can introduce strand breaks directly (e.g., metal chelates and ellipticine) or indirectly (e.g., photosensitizers) after alkaline treatment. Metal complexes such as Fe^{2+}-EDTA, Cu^{+}-phenanthroline and Fe^{2+}-porphyrin can also produce hydroxyl radicals (OH) that can oxidize* *the sugar and subsequently result in phosphodiester bond cleavage. Phosphodiester internucleoside linkages may be substituted with phosphotriester,[3–5] methylphosphonate, phosphorothioate or phosphoroselenoate. Oligomers that are modified to phosphorothioates and phosphoroselenoates are resistant to nuclease (RNase H) and can induce cleavage of the target RNA. For this reason, phosphorothioate analogs have been widely used in cultured cells.* ***Oligomer size:*** *15- to 30-mer, depending on the particular nucleotide sequences.* ***Oligomer dose:*** *10 to 100* μM *in total, depending on the particular experiment.*

2. Based on the design, synthesize and purify appropriate antisense oligonucleotides by a DNA synthesizer according to the instructions.

Treatment of Cultured Cells with Antisense Oligomers and Determination of Optimum Dose of Oligomers by Western Blot Analysis

1. Under sterile conditions, culture the cells of interest in prewarmed, appropriate medium such as modified DMEM containing 10 to 15% (v/v) heat-inactivated fetal bovine serum (FBS) or 5 to 10% (v/v) heat-inactivated newborn calf serum (NCS) with appropriate antibiotics such as 50 μg/ml of penicillin and 100 μg/ml streptomycin. Culture the cells at 37°C in 5% CO_2 and 95% air incubator. Allow cells to grow to 70 to 80% confluence.
2. Trypsinize (for attaching cell lines) and spin down the cells at 1000 rpm for 5 min under sterile conditions. In a sterile laminar flow hood, carefully aspirate the supernatant and resuspend the cells in a minimum volume of fresh medium. Determine the viability and density of the cells by counting Trypan blue, excluding cells using a hemocytometer.
3. Perform a dose-response experiment using the synthetic antisense oligomers in order to optimize the amount of the oligomers to use in order to block the expression of the gene of interest efficiently as follows:
   a. Seed 2×10^6 cells in each well of 24-well culture plates or T-flasks containing 2 to 5 ml of culture medium. Prepare seven wells or T25 flasks with equal density of cells.
   b. Add the oligomers to the medium as follows:
   No treatment: no antisense oligomer used as a negative control
   Treatment #1: 6 μ*M* antisense oligomers
   Treatment #2: 4 μ*M* antisense oligomers
   Treatment #3: 2 μ*M* antisense oligomers
   Treatment #4: 1 μ*M* antisense oligomers
   Treatment #5: 0.5 μ*M* antisense oligomers
   Treatment #6: 0.25 μ*M* antisense oligomers
   Note: The initial concentration of the appropriate oligomers varies with a particular experiment.
   c. Incubate the cells for 2 to 6 days, adding the same concentration of the oligomers to the cells every day and gently mixing in the medium.

d. Harvest the cells for protein analysis. If the protein of interest is a secretory protein, the supernatant should be used for extraction of the proteins. If not, the pellet of cells from each treatment should be washed with an appropriate buffer, lysed and extracted for protein analysis. The detailed protocols for protein extraction from cultured cells are described in Chapter 11. The protein of interest may be purified from the total proteins with specific antibodies against the protein by the immunoprecipitation method. We recommend using total proteins for inhibition analysis with the antibodies against an appropriate housekeeping gene product as a normalization control.

e. Analyze the inhibition of protein expression by western blotting. The detailed protocol for western blot hybridization is described in Chapter 11. The major task here is to focus on analysis of western blot data, which are diagrammed in Figure 9.1. Assuming that the protein to be inhibited is 60 kDa and that monoclonal antibodies are used in western blotting, as shown in the diagram, the oligomer at 4 μ*M* is the optimal concentration that sufficiently inhibits the expression of the protein. Antibodies against actin or tubulin should be used as controls. The detected signals for actin in each of the samples are very similar, indicating that the amount of total protein from each of the cell extracts is equally loaded into SDS-PAGE.

Treatment of Cultured Cells Using an Optimum Dose of Oligomers

1. Seed 2×10^6 cells in each T-flask containing 5 ml of culture medium. Prepare three replicates for each of the samples.
2. Add the optimum dose of antisense oligomers (4 μ*M*) to the medium for each of the samples and culture for appropriate periods of time as follows:
   Sample #1: no oligomer used as a control for 7 days
   Sample #2: 4 μ*M* antisense oligomers for 0.5 days
   Sample #3: 4 μ*M* antisense oligomers for 1.5 days
   Sample #4: 4 μ*M* antisense oligomers for 2 days

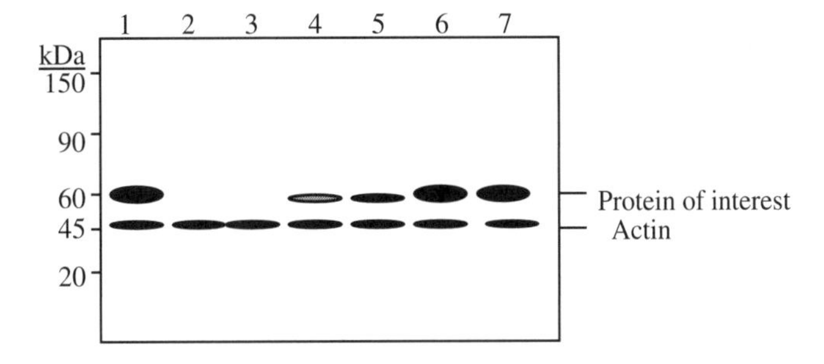

FIGURE 9.1 Protein analysis by western blotting after antisense oligonucleotide treatment. Lane 1: control cells. Lane 2 to lane 7: underexpression of the 60-kDa protein after treatment of oligomers at 6, 4, 2, 1, 0.5 and 0.25 μ*M*, respectively.

Sample #5: 4 μ*M* antisense oligomers for 2.5 days
Sample #6: 4 μ*M* antisense oligomers for 3 days
Sample #7: 4 μ*M* antisense oligomers for 3.5 days
Sample #8: 4 μ*M* antisense oligomers for 4 days
Sample #9: 4 μ*M* antisense oligomers for 5 days
Sample #10: 4 μ*M* antisense oligomers for 6 days
Sample #11: 4 μ*M* antisense oligomers for 7 days

3. Incubate the cells for 6 days, adding the same amount of the oligomers to the cells every day and gently mixing in the medium.

Analysis of Inhibition of Gene Expression by Western Blotting

1. Harvest the cells at step 3 in the preceding section for protein analysis. If the protein of interest is a secretory protein, the supernatant should be used for extraction of proteins. If not, the pelleted cells from each treatment should be washed with an appropriate buffer, lysed and extracted for protein analysis. The detailed protocols for protein extraction from cultured cells are described in Chapter 11. The protein of interest may be purified from the total proteins with specific antibodies against the protein by the immunoprecipitation method. We recommend using total proteins for the inhibition analysis with the antibodies against an appropriate housekeeping gene product as a normalization control.
2. Analyze the inhibition of protein expression by western blotting. The detailed protocol for western blot hybridization is described in Chapter 11. The major task here is to focus on analysis of western blot data, an example of which is diagrammed in Figure 9.2. Assuming that the protein to be inhibited is 60 kDa and that monoclonal antibodies are used in the western blotting, as shown in the figure, the level of the 60-kDa protein gradually decreases during the course of oligomer treatment. Antibodies against actin are used as the control. The detected signals for actin in each of the samples are very similar, indicating that the amount of total protein from each of the cell extracts is equally loaded into the SDS-PAGE.

DESIGN AND SELECTION OF PLASMID-BASED EXPRESSION VECTORS

Use of antisense oligonucleotides, described earlier, provides an effective and fast strategy for down-regulation of the expression of a protein of interest. Nonetheless, this approach gives very temporary underexpression. To inhibit the expression of a specific gene stably, a genetic approach is necessary. In fact, to use stably transfected cell lines or to make transgenic animals to study the function of a gene, it may be necessary to choose the strategy of stable genetic alteration of the expression of the gene of interest. Detailed descriptions for the design and selection of eukaryotic expression vectors are given in Chapter 8.

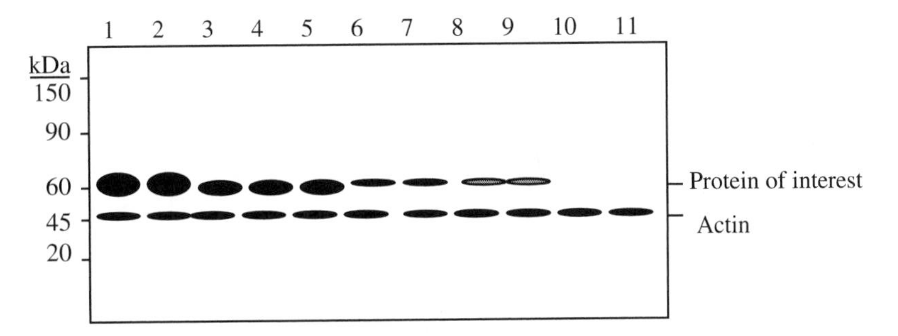

FIGURE 9.2 Protein analysis by western blotting after antisense oligonucleotide treatment. Lane 1: control cells. Lane 2 to lane 11: time course of underexpression of the 60-kDa protein after treatment of oligomers at 4 μ*M* for 0.5, 1, 1.5, 2, 2.5, 3, 3.5, 4, 5, 6, and 7 days, respectively.

PREPARATION OF PLASMID ANTISENSE cDNA CONSTRUCTS

Chapter 3 describes cDNA isolation and characterization. Restriction enzyme digestion of DNA, ligation, bacterial transformation, and purification of plasmid DNA are given in Chapter 4. This section emphasizes special notes regarding the preparation of antisense cDNA constructs.

1. **Length of cDNA to be cloned:** Unlike the gene overexpression described in Chapter 8, underexpression of a gene of interest does not need full-length cDNA, nor is the entire ORF required. In our experience, the 5′UTR and the 5′-portion of the ORF have high efficiencies of translational inhibition.
2. **Orientation of the cDNA:** It is important that the cDNA be placed in the antisense orientation downstream from the driving promoter. In other words, the poly(A) strand, or (+)strand of the cDNA, should be 3′ → 5′ downstream from the promoter in order to make 5′ → 3′ poly(T) RNA or antisense RNA (Figure 9.3).
3. **Start or stop codon:** Unlike gene overexpression in Chapter 8, the presence of a stop or start codon between the driving promoter and the first-start codon ATG of the cDNA is acceptable. Remember, antisense RNA will primarily hybridize with sense RNA. As a result, mRNA will undergo degradation or the translation process is blocked.

TRANSIENT TRANSFECTION OF CULTURED CELLS WITH ANTISENSE CONSTRUCTS

Methods and detailed protocols are described under Transient Transfection in Chapter 8.

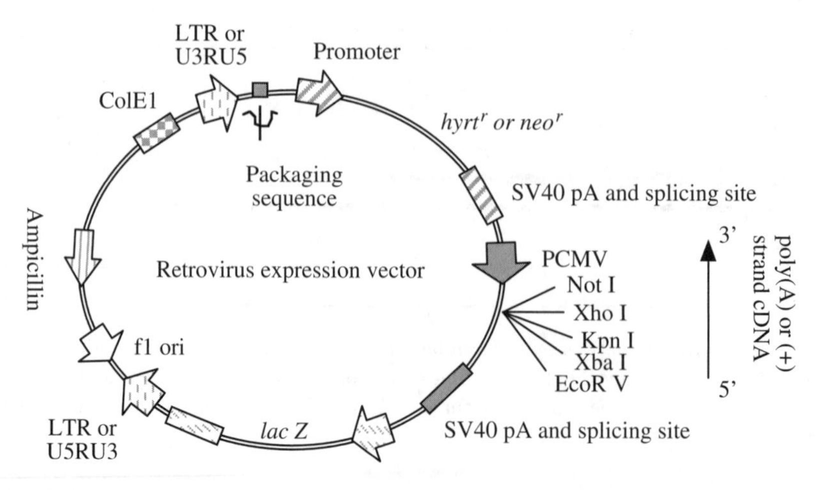

FIGURE 9.3 Diagram of a retrovirus expression vector.

STABLE TRANSFECTION OF CULTURED CELLS WITH ANTISENSE DNA CONSTRUCTS

Method A. Transfection by Liposomes

Liposome-mediated transfection is relatively safe and has a high efficiency of transfection in cultured cells. It is suitable for both transient and stable transfection of tissues or cells. It mandates smaller amounts of DNA for transfection compared with electroporation and other methodologies. Lipofectin reagents are commercially available (e.g., Gibco BRL Life Technologies). Liposomes are composed of polycationic lipids and neutral lipids at an appropriate ratio (e.g., 3:1). The primary principles of this method are: (1) the positively charged and neutral lipids can form liposomes that can complex with negatively charged DNA constructs; (2) the DNA–liposome complexes are applied to cultured cells and are uptaken by endocytosis; (3) the endosomes undergo breakage of membranes, releasing the DNA constructs; and (4) the DNA enters the nucleus through the nuclear pores and facilitates homologous recombination or integration into the chromosomes of the cell.

The primary disadvantage of the lipo-transfection method is that if the ratio of lipids–DNA and the cell density are not appropriate, severe toxicity will be generated, killing quite large numbers of cells. Embryonic stem (ES) cells are more sensitive to toxicity compared with other fibroblasts such as NIH 3T3 and mouse STO cells.

1. Harvest cells by gentle trypsinization as described in Chapter 16. Centrifuge the cells at 900 rpm for 5 min and resuspend the cells well in normal culture medium at approximately 10^6 cells/ml.

Note: Make sure that the cells are suspended gently but well prior to transfection because aggregation of the cells may cause difficulty in the isolation of stably transfected cell clones.

2. In a 60-mm tissue culture dish, seed the suspended cells at about 1×10^6 cells in 5 ml of normal culture medium. Allow the cells to grow under normal conditions for 12 h. Proceed to transfection.

Note: Mouse ES cells grow fast and form multiple layers of clones instead of the monolayers seen in other cell lines. Because of these features, it is not recommended to overculture ES cells prior to transfection because multilayers of ES cells may result in a low efficiency of transfection and a high percentage of potential cell death. Therefore, once a relatively high density of cells has been obtained, transfection can be carried out as soon as the ES cells attach to the bottom of the dish and grow for 6 to 12 h.

3. While the cells are growing, DNA constructs can be linearized to be transferred by the use of a unique restriction enzyme. Protocols for appropriate restriction enzyme digestion are described in Subcloning of Gene and DNA Fragments in Chapter 4.
4. In a sterile laminar flow hood, dissolve 10 to 20 μg DNA in 0.1 ml of serum-free medium (free of antibacterial agents) and dilute the lipofectin reagents two- to fivefold in 0.1 ml of serum-free medium (free of antibacterial agents). Combine and gently mix the DNA–lipids media in a sterile tube. Allow this mixture to incubate at room temperature for 15 to 30 min. A cloudy appearance of the medium indicates the formation of DNA–liposome complexes.
5. Rinse the cells once with 10 ml of serum-free medium. Dilute the DNA–liposome solution to 1 ml in serum-free medium (free of antibacterial agents) and gently overlay the cells with the diluted solution. Incubate the cells for 4 to 24 h under normal culture conditions. During this period, transfection takes place.
6. Replace the medium with 5 ml of fresh normal culture medium with serum every 12 to 24 h. Allow the cells to grow for 2 to 3 days before drug selection.
7. Trypsinize the cells and plate them at a density of about 3000 cells/100-mm dish or T25 flask culture for 24 h at 37°C with 5% CO_2. Proceed to drug selection.

Method B. Transfection by Microinjection

Microinjection is one of the highest efficiencies of transfection, although the number of cells that can be injected within a given time is usually limited, depending on personal experience. The equipment for microinjection should include: (1) an injection microscope on a vibration-free table (Diaphot TMD microscope with Nomarski optics is available from Nikon Ltd); (2) an inverted microscope for setting up the injecting chamber and holding the pipette and needle (SMZ-2B binocular microscope is available from Leitz Instruments Co.); (3) two sets of micromanipulators (Leitz Instruments Ltd.); (4) Kopf needle puller model 750 (David Kopf Instruments); (5) a finer needle and a pipette holder; and (6) a Schott-KL-1500 cold light source (Schott).

1. Thoroughly clean a depression slide with teepol-based detergent and extensively rinse it with tap water and distilled water. Rinse it with ethanol and air-dry. Add a drop of culture medium to the depression slide and a drop of liquid paraffin on top of a drop of medium. This is preparation for suspension cells but not fibroblast cells attaching on the bottom of culture dishes or flasks.
2. Assemble all parts necessary for the injection according to the manufacturer's instructions. Carry out pretesting of the procedure prior to injection of the DNA sample.
3. For cultured fibroblast cells attached to the bottom, transfer the cultured cells onto the stage under the microscope. For cell suspensions, carefully transfer the cells to the medium drop using a handling pipette that can take up the cells by suction under the microscope.
4. Prepare microinjection needles with a Kopf needle puller (model 750). Slowly take the DNA sample up into the injection needle by capillary action.
5. Under the microscope, hold the suspended cells with the holding pipette and use the micromanipulator to insert the needle individually into the cell. Slowly inject the DNA sample into the nucleus. The pressure can be maintained on the DNA sample in the needle by a syringe connected to the needle holder. As long as nucleus swelling occurs, immediately remove the needle a little bit away from the cell. The volume injected is approximately 1 to 2 pl. For cultured fibroblast cells attached to the bottom of a culture dish or flask, no holding of the cell with a pipette is needed. Injection can be directly performed into the nucleus of the cell.
6. Transfer the injected cells to one side and repeat injections for other cells. About 40 to 60 cells can be injected in 1 to 2 h, depending on an individual's experience.
7. Carefully transfer the injected cells back to the incubator. Allow the cells to grow for 2 to 3 days prior to harvesting or drug selection.

Method C. Transfection by Electroporation

Detailed protocols are given in Chapter 8.

Method D. Transfection by Retrovirus Vectors

Detailed protocols are described in Chapter 8.

SELECTION OF STABLY TRANSFECTED CELL LINES WITH APPROPRIATE DRUGS

Detailed descriptions are given in Chapter 8.

CHARACTERIZATION OF STABLY TRANSFECTED CELL CLONES

This characterization will provide information about whether or not the developed clones contain and express antisense-induced inhibition in the cells.

Analysis of Gene Underexpression at the Protein Level by Western Blotting

The detailed protocols for protein extraction and western blot hybridization are described in Chapter 11, Analysis of Gene Expression at the Protein Level. The major focus of this section is on how to analyze western blot data. Assuming that the product or protein of the cDNA is 80 kDa and that monoclonal antibodies have been raised to be against the 80-kDa protein (shown in Figure 9.4), a strong 80-kDa band is detected in control cells (lane 1). In contrast, putative antisense clones show a significant reduction of the 80-kDa protein (lane 2 to lane 4) compared with the control cells.

Examination of Expression of Antisense RNA by Northern Blotting

The detailed protocols for RNA isolation and northern blot hybridization are described in Chapter 11. The major focus of this section is on how to analyze northern blot data. Assuming that the antisense RNA of the gene is 0.8 kb and that a specific sense RNA probe is used (shown in Figure 9.5), no band is detected in control cells (lane 1). However, four stably transfected clones clearly show expression of antisense RNA (lane 2 to lane 5).

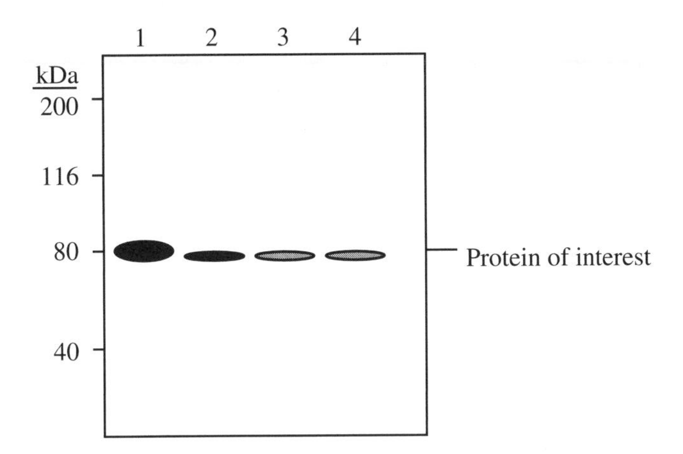

FIGURE 9.4 Protein analysis of antisense clones by western blotting. Lane 1: nontransfected cells. Lane 2 to lane 4: underexpression of the 80-kDa protein in different antisense cell clones.

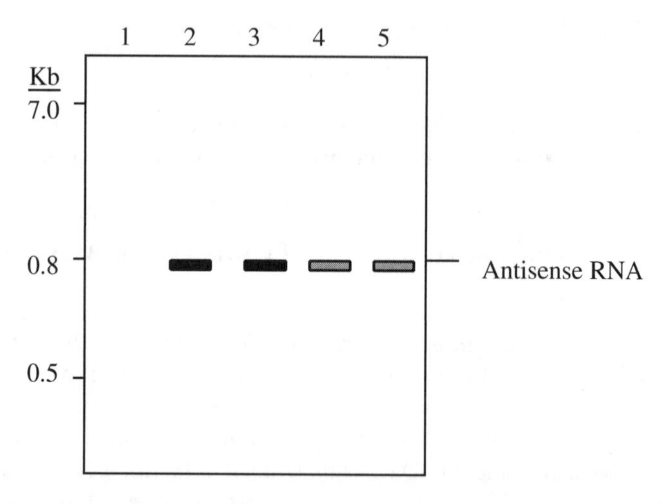

FIGURE 9.5 Antisense RNA analysis by northern blotting using a sense RNA probe. Lane 1: nontransfected cells. Lane 2 to lane 5: expression of antisense RNA (0.8 kb) in four antisense clones.

Determination of Integration Copy Number by Southern Blot Analysis

Chapter 7 describes the detailed protocols for Southern blot hybridization. Assuming that genomic DNA is digested with a unique restriction enzyme that cuts outside the endogenous gene or cDNA introduced and that the cDNA is labeled as the probe for hybridization (shown in Figure 9.6), lane 1 is the control cells and contains one band. Lane 2 and lane 4 have one copy of integration of the introduced cDNA. Lane 3 is a clone that shows two copies of insertion of the cDNA.

Expression Assay of Reporter Genes

Detailed protocols are described in Chapter 8.

GENERATION OF TRANSGENIC MICE

There are two methods for the production of transgenic mice. One is to develop stably transfected mouse ES cells with antisense cDNA constructs; these cells can be injected into female mice to generate genetically altered animals. The other method is to inject antisense cDNA constructs directly into mouse oocytes by microinjection. The injected oocytes are then transferred into the oviduct of a female mouse to produce transgenic mice.

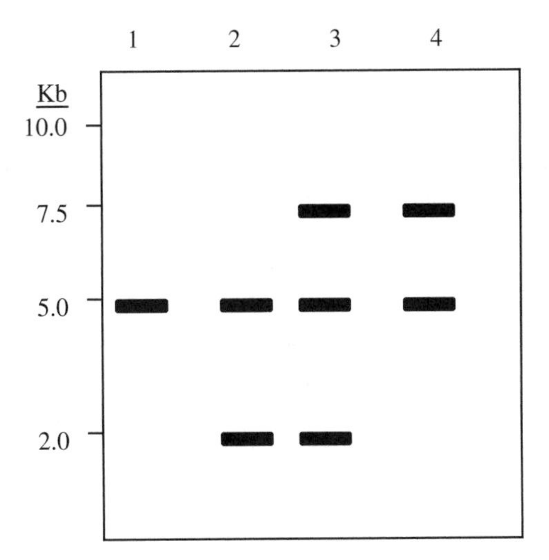

FIGURE 9.6 Antisense DNA analysis by Southern blotting using the introduced cDNA as a probe. Lane 1: nontransfected cells. Lane 2 and lane 4: single-copy integration of antisense cDNA. Lane 3: two copies of insertion of antisense cDNA.

Method A. Production of Transgenic Mice from Stably Transfected ES Cells

Selection of C57BL/6J Estrous Females

The C57BL/6J mouse is a well-established strain widely chosen as the best choice of host embryo for ES cell chimeras and germline transmission. These mice have a coat color different from 129 SVJ mice that can be used as a coat color selection marker. In order to synchronize the estrous cycle relatively and simplify female selection, it is recommended that female mice be caged together without a male. Their estrous cycle is approximately 3 to 4 days and ovulation usually occurs at about the midpoint of the dark period in a light/dark cycle that can be altered to meet a schedule. For example, the cycle may be adjusted to 14 h of light and 10 h of darkness. In this way, most strains of mice produce blastocysts that can be utilized for injection in the late morning of the third or fourth day.

The question here is how to identify pre-estrous or estrous females for mating. The basic criteria are the external vaginal epithelial features. At the pre-estrous stage, the vaginal epithelium is usually characterized by being moist, folded, pink/red and swollen. At the estrous stage, the vaginal epithelium, in general, is dry but not wet, pink but not red or white, wrinkled on upper and lower vaginal lips, and swollen. After the females have been caged for a while and when they reach the prediction of pre-estrous stage, they should be monitored on a daily basis. Once they are at the estrous stage, mating should be allowed to occur.

An alternative way to promote the estrous cycle is through hormonal regulation of ovulation with exogenous hormones — a procedure termed superovulation. Injection of pregnant mice with mare serum gonadotropin (PMSG, Sigma) can stimulate and induce immature follicles to develop to the mature stage. Superovulation can be achieved by further administration of human chorionic gonadotropin (HCG, Sigma). The procedure is outlined as follows:

1. Three days or 72 to 75 h prior to mating, induce superovulation of 10 to 15 female mice (3 to 4 weeks old) by injecting 5 to 10 units PMSG intraperitoneally into each female between 9:00 a.m. and noon on day 1.
2. Two days or 48 to 50 h later, inject 5 to 10 units of HCG intraperitoneally into each female between 9:00 a.m. and noon on day 3. After injection, cage the females and proceed to mate them with sterile males.

Tip: Superovulation will occur about 12 h following HCG administration. The injection time at step (b) will coincide with the midpoint of the dark cycle.

Preparation of a Bank of Sterile Males by Vasectomy

While females undergo their estrous cycle, prepare a stock of sterile males to mate with the estrous female chosen as the embryo transfer host so as to generate pseudopregnancy. The following procedures are recommended to be performed in conditions as sterile as possible.

1. Weigh a mouse and use 0.2 to 25 ml of tribromoethanol to anesthethize it. Swab the abdomen with 70% ethanol using cotton.
2. Lift the skin free of the peritoneum with forceps and carefully make a 1-cm transverse skin incision at approximately 1 cm rostral to the penis. Carefully make another similar incision through the peritoneum.
3. Use blunt forceps and reach laterally into the peritoneum and carefully grasp the testicular fat pad located ventral to the intestine and attached to the testis. Through the incision, gently pull out the pad together with the testis.
4. Distinguish the vas deferens from the corpus epididymis and carefully separate the vas deferens from the mesentery by inserting the closed tips of the iridectomy scissors through the mesentery. Then, open the scissors and leave them at this position to support the isolation of a length of the vas deferens.
5. Insert two pieces of suture under the vas deferens with forceps, tie up the vas deferens and, with the forceps, at two sites approximately 2 cm apart, cut out a piece of the vas deferens (about 1 cm in length) between the two knots.
6. Return the testis to the peritoneum and perform the same procedures on the other side.
7. Suture the peritoneal wall and clip the skin incision with a couple of small wound clips. Leave the vasectomized male for 10 to 14 days to clear the tract of viable sperm and to ensure that he is sterile prior to use for mating.

Materials Needed

Small blunt, curved forceps
Three pairs of watchmaker's forceps
Small sharp scissors
Iridectomy scissors
Note: These surgical instruments can be wrapped in foil and kept in a 120 to 140°C oven for at least 6 hours prior to use.
Surgical gloves
Small wound clips
Tribromoethanol anesthesia
70% Ethanol and cotton
Surgical silk suture and needle

Pairing Estrous Females and Sterile Males

1. Once the estrous females and stud males are ready for mating, cage them together overnight.
2. Check for vaginal plugs in the morning on day 4 and record that this is the first day of pregnancy.

Tips: If mating takes place, a white or yellow vaginal plug composed of male secretions from the vesicular and coagulating glands should be present within 12 h following mating. The plug may be visible at the surface or out of sight deep in the vagina. Therefore, a narrow, clean steel spatula or its equivalent (3 to 4 mm wide) may be used to detect the plug in the vagina, which feels harder than the surrounding soft tissue. Thus, pseudopregnancy has been induced because the males are sterile.

Preparation of Blastocyst-Stage Embryos from Pseudopregnant Mice

Approximately 4 days after the presence of vaginal plugs, the blastocysts usually present in the uterine lumen can be removed by flushing the uterine horns with embryo culture medium. Surgical gloves should be worn during this procedure.

1. Kill the pregnant females on the fourth day (84 to 96 h) of pregnancy by cervical dislocation.
2. Briefly clean the abdomen with 70% ethanol and make a small, transverse incision in the middle area of the abdomen. Remove the cut hairs with alcohol-soaked cotton.
3. At the incision site, tear the skin and pull it away to the front and hind legs, respectively. At this point, the peritoneum should be exposed.
4. Make a large, transverse incision in the peritoneum to expose the abdominal cavity.
5. Identify the reproductive tract by pushing aside the intestines and use blunt forceps to grasp the point of bifurcation of the uterus near the base of the bladder.

6. Carefully lift the uterus with blunt forceps, remove the supporting mesentery on both sides of the uterus, and cut across the cervix at the base.
7. Remove the entire tract by cutting it between the oviduct and the ovary and place it in a sterile Petri dish.
8. Under the low power of a dissecting microscope, carefully remove any mesentery and fat. At the utero–tubal junction site, cut off the oviduct and make an approximately 0.5 cm longitudinal clip at the end of each horn. Finally, cut across the cervix, exposing the entrances to both uterine horns.
9. Rinse the tract with embryo culture medium and place it in a sterile Petri dish. Take up 0.5 ml of embryo culture medium using a small Pasteur pipette or a pipette attached to a sterile 1-ml tip, insert the tip into the cervical end and gently flush the lumen a few times. The uterus should be slightly inflated and the medium containing embryos should flow through the tract.
10. Under the low power of a dissecting microscope, quickly transfer the blastocyst-stage embryos that have settled to the bottom of the flush dish into a micromanipulation chamber or dish using a sterile Pasteur pipette or pipette tip.

Materials Needed

Small blunt, curved forceps
Watchmaker's forceps
Small sharp scissors
Iridectomy scissors
Surgical gloves
70% ethanol and cotton

Embryo Culture Medium (1 l)

$Na_2HPO_4 12H_2O$, 3.0 g
NaCl, 6.2 g
KCl, 0.2 g
KH_2PO_4, 0.2 g
$CaCl_2.2H_2O$, 0.1 g
$MgCl_2$, 0.1 g
Na pyruvate, 0.05 g
Penicillin, 62 mg
D-Glucose, 1.04 g
Phenol red, 100 mg

Dissolve well after each addition in 1000 ml distilled water. Add 10% (v/v) of heat-inactivated fetal calf serum to the medium. Filter-sterilize and aliquot prior to use. The medium can be stored at 4°C for up to 5 weeks.

Preparation of Micromanipulation Apparatus

Micromanipulation can be set up with a micromanipulator system, including (a) an inverted microscope or a standard microscope (Leitz Instruments, Ltd.); (b) two sets of micromanipulators (Leitz Instruments Ltd.); (c) Kopf needle puller model 750 (David Kopf Instruments); (d) a finer needle and a pipette holder; and (e) a Schott-KL-1500 cold light source (Schott).

1. **Holding pipette:** To hold and stabilize a blastocyst during the injection of ES cells, a holding pipette needs to be prepared as follows:
   a. Pull out a length of glass capillary with a pipette puller according to the manufacturer's instructions or by hand over a low flame.
   b. Break the pipette at the point of 70 to 90 μm with a microforge according to the manufacturer's instructions. Fire-polish the tip of the pipette and close it down to a diameter of approximately 15 μm. The pipette may be further bent so that it can be placed horizontally in the microscope field.
2. **ES cell injection pipettes:** These pipettes will be used to inject ES cells into the blastocyst. The procedures are similar to those described for the holding pipette except that the tip should be broken off to generate about 12 to 15 μm for ES cells to pass through. No fire-polishing and bending are necessary.
3. **Blastocyst injection pipettes:** This type of pipette will be utilized to transfer blastocyst embryos for injection of ES cells into the embryos. The procedures are similar to those described for holding pipettes except that the tip should be broken off to generate a relatively large inside diameter (about 130 μm) for the blastocyst to pass through and fire-polishing should be very brief to eliminate the sharp edges.
4. **Embryo transfer pipettes:** These pipettes will be employed to transfer the blastocyst embryos containing the injected ES cells back to the uterus of the host. The procedures are similar to those described for holding pipettes except that the tip should be broken off to generate a relatively large inside diameter (about 130 μm) for the blastocyst to pass through freely and fire-polishing should be very brief to eliminate the sharp edges.

Injection of ES Cells into Blastocysts

1. Set up a suitable schedule so that ES cells may be trypsinized and gently suspend them to maintain them as a single cell suspension in embryo culture medium at 4°C prior to use; the blastocysts may be prepared in the morning.
2. Take up a tiny volume of light liquid paraffin (optional) by positive pressure and then take up blastocysts or ES cells into an embryo transfer pipette by capillary suction. Under the microscope, introduce the blastocysts into drops of embryo culture medium in a chamber, dish or depression slide.
3. Assemble all parts necessary for the injection according to the manufacturer's instructions.
4. Position the pipette tips so that they are parallel in the field using the coarse adjustments of the manipulator. The holding pipette and ES cell injection

pipette should be placed in opposite directions (e.g., north and south). Focus and place them in the same plane, utilizing the fine adjustments.
5. Carefully pull the pipettes far enough out to allow the introduction of a chamber, dish or slide containing the blastocysts onto the stage.
6. Carefully reposition the pipettes and focus. Position and place the relatively dark side of the blastocysts in the center of the field, then carefully bring the ES cell pipette closer to the blastocyst.
7. Carefully insert the ES cell pipette and expel one to five ES cells into the blastocoelic cavity (relatively light part of the blastocyst). Slightly remove the ES cell pipette and put the injected blastocyst aside.
8. Repeat the procedure for other blastocysts.

Reimplantation of Injected Blastocysts into Uterus of Recipient Female

1. Prepare the third- or fourth-day pseudopregnant females as described previously. It is recommended that these mice be prepared 1 day later than the mice prepared as the blastocyst donors.
2. Weigh and anaesthetize a couple of females as described earlier.
3. Clean the back of a female with 70% ethanol and make an approximately 1-cm transverse incision in the skin in the area of the first lumbar vertebra. Clean the cut hairs with cotton wetted in alcohol.
4. Carefully slide the skin incision to both sides until the left ovarian fat pad and the ovary can be seen through the peritoneal wall.
5. Make a 4-mm incision through the peritoneum. Locate the ovary, oviduct and uterus using blunt forceps. Grasp the fat pad and carefully pull it out together with the uterus through the opening.
6. Under a dissecting microscope, hold a pipette loaded with the embryos and a hypodermic needle in the right hand, and carefully hold the uterus with forceps in the left hand. Insert the needle through the muscle layers of the uterus and position the needle parallel to the long axis of the uterus in order to allow the tip to enter the lumen.
7. Pull out the needle and insert the tip of the embryo transfer pipette through the hole made by the needle into the lumen of the uterus. Carefully expel all of the blastocysts into the lumen and remove the pipette. Bubbles should be avoided as much as possible.
8. Replace the uterus and perform the same procedure for the other side if bilateral transfer is necessary.
9. Seal the incision or close it with a small wound clip.
10. Recover the foster mothers by briefly warming them under an infrared lamp (do not overheat), and house two or three in one cage. Live offspring are usually born 18 days after the surgery. The initial offspring are called the founder or GO animals in terms of genetics. The offspring can be allowed to breed to maintain the animals.

METHOD B. PRODUCTION OF TRANSGENIC MICE FROM OOCYTES

Preparation of Oocytes

1. Four days (94 to 96 h) before removal of the eggs, induce superovulation of 10 to 15, 3- to 4-week-old female mice by injecting 5 units per female of pregnant mare's serum (PMS) as a stimulating hormone.
2. Inject 5 units per female of human chorionic gonadotrophin (HCG) 75 to 76 h following the injection of PMS. This injection should be made 19 to 20 h prior to harvesting the eggs.
3. After the injection of HCG, cage the females with males overnight by maintaining them on a 12-h light/12-h dark cycle.
4. The next morning, check the females for vaginal plugs. The mated females should have vaginal plugs within 24 h after mating.
5. Kill the superovulated females that have visible, vaginal plugs by cervical dislocation.
6. Clean the body surface with 70% ethanol, dissect the oviducts and transfer them to disposable Petri dishes containing 30 ml of medium A. Carefully transfer the oocytes from the oviducts to the medium.

Note: Under a dissecting microscope, the fertilized eggs can be identified as containing two nuclei that are called pronuclei. The male pronucleus is bigger than that of the female.

7. Aspirate medium A, carefully transfer the oocytes to a fresh Petri dish and incubate them with hyaluronidase for an appropriate time period to remove the cumulus cells.
8. Rinse the oocytes extensively with four changes of medium A and carefully transfer them to a fresh Petri dish containing 10 to 30 ml of medium B. Place the dish in a 37°C incubator with 5% CO_2.

Reagents Needed

Medium A

4.78 m*M* KCl
94.66 m*M* NaCl
1.19 m*M* KH_2PO_4
1.19 m*M* $MgSO_4$
1.71 m*M* $CaCl_2$
4.15 m*M* $NaHCO_3$
23.38 m*M* Sodium lactate
0.33 m*M* Sodium pyruvate
5.56 m*M* Glucose
20.85 m*M* Hepes, pH 7.4 (adjust with 0.2 *N* NaOH)
BSA, 4 g/l
Penicillin G (potassium salt), 100,000 units/l
Streptomycin sulphate, 50 mg/l

Phenol red, 10 mg/l
Dissolve well after each addition. Filter sterilize and store at 4°C. Warm the medium to 37°C prior to use.

Medium B

4.78 m*M* KCl
94.66 m*M* NaCl
1.71 m*M* $CaCl_2$
1.19 m*M* KH_2PO_4
25 m*M* $NaHCO_3$
1.19 m*M* $MgSO_4$
23.38 m*M* Sodium lactate
0.33 m*M* Sodium pyruvate
5.56 m*M* Glucose
BSA, 4 g/l
Penicillin G (potassium salt), 100,000 units/l
Streptomycin sulphate, 50 mg/l
Phenol red, 10 mg/l
Dissolve well after each addition. Filter-sterilize and store at 4°C. Warm up to 37°C prior to use.

Microinjection of DNA Constructs into Oocytes

The equipment used for microinjection includes: (1) an injection microscope on a vibration-free table (Diaphot TMD microscope with Nomarski optics is available from Nikon Ltd.); (2) an inverted microscope for setting up the injecting chamber, and holding pipette and needle (A SMZ-2B binocular microscope available from Leitz Instruments Co.); (3) two sets of micromanipulators (Leitz Instruments Ltd.); (4) Kopf needle puller model 750 (David Kopf Instruments); (5) a finer needle and a pipette holder; and (6) a Schott-KL-1500 cold light source (Schott).

1. Thoroughly clean a depression slide with teepol-based detergent and extensively rinse it with tap water and distilled water. Rinse it with ethanol and air-dry. Add a drop of culture medium to the depression slide and a drop of liquid paraffin on top of the medium drop.
2. Assemble all parts necessary for the injection according to the manufacturer's instructions. Carry out a pretest of the procedure prior to injection of the DNA sample.
3. Carefully transfer the eggs into the medium drop using a handling pipette that can take up the cells by suction under the microscope.
4. Prepare microinjection needles with Kopf needle puller model 750. Slowly take up the DNA sample into the injection needle by capillary action.
5. Under the microscope, hold the suspended eggs with the holding pipette and use the micromanipulator to insert the needle individually into the male pronucleus, which is larger than the female pronucleus. Slowly inject the DNA sample into the pronucleus. The pressure can be maintained on

the DNA sample in the needle by a syringe connected to the needle holder. After swelling of the pronucleus occurs, immediately remove the needle a short distance from the cell. The volume injected is approximately 1 to 2 pl.

6. Transfer the injected eggs to one side and repeat injections for others. About 40 to 60 eggs can be injected in 1 to 2 h, depending on an individual's experience.
7. Transfer the injected eggs into a fresh dish containing medium B and place it in the incubator.

Reimplantation of Injected Eggs into Recipient Female Mice and Generation of Founder Mice

The injected eggs can be reimplanted into recipient female mice right after injection (one-cell stage embryo) or after being incubated in the incubator overnight (two-cell stage embryo). It is recommended that the injected eggs be transferred at the one-cell stage so that the introduced DNA integrates into chromosomes at the same site. Thus, all of the cells in transgenic animals contain the injected DNA. In general, the introduced DNA integrates randomly at a single site. In some cases, however, multiple copies of DNA insertion may occur.

1. Prepare pseudopregnant females by caging four to eight females (6 to 9 weeks old) with vasectomized males (two females per male) on the day they are needed. Thus, the females will be in the correct hormonal stage to allow the introduced embryos to be implanted but none of their own eggs will form embryos. The pseudopregnant females can be distinguished by inspecting the vaginal area for swelling and moistness (see Method A).
2. Carefully make a small incision in the body wall of a half-day pseudopregnant female (e.g., C57B1/CBA F1), and gently pull out the ovary and oviduct from the incision. Hold the oviduct in place with a clip placed on the fat pad attached to the ovary.
3. Under a binocular dissecting microscope, place the injected eggs or embryos into the infundibulum, using a sterile glass transfer pipette. Generally, 10 to 15 injected eggs can be transferred into an oviduct.
4. Carefully place the oviduct and ovary back in place and seal the incision.
5. Recover the foster mothers by briefly warming them under an infrared lamp (do not overheat), and keep two or three in one cage. Live offspring are usually born 18 days after the surgery. These initial offspring are named the founder animals or GO animals in terms of genetics.
6. Allow the offspring to breed in order to maintain the transgenic mice.

CHARACTERIZATION OF TRANSGENIC MICE

Detailed protocols for isolation of DNA, RNA, extraction of proteins and Southern, northern and western blot hybridizations are described in Chapter 10, Chapter 11 and Chapter 15, respectively. Data analyses are very similar to cellular characterization protocols described in this chapter.

TROUBLESHOOTING GUIDE

1. **Low efficiency of transfection of cells by liposomes, electroporation and calcium phosphate methods.** If cells are normal and the cell density is appropriate, it is likely that the amounts of DNA constructs used and parameters applied in transfection are not optimal. Make sure to optimize the conditions for transfection of the cells, including DNA concentration, lipids ratio or voltage pulse. In the case of stable transfection, ensure that the concentration of drug used in selection is not too high. A killing curve should be established.
2. **Low titer of retroviruses**. The virus stock is overdiluted or density of retrovirus particles is low due to low efficiency of stable transfection of packaging cells with modified virus DNA constructs. Try to use a good packaging cell line and optimize the parameters used for transfection. Reinfection may be performed to amplify the virus titer.
3. **Low efficiency of transfection with frozen high-titer viruses.** Assuming transfection conditions were optimal, it is very likely that the viruses lost significant activity. It is recommended that high-titer viruses be used in a couple of hours. Frozen aliquots should not be frozen and thawed.
4. **No antisense RNA is expressed in transfected cells.** This could be due to the orientation of the introduced cDNA. Make sure that the poly(A) strand of the cDNA is inserted downstream from the driving promoter in 3′ to 5′ orientation in order to make the antisense RNA or poly(T) strand. Try to check the mRNA level in the cells. If the level is equivalent to that of the control cells, the wrong orientation of the introduced cDNA is the case.
5. **Low reduction of the protein of interest.** This is obviously low efficiency of antisense inhibition. Try to run a northern blot analysis to check for antisense RNA expression using labeled sense RNA as a probe. Very little antisense RNA or not much reduction in mRNA level compared with nontransfected cells is exactly the cause of a low-percentage reduction of the protein. Additionally, it may be related to the activity of the promoter driving the expression of the introduced cDNA. If it is a constitutive promoter, its activity is low. Try to use a stronger promoter. If the driving promoter is inducible, try to optimize the concentration of the inducer.
6. **Some of drug-selected stably transfected cell clones show 100% death during subsequent expanding or culture in a selection medium.** If these clones are not frozen, it is certain that these isolated clones are false stably transfected cells. It is likely that during trypsinization and resuspension, the cells were not well suspended; instead, they aggregated. After transfection, some of cells within the aggregated colonies received a smaller amount of drug and still underwent proliferation, forming false colonies that were picked up. During the trypsinizing and picking-up,

these cells were loosened by physical force. Because they were not particularly stable colonies, they were killed in drug selection medium during subsequent expanding or culture.

REFERENCES

1. Wu, W. and Welsh, M.J., Expression of HSP27 correlates with resistance to metal toxicity in mouse embryonic stem cells transfected with sense or antisense HSP27 cDNA, *Appl. Toxicol. Pharmacol.,* 141, 330, 1996.
2. Robert, L.S., Donaldson, P.A., Ladaigue, C., Altosaar, L., Arnison, P.G., and Fabijanski, S.F., Antisense RNA inhibition of β-glucuronidase gene expression in transgenic tobacco can be transiently overcome using a heat-inducible β-glucuronidase gene construct, *Biotechnology,* 8, 459, 1990.
3. Rezaian, M.A., Skene, K.G.M., and Ellis, J.G., Antisense RNAs of cucumber mosaic virus in transgenic plants assessed for control of the virus, *Plant Mol. Biol.,* 11, 463, 1988.
4. Hemenway, C., Fang, R.-X., Kaniewski, W.K., Chua, N.-H., and Tumer, N.E., Analysis of the mechanism of protection in transgenic plants expressing the potato virus X coat protein or its antisense RNA, *EMBO J.,* 7, 1273, 1988.
5. Dag, A.G., Bejarano, E.R., Buck, K.W., Burrell, M., and Lichtenstein, C.P., Expression of an antisense viral gene in transgenic tobacco confers resistance to the DNA vu-us tomato golden mosaic virus, *Proc. Natl. Acad. Sci. USA,* 88, 6721, 1991.
6. Iwaki, T., Iwaki, A., Tateishi, J., and Goldman, J.E., Sense and antisense modification of glial αB-crystallin production results in alterations of stress fiber formation and thermoresistance, *J. Cell Biol.,* 15, 1385, 1994.
7. Yang, N.S., Burkholder, J., Roberts, B., Martinell, B., and McCabe, D., *In vitro* and *in vivo* gene transfer to mammalian somatic cells by particle bombardment, *Proc. Natl. Acad. Sci. USA,* 87, 9568, 1990.
8. Wu, W., Gene transfer and expression in animals, in *Handbook of Molecular and Cellular Methods in Biology and Medicine,* Kaufman, P.B., Wu, W., Kim, D., and Cseke, L., CRC Press, Boca Raton, FL, 1995.
9. Zelenin, A.V., Titomirov, A.V., and Kolesnikov, V.A., Genetic transformation of mouse cultured cells with the help of high-velocity mechanical DNA injection, *FEBS Lett.,* 244, 65, 1989.
10. Hasty, P. and Bradley, A., Gene targeting vectors for mammalian cells, in *Gene Targeting: A Practical Approach,* Joyner, A.L., Ed., IRL Press, Oxford University Press, Inc., New York, 1993.

10 Analysis of Gene Expression at the Functional Genomic Level

CONTENTS

INTRODUCTION

Northern blot hybridization is a procedure in which different sizes of RNA molecules are separated in an agarose gel, immobilized onto a solid support of either nylon or nitrocellulose membranes, and then subjected to hybridization with a labeled DNA or RNA probe. Unlike the Southern blotting technique that involves DNA–DNA hybridization, northern blot hybridization refers to RNA–DNA or RNA–RNA hybridization. Technically, northern is not a person's name; nor is it due to a geographic location. As a matter of fact, this technique was developed later than the Southern blot method, and thus named the northern blot method.

The northern blot technique is one of the most fundamental and powerful tools used in analysis of gene expression.[1] It is a sensitive, reliable and quantitative method widely employed in the characterization of the steady-state level of RNA transcripts. RNA analysis is a powerful approach to monitoring the activity of an endogenous or introduced gene in specific cell lines or tissues. In spite of the fact that other methods such as RNase protection and dot/slot blot hybridization can be used for

RNA analysis and are relatively more sensitive, the northern blot method has several advantages. As has been demonstrated previously,[2–4] expression patterns of genes are often complex. Multiple RNA molecules can be expressed from a single gene and thousands of RNA species are generated from a single cell, tissue or organism. Northern blot analysis can simultaneously provide information on the species, sizes and expression levels of diversity of RNAs that cannot be displayed by alternative techniques such as the dot/slot blot and RNA protection assays. Also, membrane filters containing a record of different RNA species can be reused for analysis of multiple RNAs' transcripts from several genes in the same RNA sample. Therefore, the value and applications of northern blot technology are immeasurable.

This chapter offers detailed protocols for standard northern blot hybridization and for semiquantitative PCR and dot blot analysis of mRNA expression.[1,2,4–9]

PRINCIPLES AND GENERAL CONSIDERATIONS

The primary principles and procedures of northern blot hybridization are illustrated in Figure 10.1. Different RNAs are first denatured by formaldehyde and then subjected to separation according to their molecular weights by standard agarose gel electrophoresis. An agarose gel serves as a molecular sieve; an electron transport in an electrical field is the driving force for migration of negatively charged RNAs during electrophoresis. The separated RNAs are then blotted or transferred onto a nylon or nitrocellulose membrane. Generally, different positions of RNAs in the agarose gel are identically recorded on the membrane. After covalent cross-linking of RNA molecules on the membrane, specific or target sequences of RNAs can be hybridized with a DNA or RNA probe of interest, followed by detection with an appropriate method.

One important concept to know is that the principle of mRNA transcribed from its gene or DNA is the base pairing rule. That means that the sequence of an mRNA is complementary to that of the DNA template from which it was copied, except that base T is replaced by U. It is exactly the base pairing or complementary rule that makes it possible for a single-stranded DNA or RNA probe to hybridize with its target sequences of denatured mRNA by means of hydrogen bonds. With regard to probes, an RNA probe or riboprobe transcribed *in vitro* has been demonstrated to be more sensitive or hotter than a DNA probe.[1] In addition, RNA–RNA hybrids are more stable compared with RNA–DNA hybrids. Therefore, hybridization and washing can be carried out under relatively high stringency conditions, thus reducing potential development of a nonspecific background.

Because RNAs, as compared to DNA, are very mobile molecules due to potential degradation by RNases, much care should be taken to maintain the purity and integrity of the RNA. This is certainly critical for northern blot hybridization. RNase activity is not readily inactivated. Two major sources of RNase contamination are from the user's hands and from bacteria and fungal molds present on airborne dust particles. To avoid RNase contamination, gloves should be worn and changed frequently when handling RNAs. Disposable plasticware should be autoclaved. Nondisposable glass- and plasticware should be treated with 0.1% diethyl pyrocarbonate (DEPC) in d.H_2O and be autoclaved prior to use. DEPC is a strong RNase inhibitor.

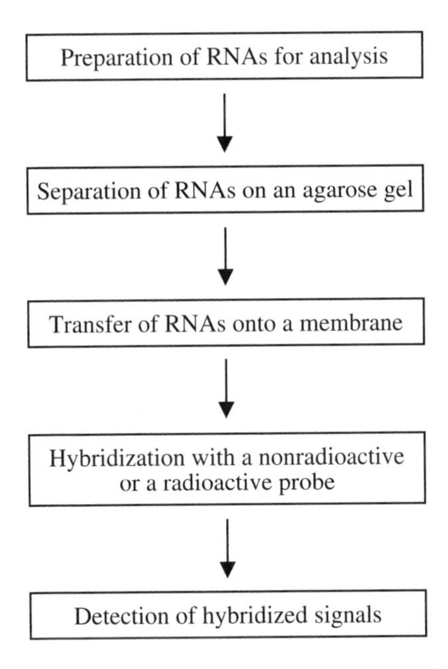

FIGURE 10.1 Scheme of mRNA analysis by northern blot hybridization.

Glassware may be baked at 250°C overnight. Gel apparatus should be soaked with 0.2% SDS overnight and thoroughly cleaned with detergent followed by thorough rinsing with DEPC-treated water. Apparatus for RNA electrophoresis, if possible, should be separated from that used for DNA or protein electrophoresis. Chemicals to be used should be of ultrapure grade and RNase-free. Gel mixtures, running buffers, hybridization solutions and washing solutions should be made with DEPC-treated water or treated with a few drops of DEPC.

Similar to Southern blot hybridization, a standard northern blot consists of five procedures: (1) preparation of RNAs to be analyzed; (2) agarose gel electrophoreis of RNA molecules; (3) blotting of RNAs onto a solid membrane; (4) hybridization of the RNA with a labeled DNA or RNA probe; and (5) detection of the hybridized signals. Agarose gel electrophoresis is a common tool for separating different RNAs. Agarose comes from seaweed algae and is a linear polymer basically composed of D-galactose and L-galactose. Once agarose powder is melted and then hardened, it forms a matrix that serves as a molecular sieve by which different sizes of RNAs are separated during electrophoresis.

A number of general factors should be considered when agarose gel electrophoresis of RNA is performed: an ultrapure grade agarose that is RNase-free is highly recommended for northern blotting and the concentration of agarose will influence separation of RNAs. (A general chart is shown below.) Normally, the larger the size of the RNA is, the lower the percentage of agarose that should be used to obtain sharp bands.

During electrophoresis, RNAs move through the gel matrix at a rate inversely proportional to the log of their molecular weight. Small RNA species move faster than large RNA molecules.

Agarose % (w/v)	RNA size (Kb)
0.6	1.0 to 20
0.8	1.0 to 10
1.0	0.6 to 7.0
1.2	0.5 to 5.0
1.4	0.2 to 2.0

ISOLATION OF TOTAL RNAs AND PURIFICATION OF mRNAs

Detailed protocols are described in Chapter 3.

ELECTROPHORESIS OF RNAs USING FORMALDEHYDE AGAROSE GELS

1. Thoroughly clean an appropriate gel apparatus and combs by soaking them in 0.2% SDS solution overnight and washing them with a detergent. Completely wash out the detergent with tap water and rinse with distilled water three to five times.
2. Prepare a formaldehyde agarose gel mixture in a clean bottle or a beaker as follows:

Component	Mini gel	Medium gel	Large gel
dd.H_2O	21.6 ml	54 ml	86.4 ml
Ultrapure agarose	0.3 g	0.75 g	1.2 g

Note: The percentage of agarose is normally 1% (w/v), but it can be adjusted with the sizes of the RNAs to be separated. The dd.H_2O should be DEPC treated. Because the agarose is small in amount, its volume can be ignored in calculating the total volume.

3. Place the bottle (with the cap loose) or beaker in a microwave oven and slightly heat the agarose mixture for 1 min. Briefly shake the bottle to rinse any agarose powder stuck to the glass sides. To melt the agarose completely, carry out gentle boiling for 1 to 3 min, depending on the volume of agarose mixture. Gently mix, open the cap and allow the mixture to cool to 50 to 60°C at room temperature.
4. While the gel mixture is cooling, seal a clean gel tray at the two open ends with tape or a gasket and insert the comb in place.
5. Add the following components to the cooled gel mixture:

Components	Mini gel	Medium gel	Large gel
10X MOPS buffer	3 ml	7.5 ml	12 ml
Ultrapure formaldehyde (18%, v/v) (37%, 12.3 M)	5.4 ml	13.5 ml	21.6 ml
Ethidium bromide (EtBr, 10 mg/ml)	3 µl	7.5 µl	12 µl

6. Gently mix after each addition and pour the mixture into the assembled gel tray placed in a fume hood. Allow the gel to harden for 30 min at room temperature.

Caution: Formaldehyde is toxic; both DEPC and EtBr are carcinogenic. These reagents should be handled with care. Gloves should be worn when working with these chemicals. Gel running buffer containing EtBr should be collected in a special container designated for toxic waste disposal. Formaldehyde serves to denature any secondary structures of RNAs. EtBr is used to stain RNA molecules by interlacing in the regions of secondary structure of RNAs and fluoresces orange when illuminated with UV light. An alternative way is to stain the gel after electrophoresis is completed. The gel can be stained with EtBr solution for 10 to 30 min followed by washing in distilled water for 3 to 5 min. In addition, an appropriate volume of EtBr stock solution can be directly added to the RNA sample (1 μl of 1 mg/ml per 10 μl RNA) prior to loading.

7. While the gel is hardening, prepare the RNA sample in a sterile tube as follows:
   Total RNA (10 to 35 μg/lane)
   or poly(A)+RNA (0.2 to 2 μg/lane), 6 μl
   10X MOPS buffer, 3.5 μl
   Ultrapure formaldehyde
   (37%, 12.3 M) ,6.2 μl
   Ultrapure formamide
   (50% v/v), 17.5 μl
   Add DEPC-treated dd.H_2O to 35 μl
8. Heat the tubes at 55°C for 15 min and immediately chill on ice to denature the RNAs. Briefly spin down afterwards.
9. Add 3.5 μl of 5X DEPC-treated loading buffer to the RNA sample. Proceed to sample loading.
10. Slowly and vertically pull the comb out of the gel and remove the sealing tape or gasket from the gel tray. Place the gel tray in an electrophoresis apparatus and add a sufficient volume of 1X MOPS buffer to the apparatus until the gel is covered to a depth of 1.5 to 2 mm above the gel.

Note: The comb should be slowly and vertically pulled out from the gel because any cracks occurring inside the wells of the gel will cause sample leaking when the sample is loaded. The top side of the gel where wells are located must be placed at the negative pole in the apparatus because the negatively charged RNA molecules will migrate toward the positive pole. If any of the wells contain visible bubbles, they should be flushed out of the wells using a small pipette tip to flush the buffer up and down the well several times. Bubbles may adversely influence the loading of the samples and the electrophoresis. The concentration of TBE buffer should never be lower than 0.4X. Otherwise, the gel may melt during electrophoresis. Pre-running the gel at a constant voltage for 10 min is optional.

11. Carefully load the samples, one by one, into appropriate wells in the submerged gel.

Notes: (1) A pipette tip should not be inserted all the way to the bottom of a well because this will likely break the well and cause sample leaking and contamination. The tip containing the RNA sample should be vertically positioned at the surface of the well, stabilized by a finger of the other hand and slowly loaded into the well. (2) The loading buffer should be at least 1X to final concentration; otherwise, the sample may float out of the well.

12. After all the samples are loaded, estimate the length of the gel between the two electrodes and set up the power supply at 5 to 15 V/cm of gel. Allow electrophoresis to proceed for 2 h or until the first blue dye reaches a distance of 1 cm from the end of the gel. If overnight running of the gel is desired, the total voltage should be determined empirically.
13. Stop electrophoresis and visualize the RNA bands in the gel under UV light. Photograph the gel with a Polaroid camera.

Caution: Wear safety glasses and gloves for protection against UV light.

Notes: The purpose of taking a picture is to record different RNA molecules at different positions following electrophoresis, which is useful once a hybridized signal is detected. Pictures should be taken with a fluorescence ruler placed beside the gel. An appropriate exposure is needed to obtain two relatively sharp rRNA bands and a black background. For successful RNA separation on a gel, smear mRNA and sharp rRNA bands should be visible in each lane. The two sharp rRNA bands are 28S and 18S for animal RNAs or 25S and 18S for plant RNAs.

Materials Needed

Microcentrifuge
Sterile microcentrifuge tubes (0.5 or 1.5 ml)
Pipette or pipetman (0 to 200 μl, 0 to 1000 μl)
Sterile pipette tips (0 to 200 μl, 0 to 1000 μl)
Gel casting tray
Gel combs
Electrophoresis apparatus
DC power supply
Small or medium size of RNA electrophoresis apparatus, depending on samples
Ultrapure agarose powder
Total RNA samples
Specific DNA or RNA as a probe
Nylon membranes
3MM Whatman filters
Paper towels for blotting
UV transilluminator
Hybridization oven or shaker
Water bath with temperature control
Diethylpyrocarbonate (DEPC)
Formaldehyde (37%)
Formamide

DEPC Water

0.1% (v/v) DEPC in dd.H_2O, placed at room temperature overnight with a stir bar to inactivate any potential RNases. Autoclave to remove the DEPC.

10X MOPS Buffer

0.2 *M* 3-(*N*-morpholino)propanesulfonic acid (MOPS)
80 m*M* Sodium acetate
10 m*M* EDTA (pH 8.0)
Dissolve well after each addition in DEPC-treated dd.H_2O.
Adjust the pH to 7.0 with 2 *N* NaOH.
Filter-sterilize and store at room temperature.

Ethidium Bromide (EtBr)

10 mg/ml in DEPC-treated dd.H_2O, dissolve well and keep in a dark or brown bottle at 4°C.
Caution: EtBr is extremely toxic and should be handled carefully.

5X Loading Buffer

50% Glycerol
2 m*M* EDTA
0.25% Bromphenol blue
0.25% Xylene cyanol
Dissolve well in DEPC-treated water, filter-sterilize, and store at 4°C.

BLOTTING OF RNAs ONTO NYLON MEMBRANES BY THE CAPILLARY METHOD

The general procedures are outlined in Figure 10.2 and Figure 10.3. Protocols are the same as those for DNA transfer, which are described in detail in Chapter 7.

PREPARATION OF ISOTOPIC OR NONISOTOPIC DNA/RNA PROBES

PROTOCOL A. PREPARATION OF DNA PROBES

Detailed protocols are described in Chapter 7.

PROTOCOL B. PREPARATION OF RNA PROBES BY TRANSCRIPTION *IN VITRO*

A number of plasmid vectors have been developed for subcloning cDNA. These vectors have polycloning sites downstream from powerful bacteriophage promoters, SP6, T7 or T3 in the vector. The cDNA of interest can be cloned at the polycloning site between promoters SP6 and T7 or T3, forming a recombinant plasmid. The cDNA inserted can be transcribed *in vitro* into single-strand sense RNA or antisense RNA from a linear plasmid DNA with promoters SP6, T7 or T3. During the process of *in vitro* transcription, one of the rNTPs is radioactively labeled and can be

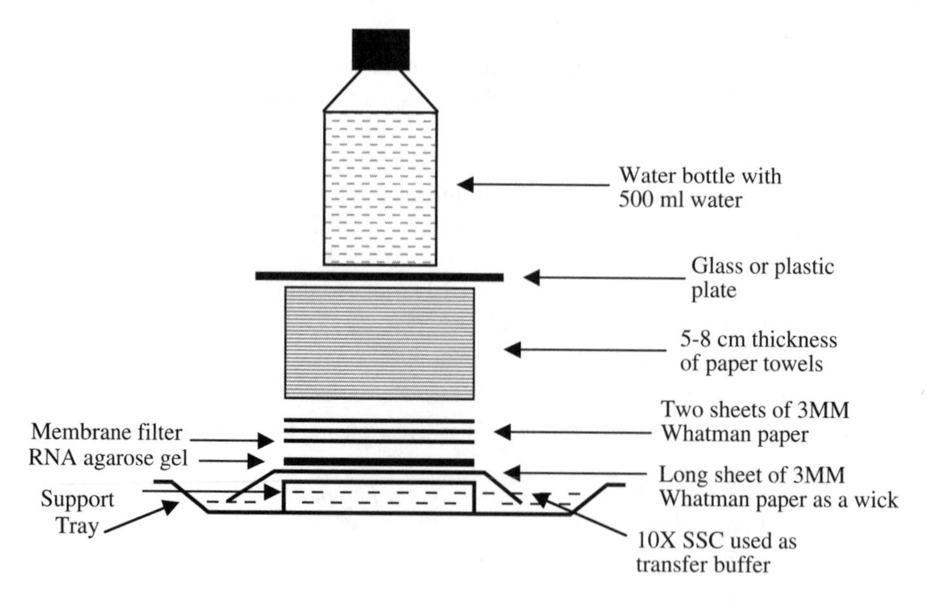

FIGURE 10.2 Standard assembly for upward capillary transfer of RNAs from an agarose gel onto a nylon membrane filter.

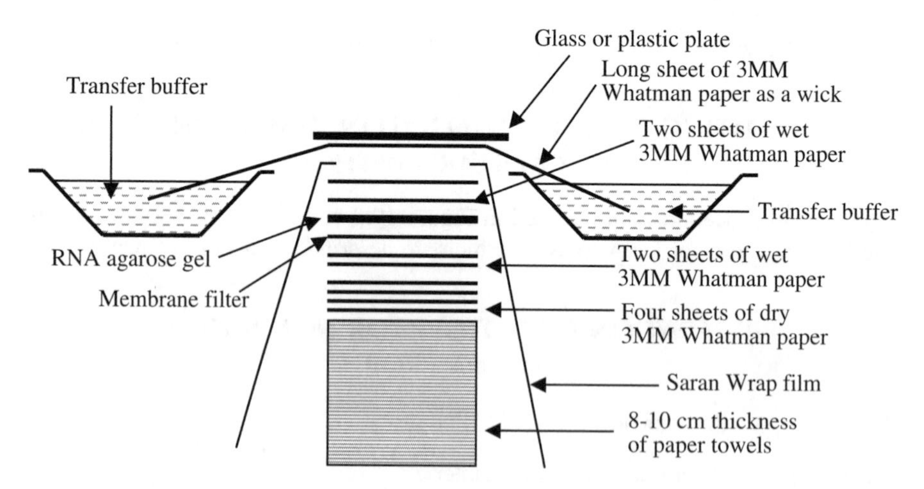

FIGURE 10.3 Standard assembly of downward capillary transfer of RNAs from an agarose gel onto a nylon membrane filter.

incorporated into the RNAs. These are the labeled RNAs. The labeled RNA probes usually have a high specific activity and are much "hotter" than ssDNA probes. Compared with DNA labeling, the yield of the RNA probe is usually very high because the template can be repeatedly transcribed. RNA probes can be easily purified from a DNA template merely by using RNase-free *DNAse* I treatment. The greatest advantage of an RNA probe over a DNA probe is that the RNA probe can produce much stronger signals in a variety of different hybridization reactions.

1. Preparation of linear DNA template for transcription *in vitro*. The plasmid is linearized by an appropriate restriction enzyme in order to produce "run-off" transcripts. To make RNA transcripts from the DNA insert, recombinant plasmid DNA should be digested by an appropriate restriction enzyme that cuts at one site that is very close to one end of the insert.
   a. On ice, set up a plasmid linearization reaction:
      Recombinant plasmid DNA (μg/μl), 5 μg
      Appropriate restriction enzyme 10X buffer, 2.5 μl
      Appropriate restriction enzyme, 3 μ/μg DNA
      Add dd.H_2O to a final volume of 25 μl.
   b. Incubate at the appropriate temperature for 2 to 3 h.
   c. Extract DNA with one volume of TE-saturated phenol/chloroform and mix well by vortexing. Centrifuge at 11,000 × *g* for 5 min at room temperature.
   d. Carefully transfer the top, aqueous phase to a fresh tube and add one volume of chloroform:isoamyl alcohol (24:1). Mix well and centrifuge as described in the previous step.
   e. Transfer the top, aqueous phase to a fresh tube. Add 0.1 volume of 2 *M* NaCl solution and two volumes of chilled 100% ethanol to the supernatant. Precipitate the DNA at –20°C for 30 min.
   f. Centrifuge at 12,000 × *g* for 5 min, aspirate the supernatant and briefly rinse the pellet with 1 ml of 70% ethanol. Dry the DNA pellet for 10 min under vacuum and dissolve the linearized plasmid in 15 μl dd·H_2O.
   g. Take 2 μl of the sample to measure the concentration of the DNA at A_{260} nm and A_{280} nm. Store the sample at –20°C until use.
2. Blunt the 3′ overhang ends using the 3′ → 5′ exonuclease activity of Klenow DNA polymerase.

Note: Although this is optional, we recommend that the 3′ protruding ends be converted into blunt ends because some of the RNA sequence is complementary to vector DNA. Therefore, enzymes such as Kpn I, Sac I, Pst I, bgl I, Sac II, Pvu I, Sfi *and* Sph I *should not be used to linearize plasmid DNA for transcription* in vitro.

3. Carry out synthesis of RNA via *in vitro* transcription.
   a. Set up a reaction as follows:
      5X transcription buffer, 8 μl
      0.1 *M* DTT, 4 μl
      rRNasin ribonuclease inhibitor, 40 units
      Linearized template DNA (0.2 to 1.0 g/μl), 2 μl
      Add dd.H_2O to a final volume of 15.2 μl.
   b. Add Klenow DNA polymerase (5 u/μg DNA) to the reaction and incubate at 22°C for 15 to 20 min.
   c. To the reaction, add the following components:
      Mixture of ATP, GTP, CTP, or UTP (2.5 m*M* each), 8 μl
      120 m*M* UTP or CTP, 4.8 μl
      [α-^{32}P]UTP or [α-^{32}P]CTP
      (50 μCi at 10 mCi/ml), 10 μl
      SP6, T7, or T3 RNA polymerase (15 to 20 u/μl), 2 μl

d. Incubate the reaction at 37 to 40°C for 1 h.
e. Remove DNA template using DNase I, using 1 u/μg DNA template. Incubate for 15 min at 37°C.
f. Extract RNA with one volume of TE-saturated phenol/chloroform. Mix well by vortexing for 1 min and centrifuging at 11,000 × *g* for 5 min at room temperature.
g. Carefully transfer the top, aqueous phase to a fresh tube and add one volume of chloroform:isoamyl alcohol (24:1). Mix well by vortexing and centrifuge at 11,000 × *g* for 5 min.
h. Carefully transfer the upper, aqueous phase to a fresh tube, add 0.5 volume of 7.5 ammonium acetate solution and 2.5 volumes of chilled 100% ethanol. Precipitate RNA at –80°C for 30 min.
i. Centrifuge at 12,000 × *g* for 10 min. Carefully discard the supernatant and briefly rinse the RNA with 1 ml of 70% ethanol and dry the pellet under vacuum for 15 min.
j. Dissolve the RNA probe in 20 to 50 μl of TE buffer and store at –20°C until use.

4. Calculate the percentage of incorporation and the specific activity of the RNA probe.
   a. Estimate the cpm used in the transcription reaction. For instance, if 50 μCi of NTP is used, the cpm is $50 \times 2.2 \times 10^6$ cpm/μCi = 110×10^6 cpm in 40 μl reaction, or 2.8×10^6 cpm/μl.
   b. Perform a TCA precipitation assay using 1:10 dilution in dd.H_2O as described previously.
   c. Calculate the percentage of incorporation: % incorporation = TCA precipitated cpm/total cpm × 100.
   d. Calculate the specific activity of the probe. For example, if 1 μl of a 1:10 dilution is used for TCA precipitation, 10 × cpm precipitated = cpm/μl incorporated. In 40 μl of reaction, 40 × cpm/μl is the total cpm incorporated. If 50 μCi of labeled UTP at 400 μCi/nmol is used, then 50/400 = 0.125 nmol of UTP is added to the reaction mixture. If there is 100% incorporation and UTP represents 25% of the nucleotides in the RNA probe, 4 × 0.125 = 0.5 nmol of nucleotides is incorporated, and 0.5 × 330 ng/nmol = 165 ng of RNA is synthesized. Then, the total ng RNA probe = % incorporation × 165 ng. For example, if 1:10 dilution of the labeled RNA sample has 2.2 × 105 cpm, the total cpm incorporated is 10 × 2.2 × 105 cpm × 40 μl (total reaction) = 88 × 106 cpm. % incorporation = 88 × 106 cpm/110 × 106 cpm (50 μCi) = 80% total RNA synthesized = 165 ng × 0.80 = 132 ng RNA. The specific activity of the probe = 88 × 106 cpm/0.132 μg = 6.7 × 108 cpm/μg RNA

Note: An alternative is to synthesize RNA without using an isotope. The synthesized RNA can be labeled using nonradioactive approaches as described in Chapter 7.

Reagents Needed

RNase-free DNase I
2 *M* NaCl solution
7.5 *M* Ammonium acetate solution
Ethanol (100%, 70%)
TE-Saturated phenol/chloroform
Chloroform:isoamyl alcohol
24:1 (v/v)
TE buffer

5X Transcription Buffer

200 m*M* Tris-HCl, pH 7.5
30 m*M* $MgCl_2$
10 m*M* Spermidine
50 m*M* NaCl

NTPs Stock Solutions

10 m*M* ATP in dd.H_2O, pH 7.0
10 m*M* GTP in dd.H_2O, pH 7.0
10 m*M* UTP in dd.H_2O, pH 7.0
10 m*M* CTP in dd.H_2O, pH 7.0

Radioactive NTP Solution

[α-^{32}P]UTP or [α-^{32}P]CTP (400 Ci/nmol)

HYBRIDIZATION AND DETECTION OF SIGNALS

The general procedures are the same as for DNA hybridization and detection, which are described in detail in Chapter 7.

ANALYSIS OF mRNA EXPRESSION BY A SEMIQUANTITATIVE PCR APPROACH

The disadvantage of northern blot hybridization or of *in situ* hybridization of mRNA is the difficulty in obtaining a desired results when analyzing nonconstitutively expressed, cell- or tissue-specific genes with a low abundance of mRNAs or when limiting amounts of mRNA are available due to their degradation. These drawbacks, however, can be overcome by using quantitative PCR. The same amount of total RNAs or mRNAs from different samples is reverse transcribed into cDNAs under the same conditions. A particular cDNA of interest in total cDNAs in each of the samples is then amplified by PCR, using specific primers. An equal amount of the amplified cDNA from each sample is then analyzed by dot blotting or Southern blotting using the target sequence between primers as a probe. In this case, different amounts of mRNA expressed by a specific gene under different conditions will generate different amounts of cDNA. The different levels of gene expression can be detected with ease by comparing the hybridized signals of different samples. Meanwhile, a control mRNA (usually an appropriate housekeeping gene transcript) such as actin mRNA is reverse transcribed and amplified by PCR, using specific primers, under exactly the same conditions in order to monitor equivalent reverse transcription and an equivalent amplification by PCR.

1. Synthesize the first-strand cDNAs from the isolated total RNAs or mRNAs using reverse transcriptase and oligo(dT) as a primer. This is described in detail in Chapter 3.
2. Design primers for amplification of specific cDNA PCR. Design oligonucleic acid primers based on conserved amino acid sequences in two different regions of a specific cDNA. For instance, two oligonucleotide primers can be designed as follows for the two amino acid sequence regions of invertase, NDPNG and DPCEW.

   Forward primer = 5′ AAC(T)GAT(C)CCIAA(C)TGGI 3′, derived from NDPNG

   Reverse primer = 3′ GGTGAGCGTCCCTAG 5′, derived from DPCEW

   Note: I stands for the third position of the codon, which can be any of TCAG. The primers can generate a product of 554 base pairs.

   For amplification of actin cDNA, the primers can be designed as follows:

   Forward primer = 5′ ATGGATGACGATATCGCTG 3′

   Reverse primer = 5′ ATGAGGTAGTCTGTCAGGT 3′

   Note: These primers can generate a product of 568 base pairs.
3. Carry out PCR amplification.
   a. In a 0.5-ml microcentrifuge tube on ice, add the following items in the order listed for one sample amplification:

      10X Amplification buffer, 10 μl
      Mixture of four dNTPs (1.25 m*M* each), 17 μl
      Forward primer (100 to 110 pmol) in dd.H_2O, 4 μl
      Reverse primer (100 to 110 pmol) in dd.H_2O, 4 μl
      cDNA synthesized = 6 μl (0.1 to 1 μg)
      Add dd.H_2O to final volume of 100 μl.

      For no cDNA control:
      10X Amplification buffer, 10 μl
      Mixture of four dNTPs (1.25 m*M* each), 17 μl
      Forward primer (100 to 110 pmol) in dd.H_2O, 4 μl
      Reverse primer (100 to 110 pmol) in dd.H_2O, 4 μl
      Add dd.H_2O to final volume of 100 μl.
   b. Add 2.5 units of high fidelity of *Taq* DNA polymerase (5 units/μl) to each of the samples and mix well.
   c. Overlay the mixture with 30 μl of light mineral oil (Sigma or equivalent) to prevent evaporation of the sample.
   d. Carry out PCR amplification for 35 to 40 cycles in a PCR cycler programmed as follows:

Note: The condition for amplification should be exactly the same for every sample — actin cDNA as well as no cDNA.

Cycle	Denaturation	Annealing	Polymerization
First	3 min at 94°C	2 min at 55°C	3 min at 70°C
Subsequent	1 min at 94°C	2 min at 55°C	4 min at 70°C
Last	1 min at 94°C	2 min at 55°C	10 min at 70°C

e. Starting after 10 cycles, carefully insert a pipette tip through the oil and remove 15 µl of reaction mixture from each of the samples and from controls every five cycles at the end of the appropriate cycle (at 70°C extension phase). Place tubes at 4°C until use.

Note: Sampling should be done in 2 min for all samples. There should be six samplings for each of the samples in 40 cycles. Oil should be avoided as much as possible.

4. Use the 15 µl cDNA from each of the samples and the controls to carry out dot blot hybridization using the appropriate ^{32}P-labeled internal oligonucleotides or the target sequence (e.g., cDNA fragment) between the two primers used as a probe.
   a. Denature DNA at 95°C for 10 to 15 min and immediately chill on ice. Spin down and add one volume of 20X SSC to the sample. An alternative is the alkaline denaturing method: add 0.2 volume of 2 *M* NaOH solution to the sample, incubate at room temperature for 15 min and add one volume of neutralization buffer containing 0.5 *M* Tris-HCl (pH 7.5) and 1.5 *M* NaCl. Incubate at room temperature for 15 min.
   b. Cut a piece of nylon or nitrocellulose membrane filter and place the filter on top of a vacuum blot apparatus. Turn on the vacuum slightly to obtain an appropriate suction that holds the membrane filter onto the apparatus.

   Tips: If the vacuum is too weak, diffusion of the loaded samples is usually a problem. However, the vacuum cannot be too strong; if the vacuum is too great, the efficiency of blotting decreases. Using an appropriate vacuum speed, a trace of each well will be visible on the membrane.
   c. Spot samples (about 2 µl/loading without vacuum or 5 µl/loading under vacuum) onto the membrane filter that is prewetted with 10X SSC and air-dried. Partially dry the filter between each spot.
   d. After spotting is complete, dry the filter under vacuum and place it onto a 3MM Whatman filter paper saturated with denaturing solution with the DNA side up. Incubate for 5 to 10 min.
   e. Transfer the membrane filter, DNA side up, to a piece of 3MM Whatman filter paper presaturated with neutralizing solution for 2 to 5 min.
   f. Air-dry the membrane filter at room temperature for 20 min.
   g. Wrap the filter with SaranWrap and place DNA side down onto a transilluminator (312 nm wavelength is recommended) for 4 to 6 min for UV cross-linking.
   h. Bake the filter in an oven at 80°C for 2 h under vacuum. Proceed to hybridization.
5. Carry out hybridization as described in Chapter 7.

6. Measure the signal intensities of dot spots with a densitometer. For the cDNA controls, no signals should be visible. For actin cDNA, the signal intensities should increase with amplified cycles, but no differences should be seen for any cell or tissue type or all developmental stages, or treated and untreated tissues. In contrast, if the expression of a specific gene is nonconstitutive but inducible with cell- or tissue-specific or chemical treatments, clear signal patterns of amplified cDNA dot blot hybridization can be seen. The signal for each sample should be stronger with amplified cycles (Figure 10.4).
7. If it is necessary to determine the size of PCR products, the cDNAs can be analyzed using 1.0 to 1.4% agarose gel electrophoresis and Southern blot hybridization using the target sequence as an internal probe (Figure 10.5).

Reagents Needed

First-Strand 5X Buffer

250 m*M* Tris-HCl, pH 8.3 (42°C)
50 m*M* $MgCl_2$
250 m*M* KCl
2.5 m*M* Spermidine
50 m*M* DTT
5 m*M* Each of dATP, dCTP, dGTP, dTTP

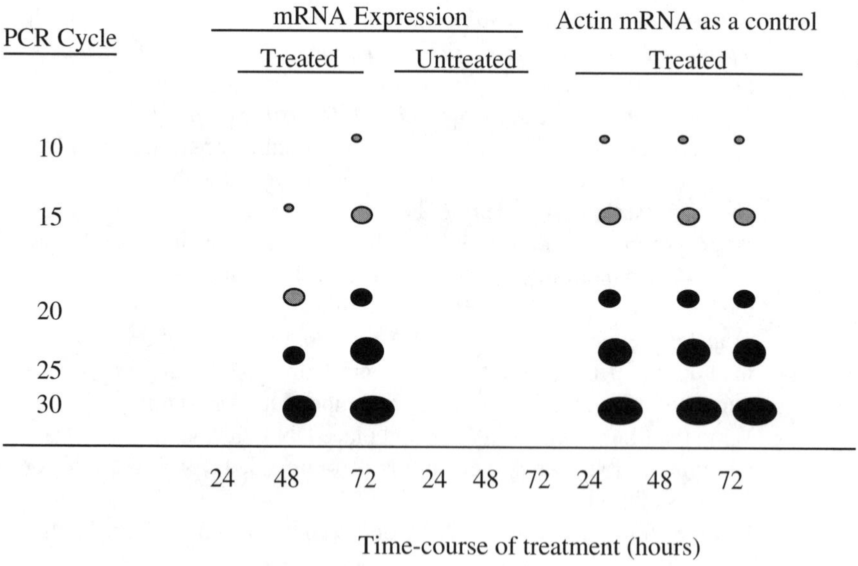

PCR Cycle
mRNA Expression
Treated
Untreated
Actin mRNA as a control
Treated
10
15
20
25
30
24 48 72 24 48 72 24 48 72
Time-course of treatment (hours)

FIGURE 10.4 Dot blot hybridization showing expected semiquantitative PCR products.

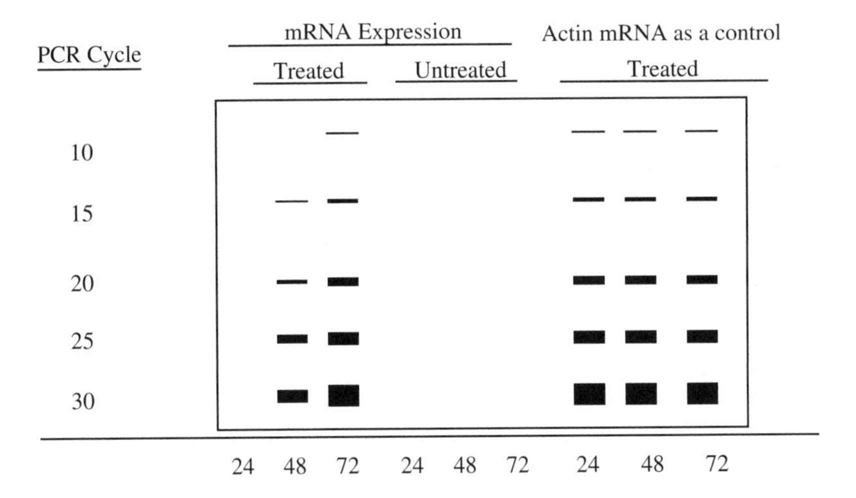

FIGURE 10.5 Southern blot hybridization showing expected semiquantitative PCR products.

10X Amplification Buffer

100 m*M* Tris-HCl, pH 8.3
500 m*M* KCl
15 m*M* $MgCl_2$
0.1% BSA

TROUBLESHOOTING GUIDE

1. **Following electrophoresis, distribution of RNA species stained by EtBr is near the very bottom in some lanes instead of a long smear ranging from the top to the bottom of the lanes.** This is due to obvious RNA degradation by RNase during RNA isolation, RNA storage or electrophoresis. Ensure that RNAs are handled without RNase contamination.
2. **Two rRNA bands are not sharp, but instead are quite diffused.** The voltage used for electrophoresis is too low and the gel is run for too long. To obtain the best results (sharp bands), a gel should run at an appropriate voltage.
3. **After transfer is complete, obvious RNA staining remains in the agarose gel.** RNA transfer is not efficient or complete. Try to set up the RNA transfer carefully and allow blotting to proceed for a longer time.
4. **The blotted membrane shows some trace of bubbles.** Obviously, this problem is due to bubbles generated between the gel and the membrane. Carefully follow instructions when assembling the blotting apparatus.
5. **No signal is detected at all.** This is the worst situation in northern blot hybridization. RNAs may not be effectively denatured or dsDNA probes

may not be denatured prior to hybridization. When this happens, it is not surprising to see zero hybridization signals. Keep in mind that ssDNA probes and denatured RNAs are crucial for hybridization.

6. **Hybridized signals are quite weak on the film or the filter.** The activities of DNA probes may be low or hybridization efficiency is not good. To overcome this, allow hybridization to proceed for a longer time or increased exposure time to the x-ray film or develop longer on the filter.
7. **Detected signals are not sharp bands; rather, long smears appear in each lane.** This problem is most likely caused by nonspecific binding. Try to carry out hybridization and washing under high-stringency conditions.
8. **Black background occurs on the x-ray film as revealed by chemiluminescent detection.** Multiple factors may be responsible for such a common problem. The membrane filter may have dried out during hybridization or during the washing process. Excess detection solution may not have been wiped out prior to exposure. Exposure time may be too long. The solution to this problem is to make sure that the filter is kept wet and that excess detection reagents are completely removed by wiping. Try to reduce exposure time for x-ray film.
9. **A highly purple/blue background occurs on the filter as shown by colorimetric detection.** Excess detection reagents may be used or the color is allowed to develop too long. Try using an appropriate amount of detection reagents and pay close attention to the color development. Once major bands become visible, stop development immediately by rinsing the filter with distilled water several times.
10. **Unexpected bands show up on the x-ray film or the membrane filter.** This problem is most likely caused by nonspecific binding between probes and the DNA species. To solve or prevent such a problem, one may increase blocking time for the blotted membrane and elevate stringency conditions for hybridization and washing processes.

REFERENCES

1. Wu, W., Electrophoresis, blotting, and hybridization, in *Handbook of Molecular and Cellular Methods in Biology and Medicine,* Kaufman, P.B., Wu, W., Kim, D., and Cseke, L., pp. 87–122, CRC Press, Boca Raton, FL, 1995.
2. Wu, W. and Welsh, M.J., Expression of the 25-kDa heat-shock protein (HSP27) correlates with resistance to the toxicity of cadmium chloride, mercuric chloride, *cis*-platinum(II)-diammine dichloride, or sodium arsenite in mouse embryonic stem cells transfected with sense or antisense HSP27 cDNA, *Toxicol. Appl. Pharmacol.*, 141, 330, 1996.
3. Welsh, M.J., Wu, W., Parvinen, M., and Gilmont, R.R., Variation in expression of HSP27 messenger ribonucleic acid during the cycle of the seminiferous epithelium and co-localization of HSP27 and microfilaments in Sertoli cells of the rat, *Biol. Reprod.*, 55, 141–151, 1996.

4. Wu, (W.)L., Song, I., Karuppiah, R., and Kaufman, P.B., Kinetic induction of oat shoot pulvinus invertase mRNA by gravistimulation and partial cDNA cloning by the polymerase chain reaction, *Plant Mol. Biol.*, 21(6): 1175–1179, 1993.
5. Wu, (W.)L., Mitchell, J.P., Cohn, N.S., and Kaufman, P.B., Gibberellin (GA_3) enhances cell wall invertase activity and mRNA levels in elongating dwarf pea (*Pisum sativum*) shoots, *Int. J. Plant Sci.*, 154(2): 278–288, 1993.
6. Wu, L., Song, I., Kim, D., and Kaufman, P.B., Molecular basis of the increase in invertase activity elicited by gravistimulation of oat-shoot pulvini, *J. Plant Physiol.*, 142: 179–183, 1993.
7. Ausubel, F.M., Brent, R., Kingston, R.E., Moore, D.D., Seidman, J.G., Smith, J.A., and Struhl, K., *Current Protocols in Molecular Biology*, Greene Publishing Associates and Wiley-Interscience, John Wiley & Sons, New York, 1995.
8. Krumlauf, R., Northern blot analysis of gene expression, in *Gene Transfer and Expression Protocols,* Murray, E.J., Ed., The Human Press Inc., Clifton, NJ, 1991.
9. Sambrook, J., Fritsch, E.F., and Maniatis, T., *Molecular Cloning: A Laboratory Manual,* 2nd ed., Cold Spring Harbor Press, Cold Spring Harbor, NY, 1989.

11 Analysis of Gene Expression at the Proteomic Level

CONTENTS

INTRODUCTION

Proteins are translated from mRNA species and considered to be the final products of gene expression. It is the protein that facilitates the specific function of a gene that makes gene expression meaningful.[1,2] Therefore, protein expression is one of the most important parts of modern molecular biology. To explore proteins, one key approach is to analyze them by electrophoresis, a process in which a net charged molecule will move in an electric field. SDS-PAGE (sodium dodecyl sulfate polyacrylamide gel electrophoresis), which was developed in the mid-1960s, is a widely used technique to separate proteins according to their net charges, sizes and shapes.[1,3–6] To identify a specific protein from a protein mixture, it may be necessary to use a powerful and commonly employed technique: western blotting.

Western blot hybridization or immunoblotting is a procedure in which different types of proteins are separated by SDS-PAGE and immobilized onto a solid support PVDF (polyvinylidene difluoride) or nitrocellulose membrane. The protein of interest on the blotted membrane is then detected by incubating the membrane with a specific antibody as a probe. The name "western" is not a person's name, nor is it due to a geographic location. This technique was developed later than the Southern blot method and thus named western blotting. Because the technique mandates specific antibodies, (monoclonal or polyclonal), it is also called immunoblotting. Western blot technique is a sensitive, reliable and quantitative method widely employed in analysis of proteins. It can simultaneously offer information about the species, sizes and expression levels of diverse proteins that cannot be obtained by other alternative techniques.[1,2,4–8]

General factors should be considered when western blot hybridization is carried out. The concentration of polyacrylamide gel will influence separation of proteins. Depending on the particular size of the protein of interest, a general chart is displayed below. Normally, the larger the size is of a protein to be detected, the lower the percentage of polyacrylamide gel that should be used in order to obtain a sharp band.

Polyacrylamid % (w/v)	Protein size (kDa)
7%	20 to 200
10%	12 to 65
11%	8.0 to 35
14%	5.0 to 27
16%	3.0 to 15

During SDS-PAGE, proteins move through the gel matrix at a rate inversely proportional to the log of their molecular weight. Small proteins migrate faster than large molecules.

Transfer of proteins from a polyacrylamide gel onto a solid nitrocellulose or nylon membrane can be completed by the electrophoretic blot method. Compared with traditional nitrocellulose membranes, PVDF membranes are durable and tear resistant. Based on our experience, we highly recommend that PVDF be used as the solid transfer support.

Western blotting requires specific antibodies as probes. The specificity and activity of an antibody are absolutely essential to the success of western blot hybridization. Two general types of antibodies are monoclonal antibodies raised in mice, and polyclonal antibodies usually raised in rabbits.[4] A monoclonal antibody has a single antibody specificity, a single affinity and a single immunoglobulin isotype. However, a preparation of polyclonal antibodies contains a variety of antibodies directed against the antigen of interest as well as nonspecific proteins. As a result, a monoclonal antibody is more specific than polyclonal antibodies. Monoclonal antibodies are produced by a monoclonal population of cells derived from one cloned cell. Hence, all the antibodies are theoretically identical.

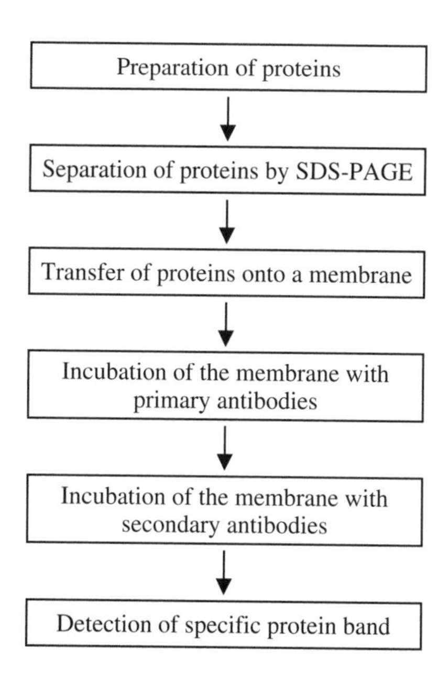

FIGURE 11.1 Scheme of detection of specific protein by western blot hybridization.

A standard western blot hybridizaiton mainly consists of four procedures: separation of proteins by SDS-PAGE, transfer of proteins onto a solid membrane, incubation of the membrane with specific antibodies and detection of hybridized signals (Figure 11.1). This chapter describes step-by-step protocols for successful western blotting.

PRINCIPLES

Knowing the principles of western blotting is extremely important for success. The primary principles of western blotting or immunoblotting start with separation of proteins by SDS-PAGE. SDS is a strong, negatively charged detergent composed of a hydrophilic head and a long hydrophobic tail (CH_3-CH_2-CH_2-CH_2-CH_2-CH_2-CH_2-CH_2-CH_2-CH_2-CH_2-CH_2-$SO_4^-Na^+$). SDS serves to denature proteins and make them negatively charged. To better understand how a protein molecule is denatured by detergents, it is necessary to review the structures of proteins briefly.

In general, functional proteins have four levels of structures. Primary structure is the amino acid sequence that is the backbone of a protein. Linear polypeptide chains undergo appropriate folding into the secondary structure of proteins by hydrogen bonds between main-chain NH and CO groups. In some proteins, disulfide bridges or bonds (-S-S-) are formed between the sulfhydryl groups of two cysteine residues. The tertiary structure of protein architecture refers to the spatial folding following the secondary structure. Tertiary structure is the highest level of structure for most proteins. However, many proteins contain multiple peptide chains or subunits. The spatial arrangement of such subunits and the nature of their connections are called quaternary structure. Proteins with multiple subunits mandate the quaternary structure for functions. These unique structures make proteins much more stable. However, for SDS-PAGE analysis of proteins, proteins need to be disrupted to their primary structures using appropriate detergents. SDS is an anionic detergent that binds to hydrophobic regions of protein molecules and causes them to unfold into extended polypeptide chains. As a result, the individual proteins are dissociated from other proteins and rendered freely soluble in the SDS solution. In addition, a reducing agent, β-mercaptoethanol (SH-CH_2-CH_2-OH), is commonly used to break any S–S bonds in proteins.

During SDS-PAGE, the anions of SDS not only denature proteins but also coat polypeptides at a ratio of about one SDS for every two amino acid residues, which gives approximately equal charge densities per unit length and makes the polypeptides negatively charged. As a result, each protein molecule binds large numbers of the negatively charged SDS molecules that overwhelm the protein's intrinsic charge. This is one of the principles of protein separation by SDS-PAGE. In an electric field, proteins wrapped with the negative SDS migrate toward the positive electrode. To enhance separation, polyacrylamide gels or PAGEs are widely used as supporting media and molecular sieves. The meshes and pore size of a PAGE can be readily made by the polymerization of acrylamide and a cross-linking reagent methylenebisacrylamide using appropriate concentrations. When a voltage is applied, proteins of the same size have the same amount of SDS molecules and the same shape due to complete unfolding by the SDS, and thus migrate at the same speed in a given pore size of a slab SDS-PAGE.

Smaller proteins move faster than larger proteins, which are retarded much more severely. As a result, proteins are fractionated into a series of discrete protein bands in order of their molecular weight (MW). There is a linear relationship between the log of the MW of a polypeptide and its R_f. R_f is the ratio of the distance from the top of the gel to a band divided by the distance from the top of the gel to the dye front. A standard curve is generated by plotting the R_f of each standard polypeptide or band marker as the abscissa and the $\log_{10}$ of its MW as the ordinate. The MW of an unknown protein band can then be determined by finding its R_f that vertically crosses on the standard curve and reading the $\log_{10}$ MW horizontally crossing to the ordinate. The antilog of the $\log_{10}$ MW will be the actual MW of the protein. Following electrophoresis, the SDS-PAGE gel can be stained with Coomassie blue solution. The major proteins or bands are then readily displayed by destaining the gel (Figure 11.2).

Inasmuch as the separated proteins are coated with negatively charged SDS, they can be readily transferred from the SDS-PAGE onto a solid membrane by electroblotting. The negatively charged proteins migrate toward the positive electrode and are retained on the membrane placed between the gel and the positive electrode (Figure 11.3). Various bands at different positions are exactly blotted at appropriate positions on the membrane. To check the efficiency of blotting, the blotted membrane can be stained by Ponceau S solution and major bands are readily seen if the transfer is successful.

The major principle of western blotting is the interaction between a specific antibody and its target protein on the membrane (Figure 11.4). To understand how the binding of antibody–antigen occurs, it is helpful to be briefly familiar with

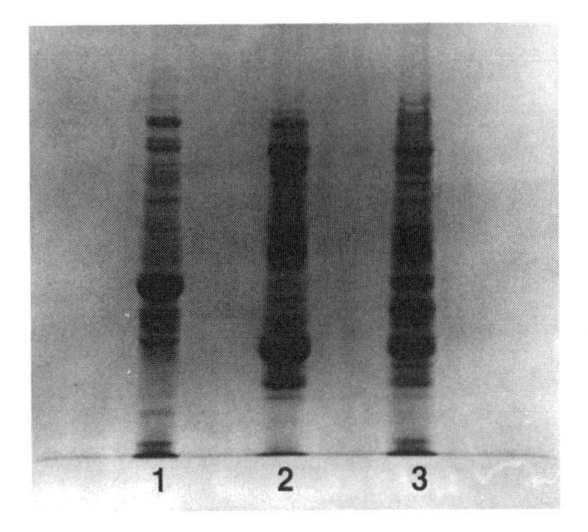

FIGURE 11.2 Coomassie blue staining and destaining of SDS-PAGE. Lane 1 to lane 3 represent proteins extracted from three different cell lines.

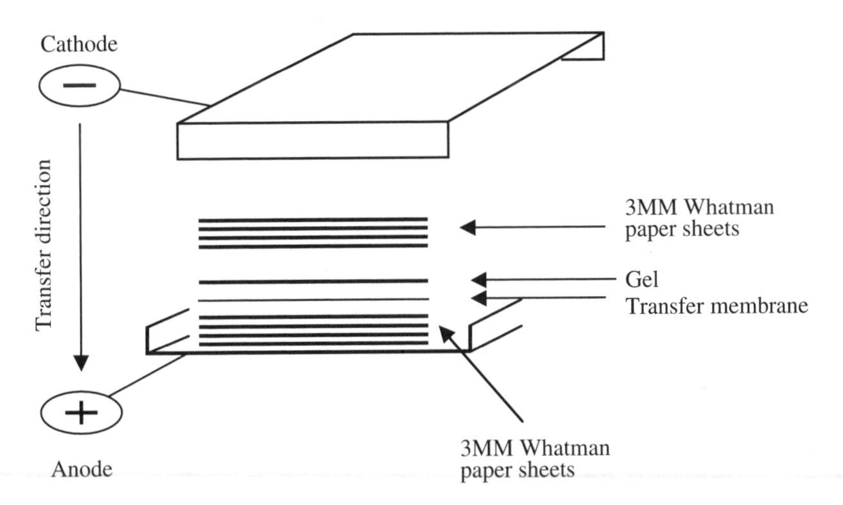

FIGURE 11.3 Diagram of semidry electric blotting of proteins onto a membrane.

structures of antibodies. An antibody is a Y-shaped molecule containing two heavy chains, two light chains and two identical antigen-binding sites or bivalent at the tip of two arms. It is important to know that the binding sites are specific for a particular antigen (e.g., a protein). The antigen-binding sites make it possible for an antibody to cross-link the antigenic determinants of the target antigen or a protein in terms of western blotting or immunoblotting. Many antigens have multiple antigenic determinants or epitopes, which enable the antigen to form linear chains or cyclic complexes with the antibody.

The antibody that directly binds to its antigen is called the primary or first antibody. In western blot hybridization, a secondary antibody is needed, which usually interacts with the primary one (Figure 11.4). The secondary antibody is

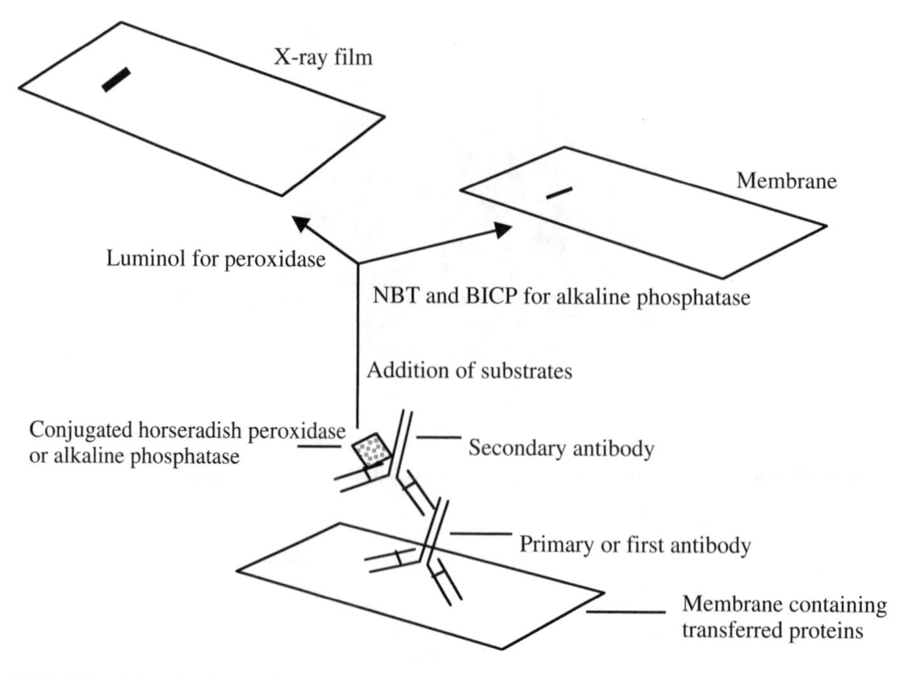

FIGURE 11.4 Principles of western blot hybridization.

frequently conjugated with an enzyme such as alkaline phosphatase or horseradish peroxidase. Once a complex of protein-primary antibody–secondary antibody is formed on a blotted membrane, it can be detected using substrates such as NBT and BCIP for alkaline phosphatase or luminol for horseradish peroxidase (Figure 11.5).

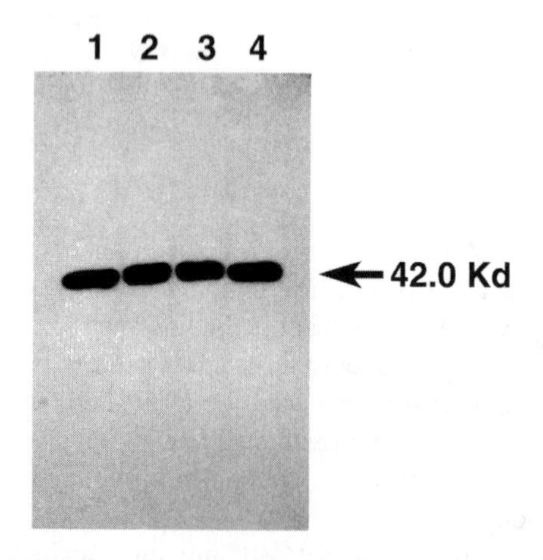

FIGURE 11.5 Detection of actin from the proteins extracted from mouse embryonic stem cells. Monoclonal antibodies against actin were used in western blotting.

EXTRACTION OF CELLULAR PROTEINS

METHOD A. EXTRACTION OF TOTAL PROTEINS USING LAUROYLSARCOSINE BUFFER

Traditionally, proteins are extracted by lysing cells with SDS gel sample loading buffer followed by boiling. This procedure rapidly inactivates proteolytic enzymes and the protein samples can be immediately used for sodium dodecyl sulfate-polyacrylamide gel electrophoresis (SDS-PAGE) or two dimensional (2-D) gel electrophoresis; however, they are not suitable for quantitative analysis of the samples' protein concentration because the samples contain SDS and 2-mercaptomethanol. In the Bradford assay, these compounds significantly decrease the O.D. at A_{595} nm, so the concentration of proteins in the extracts cannot be accurately measured. An alternative procedure is to determine the concentration of the proteins by the method of the bicinchoninic acid (BCA) assay. The major drawbacks of this approach are that these procedures are time consuming, and a portion of proteins are lost during the process of trichloroacetic acid (TCA) precipitation as well as centrifugation. As a result, the protein concentration determined by this method is lower than what truly exists in the sample.

To solve the problem, a new extraction buffer has been developed.[4] The proteins extracted are suitable for quantitative analysis using the Bradford assay and for routine analysis of proteins by SDS-PAGE, 2-D gel electrophoresis, and immunoblotting. This method provides an economical, rapid, simple and efficient approach for protein analysis. Cells in tissue culture dishes or flasks can be directly lysed with an extraction buffer containing 45 m*M* Tris-HCl (pH 6.8), 0.2% *N*-Lauroylsarcosine (sodium salt, Sigma Chemical Co., and 0.2 m*M* of phenylmethanesulfonyl fluoride (PMSF). Because of the instability of PMSF in aqueous solution, it is recommended to add this component to the buffer shortly before use.

PROTOCOL 1. PROTEIN EXTRACTION FROM CULTURED CELLS

1. When the cells reach 50 to 100% confluence, aspirate the medium and add an appropriate volume of extraction buffer (1 to 1.5 ml/100-mm dish).
2. To lyse the cells, incubate for 5 min at room temperature with occasional agitation by hand. To confirm cell lysis, the dish or flask containing the cells can be observed by phase contrast microscopy.
3. Transfer the lysed mixture into a microcentrifuge tube (Eppendorf) and store at –20°C until use. At this point, the sample can be utilized, without further processing, for protein concentration measurement by the Bradford assay and for quantitative analysis of proteins by 1D or 2D SDS-PAGE and immunoblotting.

Protocol 2. Protein Extraction from Tissues

1. Completely homogenize 1 g tissues of interest in 2 ml of extraction buffer.
2. To lyse the cells, incubate the homogenate for 10 min at room temperature with occasional agitation by hand.
3. Centrifuge at 10,000 × g for 2 min and carefully transfer the supernatant into a microcentrifuge tube and store at –20°C until use. At this point, the sample can be utilized, without further processing, for protein concentration measurement by the Bradford assay and for quantitative analysis of proteins by 1D or 2D SDS-PAGE and immunoblotting.

Method B. Determination of Protein Concentration Using the Bradford Assay

1. Label an appropriate number of cuvettes for standard curve and protein samples.
2. To the cuvettes used for standard curve, add the following components in the order shown.
   a. BSA stock solution (1 mg/ml in dH_2O): 0, 5, 10, 15, 20, 25, 30 to 40 μl, respectively.
   b. The protein extraction buffer: 20 to 40 μl for every cuvette, depending on the volume of protein sample used for measurement.
   c. dH_2O: add an appropriate volume to each cuvette to give a total volume of 0.1 ml.
   d. Bradford reagent solution: 0.9 ml to every cuvette, producing a total volume of 1 ml.
3. Transfer each protein sample (20 to 40 μl) extracted from cultured cells or from tissues to a plastic cuvette. Add $dd.H_2O$ to 0.1 ml, and then 0.9 ml of the Bradford reagent solution.
4. Carry out protein measurement at a wavelength of 595 nm in a UV-Vis spectrophotometer according to the manufacturer's instructions.

ANALYSIS OF PROTEINS BY SDS-PAGE AND WESTERN BLOTTING

Protocol 1. Preparation of the SDS Separation Gels

1. Thoroughly clean the glass plates and spacers with detergent and wash them in tap water. Rinse the plates and spacers with $dd.H_2O$ several times and air-dry.
2. Wear gloves and, using clean paper prewet with 100% ethanol, wipe dry the glass plates and spacers. Assemble the vertical slab gel unit such as the Protein II (BioRad) or the SE 600 vertical slab gel unit (Hoeffer) in the casting mode according to the instructions. Briefly, place one spacer (1.5 mm thick) on each side between the two glass plates and fix in place with clamps, forming a sandwich. Prior to casting the sandwich tightly,

vertically place the sandwich on a very level bench for an appropriate adjustment. Make sure that the glass plates and spacers are well matched to prevent any potential leaking. Repeat for the second sandwich if needed. Tightly cast the sandwich in the casting mode.

Note: The glass plates should be in very good shape with no damage at the edges. If necessary, spray a little grease or oil on the spacer areas at the top and bottom ends of the sandwich to prevent leaking.

3. Test potential leaking by filling an appropriate volume of dd.H_2O into the sandwich. Drain away the water by inverting the unit.
4. Prepare the separation gel mixture in a clean 100-ml beaker or flask in the order shown. This is sufficient for two pieces of standard size gel:
   20 ml of 30% monomer solution
   15 ml of separating gel buffer
   0.6 ml of 10% SDS solution
   24.1 ml of dd.H_2O
   Mix after each addition.

Caution: Acrylamide is neurotoxic. Gloves should be worn when handling the gel mixture.

5. Add 0.25 ml of freshly prepared 10% ammonium presulfate (AP) solution and 20 μl TEMED to the gel mixture. Briefly mix by gently swirling or using a pipette.
6. Immediately and carefully fill the mixture into the assembled sandwich up to about 4 cm from the top edge using a 10- to 15 ml-pipette or a 50-ml syringe.
7. Take up approximately 1 ml of dd.H_2O into a 1-ml syringe equipped with a 22-gauge needle and carefully load 0.5 ml of the water, starting from one top corner near the spacer, onto the surface of the acrylamide gel mixture. Load 0.5 ml of the water from the other top corner onto the gel mixture. The water will overlay the surface of the gel mixture to make the surface very even. Allow the gel to be polymerized for about 30 min at room temperature. A very sharp gel–water interface can be seen once the gel has polymerized.
8. Drain away the water layer by inverting the casting unit and rinse the surface once with 1 ml of gel overlay solution. Replace the solution with 1.5 ml of fresh gel overlay solution for at least 5 min. The gel can be kept at this point for 2 h to overnight prior to the next step.

Protocol 2. Preparation of Stacking Gels

1. Prepare the stacking gel mixture in a clean 50-ml beaker or a 50-ml flask as follows:
   2.66 ml of 30% monomer solution
   5 ml of stacking gel buffer
   0.2 ml of 10% SDS
   12.2 ml of dd.H_2O
   Mix after each addition.

2. Drain away the overlay solution on the separating gel by tilting the casted unit and rinse the surface with 1 ml of the stacking gel mixture prepared at the previous step. Drain away the stacking gel mixture and insert a clean comb (1.5 mm thick with 14 or 30 wells) from the top into the glass sandwich.
3. Add 98 μl of 10% AP solution and 10 μl of TEMED to the stacking gel mixture. Gently mix and add the gel mixture into the sandwich from both upper corners next to the spacers using a pipette or an equivalent.

Note: The stacking gel mixture may be filled into the sandwich prior to inserting the comb. However, air bubbles are readily generated and trapped around the teeth of the comb. It is recommended that the comb be inserted into the sandwich prior to filling the stacking gel mixture.

4. Allow the stacking gel to polymerize. It usually takes approximately 30 min at room temperature.

Protocol 3. Electrophoresis

Loading Protein Samples and Protein Standard Markers onto the Gel

1. While the stacking gel is polymerizing, prepare samples and standard markers. To denature proteins completely, add one volume of 2X sample loading buffer to each sample and standard marker in individual microcentrifuge tubes. Tightly cap the tubes and place them in boiling water for 4 min. Briefly spin down.

Note: The loading buffer contains SDS and mercaptoethanol, which are strong detergents for denaturation of proteins. The amount of proteins loaded into one well should be 5 to 30 μg for Coomassie blue staining and western blotting. The amount of protein standard markers for one well should be 2 to 10 μg.

2. Carefully pull the comb straight up from the gel, and rinse every well with running buffer using a pipette. Drain away the buffer by tilting the cast sandwich. Repeat rinsing once.
3. If the apparatus Protein II (BioRad) is used, the sandwich can be directly cast in place according to the manufacturer's instructions. An upper buffer chamber will be readily formed. If the SE 600 Vertical Slab Gel Unit (Hoeffer) is utilized, carefully cast the upper buffer chamber in place according to the instructions. Remove lower cams and cam the sandwich tightly to the bottom of the upper buffer chamber.

Note: Apply a little grease or oil on the spacer areas at the top corners of the sandwich to prevent leaking.

4. Fill the upper chamber with an appropriate volume of running buffer.
5. Take up sample or markers into a pipette tip or a syringe equipped with a 22-gauge needle. Insert the tip or needle into the top of a well and load the sample into it. Because of the 2X loading buffer added, the sample is

heavier than running buffer and can be seen sinking to the bottom of well when loading.

Note: The volume for each well should be less than 40 µl for a standard gel or less than 10 µl for a minigel. Overloading may cause samples to float out, resulting in contamination among wells. The markers should be loaded into the very left or right well or both wells.

Carrying Out Electrophoresis

1. Carefully insert the assembled unit with loaded samples into the electrophoresis tank filled with 2 to 4 l of running buffer. Remove any bubbles trapped at the bottom of the sandwich using a glass Pasteur pipette hook or an equivalent tool.

Note: The volume of running buffer in the tank should be sufficient to cover two thirds of the sandwich. Otherwise, the heat generated during electrophoresis will not be uniformly distributed, causing distortion of the band patterns in the gel.

2. Place the lid, or cover, on the apparatus and connect to a power supply. The cathode (negative electrode) should be connected to the upper buffer chamber and the anode (positive electrode) must be connected to the bottom buffer reservoir. Proteins negatively charged by SDS will migrate from the cathode to the anode.
3. Apply a constant power or current at 25 to 30 mA/1.5 mm thick standard size gel. Allow the gel to run for several hours or until the dye reaches about 1 cm from the bottom of the gel. If an overnight run is desired, the current should be adjusted empirically.
4. Turn off the power and disconnect the power cables. Loosen and remove the clamps from the sandwiches. Place the gel sandwiches on the table or bench. Carefully remove the spacers and separate the two glass plates starting from one corner by inserting a spacer into the sandwich and simultaneously lifting the top glass plate. Using a razor blade, make a small trim at the upper left or right corner of the gel to record the orientation. One gel will be stained to determine the MW and patterns of different protein bands. The other gel will be used for western blotting.

Protocol 4. Staining and Destaining of SDS-PAGE Using a Modified Coomassie Blue (CB) Method (Figure 11.2)

1. Wear gloves and carefully transfer one of the gels into a glass or plastic tray containing 100 to 150 ml of CB staining solution and allow the gel to be stained for 30 min in a 50°C water bath with occasional agitation.
2. Replace the staining solution with 150 ml of rapid destaining solution. Allow destaining to take place for 60 to 80 min on a shaker with slow shaking. Change the solution once every 20 min.

Notes: (1) Major protein bands should be visible after 20 min of destaining; all patterns of bands with a clear background will appear following 80 min of destaining. If desired, the destained gel can be photographed or dried or kept in 50% diluted destaining solution in distilled water for up to 1 month. (2) The gel may be rapidly destained using 1 to 1.5% bleach solution in distilled water. Allow destaining to proceed for 30 min with agitating. Constantly monitor the destaining and immediately rinse the gel with a large volume of top water once clear bands are visible. Change distilled water several times until the blue color of the bands is stable. The drawback to this procedure is that, if overly destained, some bands may disappear. (3) The gel can be stained by the silver method, which is more sensitive than Coomassie blue staining. Some weak bands seen in Coomassie blue staining can become sharp bands after silver staining. However, staining and destaining procedures are quite complicated and time consuming, so silver staining is not considered to be within the scope of this lab protocol.

3. To determine the MW of specific bands accurately, measure the distance from the top of the gel to each of the bands in the protein standard marker lane and to specific bands in the protein sample lanes. Calculate the R_f for each band: divide the distance from the top of the gel to a specific band by the distance from the top of the gel to the dye front. Make a standard curve by plotting the R_f of each standard marker band as the abscissa and the $\log_{10}$ of each MW as the ordinate. The MW of an unknown band in the protein sample can readily be determined by finding its R_f that vertically crosses the standard curve and the $\log_{10}$ MW that horizontally crosses to the ordinate. The antilog of the $\log_{10}$ MW is the actual MW of the protein band.

Note: If accurate determination of protein bands is not necessary, the MW of an unknown band can be estimated by comparing the position of a specific band with those of standard marker bands.

Protocol 5. Transfer of Proteins from SDS-PAGE onto a Nitrocellulose or PVDF Membrane by Electroblotting (Figure 11.3)

Method A. Semidry Electroblotting

1. Following SDS-PAGE, soak the gel in 200 ml of transfer buffer for 10 min.
2. Cut a piece of polyvinylidene difluoride (PVDF) or nitrocellulose membrane and 10 sheets of Whatman filter paper to the same size as the gel and soak them in transfer buffer for 4 min prior to blotting. If PVDF is used, briefly soak the membrane in methanol for 30 s prior to soaking in transfer buffer.
3. Assemble a blotting system in the order shown in Figure 11.3 using a Multiphor II electrotransfer apparatus (Pharmacia Biotech).
   a. Rinse the bottom plate (the negative electrode) in distilled water and place five sheets, one by one, of the soaked 3MM Whatman paper on

the plate. Remove any bubbles underneath by rolling a pipette over the filter papers.

b. Place the soaked membrane on the top of 3MM Whatman sheets. Gently press the membrane to remove any bubbles underneath.

c. Carefully overlay the membrane with the gel, starting at one side of the gel, then slowly proceed to the other end. The marked side of the membrane should face the gel. Carefully remove any bubbles between the membrane and the gel, which can be done by gently pressing or rolling on the membrane.

d. Gently overlay the gel with five sheets, one by one, of the presoaked 3MM Whatman paper.

e. Prewet the top plate (the positive electrode) in distilled water and carefully place the plate on top of the assembled unit.
To avoid producing bubbles, do not disturb the assembled unit lying underneath.

f. Place a bottle or beaker containing approximately 1000 ml water on the top plate to serve as a weight.

4. Connect the assembled unit to a power supply (positive to positive pole, negative to negative pole). Turn on the power and set it at a constant current. Allow blotting to proceed for 1 h at 1.2 mA/cm^2 membrane at room temperature.

Notes: (1) If several gels need to be transferred simultaneously, place four sheets of 3MM Whatman paper between each gel and membrane, and adjust the current accordingly. (2) Air bubbles should be removed from each layer of the blot by carefully rolling a pipette over the surface. It is important not to move gel, membrane and blotting paper until transfer is complete. Otherwise, bubbles will be generated and band positions are likely to be changed.

Method B. Liquid Transfer Using a Hoeffer TE 42 Blotting Apparatus

1. Soak the gel, membrane and four pieces of 3MM Whatman paper sheets (relatively bigger than the gel) in 500 ml of transfer buffer for 15 min.
2. Fill a tray large enough to hold the cassette with transfer buffer to a depth of approximate 4 cm.
3. Place one half of the cassette in the tray with the hook facing up and put one Dacron™ sponge on the cassette.
4. Place two sheets of the soaked 3MM Whatman paper on the sponge and remove any bubbles.
5. Overlay the soaked membrane on the paper sheets and remove any air bubbles between the layers.
6. Carefully place the soaked gel on the membrane and overlay the gel with two sheets of presoaked 3MM Whatman paper.

Note: Gloves should be worn. Gently press the filter papers to force out any trapped air bubbles that will locally block transfer of proteins onto the membrane.

7. Overlay another Dacron sponge on the filter sheets and place the second half of the cassette on the top of the stack.
8. Hook the two halves together by sliding them toward each other so that the hooks are engaged with the opposite half.
9. Place the assembled cassette into the blotting chamber according to the instructions. Fill the chamber with transfer buffer sufficient to cover the cassette. Place a stir bar in the chamber.

Note: It is extremely important to insert the cassette in the right place so that the membrane is between the gel and the anode (positive electrode).

10. Place the whole chamber on a magnetic stirrer plate and turn on the stir bar at low speed. Connect the apparatus to the power supply, turn on the power and set the current at 0.8 to 1 A.
11. Allow the transfer to take place for 60 to 75 min at 1 to 1.5 A.

Note: Because heat will be rapidly generated, the current should be monitored to be 1 to 1.5 amps. Otherwise, it may burn the apparatus.

Protocol 6. Blocking and Hybridization of the Membrane Filter Using Specific Antibodies

1. While the transfer is taking place, prepare blocking solution (see Materials Needed).
2. After transfer is complete, stain the membrane in 1X Ponceau S solution (Sigma Chemicals) diluted in distilled water using 10 ml per 12×14 cm^2 membrane. Allow staining to take place for 5 to 8 min at room temperature and rinse the membrane in distilled water several times.

Note: A successful transfer should show red bands. Staining will not interfere with the following steps.

3. After rinsing, block the nonspecific binding sites on the membrane in blocking solution (100 ml per 12×14 cm^2 membrane) at room temperature for 50 min with slow shaking.

Note: The membrane at this stage may be kept in the blocking solution for up to 48 h at 4°C. To check the efficiency of transferring, the blotted gel may be stained with Coomassie blue and then destained. A successful blotting should have no visible bands left in the gel.

4. Replace the blocking solution with antibody solution containing primary antibodies (monoclonal or polyclonal antibodies against the protein of interest) and appropriate secondary antibodies diluted in PBS buffer. Allow incubation to proceed at room temperature for 70 to 100 min with slow shaking.

***Please pay attention to the following notes**: (1) Traditionally, the membrane is separately incubated with primary antibodies, washed several times and then incubated with secondary antibodies. In our experience, the membrane can be simultaneously incubated with first and second antibodies with high efficiency of hybridization. This saves at least 1.5 h. (2) The concentration of antibodies varies with their activity and specificity. Therefore, different* *dilutions, in case of new antibodies, must be tested in order to obtain the optimal concentration for hybridization. (3) If the primary antibodies are monoclonal antibodies from mice, the secondary antibodies should be commercial goat antimouse* IgA, IgG, *and* IgM *conjugated with alkaline phosphatase or with peroxidase or biotin labeled. If the primary antibodies are polyclonal antibodies from rabbits, the secondary antibodies should be goat antirabbit* IgG *conjugated with alkaline phosphatase or with peroxidase or biotin labeled. (4) The standard volume of antibody solution should be 30 to 40 ml per 12 × 14 cm² membrane.*

5. Transfer the hybridized membrane to a clean tray containing 200 ml/membrane of PBS. Allow washing to proceed for 60 min with slow shaking. Change PBS once every 20 min. Proceed to detection of hybridized signals.

Note: (1) The antibody solution can be stored at 4°C and be reused three to four times. (2) It is optional to add 0.05% (v/v) Tween-20 to PBS to wash the membrane.

Protocol 7. Detection of Hybridized Signals

Method A. Chemiluminescent Detection for Peroxidase-Conjugated or Biotin-Labeled Antibodies

1. Place a piece of SaranWrap film on a bench and add 0.8 ml/filter (12 × 14 cm²) of ECL detection solution (Amersham Life Science) or an equivalent to the center of the film.
2. Wear gloves and briefly dampen the filter to remove excess washing solution; thoroughly wet the protein-binding side by lifting and overlaying the filter with the solution several times.
3. Wrap the filter with SaranWrap film, leaving two ends of the film unfolded. Place the wrapped filter on a paper towel and, using another piece of paper towel, carefully press it to wipe out excess detection solution through the unfolded ends of the film. Excess detection solution will most likely cause a high background.
4. Completely wrap the filter and place it in an exposure cassette with protein-binding side facing up. Tape the four corners of the filter.
5. In a darkroom with a safe light on, overlay the filter with an x-ray film and close the cassette. Allow exposure to proceed at room temperature for 10 s to 15 h, depending on the intensity of the detected signal.
6. In a darkroom, develop and fix the film in an appropriate developer and fixer, respectively. If an automatic x-ray processor is available, development, fixation, washing and drying of the film can be completed in 2 min. If a hybridized signal is detected, it appears as a black band on the film (Figure 11.5).

Note: Multiple films may need to be exposed and processed until the signals are desirable. Exposure for more than 4 h may generate a high black background. In our experience, good hybridization and detection should display sharp bands within 1.5 h. In addition, the film should be slightly overexposed to obtain a relatively dark background that will help identify the sizes of the bands compared with the marker bands.

Method B. Colorimetric Detection for Alkaline Phosphatase-Conjugated Antibodies

1. Place the filter in color developing solution (10 ml/filter, 12 × 14 cm^2) containing 40 μl of NBT stock solution and 30 μl of BCIP stock solution. NBT and BCIP stock solutions are commercially available.
2. Allow color to develop in a dark place at room temperature for 15 to 120 min or until a desired detection is obtained. Positive signals should appear as a blue/purple color.

Congratulations on your successful western blot hydribidation!

ANALYSIS OF PROTEINS BY 2-D GEL ELECTROPHORESIS

Two-dimensional (2-D) gel electrophoresis consists of a first-dimension gel, which is IEF or isoelectric focusing gel, and second-dimension gel, or SDS-PAGE. IEF can separate proteins based on their isoelectric points (pI value); SDS-PAGE separates proteins according to their molecular weight. Generally, 2-D does not need to be performed and one-dimensional gel or SDS-PAGE can provide sufficient information for protein expression. However, a detailed characterization of a specific protein such as protein purification and sequencing may mandate 2-D gel electrophoresis. The major disadvantage of one-dimensional gel electrophoresis using SDS-PAGE is that many proteins with the same molecular weight cannot separate. Instead, these proteins are shown in a single band in the gel. Normally, up to 30 to 40 bands can be seen in a single gel. In contrast, the major advantage of 2-D gel electrophoresis can provide incredible information about the pI values, molecular weight and species of proteins in a sample. If all goes well, hundreds of protein molecules can be revealed.

Detailed and simplified instructions for performance of IEF gel are available from the Immobilize DryStrip Kit (Pharmacia). The protocols for the second dimension gel or SDS-PAGE have been described previously.

Materials Needed

Microcentrifuge
Sterile microcentrifuge tubes (0.5 or 1.5 ml)
Pipettes or pipetman (0 to 200 μl, 0 to 1000 μl)
Sterile pipette tips (0 to 200 μl, 0 to 1000 μl)
Gel casting plates
Gel combs
Electrophoresis apparatus

DC power supply
Total protein samples
Specific primary antibody
Appropriate secondary antibody
PVDF or nitrocellulose membranes
3MM Whatman filters
Multiphor II electrosemidry transfer apparatus (Pharmacia Biotech) or Hoeffer TE 42 electroliquid blotting apparatus
Shaker
Plastic trays

Monomer Solution (30% Acrylamide)

116.8 g Acrylamide
3.2 g *N,N*-Methylene-*bis*-acrylamide
Dissolve after each addition in 300 ml dd.H_2O. Add dd.H_2O to a final volume of 400 ml. Wrap the bottle with foil and store at 4°C.
Caution: Acrylamide is neurotoxic. Gloves should be worn when working with this chemical.

Separating Gel Buffer

72.6 g 1.5 *M* Tris
Dissolve well in 250 ml dd.H_2O.
Adjust pH to 8.8 with 6 *N* HCl.
Add dd.H_2O to 400 ml.
Store at 4°C.

Stacking Gel Buffer

24 g 0.5 *M* Tris
Dissolve well in 200 ml dd.H_2O.
Adjust pH to 6.8 with 6 *N* HCl.
Add dd.H_2O to 400 ml.
Store at 4°C.

10% SDS

15 g SDS
Dissolve well in 150 ml dd.H_2O.
Store at room temperature.

1% Ammonium Persulfate (AP)

0.2 g AP
Dissolve well in 2 ml dd.H_2O.
Store at 4°C for up to 5 days.

Overlay Buffer

25 ml Running gel buffer
1 ml 10% SDS solution

Add dd.H_2O to 100 ml.
Store at 4°C.

2X Denaturing Buffer

5 ml Stacking gel buffer
8 ml 10% SDS solution
4 ml Glycerol
2 ml β-Mercaptoethanol
Add dd.H_2O to 20 ml.
0.02 g Bromophenol blue
Divide in aliquots and store at –20°C.

Running Buffer

12 g 0.25 *M* Tris
57.6 g Glycine
40 ml 10% SDS solution
Add dd.H_2O to 4 l.

1% (w/v) Coomassie Blue (CB) Solution

2 g Coomassie blue R-250
Dissolve well in 200 ml dd.H_2O. Filter using 3MM Whatman paper.

Coomassie Blue Staining Solution

62.5 ml 1% CB solution
250 ml Methanol
50 ml Acetic acid
Add dd.H_2O to final 500 ml.

Rapid Destaining Solution

500 ml Methanol
100 ml Acetic acid
Add dd.H_2O to final 1000 ml.

Transfer Buffer

39 m*M* Glycine
48 m*M* Tris
0.04% SDS
20% Methanol

Phosphate Buffered Saline (PBS)

2.7 m*M* KCl
1.5 m*M* KH_2PO_4
136.9 m*M* NaCl
15 m*M* Na_2HPO_4
Adjust the pH to 7.2 to 7.4.

Blocking Solution

5% (w/v) Nonfat dry milk or 1 to 2% (w/v) gelatin or 2 to 3% (w/v) bovine serum albumin (BSA) in PBS solution. Allow the powder to be completely dissolved by shaking for 30 min prior to use.

Predetection/Color Developing Buffer

0.1 *M* Tris-HCl, pH 9.5
0.1 *M* NaCl
50 m*M* $MgCl_2$

NBT Stock Solution

75 mg/ml Nitroblue tetrazolium (NBT) salt in 70% (v/v) dimethylformamide

BCIP Stock Solution

50 mg/ml 5-Bromo-4-chloro-3-indolyl phosphate (BCIP or X-phosphate), in 100% dimethylformamide

Others

Chemiluninescent substrates
Appropriate primary antibodies
Appropriate secondary antibodies

TROUBLESHOOTING GUIDE

1. **Following electroblotting, the blotted membrane shows weak bands after Ponceau S staining, whereas the blotted gel displays clear bands stained with Coomassie blue.** Obviously, the efficiency of protein transfer is very low. Try to set an appropriate transfer and to increase transfer time.
2. **Protein bands are not sharp but are quite diffused.** Current may be changed severely. Try to set up a constant power for electrophoresis.
3. **No signal at all is evident.** This is the worst situation in western blot hybridization and multiple factors may cause the problem.
   a. Insufficient antigen (protein) was loaded into the gel. Try to increase the amount of proteins for each well.
   b. Proteins may not be transferred at all onto the membrane. Make sure to stain the blotted membrane or use color protein markers as indicators.
   c. The activity of primary or secondary antibodies, or both, is extremely low or does not work at all. Try to check the quality of antibodies by loading each in a separate well as controls in the same gel. If no signals are shown, it is most likely that western blot process has not been handled properly.
4. **Hybridized signals are quite weak.** The concentration of antigen is low or the activities of the antibodies may be low. To solve this problem, use

appropriate concentrations of proteins or antibodies, increase exposure time for x-ray film or develop the membrane filter for a longer time.

5. **A black background on x-ray film occurs when using chemiluminescent detection.** Multiple factors may be responsible for this common problem. The membrane filter may be dried out during antibody incubation or the washing process. Excess detection solution may not be wiped out prior to exposure. Exposure time may be allowed for too long. The solution is that the filter should be kept wet and excess detection reagents need to be wiped out. Try to reduce exposure time for x-ray film.
6. **A highly purple/blue background is evident on the filter using colorimetric detection.** Excess detection reagents may have been used or color developed too long. Try to apply an appropriate amount of detection reagents and pay close attention to color development. Once major bands become visible, stop development immediately by rinsing the filter with distilled water several times.
7. **Nonspecific bands show up on x-ray film or membrane filter.** This problem is most likely caused by the nonspecificity of antibodies. Make sure that the antibodies used are specific. One possibility is to increase blocking reagents and blocking time to block the nonspecific binding sites on the blotted membrane and elevate washing conditions using 0.05% Tween-20 in PBS.

REFERENCES

1. Wu, W. and Welsh, M.J., Expression of the 25-kDa heat-shock protein (HSP27) correlates with resistance to the toxicity of cadmium chloride, mercuric chloride, *cis*-platinum(II)-diammine dichloride, or sodium arsenite in mouse embryonic stem cells transfected with sense or antisense HSP27 cDNA, *Toxicol. Appl. Pharmacol.,* 141, 330, 1996.
2. Wu, W. and Welsh, M.J., A method for rapid staining and destaining of polyacrylamide gels, *BioTechniques,* 20, 3, 386–388, 1996.
3. Wu, W., Electrophoresis, blotting, and hybridization, In *Handbook of Molecular and Cellular Methods in Biology and Medicine,* Kaufman, P.B., Wu, W., Kim, D., and Leland, C., pp. 87–122, CRC Press, Boca Raton, FL, 1995.
4. Bradford, M.M., A rapid and sensitive method for the quantitation of microgram quantities of protein utilizing the principle of protein-dye binding, *Anal. Biochem.,* 72, 248–254, 1976.
5. Towbin, J., Staehlin, T., and Gordon, J, Electrophoretic transfer of proteins from polyacrylamide gels to nitrocellulose sheets: procedure and some applications, *Proc. Natl. Acad. Sci. USA,* 76, 4350, 1979.
6. Knudsen, K.A., Proteins transferred to nitrocellulose for use as immunogens, *Anal. Biochem.,* 147, 285, 1985.
7. Johnson, D.A., Gautsch, J.W., Sportsman, J.R., and Elder, J.H., Improved method for utilizing nonfat dry milk for analysis of proteins and nucleic acids transferred to nitrocellulose, *Gene Anal. Technol.,* 1, 3, 1984.

8. Kyhse-Anderson, J., Electroblotting of multiple gels: a simple apparatus without buffer tank for rapid transfer of proteins from polyacrylamide to nitrocellulose, *J. Biochem. Biophys. Methods,* 10, 203, 1984.

12 Analysis of Cellular DNA or Abundance of mRNA by Radioactivity *in Situ* Hybridization (RISH)

CONTENTS

INTRODUCTION

Traditional methods for analysis of genomic DNA or mRNA expression include Southern blot or northern blot hybridization. In spite that the ability of Southern or northern blot hybridization to provide quantitative information on amounts of DNA or RNA in specific tissues, they cannot reveal spatial information on the distribution of DNA or RNA of interest within cells. Recently, a powerful and versatile technology has been developed to fulfill such a task: *in situ* hybridization.[1,2] This technique becomes a very useful tool for scientists to simultaneously detect and localize specific DNA or RNA sequences with spatial information about their subcellular locations or locations within small subpopulations of cells in tissue samples. *In situ* hybridization can identify sites of gene expression, tissue distribution of mRNA, and identification and localization of viral infections.[3–5]

There are two major commonly used procedures for *in situ* hybridization and detection of cellular DNA or RNA.[6,7] One is radioactivity *in situ* hybridization (RISH) using a radioactive probe[6] and the other is fluorescence *in situ* hybridization (FISH), which utilizes nonradioactive probes.[7] Each has its own strengths and weaknesses. Based on our experience, the RISH approach allows performing hybridizations at high temperatures (e.g., 65°C) and to employ washing under high-stringency conditions. As a result, a good signal-to-noise ratio can be obtained. However, the morphology of cells or tissues may not be well preserved.

In contrast, with the FISH approach, good images of cellular structure can be preserved, but this method has two major disadvantages. One is that with FISH the hybridization and washing procedures cannot easily be performed at high temperatures or under highly stringent conditions. The other limitation is that during final color development background cannot be readily controlled. As a result, nonspecific background may be a concern. The present chapter describes in detail the protocols for the success of *in situ* hybridization using isotopic probes. These methods have been well tested and have worked successfully in our laboratories.

PART A. TISSUE FIXATION, EMBEDDING, SECTIONING AND MOUNTING OF SECTIONS ON SLIDES

PREPARATION OF SILANE-COATED GLASS SLIDES

High-quality glass slides play an essential role in preventing cells or tissue sections from falling off the slides during *in situ* hybridization. For this reason, regular glass slides should be precoated prior to use for *in situ* hybridization. Precoated glass slides are commercially available (e.g., Perkin–Elmer Corporation). Next, three commonly utilized methods that work well in our laboratories are described.

Method A. Silanization of Slides

1. Wash glass slides in 2 *N* HCl for 7 min using slide racks and glass staining dishes, and rinse the slides in dd.H_2O for 2 × 1 min.
2. Rinse the slides in high-grade acetone for 1.5 min and air-dry.
3. Treat the slides in 2% organosilane solution (Aldrich Chemical Co.) diluted in high-grade acetone for 1.5 min with gentle agitation followed by rinsing in high-grade acetone for 1.5 min.
4. Air-dry the slides and store them in a box at room temperature until use. The slides can be stored up to 4 years.

Method B. Treatment of Slides with Poly-L-Lysine

1. Clean glass slides as described at step 1 and step 2 in Method A.
2. Immerse the slides in freshly prepared poly-L-lysine solution (Sigma, MW > 150,000, 1 mg/ml in dd.H_2O) for 5 min.

Tip: Air bubbles should be avoided during submersion of slides.

3. Air-dry the slides and store at 4°C for up to 7 days.

Method C. Gelatin-Coating of Slides

1. Wash glass slides as described at step 1 and step 2 in Method A.
2. Coat the slides in 1% (w/v) gelatin in dd.H_2O for 5 min at 4°C.
3. Air-dry the slides overnight at room temperature and store in a box at room temperature up to 4 weeks.

PREPARATION OF FIXATION SOLUTIONS

Selection of a fixative is very critical to the preservation of cellular or tissue morphology during *in situ* hybridization. Based on our experience, 10% buffered-formalin solution (pH 7.0) is an excellent fixative, or a paraformaldehyde solution can be used as a fixative as prepared below.

Preparation of 4% (w/v) Paraformaldehyde (PFA) Fixative (1 l)

1. Carefully weigh out 40 g paraformaldehyde powder and add it to 600 ml preheated dd.H_2O (55 to 60°C) and allow paraformaldehyde to dissolve at the same temperature with stirring.

Caution: Paraformaldehyde is toxic and should be handled carefully. The preparation should be carried out in a fume hood on a heating plate. Temperature should not be above 65°C and the fixative should not be overheated.

2. Add one drop of 2 *N* NaOH solution to clear the fixative.
3. Remove from heat and add 333.3 ml (1/3 total volume) of 3X PBS (8.1 m*M* KCl, 4.5 m*M* KH_2PO_4, 411 m*M* NaCl, 45 m*M* Na_2HPO_4, pH 7.2, autoclaved).
4. Adjust pH to 7.2 with 4 *N* HCl and bring final volume to 1 l with dd.H_2O.
5. Filter the solution to remove undissolved fine particles, cool to room temperature and store at 4°C until use.

FIXATION OF CULTURED CELLS ON SLIDES

1. Briefly spin down suspended or cultured cells of interest and resuspend them in serum-free medium at a density of 10^7 cells/ml.
2. Transfer 10 to 15 μl of the cells onto a precoated glass slide and allow cells to attach to the slide in a humid chamber for 40 min.

Three to five replicates should be set up for each cell line. Alternatively, cells can be grown directly on coated glass slides or on coverslips.

3. Place the slides into slide racks and carefully immerse them in a glass staining dish containing a sufficient volume of 10% buffered-formalin or 4% PFA fixation solution.
4. Allow fixation to proceed at room temperature.

Note: Fixation is critical for the success of the in situ *hybridization procedure. Fixation time varies with cell lines, ranging from 30 min to 10 h. Therefore, the optimal fixation time should be determined empirically for each cell line.*

5. Carefully transfer the slide rack to a staining dish filled with 3X PBS and allow to incubate for 4 min to stop fixation.
6. Rinse the slides 3 × 4 min with 1X PBS in fresh staining dishes.
7. Carry out complete dehydration by transferring the slide rack through a series of 6-min treatments in ethanol dishes (50%, 70%, 95%, 100%, 100%, and 100%).
8. Completely dry the slides at room temperature and store at –80°C in a box containing desiccant until use.

Note: The slides must be completely dried prior to being placed in a freezer in order to prevent potential artifacts during freezing.

TISSUE FIXATION, EMBEDDING, SECTIONING AND MOUNTING OF SECTIONS ON SLIDES

FIXATION

1. Label 10- or 20-ml snap-cap glass vials and their caps according to the samples, and fill 2/3 volume of each vial approximately with freshly prepared 10% buffered-formalin or 4% PFA fixation solution.
2. Place freshly dissected tissues, organs or embryos in the appropriate vials and allow fixation to proceed at 4°C for an optimal time period.

Note: Tissue fixation is a key step for the success of in situ *hybridization and the desired fixation time varies with tissues. For good morphology and a good signal-to-noise ratio following* in situ *hybridization, an optimal time should be determined empirically. Based on our experience, series of fixation times ranging from 30 min to 15 h should be set up for each sample.*

3. While the samples are being fixed, melt an appropriate amount of paraffin wax in a clean glass beaker in a 60°C oven.
4. After fixation, carry out dehydration of samples by carefully removing the fixation solution from the vials and replacing it with 10 ml of 50% ethanol.
5. Perform complete dehydration by transferring the slides every 20 min through an ethanol series (50%, 70%, 95%, 100%, 100%, and 100%) and 2×15 min in xylene.

Note: Tissue dehydration must be as complete as possible. Otherwise, melted paraffin wax may not thoroughly penetrate the tissue, resulting in poor paraffin blocks.

Caution: Xylene is a toxic solvent and should be carefully handled. It is strongly recommended that any steps with xylene be carried out in a hood. Used xylene should be collected in a special container for toxic waste disposal.

Embedding

1. Following the previous step 5, carry out partial impregnation of the samples by quickly filling each vial with 4 ml of freshly mixed xylene and melted wax (1:1, v/v) using a hot glass Pasteur pipette. Allow incubation to proceed overnight at room temperature.
2. Melt the mixture in a 60°C oven and immediately replace the xylene/wax mixture with fresh melted wax. Allow impregnation of the samples to proceed at 60°C for 1 h.
3. Continue wax infiltration of the tissue samples by quickly replacing with fresh melted wax every 20 min using a hot glass Pasteur pipette. Repeat this step 10 times. The samples are ready for embedding.
4. Prepare embedding boats using aluminum foil or molds according to the manufacturer's instructions. Appropriate molds and boats are also commercially available.
5. Fill embedding molds with melted paraffin wax using a hot glass Pasteur pipette and quickly transfer samples to appropriate molds using a hot cut-off Pasteur pipette. Orient the samples within the mold with a hot drawn-out and sealed Pasteur pipette.

Note: This step should be performed gently to avoid any bubbles that may become trapped inside the blocks. Place one sample in each block. Every embedding boat or mold should be clearly labeled for sample identification.

6. Allow cast blocks to harden completely at room temperature or place boats or molds into cold water in a tray.

7. Remove the blocks from the embedding boats or molds and store the blocks in a dry place at room temperature until sectioning.

Note: It is important to store the blocks in a dry place. Otherwise, moisture may ruin the paraffin wax blocks. The blocks can be stored for years under these conditions.

Sectioning and Mounting

1. Using a razor blade, carefully cut each of the paraffin wax blocks containing the embedded samples into a trapezoidal shape and continue to trim the sectioning surface until approximate 1 to 2 mm from the sample.
2. Place the trapezoid block in the holding clamp of a microtome with the wide edge of the trapezoid surface facing the knife.
3. Set up the thickness of sectioning at 4 to 10 μm according to the manufacturer's instructions for a particular microtome and start to cut section ribbons.
4. Label previously coated glass slides according to sample blocks and add one to three drops of dd.H_2O or 0.2% (w/v) gelatin subbing solution to a slide. Place the slide onto a heating plate set at 50°C.
5. Carefully transfer the ribbon of sections (3 to 10 sections) onto the gelatin- or water-drop on the slide using a fine brush. Allow the ribbon of sections to stretch to remove any wrinkles until complete drying of the slide has occurred (Figure 12.1).
6. To attach the sections to slides firmly, place the slides at 40°C for 24 h.
7. Store the slides in a slide box with desiccant at –20°C until use.

CRYOSECTIONING

It has been demonstrated that frozen tissue sections are particularly suitable for RNA *in situ* hybridization, immunocytochemistry and enzyme histochemistry. The major advantages of cryosectioning over traditional paraffin wax sectioning are that tissue embedding and sectioning are carried out under frozen conditions, and that activities of DNase, RNase and protease are greatly inhibited. Therefore, potential degradation of DNA, RNA or proteins is minimized, which is essential to *in situ* hybridization and enzyme histochemistry. For this reason, the present section describes in detail the protocols for tissue freezing, specimen preparation, sectioning and fixation.

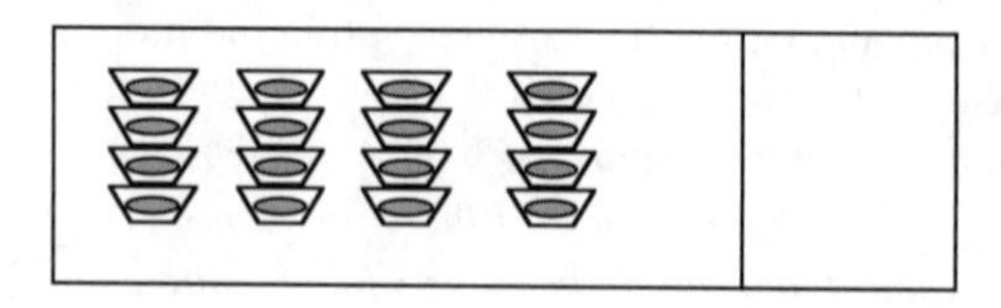

FIGURE 12.1 Diagram of section ribbons with a trapezoidal shape and mounting on a coated glass slide.

Preparation of Frozen Specimens

1. Chill all tools employed for cryosectioning in the cryostat chamber for 15 min, including razor blades, brushes and Pasteur pipettes. However, glass slides should not be prechilled.
2. Prechill a 50-ml Pyrex beaker or its equivalent in a container filled with liquid N_2 for 1 min and add a sufficient volume of CryoKwik to the beaker.
3. Label one end of a filter paper strip (3MM Whatman) with a pencil according to the specific sample and carefully place a specimen on the other end of the strip with forceps or a Pasteur pipette.
4. Carefully immerse the specimen in cold CryoKwik for 1 min and then transfer the specimen strip into a cryostat chamber. Alternatively, the specimen at this stage can be stored at –80°C until use.
5. Place a thin layer of OCT compound on the cutting chucks and place the specimen end of the strip onto the layer of OCT compound.
6. Cool the specimen/chuck mount within the cryostat with a precooled heat sink until the compound solidifies.
7. Tear off the filter paper and allow the specimen within the cryostat to reach the same temperature while the microtome is being chilled.
8. Carefully trim the specimen block to a trapezoid shape using a prechilled razor blade.

Freeze Sectioning

1. Carefully mount the chuck containing the specimen in a microtome with the trapezoid surface parallel to the knife edge. Slowly retract the chuck until it clears the knife edge.
2. Set the section thickness at 8 to 10 μm in the microtome according to the manufacturer's instructions and prepare a smooth surface on the block by rapid speed cutting (e.g., one cut per second) until it is close to the specimen.
3. Before sectioning the sample, flip the plastic roll-plate down for collecting sections. The plate should just be touching the knife and lie parallel to the knife edge, but slightly behind the cutting edge of the knife.
4. Start sectioning by turning the crank. Once sections roll up at the cutting edge of the knife, lift the plastic roll plate and carefully expand the section onto the knife using a fine brush.
5. Transfer sections onto warm prelabeled and precoated glass slides by carefully overlaying the slide onto the section (Figure 12.2).
6. Repeat step 4 and step 5 until sectioning is complete.

Fixation

Note: Frozen sections should be immediately fixed except for doing enzyme histochemistry.

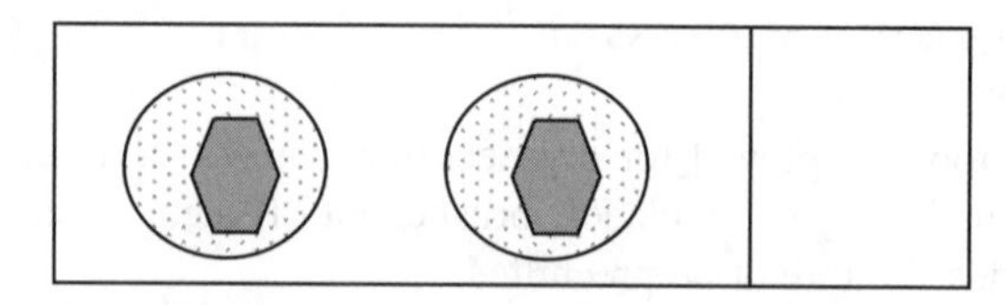

FIGURE 12.2 Diagram of thin cryosectioning ribbons and mounting on a coated glass slide.

1. Prepare a moisture chamber by adding a minimum volume of distilled water to an appropriately sized tray and placing a number of plastic or glass Pasteur pipettes on the bottom parallel to each other.
2. Place slides with sections on the pipettes in the moisture chamber and cover the sections with an appropriate volume of 4% PFA fixative or 10% buffered-formalin.
3. Cover the chamber and allow fixation to proceed for the desired time at room temperature.

Note: An optimal fixation time should be determined empirically for each tissue sample. In general, a range from 20 min to 8 h should be tested.

4. Carefully aspirate off the fixation solution and rinse with PBS for 6 min. Repeat washing with PBS three times.
5. Carry out dehydration by replacing PBS with a series of ethanol (40%, 50%, 75%, 95%, 100%, and 100%), 4 min each.
6. Air-dry the slides and store them in a slide box in an airtight container with desiccant at –80°C until use.

Note: Slides should be prewarmed to room temperature prior to use.

PART B. *IN SITU* HYBRIDIZATION AND DETECTION USING ISOTOPIC PROBES

This section describes successful techniques for *in situ* hybridization using a radioactive probe. Procedures include dewaxing and rehydration/hydration of sections, protease digestion, DNase or RNase treatment, preparation of isotopic probes, hybridization and detection of cellular DNA or RNA.

DEWAXING OF SECTIONS

Paraffin wax sections prepared in Part A must be subjected to complete dewaxing to allow treatments with protease, DNase, or RNase as well as penetration of the probe into the tissue. Dewaxing of sections can be completed readily with xylene. For unembedded cultured cells and cryosections, dewaxing is not necessary.

1. Transfer sections from the freezer to a slide rack and allow them to reach room temperature for a few minutes prior to use.
2. Fill three staining dishes with fresh xylene and two with 100% ethanol.
3. Dewax the sections by placing the slide rack into a series of three changes of xylene, prepared at step 2, for 3 min in each solution.

4. Wash the sections in two changes of 100% ethanol, prepared at step 2, for 3 min each.
5. Air-dry the slides and proceed to protease digestion.

PROTEASE DIGESTION

Protease treatment is applied to sections to be used for *in situ* hybridization of DNA or RNA but not to those sections utilized for immunohistochemistry or enzyme histochemistry. An appropriate protease digestion of sections removes some of the proteins from the cellular network to make DNA or RNA more accessible for *in situ* hybridization. It subsequently enhances the hybridized signal of interest, reducing background as well.

1. Prepare a sufficient volume of protease solution as follows:
   Proteinase K solution: 1 mg/ml in 50 m*M* Tris-HCl, pH 7.5.
   Pepsin solution: 2 mg/ml in dd.H_2O, pH ~2.0 (40 mg pepsin in 19 ml dd.H_2O and 1 ml 2 *N* HCl).
   Trypsin solution: 2 mg/ml in dd.H_2O (40 mg pepsin in 19 ml dd.H_2O and 1 ml 2 *N* HCl).

Notes: (1) Protease solutions can be stored at –20°C for up to 2 weeks. (2) Although each of the three protease solutions works equally well in our laboratories, we recommend using pepsin or trypsin solution. Proteinase K digestion is not readily controlled compared with pepsin or trypsin. As a result, overdigestion of tissue usually occurs. (3) Proteinase K is relatively difficult to inactivate or wash away from the tissue. In contrast, it is easy to inactivate and wash pepsin or trypsin by simple adjustment of the pH from 2.0 to 7.2 to 7.5, which is the standard pH of PBS.

2. Carefully overlay the tissue sections with an appropriate volume of the protease solution.

Add protease solution to the tissue sections instead of immersing the slide rack into a staining dish containing a large volume of protease solution. Otherwise, protease covers the entire surface of the slides and it is not readily washed away after treatment.

3. Carry out digestion of tissues at 37°C for 10 to 30 min.

Notes: Protease digestion is a tricky step for the success of in situ *hybridization. Insufficient protease digestion may result in a diminished or completely absent hybridization signal. If overdigestion occurs, the tissue morphology may be poor or even completely destroyed. In order to obtain a good signal-to-noise ratio and a good preservation of tissue morphology, it is recommended to determine the optimal digestion time for each tissue sample. Different tissues may have different optimal digestion times. We usually set up a series of protease treatment times (5, 10, 15 and 30 min) for tissue sections fixed for 0.5 to 2, 3 to 5, 7 to 10 and 12 to 18 h, respectively. Each test should consist of two or three replicates. It is hoped that these trial tests will provide investigators with a useful guide to determine the optimal digestion time for specific* in situ *hybridizations.*

4. After protease digestion is complete, stop proteolysis by rinsing the slides in a staining dish containing PBS (pH 7.4) with 2 mg/ml glycogen. Perform three changes of fresh PBS, 1 min each. Alternatively, wash the slides for 1 min in a solution containing 100 m*M* Tris-HCl and 100 m*M* NaCl.
5. Post-fix the specimens in freshly prepared 10% buffered-formalin solution or 4% PFA fixative for 6 min at room temperature.
6. Rinse the slides with PBS three times, 1 min for each rinse.
7. Rinse the slides in 100% ethanol for 1 min and air-dry. At this stage, the sections will be subjected to different treatments, depending on *in situ* hybridization with cellular DNA or with RNA.

DNASE TREATMENT FOR *IN SITU* HYBRIDIZATION OF RNA

This treatment serves to remove cellular DNA from tissue. We have found that genomic DNA may compete with the RNA species of interest in hybridizing with a specific probe. Subsequently, DNase digestion can reduce nonspecific background for in situ *hybridization of RNA.*

1. Overlay the tissue sections with an appropriate volume of commercially available RNase-free DNase solution (e.g., Boehrinnger Mannheim Corporation).
2. Allow digestion to proceed for 30 min at 37°C.
3. After digestion is complete, inactivate DNase by rinsing the slides in a staining dish containing a solution of 10 m*M* Tris-HCl (pH 8.0), 150 m*M* NaCl and 20 m*M* EDTA for 5 min.
4. Wash the slides in TE buffer (pH 8.0) for 3 x 2 min and 1 min in 100% ethanol and air-dry. At this stage, the slides are ready for hybridization or can be stored in a slide box with desiccant at –80°C until use.

Note: In order to prevent RNase contamination, all the tools used for in situ *hybridization of cellular RNA should be completely cleaned and rinsed with DEPC-treated dd.H_2O. All the solutions should be treated with 0.1% (v/v) DEPC or prepared using DEPC-treated water. RNase contamination may completely ruin the experiment.*

RNASE TREATMENT FOR *IN SITU* HYBRIDIZATION OF DNA

Note: This treatment functions to degradate cellular RNA in the tissue inasmuch as RNA can compete with DNA species of interest in hybridizing with a specific probe. As a result, RNase digestion can enhance the hybridized signal by reducing nonspecific background for in situ *hybridization of DNA.*

1. Overlay the tissue sections with an appropriate volume of RNase solution (DNase free).
2. Allow digestion to proceed for 30 min at 37°C.
3. After digestion is complete, inactivate RNase by rinsing the slides in a staining dish containing a solution of 10 m*M* Tris-HCl (pH 8.0), 150 m*M* NaCl and 20 m*M* EDTA for 5 min.
4. Wash the slides in TE buffer (10 m*M* Tris-HCl, 1 m*M* EDTA, pH 8.0) for 3 × 2 min and 1 min in 100% ethanol and then air-dry. At this stage, the slides are ready for hybridization or can be stored in a slide box with desiccant at –80°C if the time schedule is not convenient.

PREPARATION OF RADIOACTIVE PROBES

SYNTHESIS OF PROBES FOR DNA HYBRIDIZATION USING RANDOM PRIMER LABELING OF DSDNA

Random primer labeling kits are available from commercial companies such as Pharmacia, Promega Corporation, and Boehringer Mannhein Corporation. A kit usually includes a population of random hexanucleotides synthesized by an automated DNA synthesizer. Such a mixture of random primers contains all four bases in each position. The primers can be utilized to primer DNA synthesis *in vitro* from any denatured, closed circular or linear dsDNA as a template using the Klenow fragment of *E. coli* DNA polymerase I. Inasmuch as this enzyme lacks 5′ to 3′ exonuclease activity, the DNA product is produced exclusively by primer extension instead of by nick translation. During the synthesis, one of the four dNTPs is radioactively labeled and incorporated into the new DNA strand. By primer extension, it is relatively easy to produce a probe with extremely high specific activity (10^8 cpm/μg) because more than 70% of the labeled dNTP can be incorporated into the new DNA strand. In a typical reaction, the amount of DNA template can be as low as 25 ng.

1. Place 1 μl of DNA template (0.5 to 1.0 μg/μl in dd.H_2O or TE buffer) in a sterile microcentrifuge tube. Add an appropriate volume of dd.H_20 or TE buffer (19 to 38 μl) to the tube to dilute the DNA template to 25 to 50 ng/μl.
2. Cap the tube and denature the dsDNA into ssDNA by boiling for 6 min, and quickly chilling the tube on ice H_2O for 2 to 4 min. Briefly spin down.
3. Set up a labeling reaction by adding the following components, in the order listed:
   Denatured DNA template (25 to 50 ng), 1 μl
   5X Labeling buffer, 10 μl
   Acetylated BSA (10 mg/ml) (optional), 2 μl
   Three unlabeled dNTP mixture (500 μ*M* each), 2 μl
   [^{35}S]dCTP or [^{35}S]dATP(1000 to 1500 Ci/mmol), 2.5 to 4 μl
   Klenow enzyme, 5 units
   Add dd.H_20 to final volume of 50 μl.

Note: The standard amount of DNA template for one reaction should be 25 to 50 ng. Because the specific activity of a probe depends on the amount of the template, the lower the amount, the higher the specific activity of the probe. [^{35}S]dCTP or [^{35}S]dATP (1000 to 1500 Ci/mmol) should not be >5 µl used in the reaction to prevent potential high background.

Caution: [^{35}S]dCTP and [^{35}S]dATP are dangerous and should be handled carefully using an appropriate Plexiglas protector shield. Any gloves and waste buffer/solutions containing radioactive materials should be placed in special containers for isotopic disposal.

4. Mix well and incubate the reaction at room temperature for 60 min.
5. Stop the reaction by adding 5 µl of 200 m*M* EDTA and place on ice for 2 min.
6. Determine the percentage of [^{35}S]dCTP incorporated into the DNA and purify the probe from unincorporated nucleotides using the protocols described next.

Protocol 1. DE-81 Filter-Binding Assay

1. Dilute 1 µl of the labeled mixture in 99 µl (1:100) of 200 m*M* EDTA solution (pH 8.0). Spot 3 µl of the diluted sample onto two pieces of Whatman DE-81 filter papers (2.3 cm diameter). Air-dry the filters.
2. Rinse one filter with 50 ml of 0.5 *M* sodium phosphate buffer (pH 6.8) for 5 min to remove unincorporated cpm. Repeat washing twice. The other filter will be utilized directly to obtain the total cpm of the sample.
3. Add an appropriate volume of scintillation fluid (~10 ml) to each vial containing one of the filters. Count the cpm in a scintillation counter according to the manufacturer's instructions.

Protocol 2. TCA Precipitation

1. Dilute 1 µl of the labeled reaction mixture in 99 µl (1:100) of 200 m*M* EDTA solution and spot 3 µl on a glass fiber filter or on a nitrocellulose filter for determination of the total cpm in the sample. Allow the filter to air-dry.
2. Add 3 µl of the same diluted sample into a tube containing 100 µl of 100 µg/ml carrier DNA or acetylated BSA and 20 m*M* EDTA. Mix well.
3. Add 1.3 ml of ice-cold 10% trichloroacetic acid (TCA) and 1% (w/v) sodium pyrophosphate to the mixture. Incubate the tube on ice for 20 to 25 min to precipitate the DNA.
4. Filter the precipitated DNA on a glass fiber filter or on a nitrocellulose filter using vacuum. Rinse the filter with 5 ml of ice-cold 10% TCA four times under vacuum. Wash the filter with 5 ml of acetone (for glass fiber filters only) or 5 ml of 95% ethanol. Air-dry the filter.

5. Transfer the filter to two scintillation counter vials and add 10 to 15 ml of scintillation fluid to each vial. Count the total cpm and incorporated cpm in a scintillation counter according to the manufacturer's instructions.

Protocol 3. Calculation of Specific Activity of the Probe

1. Calculate the theoretical yield:

$$\text{ng theoretical yield} = \frac{\mu\text{Ci dNTP added} \times 4 \times 330 \text{ ng / nmol}}{\text{specific activity of the labeled dNTP } (\mu\text{Ci / nmol})}$$

2. Calculate the percentage of incorporation:

$$\%\text{ incorporation} = \frac{\text{cpm incorporated}}{\text{total cpm}} \times 100$$

3. Calculate the amount of DNA produced:

$$\text{ng DNA synthesized} = \%\text{ incorporation} \times 0.01 \times \text{theoretical yield}$$

4. Calculate the specific activity of the prepared probe:

$$\text{cpm / }\mu\text{g} = \frac{\text{total cpm incorporated (cpm incorporated} \times 33.3 \times 50)}{(\text{ng DNA synthesized} + \text{ng input DNA})\ 0.001\ \mu\text{g / ng}} \times 100$$

The factor 33.3 is derived from the use of 3 μl of 1:100 dilution for the filter-binding or TCA precipitation assay. The factor 50 is derived from the use of 1 μl of the total 50 μl reaction mixture for 1:100 dilution. For example, given that 25 ng DNA is to be labeled and that 50 μCi[^{35}S]dCTP (1000 Ci/mmol) is utilized in 50 μl of a standard reaction, and assuming that 4.92 × 10^4 cpm is precipitated by TCA and that 5.28 × 10^4 cpm is the total cpm in the sample, the calculations are as follows:

$$\text{Theoretical yield} = \frac{50\ \mu\text{Ci} \times 4 \times 330 \text{ ng / nmol}}{1000\ \mu\text{Ci / nmol}} = 66 \text{ ng}$$

$$\%\text{ incorporation} = \frac{4.92 \times 10^4}{5.28 \times 10^4} \times 100 = 93\%$$

$$\text{DNA synthesized} = 0.93 \times 66 = 61.3 \text{ ng}$$

$$\text{Specific activity} \frac{4.92 \times 104 \times 33.3 \times 50}{(61.3 \text{ ng} + 25 \text{ ng}) \times 0.001\ \mu\text{g / ng}} = 9.5 \times 10^8 \text{ cpm / }\mu\text{g DNA}$$

Protocol 4. Purification of a Radioactive Probe

It is recommended to carry out separation of a labeled probe from unincorporated isotopic nucleotide so as to avoid a high background (e.g., unexpected black spots on the hybridization filter). Use of chromatography on Sephadex G-50 spin columns or Bio-Gel P-60 spin columns is a very effective way for separation of a labeled DNA from unincorporated radioactive precursors such as [^{35}S]dCTP or [^{35}S]dATP. The oligomers will be retained in the column. This is particularly useful when an optimal signal-to-noise ratio with 150 to 1500 bases in length probe is generated.

1. Suspend 2 to 4 g Sephadex G-50 or Bio-Gel P-60 in 50 to 100 ml of TEN buffer and allow equilibration to occur for at least 1 h. Store at 4°C until use. Alternatively, such columns can be purchased from commercial companies.
2. Insert a small amount of sterile glass wool into the bottom of a 1-ml disposable syringe using the barrel of the syringe to tamp the glass wool in place.
3. Fill the syringe with the Sephadex G-50 or Bio-Gel P-60 suspension up to the top.
4. Place the syringe containing the suspension in a 15-ml disposable plastic tube and place the tube in a swinging-bucket rotor in a bench-top centrifuge. Carry out centrifugation at 1600 × *g* for 4 min at room temperature.
5. Repeat addition of the suspended resin to the syringe and continue centrifugation at 1600 × *g* for 4 min until the packaged volume reaches 0.9 ml in the syringe and remains unchanged after centrifugation.
6. Add 0.1 ml of 1X TEN buffer to the top of the column and recentrifuge as described previously. Repeat this step three times.
7. Transfer the spin column to a fresh 15-ml disposable tube and add the labeled DNA mixture to the top of the resin dropwise using a Pasteur pipette.
8. Centrifuge at 1600 × *g* for 4 min at room temperature. Discard the column containing unincorporated radioactive nucleotides into a radioactive waste container. Carefully transfer the effluent (about 0.1 ml) from the bottom of the decapped microcentrifuge tube or the 15-ml tube to a fresh microcentrifuge tube. The purified probe can be stored at –20°C until it is used for hybridization.

Preparation of [^{35}S]UTP Riboprobe for RNA Hybridization by Transcription *in Vitro* Labeling

Recently, a number of plasmid vectors have been developed for subcloning the cDNA of interest. These vectors contain polycloning sites downstream from powerful bacteriophage promoters T7, SP6, or T3 in the vector. The cDNA of interest can be cloned at the polycloning site between promoters SP6 and T7 or T3, generating a recombinant plasmid, or transcribed *in vitro* into single-strand sense RNA or antisense RNA from a linear plasmid DNA with promoters SP6, T7 or T3. During the process of *in vitro* transcription, one of the rNTPs is radioactively labeled and can be incorporated into the RNA strand, which is the labeled riboprobe. The probe has a high specific activity and is much "hotter" than a ssDNA probe. As compared with DNA labeling, a higher yield of RNA probe can be obtained because the template can be repeatedly transcribed. RNA probes can be easily purified from DNA template merely with treatment of DNase I (RNase free). The greatest merit of an RNA probe over a DNA probe is that it can produce much stronger hybridization signals.

1. Prepare linear DNA template for transcription *in vitro*. The plasmid can be linearized by an appropriate restriction enzyme for producing run-off transcripts. The recombinant plasmid should be digested with an appropriate restriction enzyme that cuts at one site very close to one end of cDNA insert.
   a. Set up a plasmid linearization reaction mixture on ice.
      Recombinant plasmid DNA (μg/μl), 5 μg
      Appropriate restriction enzyme 10 X buffer, 5 μl
      Appropriate restriction enzyme (10 to 20 u/μl), 20 units
      Add dd.H_2O to a final volume of 50 μl.
   b. Allow digestion to proceed for 2 h at the appropriate temperature, depending on the particular enzyme.
   c. Carry out extraction by adding one volume of TE-saturated phenol/chloroform and mix well by vortexing. Centrifuge at 12,000 × *g* for 5 min at room temperature.
   d. Transfer the top, aqueous phase to a fresh tube and add one volume of chloroform:isoamyl alcohol (24:1) to the supernatant. Mix well and centrifuge as described in the preceding step.
   e. Carefully transfer the top, aqueous phase to a fresh tube. Add 0.1 volume of 2 *M* NaCl solution or 0.5 volume of 7.5 *M* ammonium acetate (optional), and 2.5 volumes of chilled 100% ethanol to the supernatant. Precipitate the linearized DNA at –80°C for 30 min or at –20°C for 2 h.
   f. Centrifuge at 12,000 × *g* for 5 min, decant the supernatant and briefly rinse the DNA pellet with 1 ml of 70% ethanol. Dry the pellet for 10 min under vacuum and dissolve the DNA in 15 μl dd.H_2O.
   g. Use 2 μl of the sample to measure the concentration of the DNA by UV absorption spectroscopy at A_{260} and A_{280} nm. Store the DNA at –20°C until use.
2. Blunt the 3′ overhang ends by the 3′ to 5′ exonuclease activity of Klenow DNA polymerase. Even though this is optional, we recommend that the 3′ protruding end be converted into a blunt end because some of the RNA sequence is complementary to that of the vector DNA.

Note: Enzymes such as Kpn *I,* Sac *I,* Pst *I,* Bgl *I,* Sac *II,* Pvu *I,* Sfi *and* Sph *I should not be utilized to linearize plasmid DNA for transcription* in vitro.

   a. Set up an *in vitro* transcription reaction mixture on ice as described below:
      5X Transcription buffer, 8 μl
      0.1 *M* DTT, 4 μl

rRNasin ribonuclease inhibitor, 40 units
Linearized template DNA (0.2 to 1.0 g/μl), 2 μl
Add dd.H_2O to a final volume of 15.2 μl.

b. Add Klenow DNA polymerase (5 u/μg DNA) to the preceding mixture and incubate at 22°C for 15 min.
c. Add the following components to the reaction and mix well:
Mixture of ATP, GTP, UTP (2.5 m*M* each), 8 μl
120 μ*M* UTP, 4.8 μl
[35S]UTP (1000 to 1500 Ci/mmol), 10 μl
SP6 or T7 or T3 RNA polymerase (15 to 20 u/μl), 2 μl
d. Incubate the reaction at 37 to 40°C for 60 min.

3. Remove the DNA template with *DNase* I.
   a. Add *DNase* I (RNase free) to a concentration of 1 u/μg DNA template.
   b. Incubate for 20 min at 37°C.
4. Purify the RNA probe as described for DNA labeling or by phenol/chloroform extraction followed by ethanol precipitation.
5. Determine the percentage of incorporation and the specific activity of the RNA probe, which is similar to the method in the preceding DNA labeling.

Note: The specific activity of the probe can be calculated as follows. For example, if 1 μl of a 1:10 dilution is utilized for TCA precipitation, 10 × cpm precipitated = cpm/μl incorporated. In 40 μl of a reaction mixture, 40 × cpm/μl is the total cpm incorporated. If 50 μCi of labeled [35S]UTP at 1000 μCi/nmol is used, then 50/1000 = 0.05 nmol of UTP added to the reaction. Assuming that there is 100% incorporation and that UTP represents 25% of the nucleotides in the RNA probe, 4 × 0.05 = 0.2 nmol of nucleotides incorporated, and 0.2 × 330 ng/nmol = 66 ng of RNA synthesized. Then, the total ng RNA probe = % incorporation × 66 ng. For instance, if a 1:10 dilution of the labeled RNA sample has 2.2×10^5 cpm, the total cpm incorporated is $10 \times 2.2 \times 10^5$ cpm × 40 μl (total reaction) = 88×10^6 cpm; hence, % incorporation = 88×10^6 cpm/110×10^6 cpm (50 μCi) = 80%. Therefore, total RNA synthesized = 66 ng × 0.80 = 52.8 ng RNA. The specific activity of the probe = 88×10^6 cpm/0.0528 μg = 1.6×10^9 cpm/μg RNA.

IN SITU HYBRIDIZATION

After treatment of specimens with DNase (see the section on DNase treatment for *in situ* hybridization of RNA) or RNase (see the section concerning RNase treatment for *in situ* hybridization of DNA) and preparation of a specific probe (see the preceding section) are complete, one can proceed to *in situ* hybridization. In general, the procedure for DNA hybridization is similar to that for RNA hybridization except for the differences indicated at specific steps.

1. Carry out blocking of nonspecific sulfur-binding sites on tissue specimens prior to hybridization.

a. Remove the slides from the freezer and place them in a slide rack. Allow the slides to warm up to room temperature for a few minutes.
b. Immerse the slide rack in PBS buffer containing 10 m*M* DTT in a staining dish, and allow the specimens to equilibrate for 12 min at 45°C in a water bath.
c. Transfer the slide rack to a fresh staining dish containing a sufficient volume of freshly prepared blocking solution. Wrap the entire staining dish with aluminum foil and allow blocking of the specimens to proceed for 45 min at 45°C.
Note: Blocking nonspecific sulfur-binding sites on specimens can reduce the background a great deal when using [^{35}S]-labeled probes. However, it is not necessary for [^{32}P]-labeled probes.
Caution: The blocking solution contains toxic iodoacetamide and N-*ethylmaleimide, which should be handled carefully. Waste solutions should be placed in special containers for toxic waste disposal.*
d. Wash the slides 3 × 3 min in PBS buffer and 3 min in TEA buffer at room temperature.
e. Block polar and charged groups on tissue specimens by transferring the slide rack to fresh TEA buffer containing 0.5% acetic anhydride.
f. Incubate the specimens in 2X SSC for 5 min at room temperature.
g. Dehydrate the tissue specimens through a series of ethanol (50%, 75%, 95%, 100%, 100%, and 100%), 3 min each at room temperature.
h. Completely dry the slides by air-drying or in a desiccate and proceed to prehybridization or store at –80°C if the schedule is not convenient.

2. Perform prehybridization as follows:
a. Carefully overlay tissue specimens with an appropriate volume of prehybridization solution (15 to 20 μl/10 mm^2) using a pipette tip.
b. Place the slides in a high-humidity chamber and allow prehybridization to proceed for 1 h at 50°C.
c. During prehybridization, prepare an appropriate volume of hybridization cocktail (see Reagents Needed section).
d. After prehybridization is complete, carefully remove the prehybridization solution from the slides using a 3MM Whatman paper strip or a pipette tip.
e. Immediately overlay each specimen with freshly prepared hybridization cocktail containing the specific probe (15 to 20 μl/10 mm^2). Cover the cocktail with a piece of plastic coverslip slightly larger than the tissue section.
f. Simultaneously denature the probe and the target DNA or RNA by placing the slides on a hot plate for 3 to 4 min at 95 to 100°C for DNA hybridization or at 65 to 70°C for RNA hybridization.
Note: DNA or RNA in the specimens can be denatured prior to prehybridization. Labeled probes can be denatured immediately before preparation of the hybridization cocktail.
g. Quickly place the slides in a moisture chamber and carry out hybridization at 55 to 65°C for 2 to 4 h.

Notes: (1) Any bubbles underneath the plastic coverslips should be removed by gently pressing the coverslip or with a toothpick. (2) In order to prevent uneven hybridization, slides should be kept level in the chamber. (3) The humidity chamber must be tightly covered or sealed prior to being placed in a hybridization oven or an equivalent. If the specimens are dried during hybridization, a high background will occur.

h. Following hybridization, quickly place slides in a slide rack and insert the rack into a staining dish filled with washing solution containing 2X SSC and 2 m*M* β-mercaptoethanol. Wash the slides with three changes of fresh washing solution, each time for 15 min at 55°C with gentle shaking.

   Caution: Any used washing solution containing isotope should be collected in an isotope waste container for disposal. Any areas that touch isotopic solution or slides should be surveyed afterward for potential contamination by following local guidelines.

i. Continue to wash the slides twice, each session for 5 min, in a fresh washing solution containing 0.2X SSC and 2 m*M* β-mercaptoethanol at room temperature with gentle shaking.

   At this stage, unhybridized RNA in the tissue specimens can be removed by using an appropriate amount of RNase (DNase free, Boehringer Mannheim Corporation) or a mixture containing 45 μg/ml RNase A, 2 μg/ml RNase T1 (Sigma Chemicals), 5 mM *EDTA and 250* mM *NaCl in TE buffer, pH 7.5. RNase treatment should enhance the signal-to-noise ratio, thus reducing background.*

j. Dehydrate the slides through a series of ethanol (50%, 75%, 95%, 100%, 100%, and 100%), 3 min each.

k. Air-dry the slides. Proceed to emulsion autoradiography.

DETECTION OF HYBRIDIZED SIGNALS

For large organs or tissues, tape the slides to a film exposure cassette or cardboard and expose them against the film at 4°C for 1 to 10 days prior to emulsion autoradiography. In general, slides are directly subjected to autoradiographic emulsion by the following method:

1. Add an appropriate volume of working NTB-2 emulsion (prethawed in a 45°C water bath) to a slide-dipping chamber.
2. Slowly and carefully dip the slides into the dipping chamber and gently pull them out. Place the slides vertically in an appropriate rack and air-dry them for 2 h or in front of a fan for 20 min.
3. Place the slides in a light-tight slide box containing desiccant and seal the box with black tape. Thoroughly wrap the box with aluminum foil and allow exposure to proceed at 4°C for 1 to 10 days, depending on the intensity of the signal.

Notes: (1) The exposure of slides mandates a-^{32}P-free environment. Thus, make sure to perform a radioactive survey of the refrigerator or cold room *where the exposure takes place. (2) The slides must be completely dried prior to exposure because moisture results in fading of the images in the autoradiographic emulsion.*

4. In a darkroom with a safe light, add sufficient volumes of developer, water, and fixer to slide jars or staining dishes and allow the solution and water to warm up to 20°C.
5. Transfer the slide exposure box from the refrigerator or cold room to the darkroom and allow the box to warm to 20°C.
6. Turn on an appropriate safe light in a light-tight darkroom, open the exposure box and place the slides in a slide rack.
7. Carefully immerse the slides in developer for 2 to 4 min; quickly rinse the slides in water for 30 s and immediately place the slides in fixer for 3 min.
8. Wash the slides in slowly running tap water for 30 min. Turn on the normal light, remove the emulsion on the back side of the slide using a razor blade and air-dry the slides in a dust-free place.
9. Carry out slight counterstaining of the slides in order to enhance the images using the hematoxylin/eosin-staining procedure described next:
   a. Prepare a series of solutions in slide-staining dishes in the order below: 1 dish: hematoxylin stain; 2 dishes: water; 1 dish: 0.1% NH_4OH; 2 dishes: water; 1 dish: eosin stain; 1 dish: 95% ethanol; 3 dishes: 100% ethanol; 3 dishes: xylene
   b. Prewet the developed slides in water in a staining dish for 2 min.
   c. Stain the specimens in hematoxylin for 10 to 30 sec and rinse the slides in water for 2 × 2 min.
   d. Quickly dip the slides in 0.1% NH_4OH and rinse in water for 2 × 3 min.
   e. Stain the specimens in eosin 10 to 30 sec and quickly dip the slides nine times in 95% ethanol, then nine times in 100% ethanol.
   f. Dehydrate the specimens for 2 min in 100% ethanol, then 3 min in fresh 100% ethanol, followed by 2 × 2 min in xylene
   g. Place the slides flat and add a few drops of Permount™ to one end of the specimens and slowly overlay them with a clean coverslip. Carefully remove excess Permount with 3MM Whatman paper filter strips.
   h. Seal the edges of the coverslip with melted wax or allow the slides to air-dry for 1 to 2 days.
   i. Observe and photograph the specimens using darkfield microscopy according to the manufacturer's instructions.

Reagents Needed

5X Labeling Buffer

0.25 *M* Tris-HCl, pH 8.0;
25 m*M* $MgCl_2$
10 m*M* DTT, 1 m*M* HEPES buffer, pH 6.6
26 A_{260} units/ml random hexadeoxyribonucleotides

5X Transcription Buffer

30 m*M* $MgCl_2$
0.2 *M* Tris-HC, pH 7.5
10 m*M* Spermidine
50 m*M* NaCl

NTPs Stock Solution

100 m*M* ATP in dd.H_20, pH 7.0
10 m*M* GTP in dd.H_20, pH 7.0
10 m*M* UTP in dd.H_20, pH 7.0
10 m*M* CTP in dd.H_20, pH 7.0

TE Buffer (pH 8.0)

10 m*M* Tris-HCl
1 m*M* EDTA

Specimen Blocking Solution

Warm an appropriate volume of PBS to 45°C, add 0.16% (w/v) DTT, 0.19% (w/v) iodoacetamide, and 0.13% (w/v) *N*-ethylmaleimide. Mix well and cover with aluminum foil prior to use.

Note: This solution should be freshly prepared and used immediately. Because it contains toxic iodoacetamide and N*-ethylmaleimide, gloves should be worn and caution should be taken.*

PBS Buffer

2.7 m*M* KCl
1.5 m*M* KH_2PO_4
136.9 m*M* NaCl
15 m*M* Na_2HPO_4
Adjust the pH to 7.3.

20X SSC Solution

0.3 *M* Na_3Citrate (trisodium citric acid)
3 *M* NaCl
Adjust the pH to 7.0 with HCl; autoclave.

50X Denhardt's Solution:

1% (w/v) BSA (bovine serum albumin)
1% (w/v) Ficoll (Type 400, Pharmacia)
1% (w/v) PVP (polyvinylpyrrolidone)

Dissolve well after each addition; adjust to the final volume with distilled water and sterile-filter. Divide the solution into 50-ml aliquots and store at –20°C. Dilute 10-fold prior to use.

Prehybridization Solution

50% (v/v) Deionized formamide
15 m*M* Tris-HCl, pH 8.0
2 m*M* EDTA, pH 8.0
2X SSC
5X Denhardt's solution
0.2% Denatured and sheared salmon sperm DNA for DNA hybridization or 0.2 mg/ml yeast tRNA and 0.2 mg/ml poly(A) for RNA hybridization.

Hybridization Cocktail

Add [^{35}S]dCTP- or [^{35}S]UTP -labeled probe to an appropriate volume of fresh prehybridization solution.

Triethanolamine (TEA) Buffer

100 m*M* TEA in dd.H_2O; adjust the pH to 8.0.

Working Emulsion Solution

Thaw Kodak NTB-2 autoradiographic emulsion at 45°C and warm an equal volume of dd.H_2O at the same temperature. Gently mix together Kodak NTB-2 solution and water, and wrap the container containing the working emulsion solution with aluminum foil prior to use.

Others

Developer
Fixer
Hematoxylin stain
0.1% NH_4OH (freshly prepared)
Eosin stain
DNase I (RNase free)
RNase I (DNase free)
SP6, T7 or T3 RNA polymerase
15 mg/ml Yeast tRNA or poly(A) RNA as a carrier
0.5 *M* Dithiothreitol (DTT)
2 *M* NaCl solution
7.5 *M* Ammonium acetate solution
Ethanol (100%, 70%)
TE-saturated phenol/chloroform
Chloroform:isoamyl alcohol: 24:1 (v/v)
Deionized formamide, ultrapure grade

TROUBLESHOOTING GUIDE

1. **Individual sections fall apart during sectioning with no ribbons formed**. Overfixation of tissues could have occurred. Try a shorter fixation time and determine an optimal fixation time.
2. **Sections crumble and are damaged during stretching.** Underfixation of tissues has occurred. Use a longer fixation time and determine the optimal time empirically.
3. **Sections have holes.** This is due to incomplete tissue dehydration and improper impregnation with wax. Perform complete dehydration of fixed tissues followed by thorough impregnation with melted paraffin wax.
4. **Holes occur in frozen sections.** Ice crystals have been formed in the tissues during freezing or in the cryostat. Make sure to freeze the tissue correctly. Do not touch the frozen sections with fingers or relatively warm tools while mounting the specimens in the microtome or brushing the sections from the cutting knife.
5. **Sections are not of uniform thickness or have lines throughout.** The cutting surface of the specimen block is not parallel to the edge of the knife or the knife is dull. Take great care to use a sharpened knife while mounting the specimens in the microtome.
6. **Morphology of tissue is poorly preserved.** Tissues are stretched or shrunken. Make sure not to press or stretch the tissue during dissection and embedding.
7. **The specific activity of a labeled probe is $<10^5$ cpm/μg DNA or RNA.** [^{35}S]dCTP for DNA or [^{35}S]UTP for RNA has passed its half-life time. The activity of DNA or RNA polymerase is low or may be inhibited. Try to use fresh isotope and ensure that the labeling reaction is set up under the optimal conditions, including high-quality DNA
8. **A low signal is evident.** Tissue is underdigested with protease. The specific activity of the enzyme is low. Determine the optimal digestion time of the specimens. Make sure that the probe is good.
9. **A high background occurs.** Tissue is overfixed. Protease digestion of the tissues is not sufficient. Set up a series of fixation times. Try to use an optimal digestion.
10. **A good signal is evident but poor morphology occurs.** Tissue overdigestion occurs. Counterstaining the tissue is not sufficient. Tissue should be properly digested with an appropriate protease. Counterstain the sections slightly longer.

REFERENCES

1. Berger, C.N., *In situ* hybridization of immunoglobulin-specific RNA in single cells of the β-lymphocyte lineage with radiolabeled DNA probes, *EMBO J.*, 5: 85–93, 1986.

2. Zeller, R., Bloch, K.D., Williams, B.S., Arceci, R.J., and Seidman, C.E., Localized expression of the atrial natruuretic factor gene during cardiac embryogenesis, *Genes Dev.,* 1: 693–698, 1987.
3. Brahic, M. and Haase, A.T., Detection of viral sequences of low reiteration frequency by *in situ* hybridization, *Proc. Natl. Acad. Sci. USA,* 75: 6125–6129, 1978.
4. Lawrence, J.B. and Singer, R.H., Quantitative analysis of *in situ* hybridization methods for the detection of actin gene expression, *Nucl. Acids Res.,* 13: 1777–1799, 1985.
5. Singer, R.H., Lawrence, J.B., and Villnave, C., Optimization of *in situ* hybridization using isotopic and nonisotopic methods, *BioTechniques,* 4: 230–250, 1986.
6. Trembleau, A., Roche, D., and Calas, A., Combination of nonradioactive and radioactive *in situ* hybridization with immunohistochemistry: a new method allowing the simultaneous detections of two mRNAS and one antigen in the same brain tissue section, *J. Histochem. Cytochem.,* 41: 489–498, 1993.
7. Zeller, R., Watkins, S., Rogers, M., Knoll, J.H.M., Bagasra, O., and Hanson, J., *In situ* hybridization and immunohistochemistry, in *Current Protocols in Molecular Biology,* Ausubel, F.M., Brent, R., Kingston, R.E., Moore, D.D., Seidman, J.G., Smith, J.A., and Struhl, K., Eds., John Wiley & Sons, New York, pp. 14.0.3–14.8.24, 1995.

13 Localization of DNA or Abundance of mRNA by Fluorescence *in Situ* Hybridization (FISH)

CONTENTS

INTRODUCTION

Fluorescence *in situ* hybridization (FISH) is a very current technology used for the detection of DNA or RNA of interest within cells.[1–3] The basic principle of FISH is that DNA or RNA in the prepared specimens is hybridized with a probe labeled nonisotopically with biotin, digoxigenin or a flurorescent dye. The hybridized signals are then detected by fluorescence or enzymatic methods using a fluorescence or light microscope. The detected signal and image can then be recorded on light-sensitive film. The striking feature of using a fluorescence probe is that the hybridized image can be readily analyzed using a powerful confocal microscope or an appropriate image analysis system with a CCD camera.

Compared with the radioactivity *in situ* hybridization (RISH) described in Chapter 11, FISH offers an increase in sensitivity over RISH. In addition to offering positional information, FISH makes it possible to obtain better morphology of cells or tissues than RISH does. The major drawback for RISH is that it depends on the use of radioactive probes, which are well known for their toxicity to human beings. To carry out RISH, it is necessary to be very cautious in handling radioactive materials, including toxic waste disposal. In other words, one is under some stress during the performance of RISH. As a result, a nonradioactive approach such as FISH has become widely used for the localization of a specific DNA or mRNA in a specific cell or tissue type.[2–9]

This chapter offers detailed protocols for FISH technology. There are similarities between FISH and RISH; in fact, many procedures are exactly the same. These identical procedures are not described again in this chapter but simply referred to in specific sections in Chapter 12. In general, techniques described for the use of *in situ* hybridization have been successfully used in our laboratories. Excellent results can be obtained once the optimal conditions are determined.

CELL OR TISSUE FIXATION, TISSUE EMBEDDING, SECTIONING AND MOUNTING

These procedures are described in Part A of Chapter 12.

DEWAXING OF SECTIONS, PROTEASE DIGESTION, AND DNASE OR RNASE TREATMENT OF SPECIMENS FOR *IN SITU* HYBRIDIZATION

These procedures are described in Part B in Chapter 12.

PREPARATION OF NONISOTOPIC PROBES USING ONE OF THE FOLLOWING METHODS, DEPENDING ON SPECIFIC PROBE AND DETECTION STRATEGIES

METHOD A. RANDOM PRIMER DIGOXIGENIN LABELING OF DSDNA

The basic principle of biotin or digoxigenin labeling is that, in a random labeling reaction, random hexanucleotide primers anneal to denatured DNA template. A new strand of DNA complementary to the template DNA is catalyzed by the Klenow enzyme. During the incorporation of four nucleotides, one of them is prelabeled by digoxigenin such as biotin-dUTP or DIG-dUTP. After hybridization of the probe with the target DNA or RNA sequence in the specimen, an antidigoxigenin antibody conjugated with an alkaline phosphatase will bind to the bound biotin-dUTP or digoxigenin-dUTP. The hybridized signal will then be detected using colorimetric substrates NBT and BCIP or X-Phosphate, which give a purple/blue color.

1. Add an appropriate amount of dsDNA of interest as a template into a microcentrifuge tube. Denature the DNA by boiling for 10 min and quickly chill the tube on ice for 4 min. Briefly spin down and place on ice until use.
2. Set up a standard labeling reaction as follows:
   Denatured DNA template (50 to 10 μg), 5 μl
   10X Hexanucleotide primers mixture, 2.5 μl
   10X dNTPs mixture, 2.5 μl
   dd.H_2O to 13.5 μl
   Klenow enzyme (2 units/μl), 1.5 μl
   Total volume of 25 μl
3. Incubate at 37°C for 2 to 12 h.

Note: The amount of labeled DNA depends on the amount of DNA template and on the length of incubation at 37°C. Based on our experience, the longer the incubation within a 12-h period, the more DNA is labeled.

4. After the labeling is complete, add 2.5 μl of 0.5 *M* EDTA solution to stop the reaction.
5. Add 0.15 volumes of 3 *M* sodium acetate buffer (pH 5.2) and 2.5 volumes of chilled 100% ethanol to the mixture. Allow DNA to precipitate at –80°C for 1 h.
6. Centrifuge at 12,000 × *g* for 6 min at room temperature and carefully aspirate the supernatant. Briefly rinse the DNA pellet with 1 ml of 70% ethanol and dry the pellet under vacuum for 10 to 15 min.
7. Dissolve the labeled DNA in 50 μl dd.H_2O or TE buffer. Store at –20°C prior to use.

Notes: (1) Labeled dsDNA must be denatured prior to use for hybridization. This can be done by boiling the DNA for 10 min then quickly chilling it on ice for 3 min. (2) Following hybridization, the probe contained in the hybridization buffer can be stored at –20°C and reused up to four times. (3) The yield of the labeled DNA can be estimated by dot blotting of serial dilutions of commercially labeled control DNA and labeled sample DNA on a nylon membrane filter. After hybridization and detection, the spot intensities of the control and the labeled DNA can be compared.

Method B. Nick Translation Labeling of dsDNA with Biotin-11-dUTP or Digoxigenin-11-dUTP

Nick translation labeling is also a widely used method for labeling DNA probes. The primary principle of nick translation labeling is that single-stranded nicks in dsDNA are first created by DNase I. *E. coli* DNA polymerase I with 5′ to 3′ exonuclease activity removes stretches of ssDNA in the direction of 5′ to 3′. Almost simultaneously, the DNA polymerase I catalyzes the synthesis of a new strand of DNA from 5′ to 3′, in which a prelabeled deoxyribonucleotide such as DIG-11-dUTP or biotin-11-dUTP is incorporated into the new strand of DNA, producing DNA probes.

1. Set up a labeling reaction mixture on ice as follows:
   dsDNA template (2 to 10 μg), 2 to 5 μl
   10X biotin or DIG DNA labeling mixture, 2.5 μl
   10X reaction buffer, 2.5 μl
   DNase I/DNA polymerase I mixture, 2.5 μl
   Add dd.H_20 to a final volume of 25 μl.
2. Incubate at 15°C for 70 min. Stop the reaction by adding 2.5 μl of 0.4 *M* EDTA solution to the reaction and heating to 65°C for 10 min.
3. Precipitate and dissolve the labeled DNA as described in Method A.

METHOD C. RIBOPROBE OR RNA LABELING

A cDNA of interest can be cloned at the polycloning site between promoters SP6 and T7 or T3 of an appropriate plasmid or lambda vector. The cDNA can be transcribed *in vitro* into single-strand antisense RNA from a linear DNA template with promoters SP6, T7 or T3. During the process of *in vitro* transcription, one of the rNTPs, nonradioactively prelabeled as biotin-11-UTP or DIG-11-UTP, can be incorporated into the RNA strand, generating the labeled riboprobe. Compared with DNA labeling, a higher yield of RNA probe can be obtained because the template can be repeatedly transcribed. RNA probes can be easily purified from DNA template merely by treatment of DNase I (RNase-free) treatment. Riboprobes are now widely utilized for mRNA *in situ* hybridization.

1. Linearize plasmid DNA by an appropriate restriction enzyme for producing run-off transcripts.
   a. Set up a plasmid linearization reaction as follows.
      Recombinant plasmid DNA, 10 μg
      Appropriate restriction enzyme 10X buffer, 2.5 μl
      Acetylated BSA (1 mg/ml) (optional), 2.5 μl
      Appropriate restriction enzyme (10 to 20 u/μl), 10 units
      Add dd.H_2O to a final volume of 25 μl.

   Note: The recombinant plasmid DNA containing cDNA of interest should be digested with an appropriate restriction enzyme that cuts at one site very close to one end of the cDNA insert.

   b. Allow digestion to proceed for 2 h at the appropriate temperature, depending on the specific enzyme used.
   c. Extract DNA by adding one volume of TE-saturated phenol/chloroform and mix well by vortexing. Centrifuge at 12,000 × *g* for 5 min at room temperature.
   d. Transfer the top, aqueous phase to a fresh tube and add one volume of chloroform:isoamyl alcohol (24:1) to the supernatant. Mix well and centrifuge as described in the preceding step.
   e. Carefully transfer the top, aqueous phase to a fresh tube.
      Note: Phenol/chloroform extraction step (c) to step (e) can be replaced by simple filtration using a protein-binding filter (e.g., Millipore Corporation). Transfer the linearized mixture into a filter cup with a PVDF

membrane fixed at the bottom. Place the cup into a sterile microcentrifuge tube and centrifuge at 12,000 × g for 30 s. Proteins are retained on the membrane but DNA in the supernatant is filtered to the bottom of the centrifuge tube. The filter cup can be discarded.

f. Add 0.1 volume of 2 *M* NaCl solution (optional) and 2.5 volumes of chilled 100% ethanol to the supernatant. Precipitate the linearized DNA at –80°C for 30 min or at –20°C for 2 h.

g. Centrifuge at 12,000 × *g* for 5 min, decant the supernatant and briefly rinse the DNA pellet with 1 ml of 70% ethanol. Dry the pellet for 10 min under vacuum and dissolve the DNA in 10 µl dd.H_2O.

2. Blunt the 3′-overhang ends of the DNA by the 3′ to 5′ exonuclease activity of Klenow DNA polymerase (optional).

Note: Enzymes such as Kpn I, Sac I, Pst I, Bgl I, Sac II, Pvu I, Sfi *and* Sph I *should not be utilized to linearize plasmid DNA for transcription* in vitro.

3. Carry out *in vitro* transcription and labeling as follows:
   a. Set up a reaction mixture on ice:
      Purified and linearized plasmid DNA template containing the insert of interest, 2 µg
      10X NTP labeling mixture with biotin- or DIG-UTP, 4 µl
      10X Transcription buffer, 2.5 µl
      RNA polymerase (T7, or SP6, or T3), 4 µl
      Add DEPC-treated dd.H_20 to a final volume of 25 µl.
   b. Incubate at 37°C for 2 h.
   c. To remove the DNA template, add 20 units of DNase I (RNase-free) and incubate at 37°C for 15 min. Inactivate DNase I by adding 4 µl of 0.5 *M* EDTA solution.
   d. Precipitate and dissolve the RNA probe as described in Method A.

Reagents Needed

10X Random Hexanucleotide Mixture

0.5 *M* Tris-HCl, pH 7.2
1 m*M* Dithioerythritol (DTE)
100 m*M* $MgCl_2$
2 mg/ml BSA
62.5 A_{260} units/ml of random hexanucleotides

10X Labeling dNTP Mixture

1 m*M* dATP
1 m*M* dCTP
1 m*M* dGTP
0.60 m*M* dTTP
0.35 m*M* Biotin-dUTP or DIG-dUTP
pH 6.5

Klenow DNA Polymerase I

2 units/µl, labeling grade

10X Reaction Buffer

0.5 *M* Tris-HCl, pH 7.5
0.1 *M* $MgCl_2$
10 m*M* DTE

DNase I/DNA Polymerase I Mixture

0.08 milliunits/µl DNase I/0.1 units/µl DNA polymerase I in 50 m*M* Tris-HCl, pH 7.5
10 m*M* $MgCl_2$
1 m*M* DTE
50% (v/v) Glycerol

TE Buffer

10 m*M* Tris-HCL, pH 8.0
1 m*M* EDTA

3 M Sodium Acetate Buffer, pH 5.2

Ethanol (100%, 70%)

10X Transcription Buffer

0.4 *M* Tris-HCl, pH 8.0
100 m*M* DTT
60 m*M* $MgCl_2$
20 m*M* Spermidine
0.1 *M* NaCl
1 unit of RNase inhibitor

10X NTP Labeling Mixture

0.1 *M* Tris-HCl, pH 7.5
10 m*M* ATP
10 m*M* CTP
10 m*M* GTP
6.5 m*M* UTP
3.5 m*M* Biotin-UTP or DIG-UTP

DEPC-Treated Water

0.1% Diethylpyrocarbonate (DEPC) in dd.H_20. Incubate at 37°C overnight followed by autoclaving.

EDTA Solution

0.5 *M* EDTA in dd.H_20, pH 8.0

RNA Polymerase

20 units/µl of T7, SP6, or T3

IN SITU HYBRIDIZATION OF SPECIMENS

Once treatment of specimens with DNase or RNase and preparation of a specific probe are complete, one can proceed to *in situ* hybridization. In general, the procedure for DNA hybridization is similar to that of RNA hybridization except for differences indicated at specific steps.

1. Carry out prehybridization as follows:
   a. Carefully overlay tissue specimens with an appropriate volume of prehybridization solution (15 to 20 μl/10 mm^2) using the tip of a pipette or a plastic coverslip cut slightly larger than the tissue section.
   b. Place the slides in a moist chamber and allow prehybridization to proceed for 1 h at 37°C.
   c. During prehybridization, prepare hybridization cocktail (see Reagents Needed).
   d. After prehybridization is complete, carefully remove the prehybridization solution from the slides with a 3MM Whatman paper strip or the tip of a pipette.
2. Perform hybridization and washing as follows:
   a. Immediately overlay each specimen with freshly prepared hybridization cocktail containing the specific probe (15 to 20 μl/10 mm^2). Cover the cocktail with a plastic coverslip cut slightly larger than the tissue section.
   b. Simultaneously denature both probe and target DNA or RNA by placing the slides on a hot plate or an equivalent for 3 to 4 min at 95 to 100°C for DNA hybridization or at 55 to 65°C for RNA hybridization. *Notes: (1) DNA or RNA in the specimens can be denatured prior to prehybridization. (2) Labeled probes can be denatured immediately before preparation of the hybridization cocktail. (3) The concentration of the probe is critical for the success of* in situ *hybridization using a nonradioactive probe. A concentration of 2 μg/ml DNA probe or 1 μg/ml RNA probe is recommended. The efficiency of RNA probes is higher than that of DNA probes because some of the denatured ssDNA probes can be renatured to dsDNA under hybridization conditions. If a high background occurs, the concentration of the probe can be reduced 5- to 10-fold.*
   c. Quickly place the slides in a high-humidity chamber and carry out hybridization at 37°C for 2 to 4 h.
      Notes: (1) Any bubbles underneath the plastic coverslips should be removed by gently pressing the coverslip or by using a toothpick. (2) In order to prevent uneven hybridization, slides should be kept level in the chamber. (3) The humidity chamber must be tightly covered or sealed prior to being placed in a hybridization oven or an equivalent. If the specimens are dried during hybridization, a high background will occur. (4) The relatively high concentration of formamide (50%) and low salt concentration (e.g., 2X SSC) in the hybridization cocktail solution

are essential and have two functions in general. One is to facilitate denaturing of target DNA or RNA and probe at 100°C, about 35 to 40°C above the Tm *of homologous hybridized DNA. The other is to prevent potentially nonspecific binding of the probe while not interfering with probe-target annealing at 100°C. (5) Inasmuch as RNA–RNA hybrids have a relatively higher* Tm *than DNA–DNA hybrids, it is recommended that, prior to hybridization, the RNA probe and the target RNA be heated at 65 to 70°C to denature any secondary structure in RNA molecules. In addition, hybridization temperature should be set up at 45 to 50°C for RNA–mRNA* in situ *hybridization, and 40 to 45°C for DNA–RNA* in situ *hybridization.*

d. Following hybridization, quickly place slides in a slide rack and insert the rack into a staining dish filled with washing solution containing 2X SSC and 2.5% BSA. Wash the slides with three changes of fresh washing solution, 15 min each at 45°C with gentle shaking.
e. Continue to wash the slides twice, 5 min each time, in fresh washing solution containing 0.2X SSC, 2.5% BSA and 0.05% Tween 20 in PBS at room temperature with gentle shaking.

   Note: Do not allow the specimens to become dry prior to detection. At this stage, unhybridized RNA of the tissue specimens can be removed using an appropriate amount of RNase (DNase free) or a mixture containing 45 μg/ml RNase A, 2 μg/ml RNase T1 (Sigma Chemicals), 5 mM *EDTA and 250* mM *NaCl in TE buffer, pH 7.5. RNase treatment could enhance the signal-to-noise ratio, reducing the background.*

ENZYMATIC DETECTION OF HYBRIDIZED SIGNALS USING COLORIMETRIC SUBSTRATES NBT AND BCIP

1. Overlay the specimens with an appropriate volume of streptavidin–alkaline phosphatase (AP) conjugate solution (for biotin probes) or anti-DIG–alkaline phosphatase solution (for DIG probes) at an appropriate dilution.
2. Place the slides in an aluminum foil-wrapped high-humidity chamber and incubate for 20 to 30 min at 37°C or 1 h at room temperature.
3. Carefully remove the solution from the specimens and carry out washing in PBS containing 0.05% Tween 20 in a slide dish for 3 × 5 min at 42°C with gentle agitation.
4. Place the slides in a slide rack and wash them in an appropriate volume of enzymatic predetection solution for 2.5 min.
5. Replace predetection solution with fresh enzymatic detection solution containing NBT and BCIP as substrates.
6. Place the slides in a container and wrap it with aluminum foil. Allow color to develop at room temperature for 20 to 120 min. Check results once every 4 min under a microscope.

Note: Color development should be time controlled. Overdevelopment may bring up higher background. For this reason, it is recommended to stop the color development once the primary color is visible, assuming that the probe-target binding that results in the strongest signal is the primary color.

7. Wash the slides in PBS for 3 × 3 min and counterstain the specimens in nuclear fast red solution for 5 min if desired.

Note: Nuclear staining is omitted for RNA–RNA in situ *hybridization if the specimens are pretreated with DNase I.*

8. Partially dry the slides and mount the specimens with glass coverslips using Permount™.
9. View and photograph with a phase-contrast microscope.

FLUORESCENCE DETECTION OF HYBRIDIZED SIGNALS

1. Following step (e) in *in Situ* Hybridization of Specimens, carefully drain excess washing solution from the specimens but do not air-dry the slides.
2. According to the specific fluorescent probe used in hybridization, add biotin detection solution or a biotin–digoxigenin detection solution onto the hybridized slide (30 μl/specimen). Cover the solution with a piece of Parafilm cut slightly larger than the specimen.
3. Place the slides in a container and wrap it with aluminum foil. Place the wrapped container in a moist chamber at 37°C for 50 min.
4. Wash the slides in 2X SSC in a dark place for 2 × 10 min.
5. If desired, carry out counterstaining of nuclear DNA by adding DAPI staining solution to the slides (30 μl/specimen). Cover the solution with a piece of Parafilm cut slightly larger than the specimen.
6. Allow staining to occur for 5 min at room temperature and briefly rinse the specimens in 2X SSC.
7. View and photograph the specimens using a fluorescence microscope with an appropriate filter.

Note: Ektar-1000 and Ektachrome-400 color films are recommended for prints and slides, respectively. Kodak Technical Pan 2415 film can be used for black and white prints. If dual- or triple-band-pass filter sets are utilized, the exposure time may be set at 20 to 100 s and 2 to 10 s, respectively.

Reagents Needed

20X SSC Solution

0.3 *M* Na_3Citrate (trisodium citric acid)
3 *M* NaCl
Adjust the pH to 7.0 with HCl. Autoclave.

Prehybridization Solution

50% (v/v) Deionized formamide

15 m*M* Tris-HCl, pH 8.0
2 m*M* EDTA, pH 8.0
2 mg/ml BSA (nuclease free)
20% (w/v) Dextran sulfate (MW 500,000)
2X SSC

Hybridization Cocktail

Add an appropriate amount of nonradioactively labeled probes to an appropriate volume of fresh prehybridization solution.

Enzymatic Predetection Solution

100 m*M* Tris-HCl, pH 9.5
100 m*M* NaCl
50 m*M* $MgCl_2$ (freshly added)

PBS Buffer

2.7 m*M* KCl
1.5 m*M* KH_2PO_4
136.9 m*M* NaCl
15 m*M* Na_2HPO_4,
Adjust the pH to 7.3.

Nitroblue Tetrazolium (NBT) Stock Solution

75 mg/ml in dimethylformamide
Dissolve well and store in a dark place.

5-Bromo-4-Chloro-3-Indolyl Phosphate (BCIP) Stock Solution

50 mg/ml in dimethylformamide
Dissolve well and store in a dark place.

Enzymatic Detection Solution Containing NBT/BCIP

100 m*M* Tris-HCl, pH 9.5
100 m*M* NaCl
50 mM $MgCl_2$ (freshly added)
330 μg/ml of NBT (freshly diluted from NBT stock solution)
170 μg/ml of BCIP (freshly diluted from NBT stock solution)

Streptavidin–Alkaline Phosphatase Conjugate Solution

1% (w/v) BSA (nuclease free)
0.05% (v/v) Tween 20
3.5 μg/ml streptavidin–alkaline phosphatase conjugate (freshly added)
Dissolve well after each addition in PBS.

DAPI Stock Solution

Dissolve 0.1 mg/ml of 4′,6-diamidino-2-phenylindole (DAPI) in distilled water containing 4 μg/ml methanol. Store the stock solution in an aluminum-wrapped tube at –20°C up to 1 year.

DAPI Staining Solution

Freshly dilute DAPI stock solution for 1000X in PBS and store in a dark place prior to use.

Biotin Detection Solution

Dilute fresh rhodamine-avidin D or fluorescein-avidin DCS to a final concentration of 2 μg/ml in 2X SSC containing 1% (w/v) BSA.

Biotin/Digoxigenin Detection Solution

Freshly dilute fluorescein-avidin DCS and rhodamine-conjugated Fab fragment of sheep antidigoxigenin to a final concentration of 2 μg/ml each in 2X SSC containing 1% (w/v) BSA.

TROUBLESHOOTING GUIDE

1. **Individual sections fall apart during sectioning with no ribbons formed**. Overfixation of tissues could have occurred. Try a shorter fixation time. An optimal fixation time should be determined.
2. **Sections crumble and are damaged during stretching.** Underfixation of tissues has occurred. Use a longer fixation time and determine the optimal time empirically.
3. **Sections have holes.** This is due to incomplete tissue dehydration and improper impregnation with wax. Perform complete dehydration of fixed tissues followed by thorough impregnation with melted paraffin wax.
4. **Holes occur in frozen sections.** Ice crystals are formed in the tissues during freezing or in the cryostat. Make sure to freeze the tissue correctly. Do not touch the frozen sections with fingers or relatively warm tools during mounting in the microtome or during brushing from the cutting knife.
5. **Sections are not of uniform thickness or have lines throughout.** The cutting surface of the specimen block is not parallel to the edge of the knife. The knife is dull. Take great care when mounting the specimens in the microtome and use a sharpened knife.
6. **Morphology of tissue is poorly preserved.** Tissues are stretched or shrunken. Make sure not to press or stretch the tissue during dissection and embedding.
7. **No hybridization signal is shown in the specimens.** Probes or target DNA or RNA sequences are not really denatured prior to hybridization. Make sure to denature the probe and the target DNA or RNA in the specimens completely.

8. **The intensity of fluorescence signal in specimens is too bright with a low signal-to-noise ratio.** The amount of fluorescein probes or biotin- or digoxigenin-detection solution is overloaded. Overhybridization thus occurs. Try to reduce the probe and detection solutions to 5- to 10-fold. Control the hybridization time to 2 to 4 h.
9. **A high background is shown in enzymatic detection sections. Inasmuch as color can be seen everywhere in specimens, it is difficult to tell which signal is the correct signal.** Color development goes for too long or the specimens may be partially or completely dried during hybridization or washing procedures. Try to monitor color development once every 3 to 4 min and terminate color formation as soon as the primary color signal appears. Probe-target sequence binding should reveal first visible color before potential background develops later. Make sure not to allow specimens to become dry.

REFERENCES

1. Berger, C.N., *In situ* hybridization of immunoglobulin-specific RNA in single cells of the β-lymphocyte lineage with radiolabeled DNA probes, *EMBO J.,* 5: 85–93, 1986.
2. Zeller, R., Bloch, K.D., Williams, B.S., Arceci, R.J., and Seidman, C.E., Localized expression of the atrial natruuretic factor gene during cardiac embryogenesis, *Genes Dev.,* 1: 693–698, 1987.
3. Brahic, M. and Haase, A.T., Detection of viral sequences of low reiteration frequency by *in situ* hybridization, *Proc. Natl. Acad. Sci. USA,* 75: 6125–6129, 1978.
4. Lawrence, J.B. and Singer, R.H., Quantitative analysis of *in situ* hybridization methods for the detection of actin gene expression, *Nucl. Acids Res.,* 13: 1777–1799, 1985.
5. Singer, R.H., Lawrence, J.B., and Villnave, C., Optimization of *in situ* hybridization using isotopic and nonisotopic methods, *BioTechniques,* 4: 230–250, 1986.
6. Lichter, P., Boyle, A.L., Cremer, T., and Ward, D.D., Analysis of genes and chromosomes by nonisotopic *in situ* hybridization, *Genet. Anal. Techn. Appl.,* 8: 24–35, 1991.
7. Speel, E.J.M., Schutte, B., Wiegant, J., Ramaekers, F.C., and Hopman, A.H.N., A novel fluorescence detection method for *in situ* hybridization, based on the alkaline phosphatase-fast red reaction, *J. Histochem. Cytochem.,* 40: 1299–1308, 1992.
8. Trembleau, A., Roche, D., and Calas, A., Combination of nonradioactive and radioactive *in situ* hybridization with immunohistochemistry: a new method allowing the simultaneous detection of two mRNAS and one antigen in the same brain tissue section, *J. Histochem. Cytochem.,* 41: 489–498, 1993.
9. Zeller, R., Watkins, S., Rogers, M., Knoll, J.H.M., Bagasra, O., and Hanson, J., *In situ* hybridization and immunohistochemistry, in *Current Protocols in Molecular Biology,* Ausubel, F.M., Brent, R., Kingston, R.E., Moore, D.D., Seidman, J.G., Smith, J.A., and Struhl, K., Eds., John Wiley & Sons, New York, pp. 14.0.3–14.8.24, 1995.

14 *In Situ* PCR Hybridization of Low Copy Genes and *in Situ* RT-PCR Detection of Low Abundance mRNAs

CONTENTS

INTRODUCTION

In recent years, detection in combination with localization of DNA or RNA by *in situ* hybridization has been demonstrated.[1–3] As described in Chapter 12 and Chapter 13, the power of *in situ* hybridization allows DNA or RNA of interest to be detected in intact cells or specific tissues. Unfortunately, this technique has a sensitivity limitation. As a matter of fact, it merely works well for analysis of high abundance nucleic acids, but can barely detect low copy gene/DNA or low abundance RNA. In general, traditional Southern or northern blot hybridization also does not detect low abundance nucleic acids. To overcome this limitation, *in situ* PCR and *in situ* RT-PCR methods have been developed for the detection of low copy genes/DNAs and low abundance mRNAs, respectively.[4–6] These technologies are a virtual combination of PCR and *in situ* hybridization, which have become very powerful tools and are widely used in many disciplines.[3–12]

Clearly, there is a major difference between *in situ* PCR and regular solution PCR, even though the basic mechanism of DNA amplification is similar. Solution PCR is carried out in a tube and requires extraction of DNA/RNA from cells or tissues to be tested. As a result, one cannot correlate PCR data with the histological or pathological features of the sample being examined. In contrast, *in situ* PCR can directly amplify target DNA sequences within intact cells or specimens, which enables one to know exactly which particular cell or tissue type contains the amplified DNA or cDNA. With *in situ* PCR hybridization techniques, a virus infection or a few copies of mRNA can be rapidly detected and localized.

The primary principles of *in situ* PCR hybridization are that low copy DNA or RNA is first amplified within intact cells by the extremely sensitive polymerase chain reaction (PCR), and that the amplified signals are then readily detected following regular *in situ* hybridization using a specific probe. With regard to *in situ* RT-PCR hybridization, the basic difference is that mRNA is reversely transcribed into cDNA prior to proceeding with PCR amplification using specific primers. It is important to realize how PCR products are precisely retained inside the cells with a little or no diffusion. It has been speculated that trapping amplified DNA or cDNA in cellular compartments depends on the fixative used for fixation.[7–8] Perhaps both ethanol and acetone fixation cannot induce cross-linking between proteins and nucleic acids, resulting in migration of PCR products out of cells. Formalin (i.e., formaldehyde), however, can extensively polymerize proteins and cross-link nucleic acids. Following appropriate formalin fixation, a barrier may be created to prevent the amplified products from migrating out of the cell. In addition, it is a possibility that the positively charged amino acids contained in the fixed protein meshwork may interact with the negatively charged PCR products, which inhibit the amplified

DNA or cDNA from diffusion from its sites of origin. Nonetheless, not 100% PCR products can remain in their original sites.[3,6,8–17]

The present chapter describes detailed protocols for successful *in situ* PCR hybridization. It consists of three parts: *in situ* PCR hybridization, *in situ* RT-PCR hybridization and direct *in situ* PCR with incorporation of DIG-11-dUTP or biotinylated-dUTP. The methods covered here are improved or modified from previous methodologies. All the procedures have been well tested and work successfully in our laboratories.

PART A. DETECTION OF A LOW COPY GENE BY *IN SITU* PCR HYBRIDIZATION

As outlined in Figure 14.1, these general procedures include design of primers; cell or tissue fixation; tissue embedding, sectioning and mounting on slides; protease digestion; RNase treatment; *in situ* PCR amplification; and *in situ* hybridization and detection.

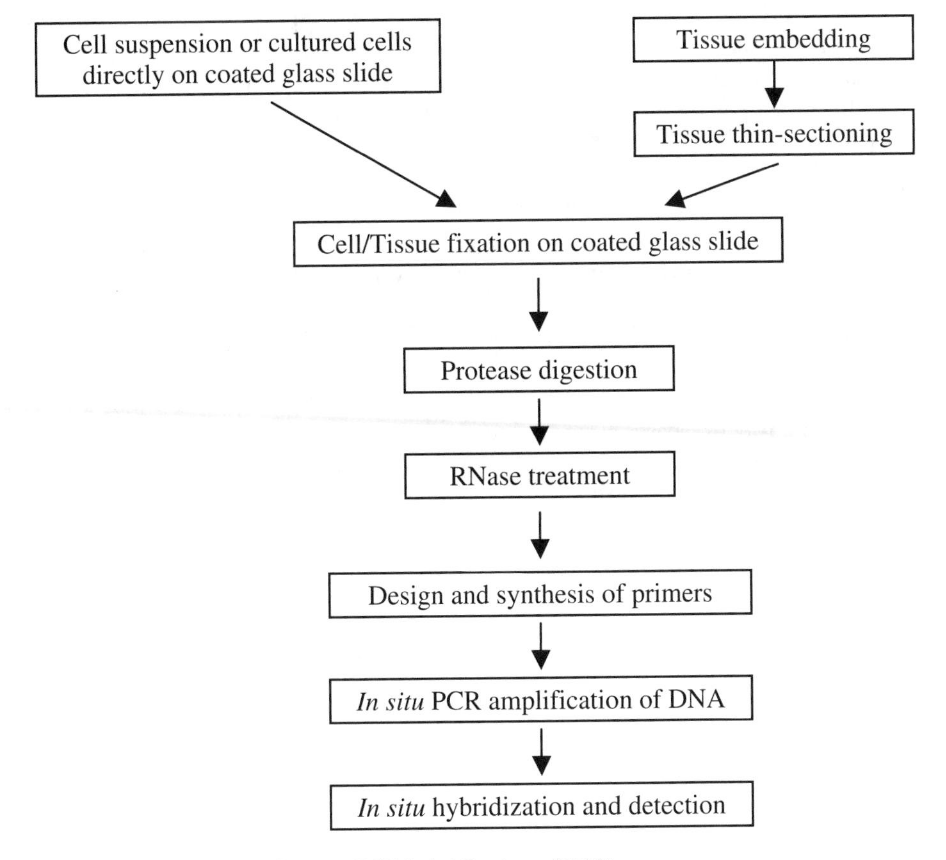

FIGURE 14.1 Scheme of *in situ* PCR hybridization of DNA.

Design of Specific Primers for *In Situ* PCR Amplification of Target DNA Sequences

Like solution PCR in a tube, the design of primers is essential for the success of *in situ* PCR in cells or tissues. The length, Tm value, GC content and specificity of the primers mandate careful planning and analysis. The DNA sequences to be amplified can be determined by searching for appropriate publications or databases from GenBank. To amplify a target DNA sequence, two primers should be designed and synthesized. One is sense or forward primer; the other is called antisense or reverse primer. The primers are designed so that one is complementary to each strand of the DNA double helix; they are located at opposite sides of the region to be amplified. In other words, forward primer directs the synthesis of a DNA strand towards the reverse primer, and vice versa, generating double-stranded DNA flanked by the two primers (Figure 14.2).

The authors routinely use the computer program Oligo™ version 4.0 or Cprimer f (Internet) to design and analyze oligonucleotide primers for amplification of partial- or full-length HSP27 cDNA. The major advantages of these programs are that they: (1) allow a search for the most suitable regions in the target DNA sequences for the design of primers; (2) verify the specificity of the primers; (3) display the AT and GC contents and *Tm* values of the primers; and (4) allow checking for potential intrahairpin structure and interstrand base pairing of the primers (Figure 14.3).

In our experience, the following points should be noted when designing primers for *in situ* PCR and *in situ* RT-PCR:

- DNA target length to be amplified should be a relatively small fragment (0.3 to 2.0 kb). Although some new *Taq* polymerases such as the Expand™ PCR System (Boehringer Mannheim Corporation) enable DNA >10 kb to be amplified, it is suitable for solution PCR in a tube but not desirable for *in situ* PCR. Often, the larger the fragment to be amplified, the more potential there is to create artifacts of PCR products.
- The length for forward/sense and reverse/antisense primers should be 17- to 25-mer or bases. Primers less than 14-mer may have a low efficiency of annealing to template or they may cause failure of PCR amplification. On the other hand, primers >30-mer in length may bring about nonspecific annealing to templates, resulting in unexpected PCR products.

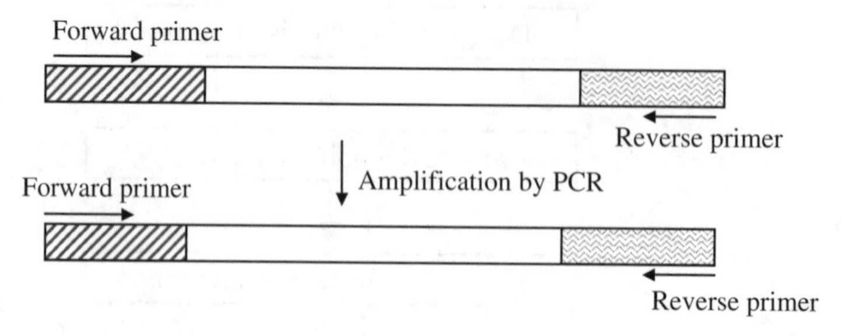

FIGURE 14.2 Diagram of DNA amplification by PCR.

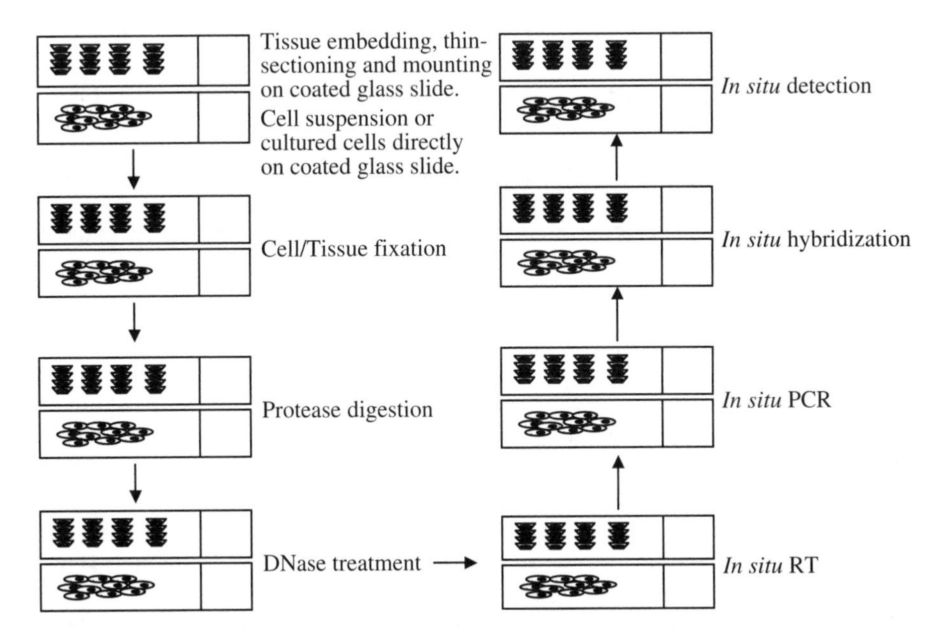

FIGURE 14.3 Scheme of *in situ* RT-PCR.

- The GC content of a primer should be 45 to 58%. A GC-rich primer (>60%) will significantly increase the *Tm* value of the primer, which will affect the annealing temperature.
- To facilitate complementary-strand annealing between the primer and the DNA template efficiently, a primer should be designed so that it contains two to three GC-type bases at its 3′ end and one G or C base at its 5′ end. Bases C and G have three hydrogen bonds compared with A or T that contains two hydrogen bonds.
- Perhaps the most critical caution when designing primers is whether an intrastrand hairpin structure or interstrand complementary base pairs will be formed. According to the computer programs Oligo™ version 4.0 and CPrimer f (Internet) and based on our experience, no more than three intraprimer base pairs or no more than four interprimer base pairs are allowed. Particularly, the 3′ ends of forward and reverse primers must not be complementary to each other. Any intraprimer secondary structures or primer–primer dimers will cause a failure in the annealing between primers and DNA templates. In this case, there will be no PCR products. Therefore, it is absolutely important to design primers that contain no secondary structures and no interstrand primer–dimers.
- In general, the melting temperature or *Tm* of a primer should be <68°C. Primers with *Tm* >75°C are usually difficult to anneal to a DNA template or difficult to melt once secondary structures are formed. If the computer program CPrimer f, Oligo™ version 4.0 or Cprimer f is used, the *Tm* of primers can be calculated and displayed by the programs. If the programs

are not available, the *Tm* can be estimated by the formula: *Tm* of a primer = (2°C × AT bases)+ (4°C × GC bases)

More accurately, one can consider the salt concentration when calculating *Tm* using the following formula:

$$\text{Tm of a primer} = 81.5°\text{C} + 16.6\ (\text{Log M}) + 0.41\ (\%\text{GC}) - 500/\text{n}$$

where *Tm* is the melting temperature of the primer, M is the molarity of the salt in the buffer and n is the length of the primer.

Synthesis and Testing of the Designed Primers

Once specific primers have been properly designed, they can be synthesized by an appropriate DNA oligonucleotide synthesizer. Purification of the primers is optional. We routinely use synthesized primers and obtain successful PCR products. At this point, it might be assumed that the primers can be directly applied to *in situ* PCR. To ensure the quality of primers, however, the authors strongly recommend that the designed primers be tested for the specificity and integrity of PCR products simultaneously or prior to *in situ* PCR. The testing can be done by regular solution PCR in a tube using genomic DNA from the cells or tissues to be used for *in situ* PCR. The PCR products should be analyzed by agarose gel electrophoresis.

A good pair of primers should generate one band with expected size. No bands or multiple bands with unexpected sizes indicate that the designed primers are not suitable for *in situ* PCR and should be redesigned and synthesized. In our laboratories, even though the expected PCR products are obtained, we sometimes partially or completely sequence the PCR products to make sure that the right target sequences have been amplified. Inasmuch as nonspecific or mispaired primers is a common potential phenomenon, PCR products with any artifacts will cause misinterpretation of PCR results. Once the quality of primers and the integrity of PCR products have been verified, one can proceed to *in situ* PCR.

Carrying Out Fixation of Cells or Tissues, Tissue Embedding, Sectioning and Mounting of Sections on Slides

These activities are described in Part A of Chapter 12.

Note: For in situ *PCR hybridization, three separate drops of cell suspension or tissue sections should be placed on the same slide. One drop of cells or tissue section will be subjected to DNase and RNase digestion prior to* in situ *PCR. Following* in situ *PCR hybridization, this will serve as a negative control because all cells should show no PCR products. Another section or a drop of cells will not be subjected to DNase and RNase treatment. This will serve as a positive control because all cells should have PCR products and hybridization signals. The third drop of cells or section will serve as target-specific test, which will be digested with RNAse but not with DNase.*

Dewaxing of Sections (if Necessary), Protease Digestion and DNase or RNase Treatment of Specimens

These activities are described in Part B of Chapter 12.

Preparation of Radioactive Probes for RISH or Nonradioactive Probes for FISH

RISH preparation is described in Part B of Chapter 12. FISH preparation is described in Chapter 13.

Performance of *In Situ* PCR

1. Prepare an appropriate volume of PCR reaction cocktail mixture (50 μl per section). The following cocktail is a total of 500 μl for 10 sections.
   10X Amplification buffer with $MgCl_2$, 50 μl
   Mixture of four dNTPs (2.5 m*M* each), 50 μl
   Sense or forward primer (20 μ*M* in dd.H_2O), 25 μl
   Antisense or reverse primer (20 μ*M* in dd.H_2O), 25 μl
   Taq DNA polymerase, 50 units
   Add dd.H_2O to final volume of 500 μl.

Note: PCR kits containing 10X amplification buffer, dNTPs and Taq polymerase are available from Boehringer Mannheim and Perkin–Elmer Corporation. $MgCl_2$ concentration plays a role in the success of PCR amplification, which should be adjusted accordingly. In our experience, 4.0 to 4.5 mM *$MgCl_2$ is an optimal concentration to obtain an intense signal.*

2. Add the PCR cocktail to the slides prepared at step 4 (50 μl/section). Carefully cover each slide and overlay cells or tissue sections with the cocktail using an appropriate size of coverslip. If necessary, add mineral oil to seal all the edges of the coverslip to prevent evaporation during the PCR process.

Note: The assembly of slides for in situ *PCR varies with the specific instrument. The procedures should be carried out according to the manufacturer's instructions. We use the GeneAmp* in Situ *PCR System 1000 (Perkin–Elmer Corporation) that can simultaneously carry out 10 slides for* in situ *PCR without using mineral oil. Brief instructions for assembly of slides in the correct order are:*

- Place the slides with cells or tissue sections on the platform of the assembly tool.
- Fully expand the AmpliCover clip and put it against a magnetic holder in the arm of the assembly tool provided.
- Place the AmpliCover disc in the clip, and add the PCR cocktail to the center of the cell sample or tissue section on a glass slide (50 μl/section).

  Note: To prevent formation of bubbles, the cocktail should be placed in a single drop instead of evenly distributed on the whole cell sample or tissue section. In addition, do not touch the cells or tissue section with a pipette tip when loading the PCR cocktail so as to avoid any damage to the tissue sections.

- Place the arm down, turn the knob and simultaneously clamp the AmpliCover clip onto the slide according to the instructions.
  Note: The cells or tissue sections overlaid with the PCR cocktail should now be completely sealed. To eliminate mispriming between primers and DNA template, we recommend using a hot-start approach at 70°C. The cocktail should not be added onto the sections until a slide temperature of 70°C has been reached. Briefly preheat the assembly tool to 70°C. Place the slide with sections on the tool for about 1 min prior to adding the PCR cocktail to the tissue sections, or the Taq *polymerase may not be added into the PCR cocktail. After the cells or tissue sections on slides are overlaid with the PCR cocktail (about 2/3 of the total volume) lacking* Taq *polymerase, ramp to 70°C or 3 to 5°C above the Tm of primers, quickly add an appropriate amount of* Taq *polymerase diluted in dd.*H_2O *(about 1/3 of the total volume) onto the sample and assemble the slides for* in situ *PCR according to the instructions.*
- Load the assembled slides, one by one, into the slide block of the GeneAmp *in Situ* PCR System 1000[R].

3. Carry out 25 cycles of PCR amplification, preprogrammed according to the manufacturer's instructions, as follows:

Cycle	Denaturation	Annealing	Extension
First	94°C, 3.5 min	55°C, 2 min	72°C, 2.5 min
2 to 24	94°C, 1 min	55°C, 2 min	72°C, 2.5 min
25	94°C, 1 min	55°C, 2 min	72°C, 7.0 min

Note: For the final step, hold at 4°C until the slides are removed.

Tips: (1) This PCR profile may be appropriately adjusted according to specific primer conditions. The annealing temperature should be approximately 2°C above the Tm of the primers.. Annealing primers and DNA template at a temperature too far below the Tm will generate unexpected PCR products due to false priming. However, annealing at a temperature too far above the Tm may cause no priming and no results. (2) Special tips are recommended for a high fidelity of amplification of large DNA fragments (8 to 25 kb). Primers with at least 30 bases or 30-mer in length should be designed. High quality of Taq *polymerase with proofreading capability, which is commercially available, should be utilized. The PCR profile should be programmed as follows:*

Cycle	Denaturation	Annealing	Extension
First	94°C, 3.5 min	60°C, 1 min	70°C, 4.5 min
2 to 24	94°C, 40 s	60°C, 1 min	70°C, 4.5 min with an extension of 10 s per cycle
25	94°C, 40 s	60°C, 1 min	70°C, 15 min

Note: For the final step, hold at 4°C until the slides are removed.

To avoid mispriming or artifacts such as base misreading/miscomplementing, it is strongly recommended that primers be 17 to 24 bases, and that the DNA length to be amplified should not be longer than 2.0 kb. In our experience, amplification of large DNA molecules is not necessary with regard to in situ *PCR or RT-PCR. (3) Also pay attention to the correct linking between each of the files or cycles prior to performing* in situ *RT-PCR of cell samples or tissue sections, particularly when using a new PCR instrument such as GeneAmp* in Situ *PCR System 1000 (Perkin–Elmer Corporation). It is extremely important to check through each smooth linking between each of the files before using samples. The easiest way is to program the instrument as follows for testing:*

Cycle	Denaturation	Annealing	Extension
First	94°C, 5 s	60°C, 6 s	70°C, 7 s
2 to 4	94°C, 5 s	60°C, 6 s	70°C, 7 s
5	94°C, 5 s	60°C, 6 s	70°C, 10 s

Note: For the final step, hold at 4°C.

4. Once programming is complete, the running mode can be tested. If each step is linked smoothly, the system should be fine with actual samples; therefore, programming of the instrument may begin at the appropriate temperatures and cycles. The GeneAmp *in Situ* PCR System 1000 (Perkin–Elmer Corporation) has four files for programming. We usually utilize file 1 and file 4. File 1 is a soaking file used for setting up a temperature indefinitely or for a specified time period. File 4 is for programming at different temperatures, incubation times and cycles. For hot-start, the author routinely uses file 1 to set up a temperature at 70°C without specifying a time so that this soaking temperature can be terminated once all the slides have been placed inside the instrument. A soaking file can be programmed by turning on the instrument, pressing **File**, pressing 1 and then pressing **Enter**. The next step is to press **Step** to edit the file: set up a temperature (e.g., 70°C) followed by pressing 0 for other temperatures or time parameters except to store the file by giving a specific number and remembering that number. If specifying a soaking time (e.g., 5 min) is desirable, the soaking file number can be automatically linked to the first file number (e.g., file 10) programmed for PCR. If the soaking file is not linked to any file numbers, the soaking file can be terminated by quickly pressing **Stop** twice, pressing **File**, pressing the first file number (e.g., 10), pressing **Enter** and then pressing **Cycle** to start running PCR of samples.

Now, let us program files for PCR with **Bold Font Style** according to the order shown below. Remember that if the link file number is N, the store file number should be assigned as N-1. Let us assume the first file number to be stored is 10. If we wish to set up 70°C, 5 min for hot-start, 94°C, 3 min for the first denaturation temperature and time, 24 cycles of 94°C, 40 s (denaturation), 55°C, 1 min (annealing), 72°C, 2 min (extension), and 4°C hold until the samples are taken out from the instrument, we can program and link the file number as **10→11→12→13→14**. For running, we simply press **File→10→Enter→Cycle**.

Step 1: press **File→4→Enter→Step** to edit file→**70**°C, **5** min→**0** to other temperature and time parameters→cycle number? **No** extension→link to file? 11→run-store-print? **Store→Enter**→store file number? **10→Enter**.

Step 2: press **File→ 4→Enter→Step** to edit file→**94**°C, **3** min for the first temperature and time in deg. 1→**0** to other temperature and time parameters→**No** extension→**1** for cycle number→link to file number? **12→Enter**→run-store-print? **Store→Enter**→store file number? **11→Enter**.

Step 3: press **File→4→Enter→Step** to edit file→**94, 40** s for the first temperature and time in seg. 1→**55**°C, **1** min for seg. 2→**72**°C, **2** min→**0** to other temperature and time parameters→**No** extension→**24** for cycle numbers→link to file number? **13→Enter**→run-store-print? **Store→Enter**→store file number? **12→Enter**.

Step 4: press **File→4→Enter→Step** to edit file→**72**°C, **7** min for the first temperature and time in seg. 1→**0** to other temperature and time parameters→**No** extension→**1** for cycle number→link to file number? **14→Enter**→run-store-print? **Store→Enter**→store file number? **13→Enter**.

Step 5: press **File→1→Enter→Step** to edit file→**4**°C→**0** to other temperature and time parameters→run-store-print? **Store→Enter**→store file number? **14→Enter**.

5. After PCR amplification is complete, remove the clips and cover discs or coverslips. Place the slides in xylene for 5 min, in ethanol for 5 min, and then air-dry. Proceed to the next step.

Performance of *in Situ* Hybridization and Detection by RISH or by FISH

RISH is described in Part B in Chapter 12 and FISH is described in Chapter 13.

Materials Needed

Thermal Cycler

GeneAmp *in Situ* PCR System 1000 (Perkin–Elmer Corporation) or the PTC-100 (MJ Research), which can simultaneously perform solution PCR in a tube and *in situ* PCR on glass slides, or other equivalent thermal cyclers.

PCR Kits

Available from Boehringer Mannheim and Perkin–Elmer Corporation.

10X Amplification Buffer

0.1 *M* Tris-HCl, pH 8.3

0.5 *M* KCl
25 m*M* $MgCl_2$
0.1% BSA

Stock dNTPs
dATP10 m*M*
dTTP10 m*M*
dATP10 m*M*
dTTP10 m*M*

Working dNTPs Solution
dATP (10 m*M*)10 µl
dTTP (10 m*M*)10 µl
dATP (10 m*M*)10 µl
dTTP (10 m*M*)10 µl

Taq *Polymerase*

PART B. DETECTION OF LOW ABUNDANCE mRNA BY *IN SITU* RT-PCR HYBRIDIZATION

The greatest value of the *in situ* PCR technique is its use as a powerful tool for detection of rare copy or low abundance mRNA, which is not possible with traditional northern blotting or standard *in situ* hybridization. In general, mRNA with few copies can be transcribed into cDNA using a reverse transcriptase. The cDNA is then amplified by PCR followed by standard *in situ* hybridization and detection of signals by appropriate methods. This entire procedure is referred to as *in situ* RT-PCR hybridization. In other words, the expression of extremely low abundance mRNA species can first be amplified at their original sites and then readily detected in intact cells or tissues. The general procedures and principles of *in situ* RT-PCR hybridization are outlined in Figure 14.4 and Figure 14.5, respectively.

In situ RT-PCR is similar to the *in situ* PCR described in Part A except that a reverse transcription or RT is required at the first step. Because the very mobile RNA molecules are involved in RT-PCR, the procedure should be carried out with much care. Solutions, buffers, dd.H_2O, glass slides, coverslips, and plastic- and glassware need to be treated with 0.1% DEPC or made from DEPC-treated water. Chemicals should be RNase free and gloves should be worn when performing RT-PCR. This section describes step-by-step protocols for the success of *in situ* RT-PCR hybridization with emphasis on the RT process.

Carrying Out Cell or Tissue Fixation, Tissue Embedding, Sectioning and Mounting of Sections on Slides

These procedures are described in Part A in Chapter 12.

Note: For in situ *RT-PCR hybridization, three separate drops of cell suspension or tissue sections should be placed on the same slide. The first cell drop or tissue*

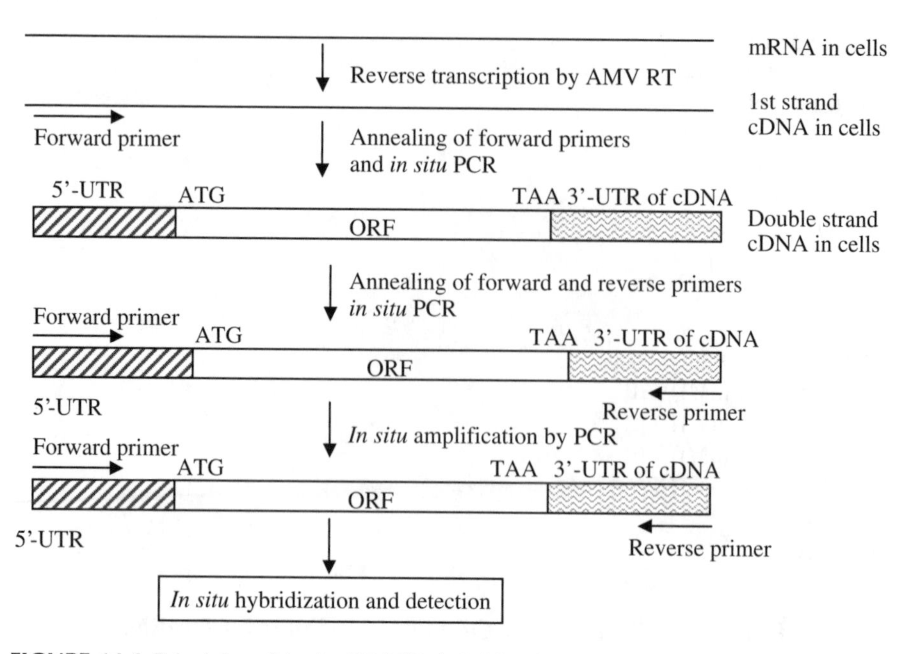

FIGURE 14.4 Principles of *in situ* RT-PCR, hybridization and detection.

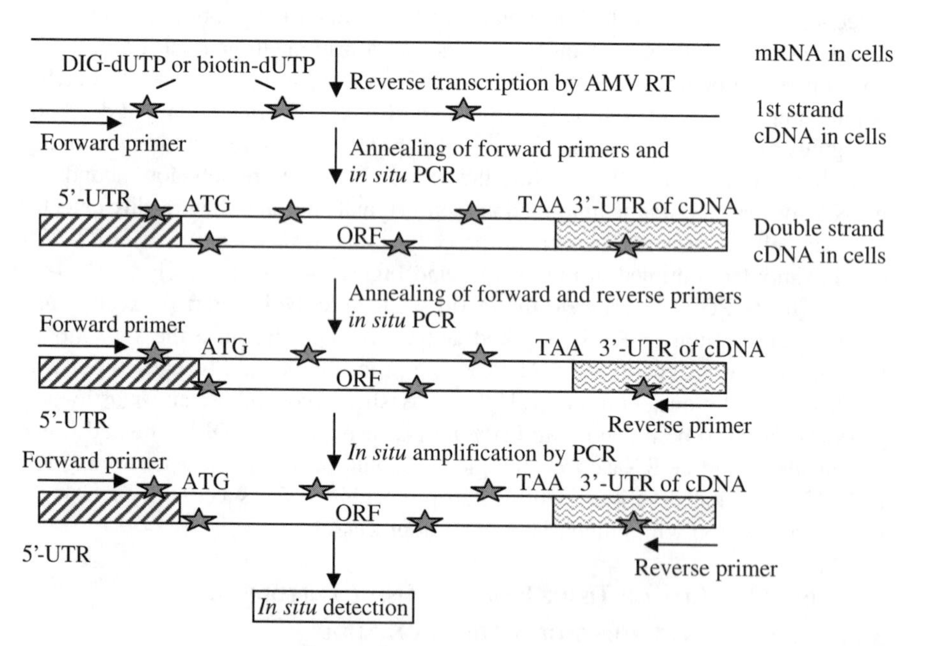

FIGURE 14.5 Principles of direct *in situ* RT-PCR and detection.

section will be subjected to DNase and RNase digestion overnight prior to in situ *RT-PCR. Following* in situ *PCR hybridization, this will serve as a negative control because no PCR products are generated. The second drop of cells or tissue section will not be digested with DNase or RNase treatment, which will serve as a positive control because all cells should have PCR products and hybridization signals. The third drop of cells or tissue section will serve as the RT reaction, which will be treated with DNase but not RNase.*

Dewaxing of Sections, Protease Digestion, and DNase or RNase Digestion of Specimens

These activities are described in Part B in Chapter 12.

Note: Digestion of specimens with DNase or RNase or both should be clearly marked or recorded according to the design stated in the note at step 1. Protease and DNase treatment will eliminate potential mispriming artifacts during the RT process.

Performance of *In Situ* Reverse Transcription (ISRT) Reaction on the Sections

Target mRNA can be selectively transcribed into cDNA using a specific primer that anneals to the 3′ portion near the target region of the mRNA. In case multiple probes are employed for multiple hybridization and detection, the authors recommend that oligo(dT) primers be used to transcribe all the mRNA species into cDNAs. Oligo(dT) primers will anneal to the 3′ poly(A) tails of mRNA molecules and facilitate the synthesis of the first-strand cDNAs. The cDNA pool allows selectively amplifying or carrying out hybridization in the sample. Avian myeloblastosis virus (AMV) reverse transcriptase and Moloney murine leukemia virus (MoMuLV) reverse transcriptase are commonly used for RT reactions. RT kits are commercially available. A standard reaction volume for one slide is 25 μl. The following cocktail is a total volume of 100 μl for four slides:

1. Prepare the RT cocktail as follows:
   10X RT Buffer for AMV or MoMuLV, 10 μl
   dNTPs mixture (2.5 m*M* each), 10 μl
   RNasin (40 units/μl) as RNase inhibitor, 2.8 μl
   Oligo(dT) primers (20 μ*M*), 4.8 μl
   AMV or MoMuLV reverse transcriptase, 50 units
   Add DEPC-treated dd.H_2O to a final volume of 100 μl.
2. Add 25 μl of the cocktail mixture to each slide and carefully overlay the cells or tissue sections with the cocktail using a clean coverslip.
3. Incubate the slides at 42°C in a high-humidity chamber for 60 min.
4. Stop the RT reaction by heating at 80°C for 2 min, ethanol wash for 5 min and partial air-drying.

Reagents Needed

10X AMV or MoMuLV Buffer

0.1 *M* Tris-HCl, pH 8.3
0.5 *M* KCl
25 m*M* $MgCl_2$

Stock dNTPs

dATP10 m*M*
dTTP10 m*M*
dATP10 m*M*
dTTP10 m*M*

Working dNTPs Solution

dATP (10 m*M*)10 µl
dTTP (10 m*M*)10 µl
dATP (10 m*M*)10 µl
dTTP (10 m*M*)10 µl

Performance of RNase (DNase-free) Digestion of Specimens to Eliminate Potential Mispriming between PCR Primers and mRNA Molecules

This is an optional procedure.

Preparation of Radioactive Probes for RISH or Nonradioactive Probes for FISH

Preparation of radioactive probes for RISH is described in Part B in Chapter 12 and preparation of nonradioactive probes for FISH described in Chapter 13.

Design of Specific Primers for *in Situ* PCR

This procedure is described in Part A of this chapter.

Carrying Out *in Situ* PCR

This procedure is described in Part A of this chapter.

Performance of *in Situ* Hybridization and Detection by RISH or by FISH

Performance of *in situ* hybridization and detection by RISH is described in Part B in Chapter 12. Performance of *in situ* hybridization and detection by FISH is described in Chapter 13.

IN SITU PCR OR RT-PCR DETECTION OF DNA OR mRNA BY DIRECT INCORPORATION OF DIGOXIGENIN-11-dUTP OR BIOTIN-dUTP WITHOUT HYBRIDIZATION

In situ PCR or RT-PCR hybridization as described earlier in Part A and Part B of this chapter mandates hybridization following PCR amplification — a relatively time-consuming and costly activity. This section, however, presents a rapid and simple approach for *in situ* PCR or *in situ* RT-PCR with direct incorporation of DIG-dUTP or biotin-dUTP. After a brief washing, the amplified sample can be directly used for detection of the target sequences, thus eliminating probes and hybridization.

Basically, the *in situ* PCR or RT-PCR procedure with direct incorporation is virtually the same as those described in Part A and Part B except that a labeled dUTP is used as a reporter molecule. Therefore, the procedures and notes mentioned in Part A and Part B will not be repeated here except for the preparation of the PCR reaction cocktail (50 µl per slide). The following cocktail is a total volume of 500 µl for 10 sections.

10X Amplification buffer, 50 µl
Mixture of four dNTPs (2.5 m*M* each), 50 µl
Digoxigenin-11-dUTP, 2.0 µl
Sense or forward primer (20 µ*M* in dd.H_2O), 5 µl
Antisense or reverse primer (2 µ*M* in dd.H_2O), 25 µl
2% Bovine serum albumin (BSA) or 0.02% gelatin solution in dd.H_2O, 25 µl
Taq DNA polymerase, 50 units
Add dd.H_2O to final volume of 500 µl.

Add the cocktail to the samples and perform *in situ* PCR as described in Part A in this chapter. For detection of the amplified sequences, anti-DIG-dUTP and anti-biotin-dUTP antibodies conjugated with alkaline phosphatase or peroxidase can be used (see the detection method for FISH in Chapter 13).

TROUBLESHOOTING GUIDE

1. **Individual sections fall apart during sectioning with no ribbons formed**. Overfixation of tissues could have occurred. Try a shorter fixation time. An optimal fixation time should be determined.
2. **Sections crumble and are damaged during stretching.** Underfixation of tissues has occurred. Use a longer fixation time and determine the optimal time empirically.
3. **Sections have holes.** This is due to incomplete tissue dehydration and improper impregnation with wax. Perform complete dehydration of fixed tissues followed by thorough impregnation with melted paraffin wax.
4. **Holes occur in frozen sections.** Ice crystals are formed in the tissues during freezing or in the cryostat. Make sure to freeze the tissue correctly.

Do not touch the frozen sections with fingers or relatively warm tools while mounting the specimens in the microtome or brushing the sections from the cutting knife.

5. **Sections are not of uniform thickness or have lines throughout.** The cutting surface of the specimen block is not parallel to the edge of the knife. The knife is dull. Take great care while mounting the specimens in the microtome and use a sharpened knife.
6. **Morphology of tissue is poorly preserved.** Tissues are stretched or shrunken. Make sure not to press or stretch the tissue during dissection and embedding.
7. **Following *in situ* PCR hybridization, no signals are detected in the cells, including the positive control. There are no problems with the primers and the *Taq* polymerase, and PCR amplification worked well.** Double-stranded DNA probes and target DNA sequences may not be denatured prior to hybridization. Both dsDNA and probes must be denatured into ssDNA before hybridization. This is a key factor for hybridization.
8. **The intensity of fluorescence signal in specimens is too bright with a low signal-to-noise ratio.** The amount of fluorescein probes or biotin- or digoxigenin-detection solution is overloaded. Overhybridization thus occurs. Try to reduce the probe and detection solutions to 5- to 10-fold. Control the hybridization time to 2 to 4 h.
9. **A high background is shown in enzymatic detection sections. Inasmuch as color can be seen everywhere in the specimens, it is difficult to tell which signal is the correct one.** Color development takes place for too long or the specimens may be partially or completely dried during hybridization or washing procedures. Try to monitor color development once every 3 to 4 min and terminate color formation as soon as the primary color signal appears. Probe-target sequence binding should reveal the first visible color before potential background develops later. Make sure not to allow specimens to become dry.
10. **Positive signals are shown in the negative control.** DNase and RNase digestion of specimens is not complete. Therefore, primers can still anneal to target nucleic acids, initiating PCR amplification. Allow DNase and RNase treatment to proceed overnight or longer.

REFERENCES

1. Haase, A.T., Retzel, E.F., and Staskus, K.A., Amplification and detection of lentiviral DNA inside cells, *Proc. Natl. Acad. Sci. USA,* 37: 4971–4975, 1990.
2. Bagasra, O. and Pomerantz, R.J., HIV-1 provirus is demonstrated in peripheral blood monocytes *in vivo:* a study utilizing an *in situ* PCR, *AIDS Res. Hum. Retroviruses,* 9: 69–76, 1993.
3. Don, R.H., Cox, P.T., Wainwright, B.J., Baker, K., and Mattick, J.S., Touchdown PCR to circumvent spurious priming during gene amplification, *Nucl. Acids Res.,* 19: 4008–4010, 1991.

4. Embleton, M.J., Gorochov, G., Jones, P.T., and Winter, G., In-cell PCR from mRBA amplifying and linking the rearranged immunoglobulin heavy and light chain V-genes within single cells, *Nucl. Acids Res.,* 20: 3831–3837, 1992.
5. Greer, C.E., Peterson, S.L., Kiviat, N.B., and Manos, M.M., PCR amplification from paraffin-embedded tissues: effects of fixative and fixation times, *Am. J. Clin. Pathol.,* 95: 117–124, 1991.
6. Heniford, B.W., Shum-Siu, A., Leonberger, M., and Hendler, F.J., Variation in cellular EGF receptor mRNA expression demonstrated by *in situ* reverse transcription polymerase chain reaction, *Nucl. Acids Res.,* 21: 3159–3166, 1993.
7. Patterson, B.K., Till, M., Ottto, P., and Goolsby, C., Detection of HIV-1 DNA and messenger RNA in individual cells by PCR-driven *in situ* hybridization and flow cytometry, *Science,* 260: 976–979, 1993.
8. Lawrence, J.B. and Singer, R.H., Quantitative analysis of *in situ* hybridization methods for the detection of actin gene expression, *Nucl. Acids Res.,* 13: 1777–1799, 1985.
9. Singer, R.H., Lawrence, J.B., and Villnave, C., Optimization of *in situ* hybridization using isotopic and nonisotopic methods, *BioTechniques,* 4: 230–250, 1986.
10. Lichter, P., Boyle, A.L., Cremer, T., and Ward, D.D., Analysis of genes and chromosomes by nonisotopic *in situ* hybridization, *Genet. Anal. Techn. Appl.,* 8: 24–35, 1991.
11. Speel, E.J.M., Schutte, B., Wiegant, J., Ramaekers, F.C., and Hopman, A.H.N., A novel fluorescence detection method for *in situ* hybridization, based on the alkaline phosphatase-fast red reaction, *J. Histochem. Cytochem.,* 40: 1299–1308, 1992.
12. Trembleau, A., Roche, D., and Calas, A., Combination of nonradioactive and radioactive *in situ* hybridization with immunohistochemistry: a new method allowing the simultaneous detections of two mRNAS and one antigen in the same brain tissue section, *J. Histochem. Cytochem.,* 41: 489–498, 1993.
13. Nunovo, G.J., Forde, A., MacConnell, P., and Fahrenwald, R., *In situ* detection of PCR-amplified HIV-1 nucleic acids and tumor necrosis factor cDNA in cervical tissues, *Am. J. Pathol.,* 143: 40–48, 1993.
14. Nuovo, G.J., RT *in situ* PCR with direct incorporation of digoxigenin-11-dUTP: protocol and applications, *Biochemica,* 11(1), 4–6, 1994.
15. Nuovo, G.J., *PCR* in Situ *Hybridization: Protocols and Applications,* Raven Press, New York, 1994.
16. Zeller, R., Watkins, S., Rogers, M., Knoll, J.H.M., Bagasra, O., and Hanson, J., *In situ* hybridization and immunohistochemistry, in *Current Protocols in Molecular Biology,* Ausubel, F.M., Brent, R., Kingston, R.E., Moore, D.D., Seidman, J.G., Smith, J.A., and Struhl, K., Eds., John Wiley & Sons, New York, pp. 14.0.3–14.8.24, 1995.
17. Staecker, H., Cammer, M., Rubinstein, R., and Van De Water, T.R., A procedure for RT-PCR amplification of mRNAs on histological specimens, *BioTechniques,* 16: 76–80, 1994.

15 Isolation and Characterization of Genes from Genomic DNA Libraries

CONTENTS

INTRODUCTION

Gene isolation plays a major role in current molecular biology.[1] To isolate the gene of interest, it may be necessary to construct a genomic DNA library. The quality and integrity of a genomic DNA library greatly influence the identification of a specific gene. An excellent library should contain all DNA sequences of the entire genome. The probability of fishing out a DNA sequence depends on the size of the fragments, which can be calculated by the following equation:

$$N = \frac{\ln(1-P)}{\ln(1-f)}$$

where N is the number of recombinants required, P is the desired probability of fishing out a DNA sequence, and f is the fractional proportion of the genome in a single recombinant. For instance, given a 99% probability to isolate an interesting DNA sequence from a library constructed with 18-kb fragments of a 4×10^9 base-pairs genome, the required recombinants are:

$$N = \frac{\ln(1-0.99)}{\ln[1-(1.8\times10^4/4\times10^9)]} = 1\times10^6$$

The average size of DNA fragments selected for cloning depends on cloning vectors. Three types of vectors are widely used to construct genomic DNA libraries: bacteriophage lambda vectors, cosmids, and yeast artificial chromosomes (YACs). Lambda vectors can hold 14- to 25-kb DNA fragments. Recombinants of $3 \times 10^{6-7}$ are usually needed to achieve a 99% probability of isolating a specific clone and the screening procedure of the library is relatively complicated. Cosmids, on the other hand, can accept 35- to 45-kb DNA fragments. Approximately 3×10^5 recombinant cosmids are sufficient for 99% probability of identifying a particular singly-copy sequence of interest. However, it is difficult to construct and maintain a genomic DNA library in cosmids compared with bacteriophage lambda vectors. YAC vectors are useful for cloning of 50- to 10,000-kb DNA fragments. This is especially powerful in cloning and isolating extra-large genes (e.g., human factor VIII gene, 180 kb in length; the dystrophin gene, 1800 kb in length) in a single recombinant YAC of a smaller library.

This chapter describes in detail the protocols for the construction and screening of a bacteriophage λ library.[1–7]

SELECTION OF LAMBDA VECTORS

Traditionally, the widely used bacteriophage λ DNA vectors are EMBL3 and EMBL4 (Figure 15.1). Both types of vectors contain a left arm (20 kb), a right arm (9 kb) and a central stuffer (14 kb) that can be replaced with a foreign genomic DNA insert (9 to 23 kb). The difference between two λ DNA vectors is the orientations of the multiple cloning site (MCS).

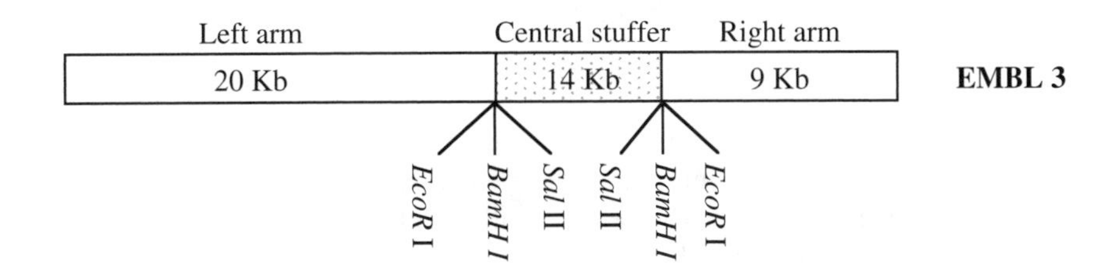

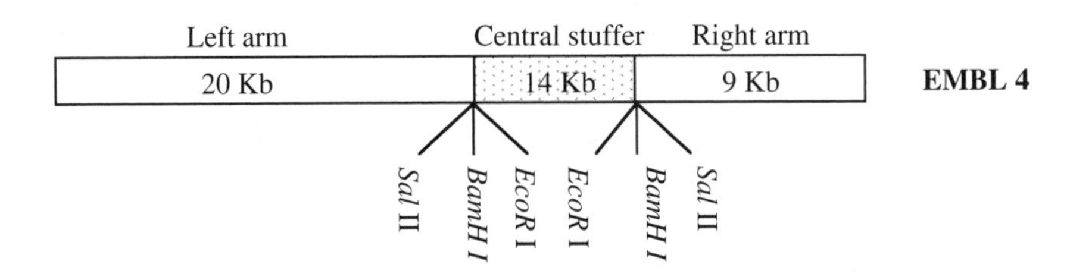

FIGURE 15.1 Structures of vectors EMBL 3 and EMBL 4.

Recently, new λ DNA vectors such as λGEM-11 and λGEM-12 have been developed (Figure 15.2); these vectors are modified from EMBL3 and EMBL4. The sizes of arms and capacity of cloning foreign DNA inserts are the same as vectors EMBL3 and EMBL4. However, more restriction enzyme sites are designed in the multiple cloning site. Additionally, two RNA promoters, T7 and SP6, are at the polylinker ends in opposite orientations.

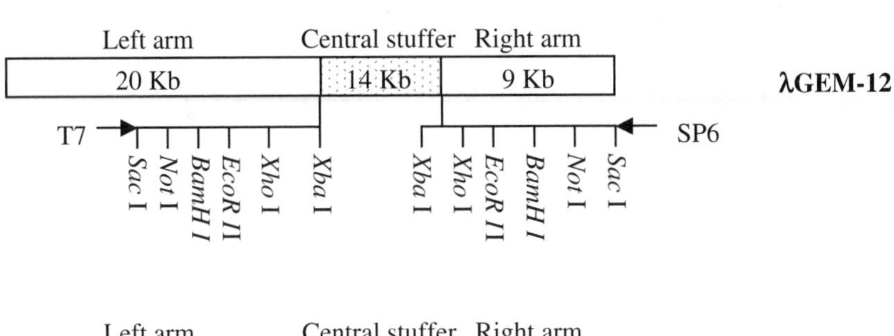

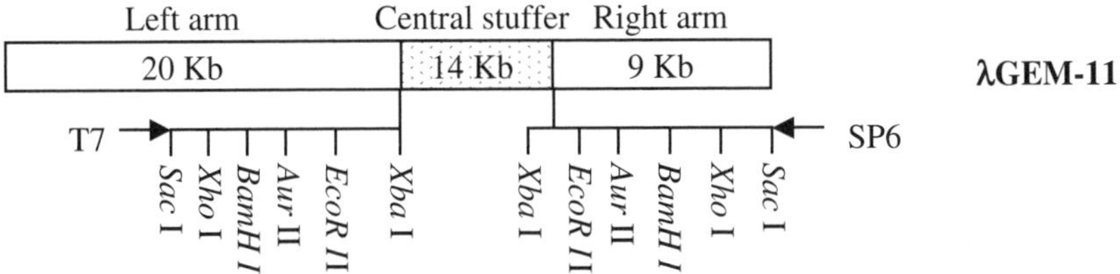

FIGURE 15.2 Structures of vectors λGEM-11 and λGEM-12.

ISOLATION OF HIGH MOLECULAR WEIGHT OF GENOMIC DNA

1. For cultured cells, aspirate culture medium and rinse the cells twice in PBS. Add 10 ml/60-mm culture plate (with approximately 10^5 cells) of DNA isolation buffer to the cells and allow cells to be lysed for 5 min at room temperature with shaking at 60 rpm. For tissues, grind 1 g tissue in liquid N_2 to fine powder and transfer the powder into a tube containing 15 ml of DNA isolation buffer. Allow cells to be lysed for 10 min at room temperature with shaking at 60 rpm.
2. Incubate at 37°C for 5 h or overnight to degrade proteins and extract genomic DNA.
3. Allow the lysate to cool to room temperature and add three volumes of 100% ethanol precooled at –20°C to the lysate. Gently mix it and allow DNA to precipitate at room temperature with slow shaking at 60 rpm. Precipitated DNA appearing as white fibers should be visible in 20 min.
4. Fish out the DNA using a sterile glass hook or spin down the DNA. Rinse the DNA twice in 70% ethanol and partially dry the DNA for 40 to 60 min at room temperature to evaporate ethanol. The DNA can be directly subjected to restriction enzyme digestion without being dissolved in TE buffer or dd.H_2O. We usually overlay the precipitated DNA with an appropriate volume of restriction enzyme digestion cocktail and incubate the mixture at an appropriate temperature for 12 to 24 h. The digested DNA is mixed with an appropriate volume of DNA loading buffer and is ready for agarose gel electrophoresis.

Reagents Needed

DNA Isolation Buffer

75 m*M* Tris-HCl, pH 6.8
100 m*M* NaCl
1 mg/ml Protease K (freshly added)
0.1% (w/v) *N*-Lauroylsarcosine

PARTIAL DIGESTION OF GENOMIC DNA USING *SAU3A I*

It is necessary to partially digest the high molecular weight of genomic DNA with a four-base cutter, *Sau3A I*, prior to ligation with appropriate vectors.

Determination of the Optimal Partial Digestion of Genomic DNA with *Sau3A I*

In order to determine the amount of enzyme used to digest the high molecular weight of DNA into 20- to 30-kb fragments, small-scale reactions, or pilot experiments, should be carried out.

1. Prepare 1X *Sau3A I* buffer on ice:
   10X *Sau3A I* buffer, 0.2 µl
   Add dd.H_2O to a final volume of 2.0 µl.
2. Perform *Sau3A I* dilutions in 10 individual microcentrifuge tubes on ice:

Tube	*Sau3A I* (3 u/µl) dilution	Dilution factors
1	2 µl *Sau3A I* + 28 µl 1X *Sau3A I* buffer	1/15
2	10 µl of 1/15 dilution + 90 µl 1X *Sau3A I* buffer	1/150
3	10 µl of 1/150 dilution + 10 µl 1X *Sau3A I* buffer	1/300
4	10 µl of 1/150 dilution + 30 µl 1X *Sau3A I* buffer	1/600
5	10 µl of 1/150 dilution + 50 µl 1X *Sau3A I* buffer	1/900
6	10 µl of 1/150 dilution + 70 µl 1X *Sau3A I* buffer	1/1200
7	10 µl of 1/150 dilution + 90 µl 1X *Sau3A I* buffer	1/1500
8	10 µl of 1/150 dilution + 110 µl 1X *Sau3A I* buffer	1/1800
9	10 µl of 1/150 dilution + 190 µl 1X *Sau3A I* buffer	1/3000
10	10 µl of 1/150 dilution + 290 µl 1X *Sau3A I* buffer	1/4500

3. Set up 10 individual small-scale digestion reactions on ice in the order listed below:

Components	Tube									
	1	2	3	4	5	6	7	8	9	10
Genomic DNA (µl)	(1 µg/µl)	1	1	1	1	1	1	1	1	1
10X *Sau3A I* buffer (µl)	5	5	5	5	5	5	5	5	5	5
dd.H_2O (µl)	39	39	39	39	39	39	39	39	39	39
Sau3A I dilution in the same order as those in step 2 (µl)	5	5	5	5	5	5	5	5	5	5
Final volume (µl)	50	50	50	50	50	50	50	50	50	50

Note: The final Sau3A *I concentration used in tubes 1 to 10 should be 1, 0.1, 0.05, 0.025, 0.015, 0.0125, 0.01, 0.0085, 0.005, and 0.0035 unit/µg DNA, respectively.*

4. Incubate the reactions at the same time at 37°C for 30 min. Place the tubes on ice and add 2 µl of 0.2 *M* EDTA buffer (pH 8.0) to each tube to stop the reactions.
5. While the reactions are being performed, prepare a large size of 0.4% agarose gel in 1X TBE buffer.
6. Add 10 µl of 5X DNA loading buffer to each of the 10 tubes containing the digested DNA prepared in step 4.
7. Carefully load 30 µl of each sample onto the wells in the order of 1 to 10. Load DNA markers (e.g., λDNA *Hind III* markers) to the left or the right well of the gel to estimate the sizes of digested DNA.
8. Carry out electrophoresis of the gel at 4 to 8 V/cm until the bromphenol blue reaches the bottom of the gel.
9. Photograph the gel under UV light and find the well that shows the maximum intensity of fluorescence in the desired DNA size range of 20 to 30 kb (Figure 15.3).

FIGURE 15.3 Partial digestion of genomic DNA using *Sau3A I*.

Large-Scale Preparation of Partial Digestion of Genomic DNA

1. Based on the optimized conditions established previously, carry out a large-scale digestion of 50 µg of high molecular weight of genomic DNA using half-units of *Sau3A I*/µg DNA that produced the maximum intensity of fluorescence in the DNA size range of 20 to 30 kb. The DNA concentration, time and temperature should be exactly the same as those used for the small-scale digestion. For instance, if tube 7 (0.01 unit of *Sau3A I* per microgram of DNA) in the small-scale digestion of the DNA shows a maximum intensity of fluorescence in the size range of 20 to 30 kb, the large-scale digestion of the same DNA can be carried out on ice as follows.

   Genomic DNA (1µg/µl), 50 µl
   10X *Sau3A I* buffer, 250 µl
   dd.H_2O, 95 ml
   Diluted *Sau3A* (0.005 u/µg) prepared as tube #9, 250 µl
   Final volume of 2.5 ml
2. Incubate the reactions at 37°C in a water bath for 30 min. Stop the reaction by adding 20 µl of 0.2 *M* EDTA buffer (pH 8.0) to the tube and place it on ice until use.
3. Add 2 to 2.5 volumes of chilled 100% ethanol to precipitate at –70°C for 30 min.
4. Centrifuge at 12,000 × *g* for 10 min at room temperature. Carefully decant the supernatant and briefly rinse the DNA pellet with 5 ml of 70% ethanol. Dry the pellet under vacuum for 8 min. Dissolve the DNA in 50 µl of TE buffer. Take 5 µl of the sample to measure the concentration at A_{260} nm and store the DNA sample at –20°C prior to use.

Reagents Needed

10X Sau3A I Buffer

0.1 *M* Tris-HCl, pH 7.5
1 *M* NaCl
70 m*M* $MgCl_2$

5X Loading Buffer

38% (w/v) Sucrose
0.1% Bromphenol blue
67 m*M* EDTA

5X TBE Buffer

54 g Tris base
27.5 g Boric acid
20 ml of 0.5 *M* EDTA, pH 8.0

TE Buffer

10 m*M* Tris-HCl, pH 8.0
1 m*M* EDTA, pH 8.0

Ethanol (100%, 70%)

PARTIAL FILL-IN OF RECESSED 3′ TERMINI OF GENOMIC DNA FRAGMENTS

Because partially filled-in *Xho* I sites of λ vectors are commercially available, the partially digested DNA fragments should be partially filled-in in order to be ligated with the vectors. The advantage of these reactions is to prevent DNA fragments from self-ligation, thus eliminating the need for size fractionation of genomic DNA fragments.

1. On ice, set up a standard reaction as follows:
   Partially digested genomic DNA, 20 µg
   10X Fill-in buffer containing dGTP and dATP, 10 µl
   Klenow fragment of *E. coli* DNA polymerase I (5 units/µl), 4 µl
   Add dd.H_2O to final volume of 0.1 ml.
2. Incubate the reaction at 37°C for 30 to 60 min.
3. Extract and precipitate as described earlier. Dissolve the DNA pellet in 20 µl of dd.H_2O. Take 2 µl of the sample to measure the concentration of DNA. Store the sample at –20°C until use.

Reagents Needed

10X Fill-in Buffer

500 m*M* Tris-HCl, pH 7.2
0.1 *M* $MgSO_4$

1 m*M* DTT
10 m*M* dATP
10 m*M* dGTP

LIGATION OF DNA INSERTS TO VECTORS

Small-Scale Ligation of Partially Filled-In Genomic DNA Fragments and Partially Filled-In λGEM-12 Arms

1. Set up on ice, sample and control reactions as follows:

Sample Reaction

Components	Tube 1	2	3	4	5
Insert DNA (0.3 μg/μl, 0.1 μg = 0.01 pmol)	0	4	3	2	0.5 (μl)
Vector DNA (0.5 μg/μl, 1 μg = 0.035 pmol)	2	2	2	2	2.0 (μl)
10X Ligase buffer	1	1	1	1	1.0 (μl)
dd.H_2O	6	2	3	4	5.5 (μl)
T4 DNA ligase (10 to 15 Weiss units/μl)	1	1	1	1	1.0 (μl)
Total volume	10	10	10	10	10.0 (μl)

Positive Control Reaction

λGEM-12 *Xho* I half-site arms (0.5 μg), 2 μl
Positive control insert (0.5 μg), 2 μl
10X Ligase buffer, 1 μl
dd.H_2O, 4 μl
T4 DNA ligase (10 to 15 Weiss units/μl), 1 μl
Total volume of 10 μl

2. Incubate the ligation reactions overnight at 4°C for *Xho* I half-site arms.

Notes: (1) A positive control is necessary in order to check the efficiency of ligation, packaging and titration. Very low efficiency (such as pfu/ml = $<10^4$) means that the vectors do not work well or that some procedures are not carried out properly. In general, the positive control is pretested by the company and pfu/ml = 10^7. (2) A ligation of vector arms only is to check the background produced by the vector arms. Usually, pfu/ml = $<10^{2-3}$. A higher pfu number, such as 10^{4-5} indicates that the vectors' self-ligation is very high, which will lower the efficiency of ligation between the vector arms and the insert DNA.

Reagents Needed

10X Ligase Buffer

0.3 *M* Tris-HCl, pH 7.8
0.1 *M* $MgCl_2$

0.1 *M* DTT
10 m*M* ATP

Packaging of Ligated DNA

1. Thaw three commercial packagene extracts (50 μl/extract) on ice.
Note: Do not thaw the extract at 37°C.
2. Immediately divide each extract into two tubes (25 μl/tube) on ice.
3. Add 4 μl of each of the ligated mixtures in the preceding step 2 into appropriate tubes, each containing 25 μl of packaging extract. To test the quality of the packaging extract, add 0.5 μg of the provided positive packaging DNA into a 25-μl packaging extract.
4. Incubate the reactions at 22 to 24°C (room temperature) for 3 to 4 h.
5. Add phage buffer to each packaged tube up to 0.25 ml and add 10 to 12 μl of chloroform. Mix well by inversion and allow the chloroform to settle to the bottom of the tube. Store the packaged phages at 4°C until use or for up to 4 weeks with a consequent several-fold drop of the titer.

Note: When the packaged phage solution is used for titration or screening, do not mix the chloroform at the bottom of the tube into the solution. Otherwise, the chloroform may kill or inhibit the growth of the bacteria.

Reagent Needed

Phage Buffer

20 m*M* Tris-HCl, pH 7.5
100 m*M* NaCl
10 m*M* $MgSO_4$

Titration of Packaged Phage on LB Plates

1. Partially thaw an appropriate bacterial strain such as *E. coli* LE 392, or KW 251 on ice. Pick up a tiny bit of the bacteria using a sterile wire transfer loop, and inoculate the bacteria on the surface of an LB plate. Invert the plate and incubate at 37°C overnight. Multiple bacterial colonies will grow up.
2. Prepare a fresh bacterial culture by picking a well-isolated colony from the LB plate using a sterile wire transfer loop and inoculate it into a culture tube containing 5 ml of LB medium supplemented with 50 ml of 20% maltose and 50 ml of 1 *M* $MgSO_4$ solution. Shake at 160 rpm at 37°C for 6 to 9 h or until the OD_{600} has reached 0.6.
3. Dilute each of the packaged recombinant phage samples to 1000X, 5000X, and 10,000X in phage buffer.
4. Add 20 μl of 1 *M* $MgSO_4$ solution and 2.8 ml of melted top agar to each of 36 sterile glass test tubes. Cap the tubes and immediately place them in a 50°C water bath for at least 30 min prior to use.

5. Mix 0.1 ml of the diluted phage with 0.1 ml of fresh bacterial cells in a microcentrifuge tube. Allow the phage to adhere to the bacteria at 37°C for 30 min.
6. Add the incubated phage–bacterial mixture into specific tubes in a water bath. Vortex gently and immediately pour onto the centers of LB plates; quickly overlay the surface of each LB plate by gently tilting the plate. Cover the plates and allow the top agar to harden for 15 min at room temperature. Invert the plates and incubate them in an incubator at 37°C overnight or for 16 h.
7. Count the number of plaques in each plate and calculate the titer of the phage (pfu) for the sample and the control.

Plaque forming units (pfu) per milliliter = number of plaques/plate × dilution times × 10

The 10 refers to the 0.1 ml, on a per-milliliter basis, of the packaging extract used for one plate. For example, if there are 200 plaques on a plate made from a 1/5000 dilution, the pfu per milliliter of the original packaging extract = $200 \times 5000 \times 10 = 1 \times 10^7$.

8. Compare the titers (pfu/ml) from different samples, determine the optimal ratio of vector arms and genomic DNA inserts and carry out a large-scale ligation and a packaging reaction using the optimum conditions.

Solutions Needed

Phage Buffer

20 m*M* Tris-HCl, pH 7.4
0.1 *M* NaCl
10 m*M* $MgSO_4$

LB (Luria-Bertaini) Medium (per Liter)

10 g Bacto-tryptone
5 g Bacto-yeast extract
5 g NaCl
Adjust to pH 7.5 with 0.2 *N* NaOH and autoclave.
20% (v/v) Maltose
1 *M* $MgSO_4$

LB Plates

Add 15 g of Bacto-agar to 1 l of freshly prepared LB medium and autoclave. Allow the mixture to cool to about 60°C and pour 30 ml of the mixture into each 85- or 100-mm diameter Petri dish in a sterile laminar flow hood. Remove any air bubbles with a pipette tip and let the plates cool for 5 min prior to covering them. Allow the agar to harden for 1 h and store the plates at room temperature for up to 10 days or at 4°C in a bag for 1 month.

TB Top Agar

Add 3.0 g agar to 500 ml freshly prepared LB medium and autoclave. Store at 4°C until use. For plating, melt the agar in a microwave oven. When the solution has cooled to 60°C, add 0.1 ml of 1 M $MgSO_4$ per 10 ml of the mixture.

S Buffer

Phage buffer + 2% (w/v) gelatin

Large-Scale Ligation of Partially Filled-In Vector Arms and Partially Filled-In Genomic DNA Fragments

1. Based on the pfu/ml of the small-scale ligations, choose the optimal conditions for large-scale ligation. For example, if tube 4 in the small-scale ligation shows the maximum pfu/ml, then a large-scale ligation can be set up as follows:
   Partially filled-in DNA inserts (0.3 μg/μl, 0.1 μg = 0.01 pmol), 10 μl
   Partially filled-in vector arms (*Xho* I half-site arms, 0.5 μg/μl, 1 μg = 0.035 pmol), 10 μl
   10X Ligase buffer, 5 μl
   dd.H_2O, 20 μl
   T4 DNA ligase (10 to 15 Weiss units/μl), 5 μl
   Total volume of 50 μl
2. Incubate the reaction at 4°C for 12 to 24 h.
3. Carry out packaging *in vitro* of ligated DNA using packagene extracts thawed on ice. Add 9 μl of the ligated mixture to each of two extracts (50 μl/extract).
4. Incubate at 22 to 24°C for 5 to 8 h.
5. Add phage buffer to each tube up to 0.5 ml. Add 25 μl of chloroform to each tube, mix well and store at 4°C until use or for up to 4 weeks. At this stage, a genomic DNA library has been established.
6. Carry out titration as described previously.

Note: A good genomic library should have a pfu number up to 10^{7-8}/ml. If the pfu is $<10^4$, it is considered a poor library that should not be used for screening.

SCREENING OF GENOMIC DNA LIBRARY AND ISOLATION OF SPECIFIC CLONES

Method A. Screening of the Genomic Library with a α-^{32}P-Labeled Probe

1. Partially thaw the specific bacterial strain, *E. coli* LE392, on ice. Pick up a tiny amount of the bacteria using a sterile wire transfer loop and streak

out the *E. coli* on an LB plate. Invert the plate and inoculate in an incubator at 37°C overnight.

2. Prepare a fresh bacterial culture by picking up a single colony from the LB plate using a sterile wire transfer loop and inoculate into a culture tube containing 50 ml of LB medium supplemented with 0.5 ml of 20% maltose and 0.5 ml of 1 *M* $MgSO_4$ solution. Shake at 160 rpm at 37°C overnight or until the OD_{600} has reached 0.6. Store the culture at 4°C until use.
3. Add 20 μl of 1 *M* $MgSO_4$ solution and 2.8 ml of melted LB top agar to each of sterile glass test tubes. Cap the tubes and immediately place them in a 50°C water bath for at least 30 min prior to use.
4. Based on the titration data, dilute the λDNA library with phage buffer. Set up 20 to 25 plates for primary screening of the genomic DNA library. For each of the 100-mm diameter plates, mix 0.1 ml of the diluted phage containing 2×10^5 pfu of the genomic library with 0.2 ml of fresh bacterial cells in a microcentrifuge tube. Allow the phage to adhere to the bacteria in an incubator at 37°C for 30 min.
5. Add the phage–bacterial mixture into the tubes in the water bath. Vortex gently and immediately pour onto the center of each LB plate, quickly overlaying the entire surface of each LB plate by gently tilting it. Cover the plates and allow the top agar to harden for 15 min at room temperature. Invert the plates and incubate in an incubator at 37°C overnight.
6. Chill the plates at 4°C for 1.5 h.
7. Transfer the plates to room temperature. Carefully overlay each plate with a dry nitrocellulose filter disk or a nylon membrane disk (prewetting treatment is not necessary) from one side of the plate slowly to the other side. Allow the phage DNA to mobilize onto the facing side of the membrane filters for 3 to 5 min. If a duplicate is needed, a second filter may be overlaid on the plate for 4 to 7 min.

Notes: (1) Air bubbles should be avoided when overlaying the membrane filters onto the LB plates. Following overlay of filters, quickly mark the top side and position of each filter in a triangular pattern by punching the filter through the bottom agar layer with a 20-gauge needle containing some India ink. (2) We recommend using positively charged nylon membrane disks because they can tightly bind the negative phosphate groups of the DNA. Nylon membranes are not easily broken, which is usually the case with nitrocellulose membranes.

8. Using forceps, carefully remove the filters and individually place them, plaque side up, on a piece of wet Whatman 3MM filter paper saturated with denaturing solution for 4 min at room temperature. Label and wrap each plate with a piece of Parafilm and store at 4°C until use.

Note: This step serves to denature the double-stranded DNA for hybridization with a probe.

9. Transfer the filters, plaque side up, onto another piece of Whatman 3MM paper saturated with neutralization solution at room temperature for 4 min.

10. Carefully transfer the filters, plaque side up, on another piece of Whatman 3MM filter paper saturated with 5X SSC at room temperature for 2 min.
11. Air-dry the filters at room temperature for about 15 min and then cover them with dry Whatman 3MM paper and bake in a vacuum oven at 80°C for 2 h or use a UV cross-linker. Wrap the filters with aluminum foil and store at 4°C until prehybridization.

Note: This step is to fix the DNA onto the filters.

12. Immerse the filters in 5X SSC for 5 min at room temperature to equilibrate the filters.

Note: Do not let the filters dry during subsequent steps. Otherwise, a high background will show up.

13. Place the filters in the prehybridization solution and carry out prehybridization for 2 to 4 h with slow shaking at 60 rpm.

Notes: (1) We strongly recommend not using plastic hybridization bags because it is not easy to get rid of air bubbles underneath the filters or they cannot be well sealed. This may cause leaking and contamination. An appropriate size of plastic beaker or tray is the best type of hybridization container. (2) The prehybridization temperature depends on the prehybridization buffer. The temperature should be set at 42°C if the buffer contains 30 to 40% (for low stringency conditions) or 50% formamide (for a high stringency condition). If the buffer, on the other hand, does not contain formamide, the temperature is set at 65°C or higher. Low stringency conditions can help identify DNA of a potential multigene family. High stringency conditions help prevent nonspecific cross-linking hybridization. (3) It is strongly recommended that a regular culture shaker with a cover and temperature control be used as the hybridization chamber. Such a chamber is easy to handle simply by placing the hybridization beaker or tray containing the filters and buffer on the shaker in the chamber. A commercial hybridization oven may be difficult to operate because the filters must be covered with a matrix and put inside the hybridization bottle; during this time, air bubbles are easily generated. (4) If many filters are used in hybridization, one beaker should not contain more than three filter disks. Too many filters in one beaker may cause weak hybridization signals. (5) The volume of prehybridization solution should be 15 ml per 100 cm^2 filter disk.

14. Prepare a DNA probe for hybridization as described in Chapter 7.
15. If the probes are double-stranded DNA, denature the probe in boiling water for 10 min and immediately chill on ice for 5 min. Briefly spin down prior to use.

Notes: This is a critical step. If the probes are not completely denatured, weak or no hybridization signals will occur. Single-strand oligonucleotide probes, however, do not require denaturation.

Caution: [α-^{32}P]dCTP is a dangerous isotope. A lab coat and gloves should be worn when working with this isotope. Gloves should be changed often and put in a special container. Waste liquid, pipette tips and papers contaminated with the isotope should be collected in labeled containers. After *finishing, a radioactive contamination survey should be performed and recorded.*

16. Dilute the purified probe with 1 ml of hybridization solution and add the probe at 2 to 10 × 10^6 cpm/ml to the hybridization buffer. Mix well and carefully transfer the prehybridized filters to the hybridization solution. Allow hybridization to proceed overnight or up to 19 h.
17. Wash the hybridized filters according to the following conditions:
    Use of High Stringency Conditions
    a. Wash the filters in a solution (50 ml/filter) containing 2X SSC and 0.1% SDS (w/v) for 15 min at room temperature with slow shaking. Repeat once.
    b. Wash the filters in a fresh solution (50 ml/filter) containing 2X SSC and 0.1% SDS (w/v) for 20 min at 65°C with slow shaking. Repeat two to four times.
    c. Air-dry the filters at room temperature for about 40 min and prepare for autoradiography.

    Use of Low Stringency Conditions
    a. Wash the filters in a solution (50 ml/filter) containing 2X SSC and 0.1% SDS (w/v) for 10 min at room temperature with slow shaking. Repeat once.
    b. Wash the filters in a fresh solution (50 ml/filter) containing 2X SSC and 0.1% SDS (w/v) for 15 min at 50 to 55°C with slow shaking. Repeat once.
    c. Air-dry the filters at room temperature for 40 min and proceed with autoradiography.
18. Wrap the filters, one by one, with SaranWrap and place in an exposure cassette. In a dark room with the safe light on, cover the filters with a piece of x-ray film and place the cassette with an intensifying screen at –80°C for 2 to 24 h prior to developing them.

Note: The film should be slightly overexposed in order to obtain a relatively dark background. This will help to identify the marks made previously.

19. To locate the positive plaques, match the developed film with the original plate by placing the film underneath the plate with the help of marks made previously.

Note: This can be done by placing a glass plate over a lamp and putting the matched film and plate on the glass plate. Turn on the light so that the positive clones can be easily identified. Make sure that the plaque-facing side exposed on the film faces down to identify the actual positive plaques. Any mismatch will cause failure of identification.

20. Remove individual positive plugs containing phage particles from the plate using a sterile pipette tip with the tip cut off. Expel the plug into a microcentrifuge tube containing 1 ml of elution buffer. Allow elution for 4 h at room temperature with occasional shaking.
21. Transfer the supernatant containing the eluted phage particles into a fresh tube and add 20 μl of chloroform. Store at 4°C for up to 5 weeks.

22. Determine the pfu of the supernatant as described previously. Replate the phage and repeat the screening procedure with the same isotopic probe several times until 100% of the plaques on the plate are positive (Figure 15.4).

Note: During the rescreening process, the plaque number used for each plate should be gradually reduced. In our experience, the plaque density for one 100-mm plate in rescreening procedures is 1000 to 500 to 300, and to 100, respectively.

23. Amplify the putative DNA clones and isolate the recombinant λDNAs by the plate or the liquid method. The purified DNA can be used for subcloning and sequencing.

Reagents Needed

Elution Buffer

10 m*M* Tris-HCl, pH 7.5
10 m*M* $MgCl_2$

Phage Buffer

20 m*M* Tris-HCl, pH 7.4
100 m*M* NaCl
10 m*M* $MgSO_4$

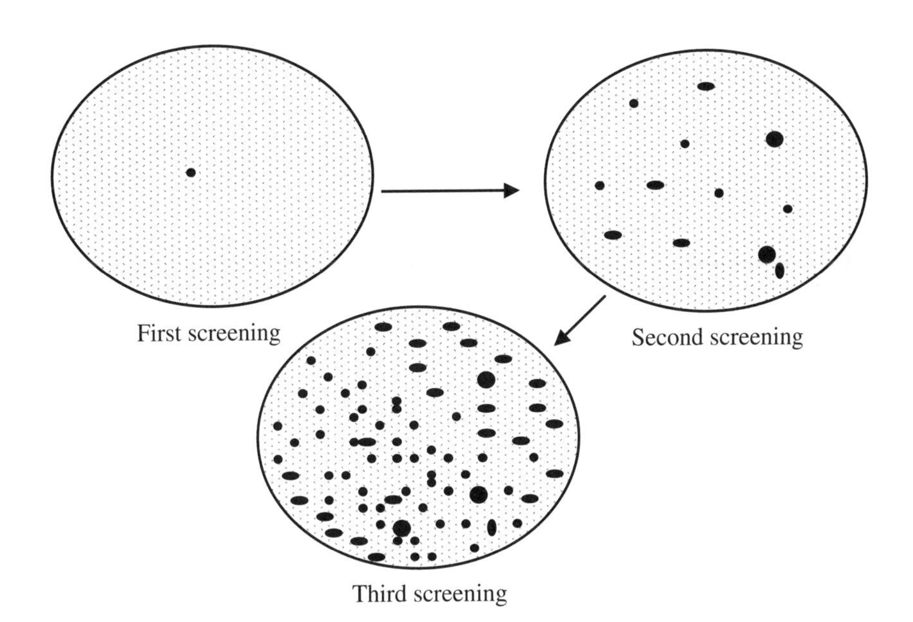

FIGURE 15.4 Diagram of rescreening and purification of positive clones from a genomic DNA library.

20X SSC Solution (1 l)

3 *M* NaCl
0.3 *M* Na_3citrate (trisodium citric acid)
Autoclave.

Denaturing solution

1.5 *M* NaCl
0.5 *M* NaOH
Autoclave.

Neutralizing Solution

1.5 *M* NaCl
0.5 *M* Tris-HCl, pH 7.4
Autoclave.

50X Denhardt's Solution

1% (w/v) BSA (bovine serum albumin)
1% (w/v) Ficoll (Type 400, Pharmacia)
1% (w/v) PVP (polyvinylpyrrolidone)
Dissolve well after each addition. Adjust to the final volume of 500 ml with distilled water and sterile filter. Divide the solution into 50-ml aliquots and store at –20°C. Dilute 10-fold into prehybridization and hybridization buffers.

Prehybridization Buffer

5X SSC
0.5% SDS
5X Denhardt's reagent
0.2% Denatured salmon sperm DNA

Hybridization Buffer

5X SSC
0.5% SDS
5X Denhardt's reagent
0.2% Denatured salmon sperm DNA
[α-^{32}P]-Labeled DNA probe

Method B. Screening of a Genomic Library Using a Nonisotope-Labeled Probe

1. Carry out step 1 to step 13 as described in Method A.
2. Perform prehybridization and hybridization, which are the same as described in Method A except that a nonradioactive probe and appropriate buffers are used.
3. Transfer the hybridized, washed filters into a clean dish containing buffer A (15 ml/filter) for 2 to 4 min.
4. Transfer the filters in buffer B (15 ml/filter) for 60 min.

5. Incubate the filters with an antibody solution, which is the anti-DIG-alkaline phosphatase (Boehringer Mannheim Biochemicals) diluted at 1:10,000 in buffer B, at room temperature for 40 to 60 min using 10 ml/filter.
6. Wash the filters (100 ml/filter) in buffer A for 20 min and repeat once using a fresh washing tray.

Note: The used antibody solution can be stored at 4°C for up to 1 month and reused five to six times without any significant decrease in the antibody activity.

7. Equilibrate the filters in buffer C for 1 to 4 min.
8. Detect the hybridized bands by one of the two methods:
   1. Using Lumi-Phos 530 (Boehringer Mannheim Biochemicals) as a substrate
      - Add 0.5 ml of the Lumi-Phos 530 to the center of a clean dish, which should be prewarmed to room temperature for 1 h prior to use. Briefly damp a filter using a forceps and completely wet the plaque facing side of the filter and both sides of the filter by slowly laying the filter down in the solution several times.
      - Wrap the filter with SaranWrap and remove the excess Lumi-Phos solution using paper towel to reduce black background.
      - Place the wrapped filter in an exposure cassette with the side facing up. Repeat steps (b), (c) and (d) until all the filters are complete.
      - Overlay the wrapped filters with an x-ray film in a dark room with the safe light on and allow the exposure to proceed, by placing the closed cassette at room temperature, for 2 min to 24 h.
      - Develop the film and proceed to identification of positive clones.

      Note: (1) Exposure for more than 4 h may produce a high black background. Based on our experience, good hybridization and detection should generate sharp positive spots with 1.5 h of exposure. (2) The film should be slightly exposed to obtain a relatively dark background that will help identify the marks made previously.
   2. Using the NBT and BCIP detection method
      - Add 40 µl of NBT solution and 30 µl of BCIP solution in 10 ml of buffer C for one filter. NBT and BCIP are available from Boehringer Mannheim Biochemicals.
      - Place the filter in the mixture prepared in the preceding step and put in a dark place for color development at room temperature for 30 min to 24 h.
      - Air-dry the filters and proceed to identification of positive clones in the original LB plates as described in Method A.

Regents Needed

Buffer A

100 m*M* Tris-HCl
150 m*M* NaCl, pH 7.5

Buffer B

2% (w/v) Blocking reagent (Boehringer Mannheim Biochemicals) or an equivalent such as BSA, nonfat dry milk, or gelatin in buffer A. Dissolve well by stirring with a magnetic stir bar.

Buffer C

100 m*M* Tris-HCl, pH 9.5
100 m*M* NaCl
50 m*M* $MgCl_2$

NBT Solution

75 mg/ml nitroblue tetrazolium salt in 70% (v/v) dimethylformamide

BCIP Solution

50 mg/ml 5-Bromo-4-chloro-3-indolyl phosphate (X-phosphate), in 100% dimethylformamide

Anti-DIG-Alkaline Phosphatase

Antidigoxigenin conjugated to alkaline phosphatase (Boehringer Mannheim Biochemicals)

ISOLATION OF ΛPHAGE DNAs BY LIQUID METHOD

1. For each 50 ml of bacterial culture, add 10 to 15 μl of 1×10^5 pfu of plaque eluate into a microcentrifuge tube containing 235 to 240 μl of phage buffer. Mix with 0.25 ml of cultured bacteria and allow the bacteria and phage to adhere to each other by incubating at 37°C for 30 min.
2. Add this mixture into a 250- or 500-ml sterile flask containing 50 ml of LB medium prewarmed at 37°C that is supplemented with 1 ml of 1 *M* $MgSO_4$. Incubate at 37°C with shaking at 260 rpm until lysis occurs.

Tips: It usually takes 9 to 11 h for lysis to occur. The medium should be cloudy after several hours of culture and then be clear upon cell lysis. Cellular debris also becomes visible in the lysed culture. There is a density balance between bacteria and bacteriophage. If the bacteria density is much over that of bacteriophage, it takes longer for lysis to occur, or no lysis takes place at all. In contrast, if bacteriophage concentration is much over that of bacteria, lysis is too quick to be visible at the beginning of the incubation and, later on, no lysis will occur. The proper combination of bateria and bacteriophage used in step 1 will assure success. In addition, careful observations should be made after 9 h of culture because the lysis is usually quite rapid after that time. Incubation of cultures should be stopped after lysis occurs. Otherwise, the bacteria grow continuously and the cultures become cloudy again. Once that happens, it will take a long time to see lysis, or no lysis will take place at all.

3. Immediately centrifuge at 9000 × *g* for 10 min at 4°C to spin down the cellular debris. Transfer the supernatant containing the amplified DNA library in bacteriophage particles into a fresh tube. Aliquot the supernatant, add 20 to 40 µl of chloroform and store at 4°C for up to 5 weeks.
4. Add RNase A and *DNase* I to the λlysate supernatant, each to a final concentration of 2 to 4 µg/ml. Place at 37°C for 30 to 60 min.

Note: RNase A functions to hydrolyze the RNAs. DNase I will hydrolyze chromosomal DNA but not phage DNA that is packed.

5. Precipitate the phage particles with one volume of phage precipitation buffer and incubate for at least 1 h on ice.
6. Centrifuge at 12,000 × *g* for 15 min at 4°C and allow the pellet to dry at room temperature for 5 to 10 min.
7. Resuspend the phage particles with 1 ml of phage release buffer per 10 ml of initial phage lysate and mix by vortexing.
8. Centrifuge at 5000 × *g* for 4 min at 4°C to remove debris and carefully transfer the supernatant into a fresh tube.
9. Extract the phage particle proteins using an equal volume of TE-saturated phenol/chloroform. Mix for 1 min and centrifuge at 10,000 g for 5 min.
10. Transfer the top, aqueous phase to a fresh tube and extract the supernatant one more time as described in the previous step.
11. Transfer the top, aqueous phase to a fresh tube and extract once with one volume of chloroform:isoamyl alcohol (24:1). Vortex and centrifuge as described in step 9.
12. Carefully transfer the upper, aqueous phase containing λDNAs to a fresh tube and add one volume of isopropanol or two volumes of chilled 100% ethanol. Mix well and precipitate the DNAs at –80°C for 30 min or –20°C for 2 h.
13. Centrifuge at 12,000 × *g* for 10 min at room temperature and aspirate the supernatant. Briefly wash the pellet with 2 ml of 70% ethanol and dry the pellet under vacuum for 10 min. Resuspend the DNA in 50 to 100 µl of TE buffer. Measure the DNA concentration (similar to the RNA measurement described previously) and store at 4°C or –20°C until use.

Reagents Needed

SM Buffer (per Liter)

50 m*M* Tris-HCl, pH 7.5
0.1 *M* NaCl
8 m*M* $MgSO_4$
0.01% Gelatin (from 2% stock)
Sterilize by autoclaving.

Phage Buffer

20 m*M* Tris-HCl, pH 7.4
100 m*M* NaCl
10 m*M* $MgSO_4$

Release Buffer

10 m*M* Tris-HCl, pH 7.8
10 m*M* EDTA

Phage Precipitation Solution

30% (w/v) Polyethylene glycol (MW 8000) in 2 *M* NaCl solution

TE Buffer

10 m*M* Tris-HCI, pH 8.0
1 m*M* EDTA

TE-Saturated Phenol/Chloroform

Thaw crystals of phenol in a 65°C water bath and mix with an equal volume of TE buffer. Allow the phases to separate for about 10 to 20 min. Mix one part of the lower phenol phase with one part of chloroform:isoamyl alcohol (24:1). Allow the phases to separate and store at 4°C in a light-tight bottle.

DNase-Free RNase A

To make DNase-free RNase A, prepare a 10 mg/ml solution of RNase A in 10 m*M* Tris-HCl, pH 7.5 and 15 m*M* NaCl. Boil for 15 min and slowly cool to room temperature. Filter if necessary. Alternatively, RNase (DNase free) from Boehringer Mannheim can be utilized.

RESTRICTION MAPPING OF RECOMBINANT BACTERIOPHAGE DNA CONTAINING GENOMIC DNA INSERT OF INTEREST

Once putative clones are isolated, the next logical step is to locate the gene of interest within the DNA insert, which will indicate how large the gene is and where it is located in the insert. It is necessary to have this information in order to subclone or directly sequence the gene. To obtain this information, restriction mapping should be carried out using three to four restriction enzymes. The selection of enzymes depends on the restriction enzyme sites in the vectors. For example, if λGEM-11 or λGEM-12 is used as the cloning vector, the mapping procedure can be performed as follows:

1. Set up, on ice, a series of single- and double-enzyme digestions.

Components	Tube 1	2	3	4	5	6
Recombinant, positive phage DNA	20 μg	20	20	20	20	20
Appropriate 10X restriction enzyme buffer	4 μl	4	4	4	4	4
EcoR I (10 units/μl)	6.7 μl	0	0	6.7	6.7	0
BamH I (10 units/μl)	0 μl	6.7	0	6.7	0	6.7
Xho I (10 units/μl)	0 μl	0	6.7	0	6.7	6.7
Add dd.H_2O to a final volume of	40 μl	40	40	40	40	40

Enzyme used for digestion	Tube
EcoR I	1
BamH I	2
Xho I	3
EcoR I + BamH I	4
EcoR I + Xho I	5
BamH I + Xho I	6

2. Incubate the reactions at an appropriate temperature for 2 to 3 h.
3. After restriction enzyme digestion, carry out electrophoresis on a 0.9% agarose gel, blot onto a nylon membrane, and perform prehybridization and hybridization with the same probe used for screening the genomic library. Washing and detection are described in detail in Southern blotting in Chapter 7.
4. To ensure that the correct results are obtained, we recommend that step 1 to step 3 be repeated. Both primary and repeated Southern blots should have identical hybridization patterns. An example of such results follows:

Restriction enzyme digestion	Bands observed on gel (kb)[a]
EcoR I	20, 9, 10, 6, 4
BamH I	20, 9, 10, 6, 4
Xho I	20, 9, 12, 6, 2
EcoR I + BamH I	20, 9, 6, 4, 2
EcoR I + Xho I	20, 9, 6, 4, 2
BamH I + Xho I	20, 9, 6, 4, 2

[a]Hybridized bands are indicated by numbers in bold typeface. Doublets or triplets are possible.

5. Identify and locate the gene in the insert according to the Southern blot hybridization results and draw a restriction map (Figure 15.5). Check out the fragments based on the following map:

Restriction enzyme digestion	Bands observed on gel (kb)[a]
EcoR I	20, 9, 10, 6, 4
BamH I	20, 9, 10, 6, 4
Xho I	20, 9, 12, 6, 2
EcoR I + BamH I	20, 9, 6, 4, 4, 4, 2
EcoR I + Xho I	20, 9, 6, 6, 4, 2, 2
BamH I + Xho I	20, 9, 6, 6, 4, 2, 2

[a]Hybridized bands are indicated by numbers in bold typeface. Doublets or triplets are possible.

In conclusion, the gene identified is 8 kb in size and located in the middle of the insert (Figure 15.5).

6. Once the size and location of the gene have been identified, purification of the gene from recombinant λGEM-11 or λGEM-12 may be carried out using appropriate restriction enzymes. The isolated gene can be subjected to subcloning and DNA sequencing.

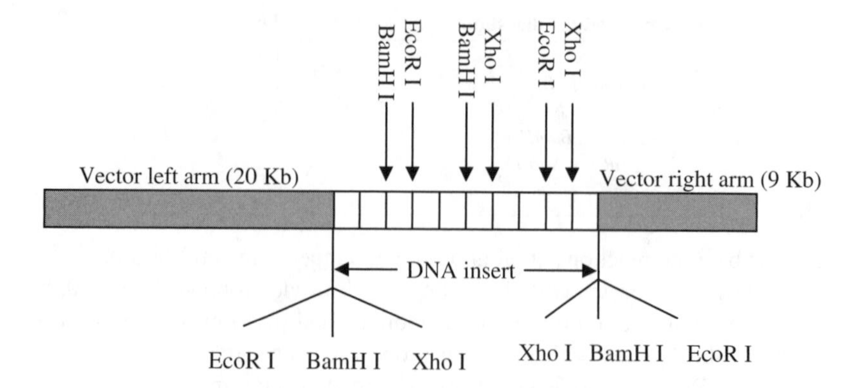

FIGURE 15.5 Restriction mapping of a gene in a λcloning vector.

SUBCLONING OF THE ISOLATED GENE OR DNA FRAGMENTS

Detailed protocols are described in Chapter 4.

CHARACTERIZATION OF ISOLATED DNA

Isolated DNA clones should be carefully characterized. For DNA sequencing and sequence analysis, detailed methods are given in Chapter 5 and Chapter 6.

TROUBLESHOOTING GUIDE

1. **Plaques are small in one area and large in other areas in the LB plates.** This is due to uneven distribution of the top agar mixture. Make sure that this mixture is spread evenly.
2. **Too many positive clones appear in the primary screening**. The specificity of the primary antibodies is low, the concentration is too high, or the blocking efficiency is low. Try to use *IgG* or affinity-column purified antibodies and carry out different dilutions of the antibodies, or increase the percentage of the blocking reagent in TBST buffer.
3. **Purple background is evident on the filter.** Color is overdeveloped. Try to stop the color reaction as soon as the desired signals appear.
4. **Unexpected larger purple spots occur.** Air bubbles may be produced when the filters are overlaid on the LB plates. Make sure that no air bubbles appear underneath the filters.
5. **No signal appears at all in any of the primary screening plates**. The pfu numbers used for each plate may be too low. For rare genes, 2×10^6 pfu should be used in primary screening of the library.
6. **Signals are weak on the filters in subsequent screening**. The antibodies have low activity or color is underdeveloped.

7. **Black background appears on the filter.** Filter is partially dried, blocking efficiency is low, or the quality of the filter is not good. Try to avoid air-drying of the filter; increase the percentage of BSA and the denatured salmon sperm DNA, and use a fresh neutral nylon membrane filter.
8. **Unexpected black spots shown on the filter.** Unincorporated ^{32}PdCTP was not effectively removed. Try to use a G-50 Sephadex column to purify the labeled DNA.

REFERENCES

1. Sambrook, J., Fritsch, E.F., and Maniatis, T., *Molecular Cloning: A Laboratory Manual,* 2nd ed., Cold Spring Harbor Press, Cold Spring Harbor, NY, 1989.
2. Wu, W., Genomic DNA libraries, in *Handbook of Molecular and Cellular Methods in Biology and Medicine,* Kaufman, P.B., Wu, W., Kim, D., and Cseke, L., pp. 165–210, CRC Press, Boca Raton, FL, 1995.
3. Tonegawa, S., Brock, C., Hozumi, N., and Schuller, R., Cloning of an immunoglobin variable region gene from mouse embryo, *Proc. Natl. Acad. Sci. USA,* 74, 3518, 1977.
4. Saiki, R.K., Gelfand, D.H., Stoffel, S., Scharf, S., Higuchi, R., Horn, G.T., Mullis, K.B., and Erlich, H.A., Primer-directed enzymatic amplification of DNA with a thermostable DNA polymerase, *Science,* 239, 487, 1988.
5 Wu, L.-L., Song, I., Kim, D., and Kaufman, P.B., Molecular basis of the increase in invertase activity elicited by gravistimulation of oat-shoot pulvini, *J. Plant Physiol.*, 142, 179, 1993.
6. Erlich, H.A., *PCR Technology: Principles and Applications for DNA Amplification,* Stockton Press, New York, 1989.
7. Jones, D.H. and Winistorfer, S.C., A method for the amplification of unknown flanking DNA: targeted inverted repeat amplification, *BioTechniques,* 15, 894, 1993.

16 Culture of Mouse Embryonic Stem Cells as a Model Mammalian Cell Line for Gene Expression

CONTENTS

INTRODUCTION

Mouse embryonic stem (ES) cells are initially derived from inner cell mass (ICM) cells of mouse blastocyst-stage embryos, and then isolated and maintained in culture conditions. In recent years, ES cells have been applied to a variety of research fields because of their advantages over other mammalian cells.[1–4] ES cells can be used as a model mammalian cell line for studies on gene cloning, manipulation, expression and cell biology[2–6] and they are pluripotent cells compared with other mammalian cell lines. This multipotent feature of ES cells offers a very powerful tool for investigators and opens a broad area of research. As a matter of fact, to investigate the potential functions of the gene of interest *in vitro* and *in vivo* in cells and animals, the ES cell line is the best choice of cell line to achieve objectives. In other words, using the ES cell line, the roles of a specific gene can be studied at the cellular level, and later transgenic animals can be generated with genetically altered expression of the gene of interest.[7–8] Therefore, ES cells provide invaluable hosts for gene over-expression, underexpression and knockout.

In spite of the fact that the general culture techniques are similar to other mammalian cell lines, the growth and maintenance of ES cells need some special care. This chapter describes in detail the appropriate procedures for ES cell culture, maintenance, freezing and thawing. We routinely employed ES cells for studying the genetic alternation of the expression of HSP27 gene. Based on our experience, we provide special tips and notes for the success of ES cell culture. Even though the present chapter focuses on ES culture, the general techniques can be adopted and applied to other mammalian cell culture.

PROTOCOL 1. CULTURE OF ES CELLS IN MEDIA CONTAINING LEUKEMIA INHIBITORY FACTOR (LIF)

Cautions: Every step should be carried out under sterile conditions to avoid bacteria or fungi contamination. Gloves should be worn and wiped periodically using 70% ethanol. Pipetteman and centrifuge tubes should be wiped outside with 70% ethanol prior to use. Pipette tips and pipettes should be autoclaved. All the solutions, buffers and media should be sterile filtered.

1. Coat the bottom of tissue culture dishes (e.g., Falcon 60 × 25 mm or 100 ×20 mm) or flasks (Falcon 25 cm^2 or 75 cm^2) by adding 3 to 6 ml of 1% gelatin solution into each dish or flask. Overlay the entire bottom surface with the solution and cap the flasks or wrap dishes. Allow coating to proceed at 4°C overnight.

Note: If culture dishes or flasks are not precoated with gelatin solution or its equivalent, ES cells cannot attach to the bottom, resulting in failure of proliferation or death.

2. Transfer the precoated containers to be used to a laminar flow hood and turn on UV light. Allow them to be warmed to room temperature under the UV light for 20 to 30 min.
3. Wipe the outside of the warmed containers with 70% ethanol and aspirate the gelatin solution. Rinse the dishes or flasks with 3 to 6 ml sterile distilled water and aspirate the water.
4. Add 3 to 6 ml of ES culture medium to each of the containers, depending on the size. Label dishes or flasks, including cell line, passage number and date.
5. Add the appropriate number of cells into each container, depending on a specific experiment. Gently mix the cells in the medium.
6. Place the containers in a cell culture incubator at 37°C with humidity, 5% CO_2 and 95% air. For flasks, the caps should be loosened a little bit so that CO_2 and air can get in.
7. Allow cells to grow to the appropriate confluence with culture medium changed once every 1 to 3 days, depending on the proliferation rate of cells.

Notes and Tips: (1) The striking feature of ES cells is that they form multiple layers of clones instead of monolayer cells. Due to the rapid proliferation/division rate of ES cells, never allow them to grow over 30% confluence on the bottom of a dish or flask. Otherwise, almost 100% of the cells will die within 1 day or so. Thus, the cells should be monitored on a daily basis. (2) If some ES cells in a flask or dish are differentiated due to some unknown reason, the entire container should be discarded. Never try to rescue some normal cells in the flask. Otherwise, it will be difficult to obtain uniform, undifferentiated ES cells. (3) Contamination is a common problem even though great care has been taken. The cells should be checked on a daily basis. To prevent potential contamination during checking, the flasks should be tightly capped and it is recommended that the stage under a microscope be wiped with 70% ethanol prior to looking at the cells under the scope. Make sure to loosen the cap of a flask before returning it to the incubator. Once a specific flask or dish is found to be contaminated with bacteria or fungi (e.g., yeast), it should be immediately treated with bleach solution and discarded.

Reagents Needed

Cell Lines

Mouse ES cell line (D3 or R1) can be obtained from the American Type Culture Collection (ATCC) (Rockville, MD).

Culture Media (GIBCO BRL, Gaithersburg, MD).

45% (v/v) Nutrient mixture Ham's F-12 with glutamine
45% (v/v) Dulbecco's modified eagle medium (D-MEM) with high glucose and glutamine
5% Fetal bovine serum (FBS)
5% Newborn calf serum (NCS)
50 µg/ml Gentamicin sulfate as an antibiotic
Sterile-filter, add 1% (v/v) of diluted 2-mercaptomethanol solution (70 µl 2-mercaptomethanol in 100 ml PBS) and store the medium at 4°C until use.

Leukemia Inhibitory Factors (GIBCO BRL, Gaithersburg, MD)

The cells can be maintained in the undifferentiated state by adding leukemia inhibitory factors (LIF) to the culture medium (10^3 units/ml).

PROTOCOL 2. PREPARATION OF MITOTICALLY INACTIVATED STO CELLS AS FEEDER LAYERS FOR GROWTH OF ES CELLS

1. Coat the bottom of each tissue culture dish or flask with 4 to 5 ml of 1% gelatin solution. Allow coating to proceed at 4°C overnight.

2. Transfer the coated containers to a laminar flow hood and allow them to be warmed to room temperature under the UV light for 20 min.
3. Rinse the containers with 3 to 6 ml sterile water and aspirate the water.
4. Add 3 to 6 ml of STO cell culture medium to each dish or flask.
5. Add the appropriate number of cells (1×10^6 cells/flask) to each container. Place the containers into a cell culture incubator at 37°C with humidity, 5% CO_2 and 95% air. For flasks, their caps should be loosened a little bit so that CO_2 and air can get in.
6. Allow the cells to grow for 5 to 7 days to form a confluent layer of cells with culture medium changed once every 3 days.
7. Replace the medium with 10 ml/flask of fresh culture medium containing 10 µg/ml of mitomycin C (Sigma Chemicals). Incubate the STO flasks for 3 to 4 h.
8. Aspirate the medium and wash the cells three times with 10 ml of PBS or fresh culture medium prior to Trypsinizing. These cells should be used within 7 days for ES cell culture.

STO Cell Culture Medium

90% (v/v) D-MEM (low glucose)
10% (v/v) FBS
50 µg/ml Gentamicin sulfate as an antibiotic
Sterile-filter and store at 4°C until use.

PROTOCOL 3. CULTURE OF ES CELLS ON FIBROBLAST FEEDER LAYERS

The ES culture described in Protocol 1 uses LIF in the culture medium to prevent cell differentiation. However, ES cells can grow on feeder cell layers such as mouse STO cells. The feeder cells can produce LIF that can keep ES cells in the proliferation stage.

1. Aspirate the medium from mitotically inactivated STO feeder cell layers and add 3 to 6 ml of ES culture medium to each dish or flask. Label dishes or flasks, including the ES cell line, passage number and date.
2. Add appropriate number of cells to each container, depending on a specific experiment.
3. Place the containers into a cell culture incubator at 37°C with humidity, 5% CO_2 and 95% air. For flasks, their caps should be loosened a little bit so that CO_2 and air can get in.
4. Allow cells to grow to appropriate confluence with culture medium changed once every 1 to 3 days, depending on the proliferation rate of cells.

Tips: The drawback of using feeder layers to provide LIF for the growth of ES cells is that if the STO cells are not 100% mitotically inactivated, they will grow very well in the ES culture medium. Hence, it may be difficult to clean them out of the ES cells.

PROTOCOL 4. TRYPSINIZING, FREEZING AND THAWING OF MAMMALIAN CELLS

TRYPSINIZING

Note: Trypsinizing procedure is only applied to those cell lines that cells usually attach to the bottom of the culture flask or dish. For floating cell culture such as some insect cells, Trypsinizing is not necessary. Instead, directly transfer cell suspension to the centrifuge tube, add an equal volume of culture medium, centrifuge and carry out step 8 to step 11 as described next.

1. Remove cell culture flasks or dishes from an incubator, tightly cover the cap of the flask and check the health and confluence/density of cells in the flask or dish.
2. Loosen caps of flasks and aspirate the medium from each container.

Note: If the medium contains toxic chemicals following toxicant treatment, the medium must be collected in a special container for toxic waste disposal.

3. Wash cells once with an appropriate volume of 1X HBSS (Gibco BRL) (3 ml/25 cm^2 flask or 4 ml/100 mm dish).

Note: HBSS solution serves to remove dead cells and serum secreted from the cells during the culture. This solution is usually stored at room temperature under sterile conditions. Serum can inhibit the activity of the Trypsine–EDTA in the next step and should be removed at this step.

4. Carefully remove the solution as described in step 2.

Note: Cells should still be attached to the bottom of the flask or dish.

5. Add an appropriate volume of 1X Trypsine–EDTA solution (Gibco BRL) to flasks or dishes (1 to 1.5 ml/25 cm^2 flask or 2 ml/100 mm dish). Tighten the caps of flasks or cover dishes and completely overlay cells with the solution by hand. Place the containers in a laminar flow hood for 7 to 10 min at room temperature or in a culture incubator at 37°C for 5 min.

Tips: In using flasks, cells can be detached from the bottom by gently patting the side walls of the flasks by hand for approximately 2 min. Trypsine–EDTA solution should be stored at 4°C to reserve its maximum activity. Its function is to detach cells from the bottom of the culture flask or dish. The mechanism, however, is not well understood. We and others believe that Trypsine–EDTA acts to depolymerize cellular microfilaments of cells, causing the cell to round up and detach from the bottom of the container.

6. Periodically check the detachment of cells under a microscope. Cells should be round and floating in Trypsine–EDTA solution.
7. Vertically place the flasks or tilt the dishes in a laminar flow hood to drain the cell solution to one side of the container. Carefully transfer the cell suspension to a sterile, plastic centrifuge tube (Falcon 15 ml or 50 ml) containing a volume of culture medium equal to the volume of Trypsine–EDTA solution added at step 5.

Note: The pipette used for transferring should be changed for each flask or dish or different cell lines in order to avoid cell–cell mixing. The centrifuge tube should be sterile and labeled according to the label of the flask or dish.

8. The centrifuge tubes should be tightly capped, balanced, and placed in an appropriate centrifuge (e.g., a swing-bucket centrifuge on a bench) and centrifuged at 1200 × *g* or 1000 rpm for 5 min.
9. Carefully transfer the tubes back to the laminar flow hood and gently wipe the outside of the tubes with Kimwipe paper wetted with 70% ethanol. Gently loosen the cap of the tube and carefully aspirate the medium from the tube.

Note: Caution should be taken at the present step. The tubes should be gently handled without vigorously disturbing the contents. The worst thing may be that if the cell pellet is loosened into the media, they will be aspirated away. Furthermore, the pipette used to aspirate the medium should never be inserted at the very bottom of tube where the cell pellet is. Otherwise, cells will be sucked away.

10. Add an appropriate volume of normal culture medium to each of the tubes; gently and thoroughly resuspend the cells using a 5- or 10-ml sterile pipette.

Note: The volume of culture medium used to resuspend cells depends on cell density. Based on our experience with mouse stem cells and STO cells, 1,000,000 cells per ml medium should be applied. To resuspend cells completely, use a 5-ml pipette to take up the medium from the tube followed by flushing the bottom of the tube. Repeat up and down suspension approximately 20 to 25 times. In this way, more than 95% of the cells are separated from each other. The efficiency of cell resuspension can be monitored under a microscope using 5 μl of the cell suspension. Pipettes used for cell suspension should be changed for each cell sample to avoid cell contamination.

11. Use the cell suspension to determine the cell density (Protocol 5), to split the cell aliquot or to freeze back the cells, depending on the particular purpose.

Freeze Back

1. Resuspend the cells at step 9 above in an appropriate volume of freezing medium at approximately 10^6 cells/ml.
2. Label sterile freezing vials: cell line name, passage number, date and person's name. Aliquot the cell suspension into the vials (0.5 to 1.0 ml/tube).
3. Place the vials in a styrofoam box precooled in a –70°C freezer.

Tip: This step is optional.

4. Store the vials in a –140 or –180°C freezer or in a liquid nitrogen tank.

Materials Needed

Sterile Freezing Vials

–70°C freezer, –180°C freezer or liquid nitrogen tank
Styrofoam box
Culture medium

Freezing Medium

70 to 75% (v/v) Culture medium
20% (v/v) FCS
5 to 10% (v/v) DMSO
Filter-sterilize.

Thawing and Reculture

1. Prepare culture dishes or flasks for growth of cells as described in Protocol 1.
2. Take a vial containing the cells from storage freezer and thaw it by holding it in a hand or in a 37°C water bath for about 3 to 6 min.
3. Once the last visible ice crystal disappears, wipe the outside of the vial with 70% ethanol and add 40 to 100 μl of the suspension to each flask or dish.
4. Culture the cells under normal conditions as described in Protocol 1.

Tip: DMSO can lower the freezing point of cells to prevent ice crystal formation inside the cells and protect the cells during freezing. However, DMSO in the freezing medium is toxic to cells under normal culture temperature. Thus, the culture medium in the flask or dish should be changed about 12 to 18 h following reculture of the frozen cells.

PROTOCOL 5. CELL COUNTING AND DETERMINATION OF CELL DENSITY

1. Trypsinize and resuspend cells as described in Protocol 2.

Note: The volume of the culture medium for suspending the cells will be considered the total volume of cell suspension.

2. Add 10 μl of cell suspension to the cross in a clean hemacytometer plate.

Tips: The cell suspension should be gently and thoroughly mixed prior to immediately taking up the 10 μl. Because cells are heavier than culture medium, they tend to sink to the bottom of the centrifuge tube. If cells are not well suspended, the counted result of the 10-μl cell suspension will not be the actual density of the cell suspension.

3. Cover the cross with a clean hemacytometer coverslip.

Note: Place the coverslip carefully so that no air bubbles are trapped underneath the coverslip. Otherwise, bubbles may affect cell counting.

4. Focus cells on the plate under an inverted phase-contrast microscope and count all of the cells in each of the four corners in the hemacytometer plate using a counter as an aid.

Tips: Counting must be accurate. Any cells located outside the four corner areas should not be counted. Only cells inside the four corners are counted. Each corner consists of 16 squares as shown below and the counting directions are recommended. Cells in any clusters should be counted as accurately as possible (Figure 16.1).

5. Calculate the cell density of cell suspension: total number of cells in four corners/4 × 10^4 = number of cells/ml cell suspension.
6. Calculate total cells of the cell suspension: cells/ml × volume of the cell suspension = total cells in the suspension.
7. In the case of cell treatment or cell splitting/subculture that mandates an equal number of cells per culture flask or dish, calculate the volume of the cell suspension containing a given number of cells for each container using the cell density determined at step 5.

Tips: For example, if total of cells in four corners is 100, the cell density will be 100/4 × 10^4 = 2.5 × 10^5 cells/ml cell suspension. If 10,000 cells are needed for each flask or dish, the volume of the cell suspension will be calculated: 2.5 × 10^5 cells/ml cell suspension = 2.5 × 10^2 cells/μl cell suspension 10,000 cells/2.5 × 10^2 cells/μl = 4 μl cell suspension per container.

8. Aliquot the volume of cell suspension calculated at step 7 into each culture flask or dish with the same size and the same volume of appropriate culture medium.

Tips: The cell suspension should be gently and thoroughly mixed and immediately take an appropriate volume of the cell suspension into each flask

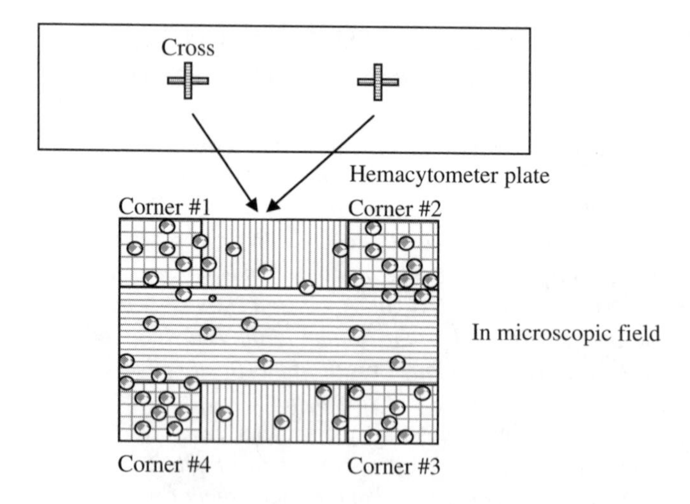

FIGURE 16.1 Diagram of cell counting in four corners of the hemacytometer plate.

as described at step 2. We strongly recommend using a Pipetteman attached with an appropriate volume of tip to aliquot the cell suspension. The tip should be changed for each aliquot to normalize the number of cells for each culture flask or dish. Because cells tend to sink to the bottom, transferring the total cell suspension into 1-, 2-, or 5-ml syringe/pipet followed by aliquot of an appropriate volume into the flask may not work well. Otherwise, the total number of cells in each of flask will be dramatically different.

TROUBLESHOOTING GUIDE

1. **ES cells do not attach to the bottom of the culture dish or flask**. One causal factor is that the container may not be well coated with gelatin solution. Unlike other mammalian cells, ES cells mandate a precoated culture container for culture. Assuming that culture conditions are normal, another possible factor is that the DMSO in the frozen cells' suspension, in the case of thawing and reculture of the cells, might not be removed. So, make sure to follow the preceding protocols to replace the old medium containing DMSO for freezing.
2. **Cells look sick or some of them are being differentiated.** Three possibilities: (a) Cell culture conditions are altered. For example, some components may be missed. Temperature in the incubator may be above 37°C and there may be no water in the water pan. (b) If culture conditions are normal, it is most likely that LIF provided may not be sufficient or its activity is gone. Try to use some fresh feeder layers or commercial LIF. (c) If (a) and (b) are not the case, check the cell passage number. If the passage number is over 50, it is recommended to use a lower passage number of ES cells.
3. **A lot of cell colonies float in the medium.** This is certainly due to overdensity or overculture of the cells. ES cells proliferate very fast. The medium should be changed once every 2 days and monitored on a daily basis. In general, if the initial cell density in a 100-mm culture dish is 10,000 cells under normal culture conditions, they should be split or frozen back after 7 days' culture. Once many floating colonies are visible, it is a good idea to discard the whole flask or dish because some ES cells have undergone differentiation and cannot be used for subsequent passages.
4. **More than 50% of cells disappear after being washed, trypsinized and centrifuged.** It is most likely that cells are washed or trypsinized for too long or, following centrifugation, the cells are kept in the suspension medium for too long prior to being frozen back or placed in the culture incubator. If that is case, a lot of cells are somehow lysed. It is good practice to handle the procedures of rinsing, trypsinizing and resuspending as fast as possible.

REFERENCES

1. Wu, W. and Welsh, M.J., Expression of the 25-kDa heat-shock protein (HSP27) correlates with resistance to the toxicity of cadmium chloride, mercuric chloride, *cis*-platinum(II)-diammine dichloride, or sodium arsenite in mouse embryonic stem cells transfected with sense or antisense HSP27 cDNA, *Toxicol. Appl. Pharmacol.*, 141, 330, 1996.
2. Capecchi, M.R., Altering the genome by homologous recombination, *Science*, 244, 1288, 1989.
3. Mansour, S.L., Thomas, K.R., and Capecchi, M.R., Disruption of the proto-oncogene *int-2* in mouse embryo-derived stem cells: a general strategy for targeting mutations to non-selectable genes, *Nature*, 336, 348, 1988.
4. Riele, H.T., Maandag, E.R., and Berns, A., Highly efficient gene targeting in embryonic stem cells through homologous recombination with isogenic DNA constructs, *Proc. Natl. Acad. Sci. USA,* 89, 5128, 1992.
5. Schorle, H., Holtschke, T., Hunig, T., Schimpl, A., and Horak, I., Development and function of T cells in mice rendered interleukin-2 deficient by gene targeting, *Nature,* 352, 621, 1991.
6. Tybulewicz, V.L.J., Crawford, C.E., Jackson, P.K., Bronson, R.T., and Mulligan, R.C., Neonatal lethality and lymphopenia in mice with a homozygous disruption of the *c-abl* proto-oncogene, *Cell,* 65, 1153, 1991.
7. Weinstock, P.H., Bisgaier, C.L., Setala, K.A., Badner, H., Ramakrishnan, R., Frank, S.L., Essenburg, A.D., Zechner, R., and Breslow, J.L., Severe hypertriglyceridemia, reduced high density lipoprotein, and neonatal death in lipoprotein lipase knockout mice, *J. Clin. Invest.*, 96, 2555, 1995.
8. Michalska, A.E. and Choo, K.H.A., Targeting and germ-line transmission of a null mutation at the metallothionein I and II loci in mouse, *Proc. Natl. Acad. Sci. USA*, 90, 8088, 1993.

17 New Strategies for Gene Knockout

CONTENTS

INTRODUCTION

The mutation of a specific gene is one of the most attractive approaches in which scientists can gain insight into the potential functions of the gene of interest.[1–5] To genetically alter the expression of a specific gene and probe its potential role, strategies for stable overexpression or underexpression of a specific gene are widely employed.[4–8] Nonetheless, up- or down-regulation of the expression of the gene of interest can only reveal the partial function of the gene. To prove the roles of an endogenous gene well, one may need to utilize another powerful tool: gene knockout or gene targeting by homologous recombination technology. This genetic approach can provide the highest level of control over producing mutations in an endogenous gene. Hence, it becomes an invaluable technique for investigators to use to make null mutations of transgenic mice to demonstrate *in vivo* gene function.[9–12] This allows using an animal model to address questions as to what a gene does in an organism, what it does for the body, and when and where in development it is mandated.

In attempt to knock out a gene of transgenic mice for evaluation of the consequences of not having the gene product, we and others have found that the previous gene targeting strategy has two obvious disadvantages.[3–6] One is that the targeted gene is usually a single knockout in one chromosome in cultured embryonic stem (ES) cells isolated from a brown mouse strain. The ES cells carrying a single-copy mutation of a specific gene are inserted by microinjection into young embryos isolated from a black female mouse strain. The black strain lacks the *agouti* gene that generates a brown coat phenotype. Hence, using the coat color as a selective marker, any newborns with brown shading intermixed with black are considered to be chimeras because they contain cells derived from two different strains of mice. In other words, these first-generation (F1) mice produced from the single-knockout clones are heterozygous mutants. To obtain homozygous mice, all resulting chimeric males must be back-crossed with black (non-*agouti*) females or heterozygous F1 mice be interbred to produce the F2 generation. The germ line transmission of F2 will be scored according to coat pigment. Any pure or solid-brown F2 mice are most likely the target mutants, with a double-knockout in both chromosomes that should be subjected to detailed characterization.[9,10] Certainly, this approach is time consuming and costly to obtain homozygous or double-knockout mice. Another major concern with previous gene knockout strategy is the potentially lethal effect of the targeted gene. In some cases, gene knockout results in early death of embryos and young animals or in morphologically and functionally abnormal offspring such as blind or otherwise handicapped animals.[9–15] These two limitations of the previous method greatly restrict the application of gene knockout techniques. As a matter of fact, many proposals for the targeting of some crucial genes have trouble winning approved due to potential lethality concerns.

To overcome these limitations, we have developed a novel strategy by which gene double-knockout can be achieved at the ES cell level instead of waiting for the F2 generation. An even more powerful argument is that, in order to address the lethality concerns, the homozygously targeted gene can be kept silent until it is activated. It is hoped that this new strategy will expand the areas of applications tremendously. The present chapter describes in detail the protocols for generating null mutant mice. These protocols and strategies can be readily adapted and applied to cells or animals other than ES cells and mice.

OVERVIEW AND GENERAL PRINCIPLES OF NOVEL STRATEGIES: GENE DOUBLE-KNOCKOUT AND TEMPORAL SILENCE

The primary principle of the gene double-knockout or homologous recombination is that a manipulated gene or fragment of genomic DNA transferred into a mammalian cell can locate and recombine with the endogenous homologous sequences in both chromosomes in cultured cells. As outlined in Figure 17.1, gene double-knockout starts with manipulation of the gene to be targeted. One approach is to mutate the cloned gene of interest by deleting one or more exons that are replaced with a selectable marker gene, such as a neomycin resistance (*neo*^{r}) or hygromycin resistance (*hyg*^{r}) gene in a targeting vector.

Another strategy is to use isogenic DNA to obtain a high efficiency of gene targeting.[15] The 5′- and 3′-end isogenic DNA fragments with deletion of the entire gene or exons of interest flank a marker gene (e.g., *neo*^{r} or *hyg*^{r}) in a targeting vector. As illustrated in Figure 17.1, each targeting vector usually contains two marker genes for drug double selections. One is the *neo*^{r} or *hyg*^{r} gene for positive selection of replacement of endogenous homologous DNA sequences. The other is the Herpes simplex virus *thymidine kinase* gene (HSV-*tk*) located immediately adjacent to the target sequence for negative selection of random insertion. The core portion of a typical targeting construct contains a specific DNA fragment A + *neo*^{r} or *hyg*^{r} gene + specific DNA fragment B + *tk* gene, forming a gene-targeting sandwich. These sandwiches containing the mutated gene or DNA fragment are then introduced into mouse ES cells (D3 or R1 cell line) followed by selection of stably transfected ES cells using the drugs G418/ganciclovir or hygromycin B/ganciclovir.

If all goes well, three groups of cells will be selected based on three specific principles. One group will be the cells that contain no insertion of gene targeting sandwiches. These cells are killed by G418 or hygromycin B during the selection because they lack *neo*^{r} or *hyg*^{r} gene. Another group of cells have a random insertion of an entire gene targeting sandwich. As a result, these cells are very sensitive to and killed by the addition of ganciclovir, the antiherpes drug, to the cell culture medium because of the *tk* gene in the sandwich. Ultimately, the survival cells are those with genomic DNA that contains the target insertion by homologous recombination. When homologous recombination occurs, the endogenous homologous sequence is replaced by the majority of the gene targeting sandwich except for the *tk* gene, which includes a specific DNA fragment A + the *neo*^{r} or *hyg*^{r} gene + the

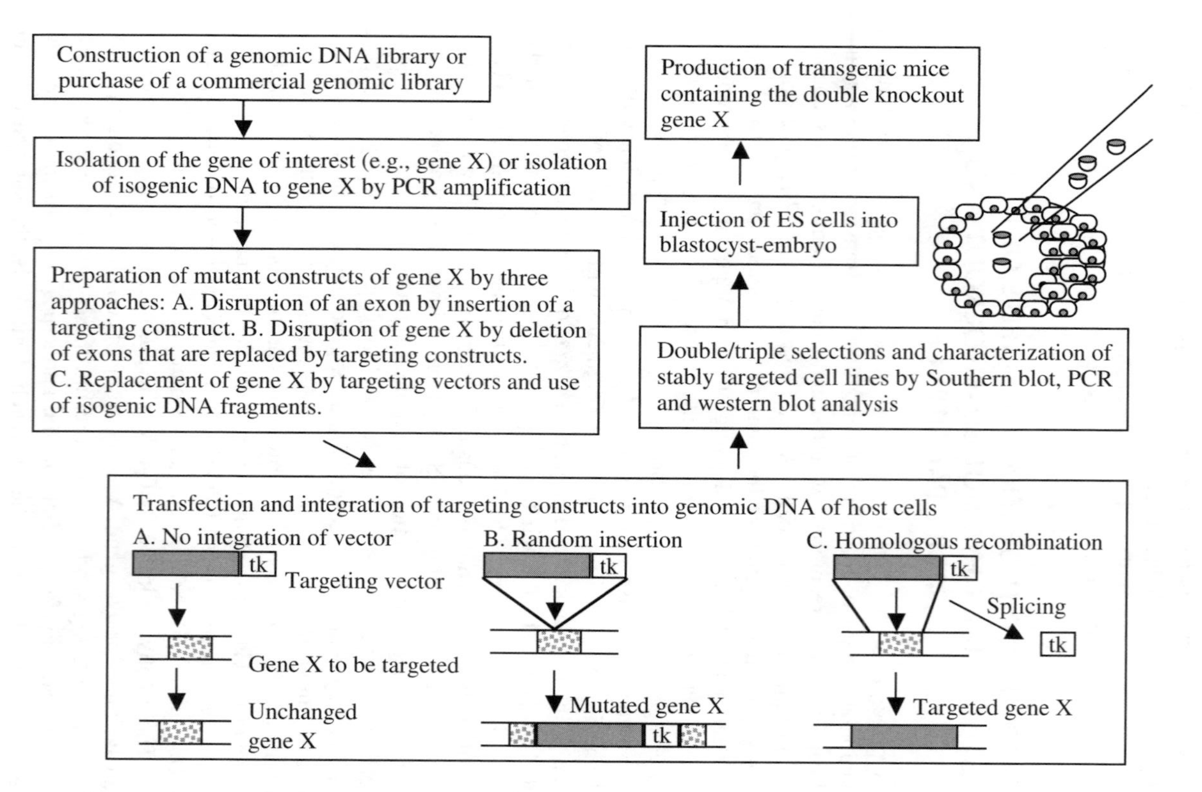

FIGURE 17.1 Diagram of gene knockout procedure in mouse ES cells and transgenic mice.

specific DNA fragment B. However, the *tk* gene, lying outside the region of matching sequences, does not integrate into the chromosome of the cell and is degraded in the cell.

According to probability analysis, the cells that carry integrated homologous DNA are extremely rare, occurring at a frequency of approximately $1/10^5$ for only a single-copy knockout. For a gene double-knockout to occur in ES cells, the stably transfected cells need to be subjected to a second round of transfection with exactly the same gene targeting sandwich or a different sandwich containing another positive selection marker gene. In other words, if the positive selection gene in the first round of transfection is *neo*^{r} or *hyg*^{r}, then, *hyg*^{r} or *neo*^{r} gene should be used as the positive selection marker in the second round. For double transfection, the survival cells after the first round of drug selection may not be individually isolated and expanded. The mixed cell populations can be used directly for retransfection. When using the same gene targeting sandwich for a second round of gene targeting, a double dose of an appropriate drug should be used for positive selection. However, the dose of ganciclovir may be the same as for the first round of selection. The primary principles of the second selection are quite similar to those for the first. If all goes well, most of the survival cells after the second selection should contain a double-knockout or homologous recombination in both chromosomes that carry both copies of the targeted gene.

The authors strongly recommend that two different gene targeting sandwiches containing two different selectable marker genes be constructed. These sandwiches (e.g., *neo*^{r} and *hyg*^{r} gene sandwiches) can be mixed at a ratio of 1:1 and simultaneously introduced into ES cells, eliminating a second round of transfection. The advantage of this approach is that the ES cells can be kept at the lowest passage number possible, and that the efficiency of generating transgenic mice will be enhanced substantially. After transfection, the stably transfected cells can be obtained by triple selections using G418, hygromycin B and ganciclovir in the culture medium. If all goes well, only cells that contain the doubly targeted gene by homologous recombination in both chromosomes can survive in a triple selection medium.

Following the drug selections, the individual cell clones or cell lines should be developed or expanded from single cells in order to generate homogeneous populations of cells. The obtained homologous recombinant cell lines are then subjected to characterization by Southern blotting, PCR screening, northern blotting, or western blotting. At this point, if the potential lethal effects of gene knockout are not of concern, some of the cells may be transferred into the cavity of blastocyst-stage embryos to generate transgenic mice. On the other hand, we offer a strategy to overcome lethality concerns even though no obvious lethal effects can be observed in cultured ES cells.

The cells with a double-knockout gene are used as the recipients for a third transfection with a DNA construct containing the entire coding region of the cDNA of the targeted gene. The expression of the cDNA is driven by an appropriate inducible promoter such as RSV-LTR enhancer linked to the dexamethasone-inducible MMTV-LTR promoter. In this way, the double-knockout gene can be temporarily rescued, or kept silent, by inducing the expression of the cDNA with dexamethasone.

However, the effects of the targeted gene can be readily activated by stopping the use of the inducer molecules. Therefore, the lethal effect of the knocked-out gene should be less or not effective until it is turned on. It should be noted that a different selectable gene such as the *Hprt* gene should be included in the inducible DNA construct so that stably transfected cells can be selected with HAT-containing medium. These cells should be characterized to ensure that the expression of the introduced cDNA occurs prior to initiating development of mice.

Once the ES cells contain the doubly knocked-out gene, they are ready to be injected into the cavities of young embryos at the blastocyst stage that are isolated from an appropriate female mouse strain (e.g., black coat strain). To keep the potentially lethal effect of gene knockout silent, an inducible chemical such as dexamethasone should be delivered into the female or the cavity at an appropriate concentration based on induction curve. The manipulated embryos are then reimplantated into recipient female mice to generate mice. F1 mice can be interbred to maintain the colonies that carry the null gene. These offspring should be characterized in detail by Southern blotting, northern blotting, western blotting and immunocytochemistry. More importantly, the potential functions of the null gene should be analyzed in the animals.

MANIPULATION OF THE SPECIFIC GENE TO BE TARGETED

For gene targeting or knockout, genomic DNA should be manipulated prior to making a gene knockout construct. This section describes in detail three protocols that have been used in our and other laboratories.[14,15]

PROTOCOL A. DISRUPTION OF AN EXON IN A SPECIFIC GENE

Disruption of an exon of the gene to be targeted is a traditional technique for targeting mutations. The basic principle of this strategy is to insert a selectable marker gene such as *neo*^{r} gene in the middle exon of the gene of interest. After homologous recombination in the genome of mouse ES cells, the disrupted protein produced by the mutated gene cannot function as the native gene product.

To disrupt an exon, a cloned DNA sequence of a chosen locus can first be digested with an appropriate restriction enzyme followed by insertion of the *neo*^{r} gene at the cut site (Figure 17.2). Detailed protocols for appropriate restriction enzyme digestion and DNA ligation are given in Chapter 4 entitled Subcloning of Genes or DNA Fragments.

PROTOCOL B. DELETION OF EXONS OF A SPECIFIC GENE

Although disruption of an exon in a gene can create a functional mutation of the gene of interest, the efficiency of disorder may not be sufficient in some cases. To overcome this limitation, deletion of one or more exons in the gene of interest has been widely utilized. The primary principle of this technique is to remove some exons from a cloned gene. In our experience, adjacent introns can also be deleted

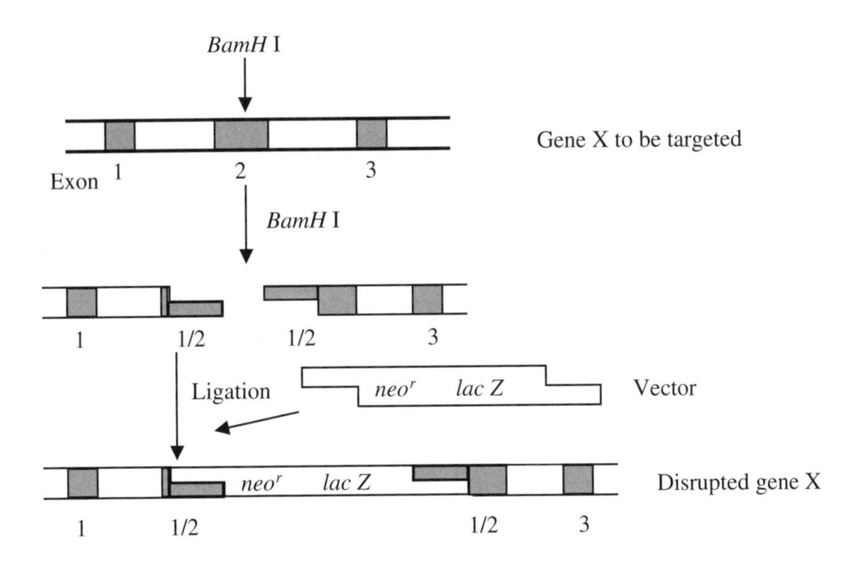

FIGURE 17.2 Mutation of gene X by insertion of a marker gene (*neo*) and a reporter gene (*lac Z*) into exon 2.

simultaneously to obtain a high efficiency of gene targeting. The deleted region is then replaced by a selectable marker gene (Figure 17.3). Refer to Subcloning of Genes or DNA Fragments in Chapter 4 for appropriate protocols for restricting enzyme digestion of DNA and for DNA ligation.

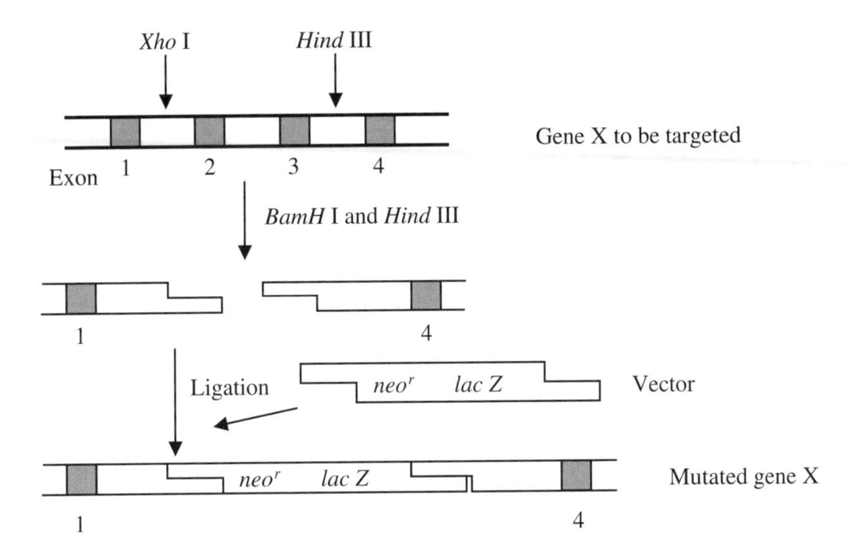

FIGURE 17.3 Mutation of gene X by deletion of exons 2 and 3 and by replacement of the exons with a marker gene (*neo*) and a reporter gene (*Lac Z*).

Protocol C. Preparation of Isogenic DNA by PCR

This is very current strategy used to obtain the highest possible efficiency of gene knockout or DNA homologous recombination. We have developed a method for preparing isogenic DNA fragments from the 129 SVJ mouse genomic library (Stratagen) by PCR techniques (Figure 17.4). Unlike the exon disruption or deletion described in Protocol A and Protocol B, the major principle of the isogenic DNA approach is the exclusion of all the exons and introns in the gene to be knocked out. The excluded exons and introns are replaced by a selectable marker gene (e.g., *neo*^{r}). During homologous recombination, exogenous isogenic DNA flanking the gene with no exons and introns will locate and recombine with endogenous isogenic DNA or genes. If all goes well, the targeted gene will express no product and become a completely null gene. Thus, the potentially functional leaking that occurred by using the exon disruption or deletion approach is avoided by using the isogenic DNA strategy.

Design of Two Pairs of Primers

As illustrated in Figure 17.4, two pairs of primers should be designed. The forward primer 1 (FP1) is designed based on the sequence in the T3 promoter region of the cloned vector. The reverse primer 1 (RP1) is based on the known sequence in the 5′ untranslated region (5′)-UTR of cDNA or the upstream region of the first exon. The design of FP2 is based on the known sequence in the 3′ untranslated region (3′-UTR) or the downstream region of the last exon of the gene to be targeted. RP2 is chosen from the T7 prompter region in the cloned vector. It is extremely important

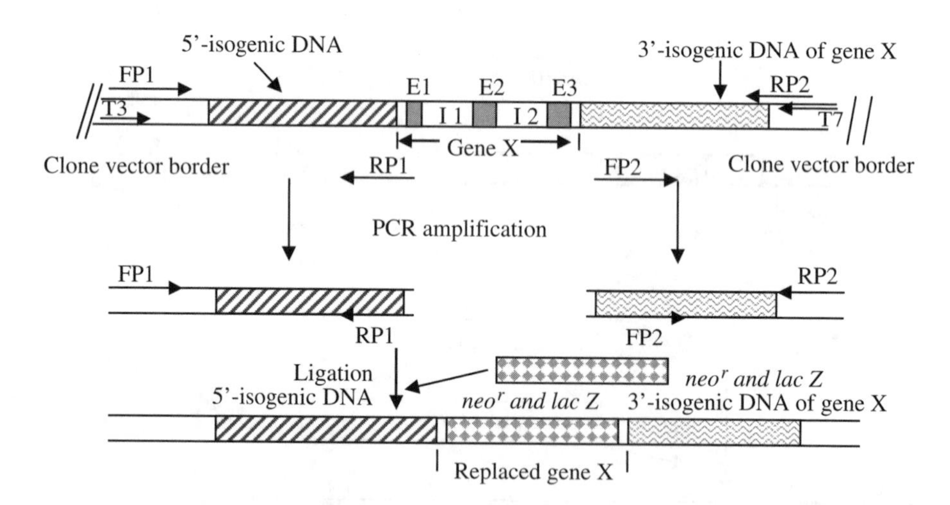

FIGURE 17.4 PCR amplification of isogenic DNA fragments for targeting of gene X by replacement. E1 to E3: exons 1 to 3; I1 and I2: introns 1 and 2; FP1: forward primer 1; RP1: reverse primer 1; FP2: forward primer 2; RP2: reverse primer 2. T3 and T7 are promoters built in the cloning vector at sites adjacent to the multiple cloning site (MCS) for insertion of genomic DNA.

that each pair of primers be carefully designed to anneal with two template DNA strands that are complementary to each other. If two primers are annealed with the same DNA strand, no PCR products will be produced. The primers should be analyzed with an appropriate computer program prior to being used in PCR. We routinely use the Oligo version 4.0 or CPrimer f to check the quality of the primers, including GC contents, *Tm* value, potential formation of intraloops/dimers and interduplexes of the primers, and potential annealing between the primers and the region flanked by the primers. Based on previous studies, the length of isogenic DNA fragments plays a significant role in the success of gene targeting.

It has been demonstrated that isogenic DNA fragments of 2 to 7 kb in length are required for gene knockout. To obtain large size DNA fragments, the length of the primers is also critical. In general, primers should be >27 bases with <60% GC contents. The first base at the 5′ or the 3′ end should be G or C followed by A or T for an efficient annealing. No more than four bases of introprimer or interprimer should be complementary. If all goes well, different sizes of 5′- or 3′-isogenic DNA fragments may be amplified because different portions of the gene to be targeted may be cloned in different vector molecules in the genomic library. These fragments should be purified by agarose gel electrophoresis. In our experiments, the following primers are designed to obtain isogenic DNA fragments for HSP25 gene targeting.

For 5′-isogenic DNA:
FP1 (33-mer): 5′-CATGGCCAGAGCTCTAATACGACTCACTATAGG-3′;
RP1 (31-mer): 5′-CTCTCGAGTTGGCGTGTCAAGGCTGAGGATC-3′

For 3′-isogenic DNA fragments:
FP2 (35-mer): 5′-GATCTAGAGTGCGCTCTTTTGATACATACATTTAC-3′;
RP2 (34-mer): 5′-GAGAATTCGCTCAATTAACCCTCACTAAAGGGAG-3′

Amplification of 5′- and 3′-Isogenic DNA Fragments by PCR

1. Place two 0.5-ml thin-walled microcentrifuge tubes (Boehringer Mannheim) on ice and set up the PCR reaction mixture by adding the following components to the appropriate tubes in the order shown:

   Tube A:
   - 10X Amplification buffer, 4 µl
   - Mixture of four dNTPs (2.5 m*M* each), 5 µl
   - Forward primer 1 (FP1) (3 µ*M* in dd.H_2O), 4 µl
   - Reverse primer 1 (RP1) (3 µ*M* in dd.H_2O), 4 µl
   - Genomic library DNA (3 to 4 µg in dd.H_2O), 2 µl (e.g., 129 SVJ mouse genomic library)
   - 3% (v/v) DMSO or formamide in case of GC-rich DNA template, 1.2 µl
   - *Taq* DNA polymerase, 4 units
   - Add dd.H_2O to a final volume of 40 µl.

   Tube B:
   - 10X Amplification buffer, 4 µl
   - Mixture of four dNTPs (2.5 m*M* each), 5 µl
   - Forward primer 2 (FP2) (3 µ*M* in dd.H_2O), 4 µl
   - Reverse primer 2 (RP2) (3 µ*M* in dd.H_2O), 4 µl

Genomic library DNA (3 to 4 μg in dd.H_2O), 2 μl (e.g., 129 SVJ mouse genomic library)
3% (v/v) DMSO or formamide in case of GC-rich DNA template, 1.2 μl
Taq DNA polymerase, 4 units
Add dd.H_2O to a final volume of 40 μl.

2. Overlay the mixture in each tube with 20 μl of light mineral oil to prevent evaporation of the sample.
3. Program a PCR thermocycler according to the manufacturer's instructions.

Cycle	Denaturation	Annealing	Extension	Additional elongation
1	94°C, 3 min	60°C, 30 s	70°C, 4 min	
2 to 34	94°C, 30 s	60°C, 30 s	70°C, 4 min	20 s per cycle
35	94°C, 30 s	60°C, 30 s	70°C, 15 min	

Finally, hold at 4°C until the sample is removed.

4. Carefully transfer the reaction mixture underneath the mineral oil using a pipette attached with a 100-μl tip. Adjust the pipette volume to 40 μl. Press the top of the pipette to its first stop position, hold it, and carefully insert the tip into the bottom of the tube. Then slowly take up the liquid underneath the oil phase. Slowly withdraw the tip from the oil layer, wipe off the outside of the tip with a clean piece of tissue, and transfer the PCR sample to a fresh tube. Alternatively, the oil layer can be removed first, leaving the PCR sample in the tube; however, the oil layer may not be completely removed. Once this is done, the amplified DNA products should be analyzed and purified by agarose gel electrophoresis.

Tips: *In addition to the points described in Design of Primers for PCR, the following points are conducive to the success of large size of PCR products.*

- Concentration of $MgCl_2$ in PCR 10X buffer: 18 m*M* for 1- to 25-kb and 22 m*M* for 15- to 30-kb PCR products.
- In case of a GC-rich sequence (GC contents >65%), 3% (v/v) of DMSO or formamide should be added to the PCR cocktail. We have found that such an addition helps a great deal.
- dNTPs concentration: 300 to 400 n*M* for 1- to 25-kb and 500 n*M* for <20-kb fragments as the final or working concentration.
- Oligonucleotide primers should be precipitated to remove any salt. The working concentration of each primer should be 250 to 300 n*M*.
- DNA template should be highly pure with a ratio of 1.85 to 2.0 at A_{260} nm/A_{280} nm. The working concentration is 350 to 400 ng.
- *Taq* polymerase plays a major role in DNA amplification by PCR. High-quality and high-fidelity *Taq* polymerase is required for obtaining long-length PCR products for gene targeting. Based on our experience, we recommend the Expand™ Long Template PCR System or Tth DNA polymerase or High Fidelity *Taq* Polymerase (Boehringer Mannheim). All of these enzymes generate single base 5′-A overhangs to PCR products.

- Denaturing temperature needs to be 93 to 94°C. Except for the first cycle, 30 s of denaturation should be used for subsequent cycles. Longer denaturation at or above 94°C may cause DNA breaks.
- Annealing temperature is recommended to be 2 to 3°C above the Tm temperature of primers.
- Extension temperature should be 5 to 8°C above the annealing temperature. Starting from the second cycle, 20 s of extra extension is recommended.

Materials Needed

Thermocycler
Thin-walled microcentrifuge tubes (0.5 ml)
Pipettes or pipetman (0 to 200 μl, 0 to 1000 μl)
Sterile pipette tips (0 to 200 μl, 0 to 1000 μl)
Taq DNA polymerase
Primers
DNA templates

10X PCR Amplification Buffer

0.5 *M* Tris-HCl, pH 9.2
18 m*M* $MgCl_2$
0.14 m*M* $(NH_4)_2SO_4$

10X PCR Amplification Buffer for GC-Rich Templates

0.5 *M* Tris-HCl, pH 9.2
18 m*M* $MgCl_2$
0.14 m*M* $(NH_4)_2SO_4$
3 % DMSO (v/v) or 3% (v/v) formamide

Electrophoresis of PCR Products

1. Clean an appropriate gel apparatus thoroughly by washing it with detergent, completely washing out the detergent with tap water and rinsing with distilled water three to five times.
2. Prepare an agarose gel mixture in a clean bottle or a beaker as follows:
   1X TBE or TAE buffer, 30 ml (minigel)
   Agarose (0.5 to 0.7%, w/v), 0.15 to 0.21 g

Note: 1X TBE or TAE buffer may be diluted from 5X TBE or 10X TAE stock solution in sterile dd.H_2O. Because the agarose is <1%, its volume can be ignored in calculating the total volume.

3. Place the bottle (with the cap loose) or a beaker in a microwave oven and slightly heat the agarose mixture for 1 min. Briefly shake the bottle to rinse any agarose powder stuck to the wall. To melt the agarose completely, carry out gentle boiling for 1 to 3 min, depending on the volume of the agarose mixture. Gently mix, open the cap and allow the mixture to cool to 50 to 60°C at room temperature.

4. While the gel mixture is cooling, seal a clean gel tray at the two open ends with tape or a gasket and insert the comb in place. Add 1 µl of ethidium bromide (EtBr, 10 mg/ml in dd.H_2O) per ml of agarose gel solution at 50 to 60°C. Gently mix and slowly pour the mixture into the assembled gel tray and allow the gel to harden for 20 to 30 min at room temperature.

Caution: EtBr is a mutagen and a potential carcinogen and should be handled carefully. Gloves should be worn when working with this material. The gel running buffer containing EtBr should be collected in a special container for toxic waste disposal.

Note: The function of EtBr is to stain DNA molecules by interlacing between the complementary strands of a double strand of DNA or in the regions of secondary structure in the case of ssDNA. It will fluoresce orange when illuminated with a UV light. The advantage of using EtBr in a gel is that DNA bands can be stained and monitored with a UV lamp during electrophoresis. During the electrophoresis, the positively charged EtBr moves towards the negative pole but negatively charged DNA molecules migrate toward the positive pole. A drawback, however, is that the running buffer and gel apparatus are likely to become contaminated with EtBr. Thus, an alternative is to stain the gel after electrophoresis is completed. The gel can be stained with EtBr solution for 10 to 30 min followed by washing in distilled water for 3 to 5 min. To minimize the amount of EtBr needed, 1 µl of EtBr (1 mg/ml in dd.H_2O) can be added into each 10 µl of DNA sample.

5. Pull the comb out of the gel slowly and vertically and remove the sealing tape or gasket from the gel tray. Place the tray in an electrophoresis apparatus and add a sufficient volume of 0.5X TBE or 1X TAE buffer to the apparatus until the gel is covered to a depth of 1.5 to 2 mm above the gel.

Note: The comb should be slowly and vertically pulled out from the gel because any cracks occurring inside the wells of the gel will cause sample leaking when the sample is loaded. The top side of the gel, where wells are located, must be placed at the negative pole in the apparatus because the negatively charged DNA molecules will migrate toward the positive pole. If any wells contain visible bubbles, they should be flushed out of the wells using a small pipette tip to flush the buffer up and down the well several times. Bubbles may adversely influence loading of the samples and electrophoresis. The concentration of TBE or TAE buffer should never be less than 0.2X; otherwise, the gel may melt during electrophoresis. Prerunning the gel at a constant voltage for 10 min is optional.

6. Add 5X loading buffer to the DNA sample and to the DNA standard markers to a final concentration of 1X. Mix and carefully load DNA standard markers (0.2 to 2 µg/well) into the left-hand or the right-hand well, or into both left- and right-hand wells of the gel. Carefully load the samples, one by one, into appropriate wells in the buffer-submerged gel.

Notes: (1) The pipette tip should not be inserted all the way down to the bottom of a well because this is likely to cause sample leaking and con-

tamination. The tip containing the DNA sample should be vertically positioned at the surface of the well and stabilized with a finger of the other hand and the sample slowly loaded into the well. (2) The loading buffer should be at least 1X to final concentration; otherwise, the sample may float out of the well.

7. After all of the samples are loaded, estimate the length of the gel between the two electrodes and set the power at 5 to 15 V/cm gel. Allow electrophoresis to proceed for 2 h or until the first blue dye reaches to 1 cm from the end of the gel. If overnight running of the gel is desired, the total voltage should be determined empirically.
8. Stop electrophoresis of the gel and visualize the DNA bands in the gel under UV light. Photograph the gel with a Polaroid camera.

Caution: Wear safety glasses and gloves for protection against UV light.

Notes: It is recommended that pictures be taken with a fluorescence ruler placed beside the gel. A short exposure is needed to obtain sharp bands and a clear background. A longer exposure is needed to visualize some very weak bands. For PCR products, white and sharp bands should be visible in a black background.

Materials Needed

Gel casting tray
Gel combs
Electrophoresis apparatus
DC power supply
Small or medium size DNA electrophoresis apparatus, depending on the number of samples to be run
Ultrapure agarose powder

5X TBE Buffer

600 ml dd.H_2O
0.45 *M* boric acid (27.5 g)
0.45 *M* Tris base (54 g)
10 m*M* EDTA (20 ml 0.5 *M* EDTA, pH 8.0)
Dissolve well after each addition. Add dd.H_2O to 1 l. Autoclave.

10X TAE Buffer

600 ml dd.H_2O
400 m*M* Tris-acetate (48.4 g tris base, 11.42 ml glacial acetic acid)
10 m*M* EDTA (20 ml 0.5 *M* EDTA, pH 8.0)
Dissolve well after each addition. Add dd.H_2O to 1 l. Sterile-filter.

Ethidium Bromide (EtBr)

10 mg/ml in dd.H_2O, dissolve well and keep in dark or brown bottle at 4°C.
Caution: EtBr is extremely toxic and should be handled carefully.

5X Loading Buffer

50% Glycerol
2 m*M* EDTA
0.25% Bromphenol blue
0.25% Xylene cyanol
Dissolve well and store at 4°C.

10X TE Buffer

100 m*M* Tris-HCl, pH 8.0
10 m*M* EDTA

Purification of PCR Bands from the Agarose Gel and Ligation of Isogenic DNA Fragments with a Selectable Marker Gene

Once the isogenic DNA fragments are separated from agarose gels, they can be eluted out from the gel matrix and ligated with a selectable marker gene as shown in Figure 17.4. Detailed protocols for DNA purification from an agarose gel, appropriate restriction digestion and DNA ligation are described in Chapter 4.

DESIGN AND SELECTION OF VECTORS AND MARKER GENES FOR GENE DOUBLE-KNOCKOUT AND TEMPORAL SILENCE

Plasmid-based vectors are commonly used for gene knockout or gene targeting. In order to achieve a double-knockout of both copies of a specific gene in two chromosomes in cultured cells and to keep the targeted gene silent, three types of vectors need to be chosen. Their structural and functional similarities and differences are described next.

TYPE A. *NEO^R* GENE VECTOR FOR FIRST COPY OF GENE KNOCKOUT

This vector is designed for single-copy knockout of the gene of interest. Its necessary components and functions are illustrated in Figure 17.5.

1. Plasmid backbone includes *f1 ori* for replication of the plasmid, ColE1 for a high copy number of the plasmid and ampicillin gene for the selection of bacterial transformants.
2. The *neo^r* marker gene with its own promoter, pA and splicing region plays a key role in selection of a single-copy targeted gene following homologous recombination.
3. The *Lac Z* gene with its own promoter, pA and splicing region is linked adjacent to the *neo* gene. This reporter gene allows staining targeted cells, embryos and tissues from adult mice.

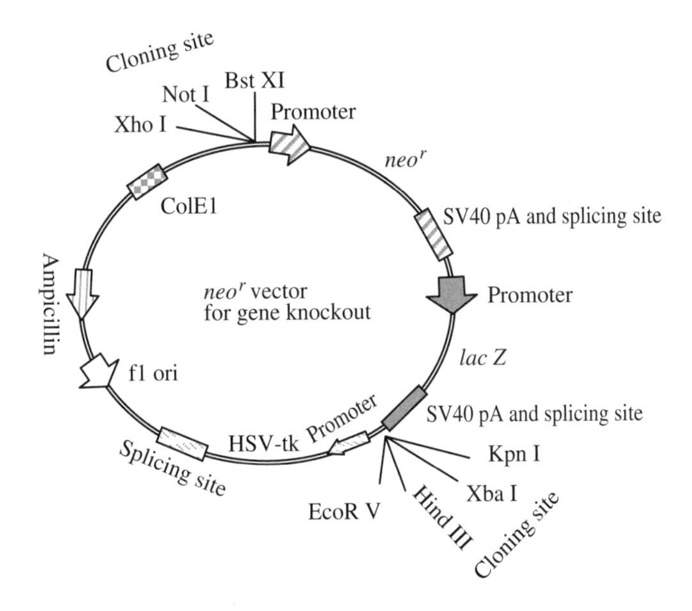

FIGURE 17.5 Diagram of a neo^r vector for gene knockout.

4. Cloning sites at both ends of the *neo-lac Z* gene cassette are used for insertion of 5′ and 3′ portions of the gene to be targeted.
5. The *hsv-tk* gene is adjacent to a cloning site. The function of this gene is to serve as a negative selection marker. If homologous recombination occurs, this gene should be spliced out and degraded in the cell. Because the cell does not contain this gene, the cell is resistant to ganciclovir-containing medium. However, if random insertion takes place instead of homologous recombination and because the cell has the *tk* gene, the cell will be sensitive to and killed by ganciclovir.

Type B. HYG^R Gene Vector for the Second Copy of Gene Knockout

Structurally, this type of vector is almost the same as type A except that the positive selection marker gene is replaced by the hyg^r gene (Figure 17.6). In case of homologous recombination, the cells should be resistant to hygromycin B and to gancicIovir in the culture medium. Nonetheless, if cells are transfected only with this type of vector with an insertion of the gene to be targeted, the homologous recombination is usually a single copy of gene knockout. However, if cells are simultaneously transfected with type A and B vectors, each with an insertion of the gene to be knocked out, the very rare homologous recombination is most likely double-knockout of the gene using triple selection with G418, hygromycin B and ganciclovir in the culture medium.

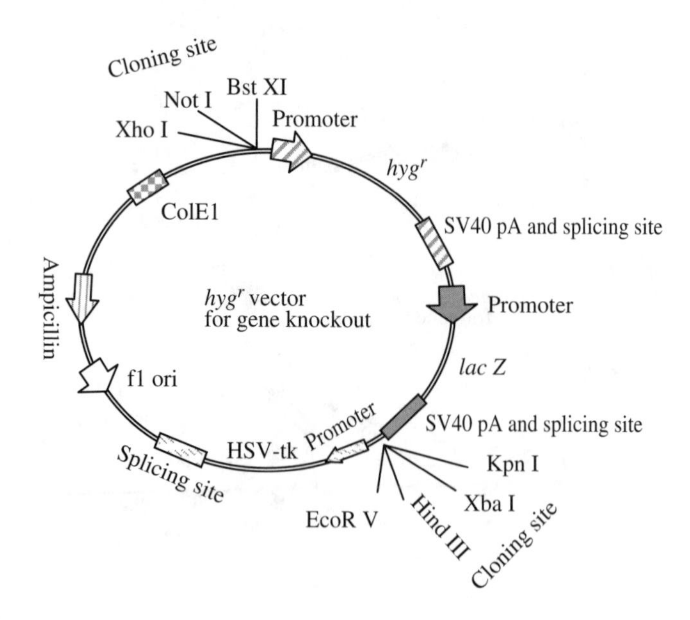

FIGURE 17.6 Diagram of a *neo*^{r} vector for gene knockout.

Type C. *hyrt*^{R} Gene Vector for Keeping the Double-Knockout Gene Silent

To avoid potential lethal effects of the targeted gene, it may be necessary to keep the double-knockout of the gene silent until it is activated. In order to achieve such an objective, the cells containing a gene double-knockout should be retransfected with a rescue DNA construct (Figure 17.7). The components and functions are described as follows.

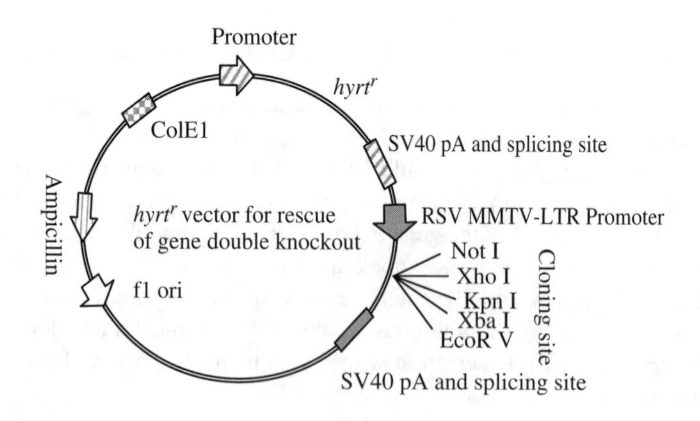

FIGURE 17.7 Diagram of a standard vector for gene knockout.

1. A plasmid backbone contains *fl ori* for replication of the plasmid, ColE1 for a high copy number of plasmids and ampicillin gene for selection of bacterial transformants.
2. An inducible promoter such as RSV MMTV-LTR that controls the expression of the cDNA to be cloned.
3. A multiple cloning site (MCS) immediately downstream of the inducible promoter will be employed for the insertion of the entire coding region of cDNA of the gene to be targeted. The expression of the cDNA will then be induced by the addition of dexamethasone to the culture medium or into the animal body.
4. An SV40 polyadenylation site and an early splicing region are placed immediately downstream of the MCS site for splicing mRNA to be expressed from the cloned cDNA.
5. The $hyrt^r$ marker gene with its own promoter, pA and a splicing region must be different from the marker genes in type A and type B. This gene will play a key role in selection of stably transfected cells. Any integration of the *hyrt* gene enables the cell to be tolerant to the HAT medium containing hypoxanthine, aminopterin and thymidine.

PREPARATION OF DNA CONSTRUCTS FOR DOUBLE-KNOCKOUT AND SILENCER

Once desirable targeting vectors and manipulated genes or isogenic DNA fragments are available, preparations of DNA constructs for gene knockout can be carried out. The gene or DNA fragments of interest should be inserted into an appropriate type of vector at the right sites in the right orientations. The general procedures and principles for gene disruption or gene targeting (homologous recombination) as well as for temporary rescue of double-knockout are illustrated in Figure 17.8 through Figure 17.10. The detailed protocols for appropriate restriction enzyme digestion and DNA ligation are described in Chapter 4.

TRANSFECTION OF ES CELLS WITH DOUBLE-KNOCKOUT AND SILENCER DNA CONSTRUCTS

This section describes two widely used methods for transfection of mammalian cells. Each method has its own principles, strengths and weaknesses.

METHOD A. TRANSFECTION BY ELECTROPORATION

This is a simple, fast and effective transfection technique. The principle underlying this method is that cells and DNA constructs to be transferred are mixed and subjected to a high-voltage electrical pulse that creates pores in the plasma membranes of the cells. This allows the exogenous DNA to enter the cytoplasm of the cell and the nucleus through nuclear pores to become integrated into the genome. This method is suitable for transient and stable transformation experiments.

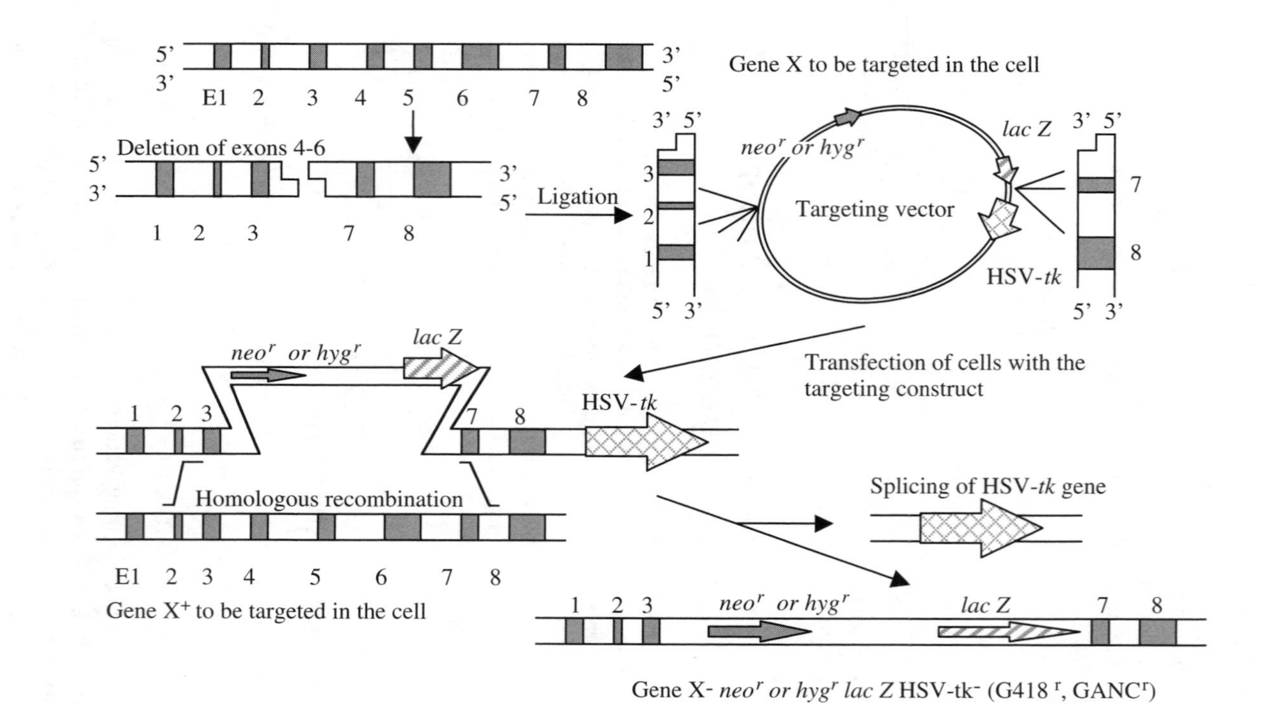

FIGURE 17.8 Diagram of the disruption of gene X by exon deletion and sequence replacement by homologous recombination in the genome of the cell.

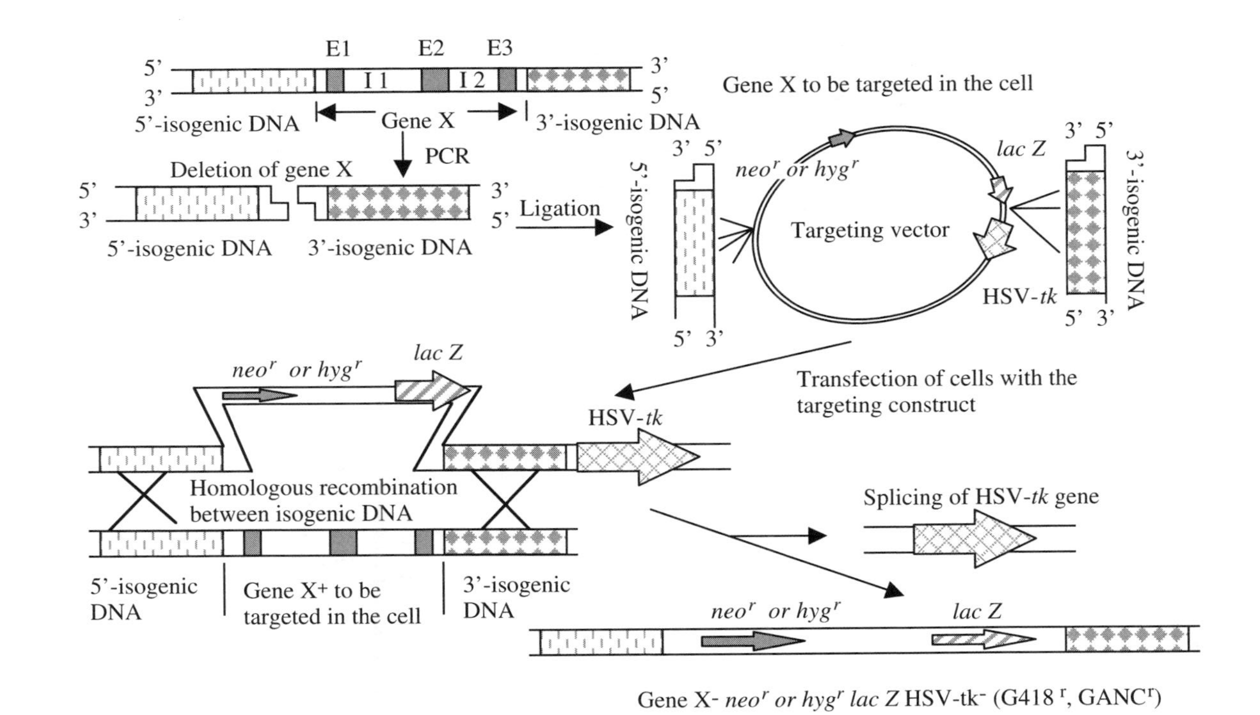

FIGURE 17.9 Targeting of gene X by deletion of the gene and sequence replacement with isogenic DNA fragments by homologous recombination in the genome of the cell.

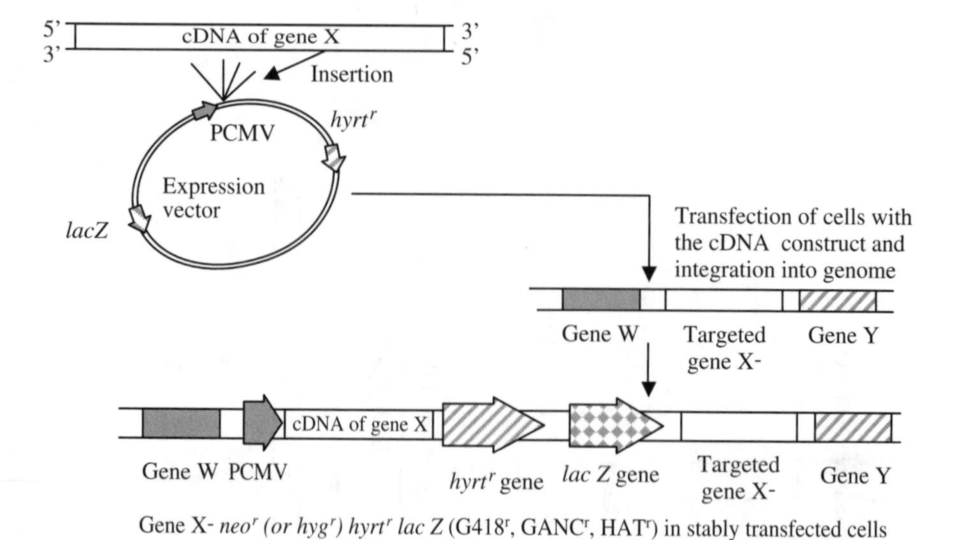

FIGURE 17.10 Temporary rescue of a double-knockout gene by induction of a cDNA expression.

However, the pulse conditions should be adjusted empirically for the particular cell type. In spite of the fact that this method has a high efficiency of transfection, its main drawback is that approximately 50% of the cells are killed by the high voltage of electrical pulse. These dead cells should be removed on a daily basis in order to reduce toxicity.

Protocol

1. Linearize the DNA constructs to be transferred by a unique restriction enzyme. Protocols for appropriate restriction enzyme digestion are described in Chapter 4.

Tips: (1) The unique restriction enzyme cut should be in the plasmid backbone area such as the ColE1, fi ori *or ampicillin gene. It is extremely important that this cut not be made in any area that will affect gene targeting by homologous recombination, as well as the activities of* lac Z *and selectable marker genes such as* neo, hyg, hyrt *and* HSV-tk. *In the case of double-knockout rescue DNA, ensure that the cut is not made in any part of the promoter, inserted cDNA and splicing site. (2) In our experience, plasmid DNA constructs do not need to be linearized for homologous recombination. In fact, plasmid DNA enhances the transfection efficiency up to 30%.*

2. Harvest the cells by gentle trypsinization as described in mammalian cell culture in Chapter 16. Centrifuge the cells at 150 × *g* for 5 min and resuspend the cells well in normal culture medium at approximately 10^6 cells/ml.

Note: Make sure that the cells are gently but well suspended prior to transfection because aggregation of cells may cause difficulty in isolation of stably transfected cell clones.

3. In a sterile laminar flow hood, gently mix 0.6 ml of the cell suspension with 15 to 30 µl of DNA constructs (25 to 50 µg) in a disposable electroporation cuvette (Bio-Rad, Hercules, CA).
4. According to the manufacturer's instructions, carry out electroporation using a gene pulser (Bio-Rad) with a pulse controller and capacitance extender at 300 V and 250 µF.
5. The electroporated cells are immediately plated into culture dishes at a density of about 3000 cells per 100-mm Petri dish and culture the cells at 37°C with 5% CO_2. Proceed to drug selection.

Method B. Transfection of Cells by Lipofectin Reagents

Lipotransfection is relatively safe and achieves a higher efficiency of transfection in mammalian cells. It is suitable for transient and stable transfection of tissues or cells and mandates a smaller amount of DNA for transfection compared with electroporation and other methodologies. Lipofectin reagents are commercially available (e.g., Gibco BRL Life Technologies). Lipsome formulation is composed of polycationic lipids and neutral lipids at an appropriate ratio (e.g., 3:1). The primary principles of this method include: (1) the positively charged and neutral lipids can form liposomes that can complex with negatively charged DNA constructs; (2) the DNA–liposome complexes are applied to cultured cells and are uptaken by endocytosis; (3) the endosomes undergo breakage of membranes, releasing DNA constructs; and (4) the DNA enters into the nucleus through the nuclear pores and facilitates homologous recombination or integration into the chromosomes of the cell.

The main disadvantage of the lipotransfection method is that if the ratio of lipids–DNA and cell density is not appropriate, severe toxicity will be generated, killing quite a number of cells. Additionally, ES cells are more sensitive to toxicity compared with other fibroblasts such as NIH 3T3 and mouse STO cells.

Protocol

1. Harvest ES cells by gentle trypsinization as described in Chapter 18. Centrifuge the cells at $150 \times g$ for 5 min and resuspend the cells well in normal culture medium at about 10^6 cells/ml.

Note: Make sure that the cells are gently but well suspended prior to transfection because aggregation of cells may cause difficulty in isolation of stably transfected cell clones.

2. In a 60-mm tissue culture dish, seed the suspended cells at about 1×10^6 cells in 5 ml of normal culture medium. Allow the cells to grow under normal conditions for 12 h. Proceed to transfection.

Note: ES cells grow fast and form multiple layers of clones instead of the monolayers seen in other cell lines. Because of these features, ES cells should not be overcultured prior to transfection. Otherwise, multilayers of

ES cells may result in a low efficiency of transfection and a high percentage of potential death. Therefore, once a relatively high density of cells is plated, transfection can be carried out as soon as the ES cells attach to the bottom of the dish and grow for 6 to 12 h.

3. While the cells are growing, DNA constructs to be transferred by a unique restriction enzyme can be linearized. Protocols for appropriate restriction enzyme digestion are described in Chapter 4.
4. In a sterile laminar flow hood, dissolve 10 to 15 μg of DNA in 0.1 ml of serum-free medium and dilute lipofectin reagents to two- to fivefold in 0.1 ml of serum-free medium. Combine and gently mix the DNA–lipids media in a sterile tube. Allow to incubate at room temperature for 15 to 30 min. A cloudy appearance of the medium indicates the formation of DNA–liposome complexes.
5. Rinse the cells once with 10 ml of serum-free medium. Dilute the DNA–liposome solution to 1 ml in serum-free medium and gently overlay the cells with the diluted solution. Incubate the cells for 4 to 24 h under normal culture conditions. During this period, transfection takes place.
6. Replace the medium with 5 ml of fresh, normal culture medium with serum every 12 to 24 h. Allow the cells to grow for 2 to 3 days before selection or analysis, in case of transient transfection.
7. Trypsinize the cells and plate them at a density of about 3000 cells per 100-mm Petri dish and culture for 24 h at 37°C with 5% CO_2. Proceed to drug selection.

SELECTION AND CHARACTERIZATION OF TARGETED ES CELL CLONES

Drug selection depends on the appropriate drug-resistant genes carried by DNA constructs. With regard to novel strategies for gene knockout described in this chapter, four different selectable marker genes are utilized. Therefore, selections can be divided into three types.

DOUBLE SELECTION OF CELLS CONTAINING A DOUBLE-KNOCKOUT GENE

If the cells were initially subjected to transfection twice with the same DNA constructs with a vector containing either *neo*^{r} and *tk* (Figure 17.5) or *hyg*^{r} and *tk* (Figure 17.6) genes, double selections (positive and negative selections) should be used. In order to identify the very rare cells that have a double-knockout gene in the genome, double-dose the positive selection drug (G418 for *neo* or hygromycin B for *hyg*).

1. The transfected cells should be plated into culture dishes at a density of 1000 cells per 100-mm dish and cultured for 36 to 48 h prior to selection. Replace the culture medium with 10 ml of drug selection medium at double the concentration. If the cells were transfected twice with *neo*^{r}–*tk*^{r} DNA constructs (Figure 17.9 and Figure 17.10), the selection medium should contain 500 to 600 μg/ml of G418 and 50 to 100 n*M* gancyclovir

(GANC). However, if the cells were transfected twice with *hyg*^{r}–*tk*^{r} DNA constructs (Figures 17.9 and 17.10), the selection medium should contain 300 to 350 μg/ml of hygromycin B and 50 to 100 n*M* gancyclovir (GANC).

Note: The concentration of drugs used for selection depends on the specific cell line. In our killing-curve experiments, 250 to 300 μg/ml of G418 can kill nontransfected ES cells. A concentration of 50 to 100 nM *GANC in the culture medium is not cytotoxic to the parental ES cells, but is sufficient to kill ES cells containing the* HSV-tk *gene. The single dose for hygromycin B is 150 to 175 μg/ml.*

2. Remove dead cells by changing the medium once every 24 h and continue the selection for 3 weeks.

Tips: Drug-resistant colonies usually begin to grow after 10 days in selection medium; within a few more days, they grow to an appropriate size for isolation.

3. To pick up stably transfected clones, remove the medium and rinse ES cell colonies with 3 ml/plate of normal culture medium (drug-free). Hold the dish under a light and, using a red marker pen, individually mark or circle well-isolated colonies outside on the bottom of the plate.
4. Place the plate in the sterile laminar flow hood and overlay each of the marked colonies with a small volume (5 μl/colony) of 1X trypsin–EDTA solution. Allow the cells to become detached from the plate during a 5-min period at room temperature without disturbing the plate. Overlay each colony with 5 μl of culture medium and gently flush the cells in each individually marked area by pipetting the minicell suspension up and down several times, using a sterile plastic 100-μl pipette tip. The purpose of doing this is to ensure detachment of cells from the culture substrate.
5. Carefully take up the cells from each individual colony without disturbing other colonies and transfer them to a 25-cm^2 culture flask containing appropriate drugs as described in step 1.
6. Allow the cells to grow for another 5 to 10 days to the appropriate confluence with medium replaced once every 3 days. Freeze back some cells from each clone (for details, see Chapter 16). The rest of cells from each of the clones can be utilized for characterization or further proliferation.

Triple Selections of Cells Containing a Double-Knockout Gene

Triple selections refer to cells simultaneously transfected with two types of DNA constructs (see Figure 17.5, Figure 17.6, Figure 17.8 and Figure 17.9). If all goes well, successfully targeted cells should contain *neo*^{r}-*tk*^{r} and *hyg*^{r}-*tk*^{r} DNA in their chromosomes. Taking advantage of the triple drug-resistant genes, triple selections can allow cells containing the double-knockout gene of interest to be isolated. The selection and development procedures for stably transfected cells are almost the same as those described in the preceding section except for a few differences. The selection medium will contain 500 to 600 μg/ml of G418, 300 to 350 μg/ml of hygromycin B and 50 to 100 n*M* GANC.

Single Selection of Cells Containing a Silent Knockout Gene

Single selection utilizes a single drug in the selection medium. Because double-knockout clones were further transfected with a rescue DNA construct containing the rescue cDNA of the targeted gene and a unique drug resistance gene (*hyrt*^{r}), the stably transfected cells should be identified with HAT medium that contains 0.1 m*M* hypoxanthine, 0.9 μ*M* animopterin and 20 μ*M* thymidine. The HAT can be prepared as a 100X stock solution and diluted to 1X in the culture medium prior to use. The procedures for isolation of colonies are the same as those described in the section on double selection.

CHARACTERIZATION OF TARGETED ES CELL CLONES

Scientifically speaking, stably transfected cell clones mandate detailed characterizations, which include an activity assay of the reporter gene, Southern blotting, PCR screening and western blot analysis. These analyses will provide crucial evidence to verify the putative cell clones containing the gene that is structurally disrupted or replaced and functionally knocked out.

β-Galactosidase Staining of Cells

This is a fast procedure used to assay the activity of the *lac Z* gene. Because the gene targeting constructs contain the *lac Z* as a reporter gene within the homologous recombination region, stably transfected cells are expected to express the *lac Z* product, or β-galactosidase. This enzyme can hydrolyze X-gal (5-bromo-4-chloro-3-indolyl-β-D-galactoside) and yield a blue precipitate, staining cells blue. Protocol is as follows:

1. Culture the stably transfected cells in a Petri dish for 1 to 2 days.
2. Aspirate the medium and rinse the cells with PBS.
3. Fix the cells in 0.2% glutaraldehyde or 4% paraformaldehyde (PFA) for 5 to 10 min at room temperature.
4. Wash the cells for 3 × 5 min.
5. Stain the cells with X-gal solution for 1 to 24 h at 37°C.
6. Wash the cells for 3 × 5 min in PBS. Any blue cells or colonies indicate that the cells express β-galactosidase, and that the cells most likely contain the targeted gene.

Reagents Needed

Potassium Phosphate Buffer (PBS)

2.7 m*M* KCl
1.5 m*M* KH_2PO_4
135m*M* NaCl
15 m*M* Na_2HPO_4
Adjust pH to 7.2, then autoclave.

Preparation of 4% (w/v) Paraformaldehyde (PFA) Fixative (1 l):

a. Carefully weigh out 40 g paraformaldehyde powder and add it to 600 ml of preheated dd.H_2O (55 to 60°C). Allow the paraformaldehyde to dissolve at the same temperature with stirring.

Caution: Paraformaldehyde is toxic and should be handled carefully. The preparation should be carried out in a fume hood on a heating plate. Temperature should not be above 65°C and the fixative should not be overheated.

b. Add one drop of 2 *N* NaOH solution to clear the fixative.
c. Remove from the heat, add 333.3 ml (1/3 total volume) of 3 × PBS (8.1 m*M* KCl, 4.5 m*M* KH_2PO_4, 411 m*M* NaCl, 45 m*M* Na_2HPO_4, pH 7.2); then autoclave.
d. Adjust pH to 7.2 with 4 *N* HCl and bring a final volume to 1 l with dd.H_2O.
e. Filter the solution to remove undissolved fine particles, cool to room temperature and store at 4°C until use.

X-Gal Solution

PBS solution containing 0.5 mg/ml X-gal, 10 m*M* $K_3[Fe(CN)_6]$. X-gal stock solution (50 mg/ml in dimethylsulfoxide or dimethylformamide). The stock solution should be kept in a brown bottle at –20°C prior to use.

Southern Blot Analysis

Data from Southern blot analysis will be the most important evidence to prove whether the stably transfected cell clones carry a double-knockout or targeted gene in their chromosomes. Detailed protocols for restriction enzyme digestion of DNA and Southern blot analysis are described in Chapter 7. This section mainly focuses on genomic DNA isolation and analysis of Southern blot data.

Isolation of Genomic DNA from Cultured Cells

1. Aspirate culture medium from cultured cells and rinse the cells twice in PBS.
2. Add 10 ml/60-mm culture plate (with approximately 10^5 cells) of DNA isolation buffer to the cells and allow them to be lysed for 5 min at room temperature with shaking at 60 rpm.
3. Incubate the plate at 37°C for 5 h or overnight to degrade proteins and extract the genomic DNA.
4. Allow the lysate to cool to room temperature and add three volumes of 100% ethanol (prechilled at –20°C) to the lysate. Gently mix and allow the DNA to precipitate at room temperature with slow shaking at 60 rpm. Precipitated DNA, appearing as white fibers, should be visible in 20 min.
5. Fish out the DNA using a sterile glass hook or spin it down. Rinse the DNA twice in 70% ethanol and partially dry it for 40 to 60 min at room temperature to evaporate the ethanol. The DNA can be directly subjected to restriction enzyme digestion without being dissolved in TE buffer or

dd.H_2O. We usually overlay the precipitated DNA with an appropriate volume of restriction enzyme digestion cocktail and incubate the mixture at an appropriate temperature for 12 to 24 h. The digested DNA is mixed with an appropriate volume of DNA loading buffer and is now ready for agarose gel electrophoresis.

Reagents Needed

DNA Isolation Buffer

75 m*M* Tris-HCl, pH 6.8
100 m*M* NaCl
1 mg/ml Proteinase K
0.1% (w/v) *N*-Lauroylsarcosine

Analysis of Southern Blot Data

Following Southern blot hybridization, it is now becomes exciting to analyze the data. Figure 17.11 illustrates the expected results. Keep in mind that disruption of gene X here refers to insertion of marker genes and *lac Z* into one of the exons of the gene to be targeted. As a result, a unique restriction enzyme (E) that originally cuts outside gene X and generates a 6-kb fragment can produce a large sized DNA fragment. For instance, insertion of *neo* and *lac Z* genes results in a 10-kb fragment. On the other hand, integration of *hyg* and *lac Z* produces a 12-kb DNA fragment. The genomic DNA is isolated from different targeted cell lines, including nontargeted cells as a control. The DNA can be digested with a unique restriction enzyme and be separated by agarose gel electrophoresis.

For Southern blot hybridization, two probes can be utilized. One is probe A, which is external to the targeting vector but internal to gene X. The other is probe B, which is internal to the marker gene (*neo* or *hyg*) but external to gene X. If probes A and B are used for hybridization, only a 6-kb band is seen in nontargeted cells (lane C). If the gene is a single-copy knockout gene, a 6-kb band in the control and a 10-kb band with the *neo* probe (lane 1) or 12-kb band with the *hyg* probe (lane 2) can be seen. In the case of a double-knockout by two different targeting constructs (Figure 17.5 and Figure 17.6), both copies of gene X in two chromosomes should be targeted. Thus, two bands should be detected: a 10-kb band corresponding to the *neo* probe and one a 12-kb band obtained with the *hyg* probe (lane 3). The 6-kb band shown in the control disappears.

Unlike exon disruption, the isogenic DNA targeting strategy described earlier in this chapter refers to completely replacing gene X to be knocked out. As shown in Figure 17.12, using 5′-isogenic DNA and 3′-isogenic DNA fragments flanking the gene X in the genome, gene X can be replaced by *neo* and *lac Z* or *hyg* and *lac Z* genes by homologous recombination at the sites indicated in the figure. If all goes well, a 20-kb fragment generated by a unique restriction enzyme (E) will become larger or smaller following homologous recombination. To detect knockout sites, probe A can be chosen to be internal to gene X and external to the marker genes. Probe B can be designed to be external to gene X but internal to the marker genes.

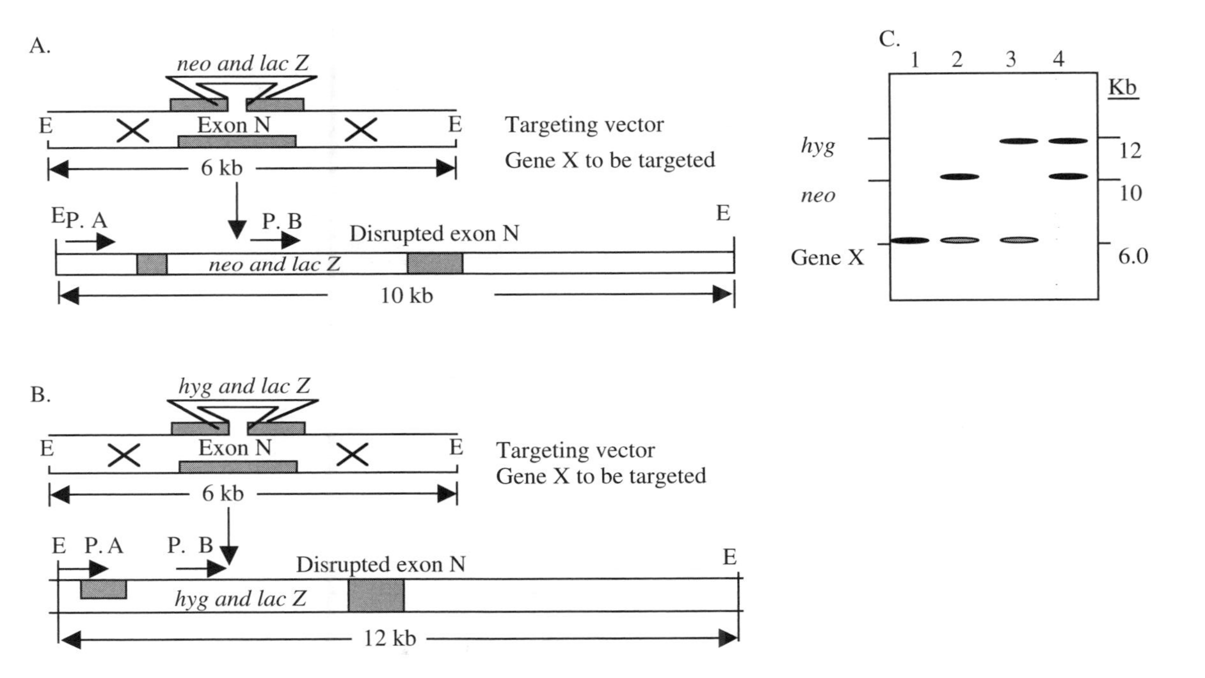

FIGURE 17.11 Southern blot analysis of genomic DNA with targeted gene X. Lane 1: nonmutant control cells. Lane 2: single-copy targeting with *neo* vector. Lane 3: single-knockout with *hyg* vector. Lane 4: double-knockout of gene X with *neo* and *hyg* vectors. P.A: probe A internal to gene X but external to the marker gene. P.B: probe B internal to marker gene but external to gene X.

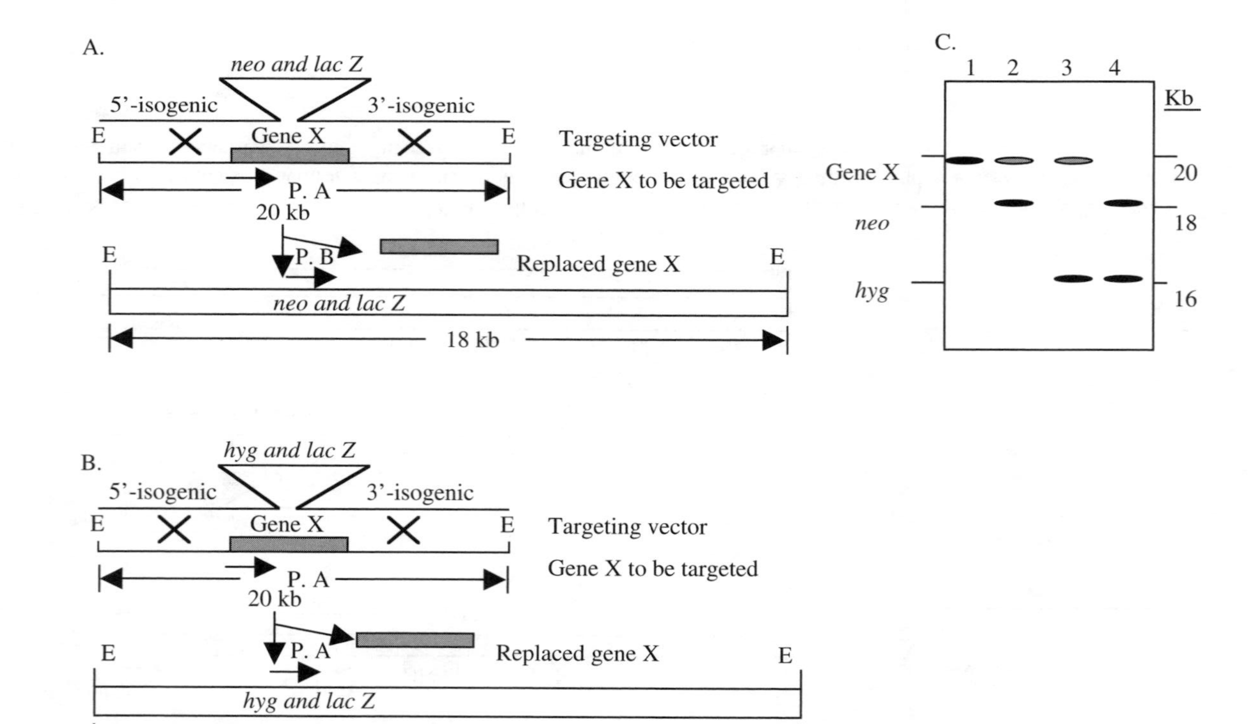

FIGURE 17.12 Analysis of double-knockout gene X with isogenic DNA. A. Use of marker *neo* gene vector. B. Use of the marker *hyg* gene vector. C. Southern blot analysis of genomic DNA with targeted gene X. Lane 1: Nonmutant control cells. Lane 2: Single-copy targeting with the *neo* vector. Lane 3: Single-knockout with the *hyg* vector. Lane 4: Double-knockout of gene X with both *neo* and *hyg* vectors. P. A: probe A internal to gene X but external to the marker gene. P. B: probe B internal to the marker gene but external to gene X.

After Southern blot hybridization, a 20-kb band should be seen in nontransfected cells as the control (lane C). If gene X is a single-copy targeted gene, a weak band of 20 kb can be seen as the control with an additional 18-kb band using the *neo* probe (lane 1) or a 16-kb band using the *hyg* probe. If both copies of gene X have been knocked out, only 18- and 16-kb bands should be detected without the 20-kb band seen in control cells (lane 3).

It should be noted that one important subject of this chapter is the rescue strategy with the double-knockout gene to avoid potential lethal effects temporarily. To analyze the success of rescue transfection after a double-knockout, Southern blot analysis should be carried out. As shown in Figure 17.13, if the unique restriction enzyme does not cut the rescue construct, after successful integration, the original 2-kb band will be expanded to 12 kb in size (Figure 17.13A). In this case, two different probes can be prepared. Probe A is internal to the marker gene (*hyrt*), but is external to the targeted gene X. Probe B, on the other hand, is internal to the rescue cDNA of gene X but external to the marker gene. Following Southern blotting, probe A will detect a 12-kb band (lane 1). In case of exon insertion by the rescue construct, probe B will detect 12- and 6-kb bands (lane 2). However, probe B should only reveal a 12-kb band after the replacement of gene X by the rescue construct (lane 3).

Western Blot Analysis

Protein analysis of the targeted gene is the bottom line of gene double-knockout. The detailed protocols for protein extraction and western blot hybridization are

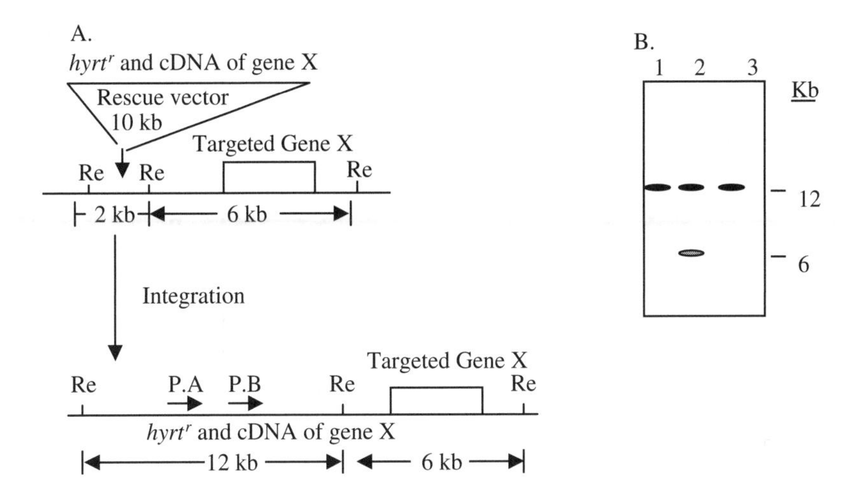

FIGURE 17.13 Analysis of integration of rescue cDNA for the targeted gene X. A. Diagram of integration. B. Southern blot analysis of genomic DNA containing the rescue cDNA for the targeted gene X. Lane 1: double-knockout control cells. Lane 2: integration into DNA containing single-copy knockout gene X. Lane 3: rescue of double-knockout of gene X. P.A: probe A internal to gene *hyrt* but external to cDNA. P.B: probe B internal to cDNA but external to *hyrt*.

described in Chapter 11. This section focuses on how to analyze western blot data. Assuming the product or protein of gene X to be targeted is 46 kDa, and that monoclonal antibodies have been raised to be against the 46-kDa protein, as shown in Figure 17.14, only a 46-kDa band is detected in wild-type (WT) or control cells. Lane 1 indicates that a single copy of gene X is targeted by a targeting construct containing the *neo* gene. Thus, a weak 46-kDa band as shown in control cells and a 90-kDa band due to the fusion protein are seen. Similar results are revealed in lane 2 if gene X is singly targeted by a construct containing the *hyg* gene (lane 2). In the case of a double-knockout by insertion into an exon of gene X, two bands larger than 46 kDa will be detected. These represent two different sizes of fusion proteins due to the insertion of two different marker genes and the *lac Z* gene (lane 3). Interestingly, if both copies of gene X are completely replaced by the double-knockout constructs, no bands should be seen (lane 4). On the other hand, the cells shown in lane 3 are rescued by gene X cDNA. In this case, three different bands will be detected, including a 46-kDa band seen in control cells (lane 5). In contrast, if the cells shown in lane 4 are rescued by gene X cDNA, only a single 46-kDa band is detected. This is the same as that from the control cells (lane 5)

Assuming that the targeted cell lines have been characterized by X-gal staining, Southern blot analysis, and western blot hybridization, and that the cell clones containing the gene double-knockout have been obtained, it is fair to say that this is a big reward for the investigators and represents more than the halfway point in the long journey to obtain transgenic mice.

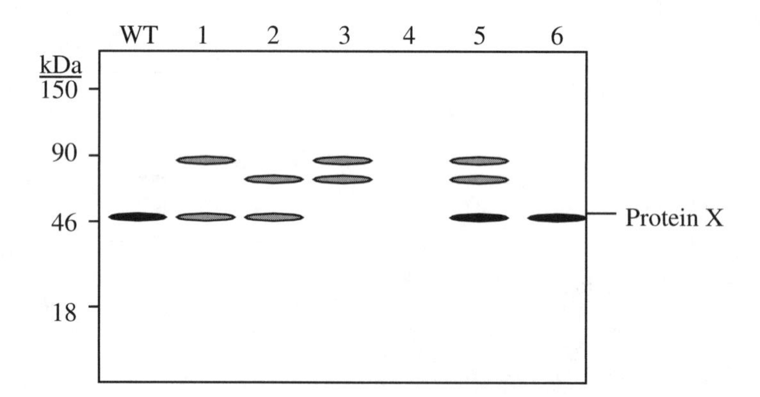

FIGURE 17.14 Protein analysis by western blotting. Assume that the product of gene X has a molecular weight of 46 kDa, and that monoclonal antibodies against the 46 kDa protein are used in western blot analysis. WT: wild-type as nontransfected cells. Lane 1: cells containing single-copy knockout of gene X by *neo* gene. Lane 2: cells containing single-copy knockout of gene X by *hyg* gene. Lane 3: double-knockout of gene X by disruption of an exon with *neo* and *hyg* genes. Lane 4: double-knockout of gene X by replacement of the gene with *neo* and *hyg* genes. Lane 5: cDNA rescue of the targeted gene X as shown in lane 3. Lane 6: cDNA rescue of the targeted gene X as shown in lane 4.

GENERATION OF TRANSGENIC OFFSPRING FROM DOUBLE-KNOCKOUT ES CELLS

C57BL/6J female mice are commercially available and widely utilized for embryo recovery and transfer. This strain of mice is different from 129 SVJ mice in coat color, which can be used as a useful marker.

PROCEDURE 1. SELECTION OF C57BL/6J ESTROUS FEMALES

C57BL/6J mouse is a well-established strain widely chosen as the best choice of host embryo for ES cell chimeras and germline transmission. These mice have a different coat color than that of 129 SVJ mice and that can be used as a coat color selection marker. In order to relatively synchronize the estrous cycle and to simplify female selection, female mice should be caged together without a male. Their estrous cycle is approximately 3 to 4 days and ovulation usually occurs at about the midpoint of the dark period in a light–dark cycle. This light and dark cycle can be altered to meet a researcher's schedule. For example, the cycle may be adjusted to 14 h of light and 10 h of dark. In this way, most strains of mice produce blastocysts that can be utilized for injection in the late morning of the third or fourth day. The key question is how to identify pre-estrous or estrous females for mating. The basic criteria are the external vaginal epithelial features. At the pre-estrous stage, the vaginal epithelium is usually characterized by being moist, folded, pink/red and swollen. At the estrous stage, the general features of the vaginal epithelium include being dry but not wet, pink but not red or white, having wrinkled epithelium on upper and lower vaginal lips, and being swollen. After the females have been caged for a while and reach the pre-estrous prediction, they should be monitored on a daily basis. Once they are at the estrous stage, mating should be set up.

An alternative way to promote the estrous cycle is by means of hormonal induction of ovulation with exogenous hormones. This procedure is termed superovulation. Injection of pregnant mare serum gonadotropin (PMSG, Sigma) can stimulate and induce immature follicles to develop to the mature stage. Superovulation can be achieved by further administration of human chorionic gonadotropin (HCG, Sigma). The procedure is outlined as follows:

1. Three days or 72 to 75 h prior to mating, induce superovulation of 10 to 15 female mice (3 to 4 weeks old) by injecting 5 to 10 units of PMSG intraperitoneally into each female between 9:00 a.m. and noon on day one.
2. Two days or 48 to 50 h later, inject 5 to 10 units of HCG intraperitoneally into each female between 9:00 a.m. and noon on day three. After injection, cage the females and proceed to mate them with sterile males.

PROCEDURE 2. PREPARATION OF A BANK OF STERILE MALES BY VASECTOMY

Detailed protocols are described in Chapter 9.

Procedure 3. Pairing of Estrous Females and Sterile Males

Detailed protocols are described in Chapter 9.

Procedure 4. Preparation of Blastocyst-Stage Embryos from Pseudopregnant Mice

Detailed protocols are described in Chapter 9.

Procedure 5. Preparation of Micromanipulation Apparatus

Detailed protocols are described in Chapter 9.

Procedure 6. Injection of ES Cells into Blastocysts

Detailed protocols are described in Chapter 9.

Procedure 7. Reimplantation of Injected Blastocysts into the Uterus of a Recipient Female

Detailed protocols are described in Chapter 9.

CHARACTERIZATION OF MUTANT MICE

Once offspring or transgenic mice have been obtained, they should be analyzed at DNA, protein, morphological and functional levels. DNA can be isolated from mouse tails and Southern blot hybridization be carried out as described in Chapter 7. Extraction of proteins from soft tissues (brain, liver) and analysis by western blotting are described in Chapter 11. Analyses of Southern and western blot data are virtually the same as those described for cell line characterization. In addition, immunohistochemical or cytochemical analyses can be performed by *in situ* hybridization (Chapter 12 to through Chapter 14).

Morphological and functional analyses can provide the most powerful evidence for the occurrence of gene double-knockout in transgenic mice. For example, if gene X controls red eyes in wild type mice. If double-knockout of gene X is successful, then, mutant mice will have white eyes. As a result, the transgenic mice may not see well as compared with wild-type mice. Other physiological or pathophysiological, toxicological and pharmacological approaches may also be employed to study transgenic mice containing targeted genes of interest.

REFERENCES

1. Capecchi, M.R., Targeted gene replacement, *Sci. Am.*, 52, 1994.
2. Capecchi, M.R., Altering the genome by homologous recombination, *Science*, 244, 1288, 1989.

3. Mansour, S.L., Thomas, K.R., and Capecchi, M.R., Disruption of the proto-oncogene *int-2* in mouse embryo-derived stem cells: a general strategy for targeting mutations to non-selectable genes, *Nature*, 336, 348, 1988.
4. Riele, H.T., Maandag, E.R., and Berns, A., Highly efficient gene targeting in embryonic stem cells through homologous recombination with isogenic DNA constructs, *Proc. Natl. Acad. Sci. USA,* 89, 5128, 1992.
5. Schorle, H., Holtschke, T., Hunig, T., Schimpl, A., and Horak, I., Development and function of T cells in mice rendered interleukin-2 deficient by gene targeting, *Nature,* 352, 621, 1991.
6. Tybulewicz, V.L.J., Crawford, C.E., Jackson, P.K., Bronson, R.T., and Mulligan, R.C., Neonatal lethality and lymphopenia in mice with a homozygous disruption of the *c-abl* proto-oncogene, *Cell,* 65, 1153, 1991.
7. Weinstock, P.H., Bisgaier, C.L., Setala, K.A., Badner, H., Ramakrishnan, R., Frank, S.L., Essenburg, A.D., Zechner, R., and Breslow, J.L., Severe hypertriglyceridemia, reduced high density lipoprotein, and neonatal death in lipoprotein lipase knockout mice, *J. Clin. Invest.*, 96, 2555, 1995.
8. Michalska, A.E. and Choo, K.H.A., Targeting and germ-line transmission of a null mutation at the metallothionein I and II loci in mouse, *Proc. Natl. Acad. Sci. USA*, 90, 8088, 1993.
9. Paul, E.L., Tremblay, M.L., and Westphal, H., Targeting of the T-cell receptor chain gene in embryonic stem cells: strategies for generating multiple mutations in a single gene, *Proc. Natl. Acad. Sci. USA*, 89, 9929, 1992.
10. Xu, L., Gorham, B., Li., S.C., Bottaro, A., Alt, F.W., and Rothman, P., Replacement of germ-line ε promoter by gene targeting alters control of immunoglobulin heavy chain class switching, *Proc. Natl. Acad. Sci. USA,* 90, 3709.
11. Hasty, P. and Bradley, A., Gene targeting vectors for mammalian cells, in *Gene Targeting: A Practical Approach,* Joyner, A.L., Ed., IRL Press, Oxford University Press, Inc., New York, 1993.
12. Wu, W., Electrophoresis, blotting, and hybridization, in *Handbook of Molecular and Cellular Methods in Biology and Medicine,* Kaufman, P.B., Wu, W., Kim, D., and Cseke, L., CRC Press, Boca Raton, FL, 1995.
13. Wu, W., Gene transfer and expression in animals, in *Handbook of Molecular and Cellular Methods in Biology and Medicine,* Kaufman, P.B., Wu, W., Kim, D., and Cseke, L., CRC Press, Boca Raton, FL, 1995.
14. Wurst, W. and Joyner, A.L., Production of targeted embryonic stem cell clones, in *Gene Targeting: A Practical Approach,* Joyner, A.L., Ed., IRL Press, Oxford University Press, Inc., New York, 1993.
15. Papaioannou, V. and Johnson, R., Production of chimeras and genetically defined offspring from targeted ES cells, in *Gene Targeting: A Practical Approach,* Joyner, A.L., Ed., IRL Press, Oxford University Press, Inc., New York, 1993.

18 Large-Scale Expression and Purification of Recombinant Proteins in Cultured Cells

CONTENTS

INTRODUCTION

Expression of large amounts of specific proteins has become ever more important to scientific research in academic fields and is invaluable to industry as well. In fact, protein labeling, protein crystalization, production of antibodies and synthesis of proteins for gene transcriptional regulation mandate a fair amount of pure proteins. Recently, DNA recombination technology, gene transfer and expression techniques have made it possible to overexpress a protein of interest at a very high level in a short period of time.[1–4] Different strategies and expression systems have been developed using prokaryotic and eukaryotic expression vectors and hosts.

Escherichia coli has been well studied for decades. It has been demonstrated that *E. coli* is the best prokaryotic system for expression of foreign genes.[1–3] Since DNA recombination technology became available, *E. coli* has been widely used as the leading host organism for broad applications of protein expression.[2–7] Compared to eukaryotic cells, *E. coli* has three major advantages. It has a very short life cycle of approximately 20 min per generation. It is easy to manipulate and large-scale cultures can be made in inexpensive media. Thus, it is not surprising that *E. coli* has been chosen as an ideal system to express large quantities of proteins encoded by cloned cDNA or genes.

One drawback, however, is that posttranslation modifications of eukaryotic proteins may not take place in *E. coli*. Fortunately, eukaryotic expression systems have been established to overcome this difficulty in processing proteins in bacteria. A well-known example is the yeast *Pichia* system from Invitrogen (San Diego, CA). Compared with other eukaryotic cell lines, yeast is easy to grow at low cost.[9] Perhaps the major advantages of using yeast as a host for protein expression are that the genome is easy for manipulation and suitable for cloning of foreign genes and overexpression and purification of proteins of interest.

The present chapter introduces and describes detailed protocols for large-scale expression, collection and purification of proteins of interest in *E. coli* and yeast systems.[4–11] Due to rapid advances in current biotechnology, the authors strongly encourage users to obtain up-to-date information or more suitable expression systems for a particular research, which may be available from biotechnology companies.

PART A. EXPRESSION AND PURIFICATION OF PROTEINS IN PROKARYOTIC SYSTEMS

This part will focus on the current strategies used for expression and purification of large amounts of specific proteins using *E. coli* as the host organism.

Design and Selection of Bacterial Expression Vectors

Expression of a specific protein in *E. coli* begins with insertion of a cDNA into an expression vector. Selection of appropriate vectors will play a major role in the success of protein expression. So far, plasmid-based vectors are widely utilized for such purposes. Different types of expression vectors are commercially available; for example, pTrcHis (A, B, C) and pRSET (A, B, C) series have been developed by Invitrogen and pET series by Novagene. All three reading frames are available in the vector series. Other expression vectors are available from Strategen and Gibco BRL.

In general, a desirable expression vector should have the features shown in Figure 18.1. If some features are not present in commercial vectors, for research purposes, appropriate sequences may be created by PCR or other subcloning approaches and the sequences inserted at appropriate positions in a vector.

1. **A strong bacteriophage T7 promoter for transcription**. Its activity is induced by endogenous T7 RNA polymerase of the host cell or by introduction of a bacteriophage T7 RNA polymerase gene into the host cell. Bacteriophage T7 RNA polymerase is a very active enzyme that synthesizes RNA much faster than *E. coli* RNA polymerase. In fact, the enzyme is so selective and active that it uses the majority of the cellular machinery for the expression of foreign proteins. Because of the high selectivity of the T7 RNA polymerase, the enzyme mainly initiates its own promoter activity but not the initiation of transcription from other sequences on *E. coli* DNA. As a result, the desired protein can be >50% of the total proteins in the cell after several hours of induction. Additionally, because the T7 promoter is controlled by T7 RNA polymerase, the expression of a cDNA, or a gene cloned downstream from the T7 promoter, can be maintained transcriptionally silent or off in the uninduced state. This is particularly useful when expression of some proteins is potentially toxic to the host cell.

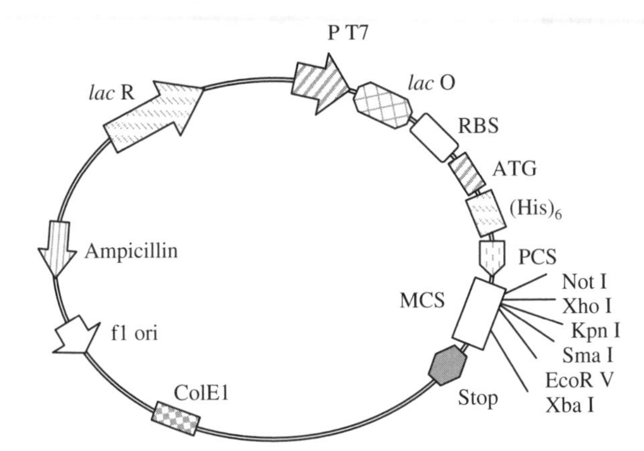

FIGURE 18.1 An ideal vector for large-scale expression and purification of specific proteins in *E. coli* strain BL21 (DE3) by IPTG induction.

2. **The *lac* operator and *lacI* gene**. Immediately downstream of the T7 promoter is the *lac* operator. Like the T7 promoter, the *lac* operator can also control the expression of the cDNA or gene cloned downstream from the operator. However, the operator is negatively controlled by the *lac* repressor encoded by the *lacI* gene that is separated from the T7 promoter and the *lac* operator. When the repressor binds to the operator, the transcription of downstream cDNA or gene is turned off. With addition of an appropriate dose of IPTG, which will bind to the repressor, the operator will be activated and the expression of a foreign gene will then be turned on. Thus, regulation between the *lac* operator and repressor can reduce basal expression of a foreign gene when the product of the gene is potentially toxic to the host cell. Besides, the *lacO* operator and the *lacI* repressor gene in the vector provide transcriptional regulation in any *E. coli* strain. In other words, the cloned cDNA or gene at the multiple cloning site (MCS) can be expressed in any strain of *E. coli* although the expression level is not as high as that controlled by the bacteriophage T7 promoter.
3. **A strong ribosome binding site (RBS).** A bacteriophage RBS consists of AAGGAG sequence lying very closely upstream from the first codon ATG, which will provide a high efficiency of translational initiation. The small subunits of ribosomes will selectively recognize and bind to the sequence, thus initiating the translation of the mRNA. Again, if the sequence is not available in the vector, the sequence immediately upstream of the ATG codon can be easily created. With the sequence, the translational rate can be enhanced several times.
4. **The ATG codon**. ATG (AUG in mRNA) is designed almost immediately following the RBS sequence. This will be the first codon for the recombinant protein. The purpose of including this first codon is due to the very next His-tag for future affinity purification. Because the first codon is fixed in the vector, a cDNA or gene to be cloned should be carefully oriented so that absolutely no stop codon and no reading frame shifts are allowed. Detailed cautions will be noted in cloning an insert into a vector.
5. **His-tag.** The His-tag sequence located immediately following the ATG codon will code for 6 to 10 consecutive histidine residues. These residues will be the N-terminal leader sequence of a protein to be expressed from the cloned cDNA or gene and will serve as the affinity tag for rapid purification of the protein with metal chelation affinity chromatography using an appropriate His-binding resin.
6. **PCS.** A sequence coding for a protease cleavage site (PCS) is designed between the His-tag and the multiple cloning site (MCS). The amino acid sequence encoded by this sequence will be the target site for an appropriate protease (e.g., thrombin, Factor Xa or enterokinase). This will allow removal of the N-terminal His-tag leader sequence *in vitro* as desired. It should be noted here that noncleaved fusion proteins will be acceptable for use as antigens. However, for biochemical, cell biology or enzyme

assays, the leader sequence may need to be removed because a native protein is better than a fusion protein.

7. **Multiple cloning site (MCS).** An MCS contains different restriction enzyme sites for insertion of a specific cDNA for expression of a specific protein. Usually, commercial vectors contain overlapping bases for adjacent restriction enzymes. Perhaps it is for ease of use that all three reading-frame vector series have been developed. This strategy allows cloning of a cDNA to be expressed at different restriction enzyme sites without causing reading frame shifts.
8. **Terminator.** A translational terminator sequence may be designed immediately downstream from the MCS. Nonetheless, if the cDNA to be expressed contains a stop codon, it will be fine without the terminator sequence built into the vector.
9. **Plasmid backbone (Ampicillin gene, F1 ori and ColE1).** A plasmid-based expression vector should contain plasmid backbone elements. An ampicillin-resistant gene or *Amp*^{r} will serve as an antibiotic marker for selection of positive transformed colonies. In theory, only bacteria carrying an *Amp*^{r} gene in the introduced plasmid DNA are resistant to killing by ampicillin. An F1 ori sequence will allow the synthesis of single-stranded plasmid DNA by infection of F′-containing host strains with helper phage (e.g., R408). The ColE1 element makes it possible to replicate the introduced plasmid DNA with a copy number as high as possible.

As mentioned earlier, the use of a bacteriophage T7 RNA polymerase/promoter system will certainly make possible the large-scale expression of specific proteins. However, for utilization of the vector type as diagrammed in Figure 18.1, the DNA constructs need to be transformed into a special *E. coli* strain such as BL21 (DE3) that contains a T7 RNA polymerase gene on the bacterial chromosome under the control of the *lac* promoter.[3] Or, two DNA constructs may need to be introduced into the same *E. coli* strain: one is the vector type of Figure 18.1 and the other is a construct containing the bacteriophage T7 RNA polymerase gene and the *lac* promoter.

An alternative approach is to design or obtain a vector that contains the features shown in Figure 18.1 and has the bacteriophage T7 RNA polymerase gene under the control of the *lac* promoter as shown in Figure 18.2. The major advantages of using this type of vector are that the DNA constructs can be introduced into any standard *E. coli* strains, and that the expression of T7 RNA polymersase can be easily induced by IPTG or by heat shock at 42°C for 30 to 40 min.

Cloning cDNA Inserts into Vectors

Detailed protocols concerning DNA digestion, preparation of vectors and insert DNA, ligation, transformation, selection and purification of plasmid DNA are described in Chapter 4. This section primarily focuses on some special precautions that should be taken when cloning a cDNA for expression.

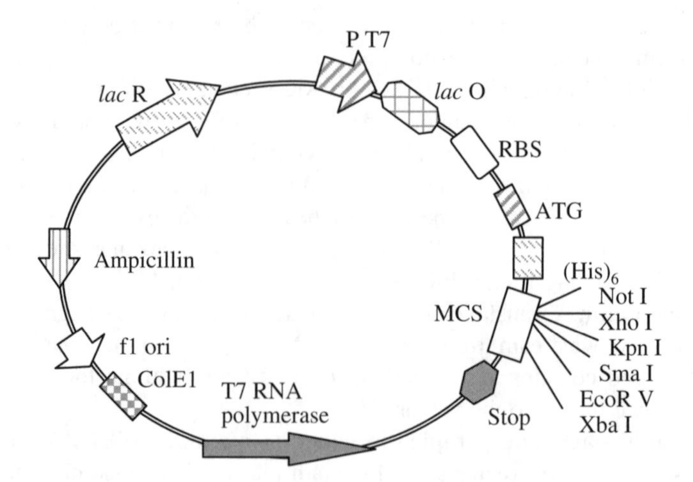

FIGURE 18.2 An ideal vector for large-scale expression and purification of specific proteins in any *E. coli* strains by heat induction.

1. Restriction enzymes used for DNA digestion should be unique so as to cut outside the entire coding region of the cDNA to be cloned at the MCS of the vector. To ensure the success of protein expression, no enzyme cuts are allowed inside the coding region of the cDNA or outside the MCS region of the vector DNA. Detailed restriction enzyme mapping should be carried out when preparing vectors and insert DNA.
2. Carefully trim out as much as possible the 5′-UTR and 3′-UTR regions of the cDNA to be expressed using appropriate restriction enzymes or PCR approaches. In our experience, long 5′-UTR or 3′-UTR regions may decrease the efficiency of protein expression.
3. Due to the orientation of the driving promoter and the first ATG codon fixed in place in the vector, it is necessary to insert the poly(T) strand instead of the poly(A) of the cDNA in $3' \rightarrow 5'$ orientation downstream from the driving promoter. In other words, the cDNA to be expressed should be cloned at the sense orientation under the control of the driving promoter. Therefore, antisense cloning done by mistake is not allowed.
4. Because the first codon ATG is fixed in place in the vector, the cDNA to be expressed must be carefully oriented so that absolutely no stop codon and no reading frame shifts are allowed. To make sure that the cDNA has been cloned at the right position with the correct reading frame starting from the ATG in the vector, it is strongly recommended that the DNA constructs obtained be carefully analyzed by appropriate restriction enzyme digestion pattern analysis and by DNA sequencing of the 5′-junction site. However, the entire cDNA does not need to be sequenced. In the authors' experience, as long as the 5′-portion of the cloned cDNA has the correct nucleotide sequence and the right reading frame with regard to the ATG codon and the leading sequence, the constructs should be fine without sequencing the whole cDNA.

Expression of Proteins of Interest in Bacteria

Once an expression construct is established, the plasmid constructs can be introduced into an appropriate *E. coli* strain for induction of protein expression. As discussed earlier, different types of vectors require different bacterial strains. For example, Protocol A and Protocol B below can be applied to the vector type shown in Figure 18.1. Protocol C and Protocol D are suitable for the vector type shown in Figure 18.2.

Protocol A. Transformation of *E. coli* BL21 (DE3) with Vector Shown in Figure 18.1 Using $CaCl_2$ Method

1. Thaw an aliquot of 0.2 ml of frozen $CaCl_2$-treated competent cells of *E. coli* BL21 (DE3) on ice.
2. Perform transformation as follows:

Components	Tube 1	Tube 2
Cells	50 μl	50 μl
DNA	100 ng in 5 μl	400 ng in 5 μl

Tip: 15 to 20 μl of ligated mixture can be directly added to 50 to 60 μl of competent cells.

3. Incubate on ice for 45 to 50 min.
4. Heat shock at 42°C for 1 to 1.5 min and quickly place the tubes on ice for 1 min.

Tip: The heat-shock treatment and chilling on ice should not exceed the time indicated. Otherwise, the efficiency of transformation will decrease 10- to 20-fold.

5. Transfer the cell suspension to sterile culture tubes containing 1 ml of LB medium without antibiotics. Incubate at 37°C for 1 to 2 h with shaking at 140 rpm to recover the cells.
6. Transfer 50 to 150 μl of the cultured cells per plate onto the center of LB plates containing 50 μg/ml ampicillin. Immediately spread the cells over the entire surface of the LB plates using a sterile, bent glass rod.
7. Invert the plates and incubate them at 37°C overnight. Selected colonies should be visible 14 h later.

Protocol B. Induction of Protein Expression with IPTG

1. Inoculate a single colony into 5 ml of LB medium containing 50 μg/ml ampicillin. Grow the cells at 37°C overnight with shaking at 200 rpm. Store the cells at 4°C prior to use.
2. Add 1 ml of the cell suspension into each flask containing 100 ml of LB medium and 50 μg/ml ampicillin in the medium and incubate with shaking

at 37°C until OD_{600} reaches 0.7 to 1.0 (usually 3 to 5 h). One sample will serve as uninduced control; the other will be subjected to IPTG induction.

3. Add IPTG from a 100-m*M* stock solution to the IPTG induction flask to a final concentration of 1 m*M*. Allow the cells to grow for another 3 to 4 h at 37°C with shaking.

Principles: During the cell culture, a large quantity of T7 RNA polymerase is expressed by the T7 RNA polymerase gene on the host chromosomal DNA. T7 RNA polymerase activates the bacteriophage T7 promoter that controls the expression of the cDNA cloned downstream from the promoter. However, there is a lac *operator immediately following the T7 promoter. The* lac *operator is usually bound by the* lac *repressor and becomes silent. Therefore, even though a lot of T7 RNA polymerases are produced, the expression of the cDNA is in an inactivated state. IPTG binds to the* lac *repressor, preventing the repressor from binding to the operator. As a result, there is no blocking door downstream of the T7 promoter. Once the T7 promoter is activated, the expression of the cloned cDNA is turned on. Using a large amount of T7 RNA polymerase, the T7 promoter can virtually convert most of the host's resources for the synthesis of foreign proteins.*

4. Place the flasks on ice for 10 min and centrifuge at 5000 rpm for 5 min at 4°C.
5. Decant the supernatant and resuspend the cells in 20 ml of Tris-HCl buffer (pH 8.0) containing 2 m*M* EDTA.
6. Carry out centrifugation as in step 4. The cells can be resuspended in an appropriate protein extraction buffer.

Reagents Needed

LB (Luria-Bertaini) Medium (per liter)

10 g Bacto-tryptone
5 g Bacto-yeast extract
5 g NaCl
Adjust to pH 7.5 with 0.2 *N* NaOH and autoclave

LB Plates

Add 15 g of Bacto-agar to 1 l of freshly prepared LB medium and autoclave. Allow the mixture to cool to about 55°C and add an appropriate amount of antibiotic stock. Pour 25 to 30 ml of the mixture into each of 100 mm Petri dishes in a sterile laminar flow hood with filtered air flowing. Remove any bubbles with a pipette tip and let the plates cool for 5 min prior to being covered. Allow the agar to harden for 2 h and partially dry at room temperature for a couple of days. Store the plates at room temperature up for 10 days or at 4°C in a bag for 1 month. Allow the cold plates to warm up at room temperature for 1 to 2 days prior to use.

Protocol C. Transformation of *E. coli* Strains HB101, JM109, TOP10 or DH5αF′ Using the Vector Type Shown in Figure 18.2

If a DNA construct is developed as shown in Figure 18.2, the constructs may be transferred into standard *E. coli* strains such as HB101, K38, JM109, TOP10 or DH5αF′, assuming that the products expressed from the cloned cDNA are not toxic to the host cell. The transformation procedure is the same as that described previously.

Protocol D. Induction of Protein Expression by Heat Shock

Cells containing the plasmid construct shown in Figure 18.2 can be induced for expression of the cloned cDNA by IPTG as described earlier or by heat shock as follows:

1. Inoculate a single colony into 5 ml of LB medium containing 50 μg/ml ampicillin. Grow the cells at 30°C overnight with shaking at 200 rpm. Store the cells at 4°C prior to use.
2. Add 1 ml of the cell suspension to each flask containing 100 ml of LB medium with 50 μg/ml ampicillin in medium and incubate with shaking at 30°C until OD_{600} reaches 0.5 to 0.7. One sample will serve as uninduced control; the other will be subjected to heat induction.
3. Raise the temperature to 42°C and allow the cells to grow at 42°C for 30 to 40 min with shaking.

Principles: During the incubation at low temperature (30°C), the expression of the cloned cDNA under the control of the T7 RNA polymerase promoter is inhibited by the lac *repressor expressed under the control of* E. coli *Plac promoter. However, at high temperature (42°C), the inhibition is inactivated, resulting in activation of the T7 promoter. Subsequently, a large quantity of T7 RNA polymerase is expressed under the control of the T7 RNA polymerase promoter. T7 RNA polymerase activates the bacteriophage T7 promoter that controls the expression of the cDNA cloned downstream from the promoter. In other words, once the T7 promoter is activated, the expression of the cloned cDNA is turned on. Using a large amount of T7 RNA polymerase, the T7 promoter can virtually convert most of the host's resources for the synthesis of foreign proteins.*

4. Reduce the temperature to 4°C and allow the cells to grow for 1.5 to 2 h with shaking.
5. Place the flasks on ice for 10 min and centrifuge at 5000 rpm for 5 min at 4°C.
6. Decant the supernatant and resuspend the cells in 20 ml of Tris-HCl buffer (pH 8.0) containing 2 m*M* EDTA.
7. Carry out centrifugation as in step 5. The cells can be resuspended in an appropriate protein extraction buffer.

Purification and Cleavage of the Expressed Proteins

Analysis of Cellular Proteins by SDS-PAGE, Dot Blot or Western Blotting

Prior to large-scale bacterial culture and protein purification, the overexpression of the proteins of interest from cultured cells should be quickly analyzed or confirmed by SDS-PAGE followed by Coomassie blue staining, or by dot blot/western blot analysis if specific antibodies are available. The bacteria containing the overexpressed proteins can be completely solublized under denaturation conditions, which can be loaded into SDS-PAGE or be used for dot blotting. The detailed protocols for protein extraction, solublization, SDS-PAGE and western blot analysis are described in great detail in Chapter 11. Here, the authors primarily focus on the assessment of the expression data.

The size of the expressed protein can be estimated based on the amino acids encoded by the reading frame of the cloned cDNA plus the leader sequence, assuming that an average of molecular weight for each amino acid is 100 KDa. However, it does not matter if it is not convenient to estimate the size of the product because, if all goes well, the band of the overexpressed proteins should be much greater than that of the uninduced protein sample; for example, if the protein band of interest is approximately 60 kDa. As shown in Figure 18.3, this band is zero or very weak in the uninduced bacterial sample but is about 50-fold higher in the induced bacterial samples (Figure 18.3A), which can be confirmed by western blotting using specific antibodies against the protein molecules (Figure 18.3B). Should the expression level of the proteins be less than fivefold as compared with control cells, the protein expression can be considered not to be successful.

Following characterization of transformed colonies, the colonies that overexpress foreign proteins should be frozen back in as many vials as possible using sterile glycerol to a final concentration of 15% (v/v) at –80°C.

Extraction of Total Proteins for Purification

In order to obtain native proteins without denaturation, a strong detergent such as SDS, 2-mercaptoethanol, dithiothreitol (DTT) should be avoided. This is particularly important for enzyme assays. However, if the expressed proteins will be used as antigens or labeling without the need of native proteins, the cells can be completely lysed and solublized using 0.1 to 0.3% (w/v) *N*-Lauroylsarcosine (sodium salt) in 45 m*M* Tris-HCl buffer (pH 6.8) containing 0.2 to 0.4 m*M* of phenylmethylsulfonyl fluoride (PMSF). PMSF should be added to the buffer shortly before use. Cells are allowed to lyse for 2 to 4 min at room temperature with occasional agitation by hand.

1. Perform a large-scale protein expression and pelletize the cells as described previously. Completely aspirate the supernatant and resuspend the cells in 5 ml of ice-cold binding buffer per 100 ml of cell culture in a tube.

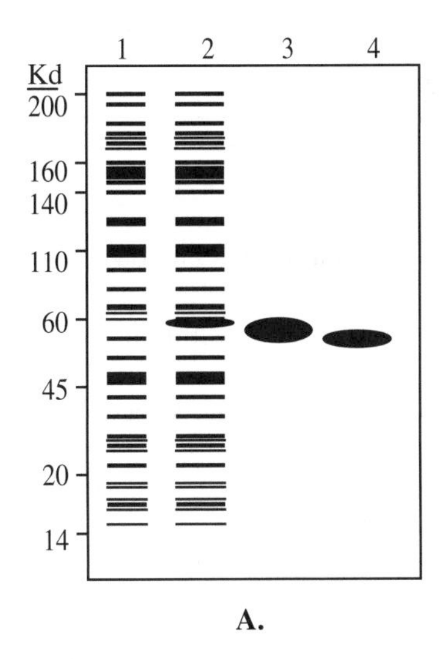

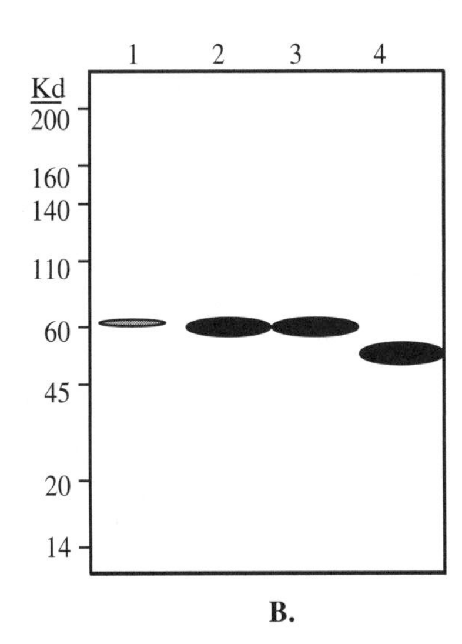

FIGURE 18.3 Analysis of protein expression in uninduced and induced bacteria. (A) Analysis of equal amounts of proteins by SDS-PAGE followed by Coomassie blue staining. (B) Analysis of protein expression by western blotting using monoclonal antibodies against the protein. Lane 1: total proteins from uninduced bacteria. Lane 2: total proteins from IPTG-induced bacteria. Lane 3: purified protein from total proteins in lane 2 with His-tag affinity chromatography. Lane 4: removal of leader sequence from the fusion protein shown in lane 3.

2. Place the tube on ice and sonicate the sample in the tube to break the cells and shear DNA using a microsonicator probe with a power setting at 20 to 40%.

Note: Do not use a high power setting and long sonication time. Otherwise, heat generated may cause denaturation of proteins. In our experience, short sonication bursts (about 5 s per burst) with cooling periods (between 2 to 3 s) are strongly recommended. Carry out sonication until the sample is no longer viscous.

3. Centrifuge the lysate at 15,000 to 20,000 × *g* for 20 min at 4°C to remove debris.
4. Filter the supernatant through a 0.45-μm membrane filter to prevent clogging of the column resin.

Binding Buffer

20 m*M* Tris-HCl, pH 7.9
0.5 *M* NaCl
5 m*M* Imidazole
0.2 m*M* PMSF

Rapid Purification of the Expressed Proteins with His-Tag Affinity Column Chromatography

Affinity columns and His-tag affinity resin are commercially available from Invitrogene and Novagene companies.

1. Prepare binding buffer, charge buffer, washing buffer and elution buffer and filter-sterile.
2. Briefly mix the commercial His-binding or His-bound resin by inverting the bottle to completely suspend the slurry. Transfer an appropriate volume of the slurry into a chromatograph column using a clean, wide-mouth Pasteur pipette or its equivalent.
3. Allow the resin to pack under gravity flow at 4°C for about 1 to 2 h. During packing, remove some liquid from the top and add more resin slurry until approximately 2.5 ml of settled resin has been reached.
4. Drain the liquid from the bottom of the column and wash the column with three volumes (to the settled bed volume) of dd.H_2O.
5. Charge the column with five volumes of charge buffer for 10 min at room temperature and drain away the liquid.
6. Equilibrate the column with three volumes of binding buffer. Close the drain valve at the bottom of the column and loosen the resin in the column using a clean needle. Allow the resin to settle at 4°C for 30 min prior to draining the liquid.
7. In a cold room, load the filtered protein sample onto the top of the column bed with the bottom drain valve closed. Loosen the resin for efficient binding of the expressed proteins to the His-binding resin and allow the binding to proceed for 30 min.

Note: If the proteins are denatured by SDS or mercaptoethanol, the procedures of chromatography can be performed at room temperature.

8. Drain the liquid and reload the collected liquid onto the column two to three times to make use of unbound proteins during the first or second loading.
9. Wash the column with 20 volumes of binding buffer.
10. Wash the column with 15 volumes of washing buffer.
11. Elute the bound proteins with six volumes of elution buffer.

Note: The column can be regenerated and reused according to the manufacturer's instructions. The eluted protein sample contains a high concentration of salts, which may need to be desalted as in step 12.

12. Carry out dialysis to desalt the protein sample as follows:
    a. Transfer the sample into a dialysis bag with a cut-off of molecular weight less than the molecular weight of the fusion proteins.
    b. Seal the bag at both ends using appropriate clips or by making knots.
    c. Place the sample bag in a beaker containing 1 l of dialysis buffer.
    d. Perform dialysis in a cold room with stirring for 1 to 2 days. Change the dialysis buffer every 12 h.

13. Concentrate the protein samples using plastic disposable microconcentrators from Amicon and centrifuge at 1000 rpm at 4°C for an appropriate period of time. An alternative and effective way is to partially freeze-dry the samples or to partially lyophilize in a speed vac system under cold conditions until a minimum volume is reached.
14. Take 20 to 100 µl of the sample to measure the concentration (see Chapter 11). Store the protein sample at –20°C until use.
15. Take 2 to 3 µg protein to analyze the purity by SDS-PAGE or western blotting (Figure 18.3).

Reagents Needed

Binding Buffer

20 m*M* Tris-HCl, pH 7.9
0.5 *M* NaCl
5 m*M* Imidazole
0.2 m*M* PMSF

Charge Buffer

50 m*M* $NiSO_4$

Washing Buffer

20 m*M* Tris-HCl, pH 7.9
0.5 *M* NaCl
60 m*M* Imidazole

Elution Buffer

20 m*M* Tris-HCl, pH 7.9
0.5 *M* NaCl
1 *M* Imidazole

Dialysis Buffer

20 m*M* Tris-HCl, pH 7.5
10 m*M* NaCl

Protease Cleavage of the His-Tag

As indicated in Figure 18.1 and Figure 18.2, each vector contains a sequence encoding a protease cleavage site that allows removal of the His-tag using an appropriate protease, depending on the specific sequence at the cleavage site. The widely used enzymatic or chemical cleavage protocols are described below. It should be noted that removal of the leader sequence is optional. As a matter of fact, the presence of the His-tag does not appear to affect the biological activity of most proteins.

Protocol 1. Enzymatic Cleavage of the Leader Sequence by Factor Xa

If a vector contains a sequence between the His-tag and the MCS region that encodes an amino acid sequence of Ile-Glu/Asp-Gly-Arg, the entire leader sequence including the His-tag and the cleavage sequence can easily be removed by treatment with Factor Xa, a mammalian serine protease that recognizes and cleaves the sequence of Ile-Glu/Asp-Gly-Arg↓.

1. **Determination of the optimum digestion conditions:**
   a. Prepare Factor Xa buffer (see Reagents Needed).
   b. Set up two pilot cleavage reactions in two microcentrifuge tubes as follows:
      Tube A:
      Fusion proteins, 30 µg in 5 to 10 µl
      10X digestion buffer, 2.5 µl
      Add dd.H2O to 29 µl.
      Factor Xa, 0.2 µg in 1 µl
      Total volume = 30 µl
      Tube B:
      Fusion proteins, 30 µg in 5 to 10 µl
      10X digestion buffer, 2.5 µl
      Add dd.H2O to 29 µl
      Factor Xa, 0.00
      Total volume = 30 µl
      Note: Tube B will serve as a control.
   c. Incubate the reactions at 4 to 37°C. Typically, the reactions can be incubated at room temperature.
   d. Transfer 5 µl aliquots from the reaction mixture into microcentrifuge tubes at each of the following time points: 6, 12, 18, 24 and 36 h following initiation of the reaction. Add 5 µl of 2X SDS sample loading buffer and place at –20°C until SDS-PAGE.
   e. Carry out the time-course analysis of protease digested samples at step (d) by SDS-PAGE followed by Coomassie blue staining. The detailed protocols are described in Chapter 11. From the stained gel, identify the optimal time-point for large-scale digestion of fusion proteins.

Reagents Needed

10X Factor Xa buffer

500 m*M* Tris-HCl, pH 8.0
1 *M* NaCl
10 m*M* $CaCl_2$

2. **Large-scale cleavage of fusion proteins.** Once the optimum incubation time is established, large-scale protease digestions can be carried out using 1 to 10 mg of fusion proteins under the optimum conditions. If all goes

well, the molecular weight of the cleaved proteins should be less than that of undigested proteins (Figure 18.3).

3. **Purification of the cleaved proteins with His-binding affinity resin.** To remove the cleaved leader sequence, the digested proteins can be readily purified by His-binding affinity matrix.
   a. In a 1.5-ml microcentrifuge tube, charge 100 to 200 µl of slurry of the His-binding resin with an equal volume of charge buffer ($NiSO_4$) and suspend it to form a good suspension.
   b. Spin down the resin at 1000 rpm for 1 min and aspirate the supernatant.
   c. Resuspend the resin in 1 ml of binding buffer. Spin down and aspirate as in the previous step.
   d. Resuspend the resin in 50 µl of binding buffer and add 100 µl of cleaved proteins to the resin. Mix well and incubate for 20 min at room temperature.
   e. Spin down at 1200 rpm for 1 min to pelletize the resin and transfer the supernatant into a fresh tube.
      Principle: The cleaved His-tag leader sequence binds to the affinity resin but the digested proteins remain in the supernatant.
   f. Analyze part of the supernatant by SDS-PAGE.
4. **Direct cleavage and purification of matrix-bound proteins in the affinity column.** The His-tag leader sequence can be cleaved for the fusion proteins after the proteins bind to the His-binding resin. Below is the protocol for small-scale simultaneous purification and digestion of His-tag in a microcentrifuge tube. This procedure, however, can be scaled up in an affinity column as desired.
   a. In a 1.5-ml microcentrifuge tube, charge 200 µl of slurry of His-binding resin with an equal volume of charge buffer ($NiSO_4$) and suspend it to form a good suspension.
   b. Spin down the resin at 1000 rpm for 1 min and aspirate the supernatant.
   c. Resuspend the resin in 1 ml of binding buffer and allow the resin to equilibrate for 30 min at room temperature. Spin down and aspirate as in the previous step.
   d. Resuspend the resin in 50 µl of binding buffer and add 100 µl of fusion protein sample to the resin. Mix well and incubate for 30 min at room temperature. Mix several times during the incubation period.
   e. Spin down at 1000 rpm for 1 min to pelletize the resin and remove the supernatant.
   f. Wash and spin down the resin with 1 ml of binding buffer twice and then 1 ml of washing buffer twice to eliminate nonspecific binding proteins from the resin.
   g. Add 200 µl of protease digestion cocktail to the resin, suspend it thoroughly and incubate the reaction at room temperature for 12 to 24 h.
   h. Spin down at 1200 rpm for 1 min to pelletize the resin and transfer the supernatant to a fresh tube.
      Principle: The cleaved His-tag leader sequence binds to the affinity resin but the digested proteins are eluted in the supernatant.

i. If necessary, concentrate the protein samples via partially freeze-drying the samples or partially lyophilizing in a speed vac system under cold conditions until a minimum volume (40 to 50 μl) is reached.
j. Analyze part of the supernatant by SDS-PAGE.

Reagents Needed

Binding Buffer

20 m*M* Tris-HCl, pH 7.9
0.5 *M* NaCl
5 m*M* Imidazole

Charge Buffer

50 m*M* $NiSO_4$

Washing Buffer

20 m*M* Tris-HCl, pH 7.9
0.5 *M* NaCl
60 m*M* Imidazole

10X Factor Xa Buffer

500 m*M* Tris-HCl, pH 8.0
1 *M* NaCl
10 m*M* $CaCl_2$

Protease Digestion Cocktail

10X Digestion buffer, 20 μl
Appropriate protease, 0.8 to 1.2 μg
Add dd.H_2O to a total volume of 200 μl.

Protocol 2. Enzymatic Cleavage of the Leader Sequence with Thrombin

If the vector contains a sequence between the His-tag and the MCS region that encodes an amino acid sequence of Leu-Val-Pro-Arg-Gly-Ser, the leader sequence (the His-tag and majority of the cleavage sequence) can be removed by thrombin with ease. Thrombin is a mammalian serine protease that recognizes and cuts the sequence between Arg and Gly: Leu-Val-Pro-Arg↓-Gly-Ser. The procedures for digestion are quite similar to those described for Factor Xa cleavage except for the use of 10X thrombin buffer.

10X Thrombin Buffer

0.5 *M* Tris-HCl, pH 7.5
1.5 *M* NaCl
25 m*M* $CaCl_2$

Protocol 3. Chemical Cleavage of Fusion Proteins with Cyanogen Bromide

If only one *M*et residue exists in each protein and the *M*et is not important to biological activity of the protein, the leader sequence fusion proteins can easily be cleaved with cyanogen bromide (CNBr). CNBr cleaves the C-terminal side of methionine residues at low pH in 70% formic acid. The general procedures are similar to those described for enzymatic digestion, including the small pilot trail reaction and the large-scale reaction. The chemical is dissolved in 70% formic acid at a concentration of 50 mg/ml. However, the CNBr can be replaced with 6 *M* guanidine-HCl/0.2 *M* HCl.

PART B. OVEREXPRESSION OF SPECIFIC PROTEINS IN EUKARYOTIC SYSTEMS

Unlike the prokaryotic systems described earlier, eukaryotic systems make it possible for the expressed proteins to undergo posttranslational modifications. Several vector systems have been developed for protein expression in eukaryotic cells, including baculovirus, mammalian and yeast systems. Due to the fact that baculovirus and mammalian systems are relatively complicated, time consuming and costly, the present chapter primarily focuses on the introduction of the yeast system, *Pichia pastoris*, which is widely used and commercially available from Invitrogen. As compared to other eukaryotic systems, it is easier to manipulate the yeast genome and to culture yeast without the need of a CO_2 incubator. The best known yeast strain is *Saccharomyces cerevisiae*. *P. pastoris* has many of the common features that *S. cerevisiae* has.

The primary principles underlying the use of *P. pastoris* as a protein expression system begin with its genome containing a gene, called *AOX1*. This gene encodes alcohol oxidase, which is a key enzyme that enables the yeast to metabolize methanol as its sole carbon source. DNA recombination technology makes it possible to manipulate the *AOX1* gene and use its promoter to drive the expression of foreign protein at a very high level.[4–6] As diagrammed in Figure 18.4, the coding region of the *AOX1* gene is deleted and its promoter (5′AOX1) and 3′ portion (3′AOX1) are utilized for the plasmid-based expression vector. The functions of each of the components of the vector are described next.

1. **A strong promoter (5′AOX1).** It facilitates DNA construct integration into the *AOX1* locus of the *Pichia* genome and drives the expression of foreign proteins.
2. **A strong ribosome-binding site (RBS).** A bacteriophage RBS consisting of AAGGAG located very close upstream to the first codon ATG will provide a high efficiency of translational initiation. The small subunits of ribosomes will selectively recognize and bind to the sequence, initiating the translation of the mRNA. With this sequence, the translational rate can be enhanced up to 10-fold.

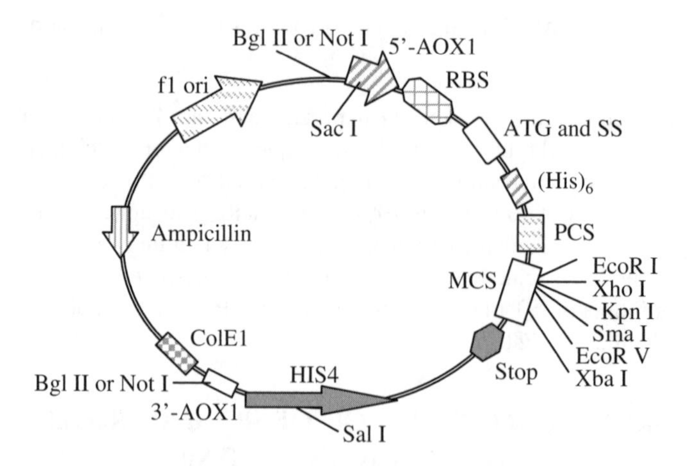

FIGURE 18.4 A diagram of a eukaryotic expression vector for large-scale expression and purification of specific proteins in the yeast strain *Pichia pastoris*.

3. **The ATG codon**. ATG (AUG in mRNA) immediately follows the RBS sequence. This will be the first codon for the recombinant protein. The purpose of including this first codon is due to the very next leader sequence, including a secretory sequence, His-tag and a cleavage sequence. Because the first codon is fixed in the vector, a cDNA or gene to be cloned should be carefully oriented so that absolutely no stop codon and no reading frame shifts are allowed. Detailed cautions will be noted in cloning of the insert into the vector.
4. **The signal sequence (SS).** This DNA sequence will code for an N-terminal leader sequence (e.g., α-factor signal sequence), which will serve to secrete the expressed proteins into culture medium. As a result, purification of the proteins will be significantly simplified.
5. **His-tag.** The His-tag sequence located immediately following the SS will code for 6 to 10 consecutive histidine residues. These His-residues will be a part of the N-terminal leader sequence of a protein to be expressed from the cloned cDNA or gene and will serve as the affinity tag for rapid purification of the protein with metal chelation chromatography using an appropriate His-binding resin.
6. **PCS.** A sequence encoding a protease cleavage site (PCS) is designed between the His-tag and the multiple cloning site (MCS). The amino acid sequence encoded by this sequence will be the target site for an appropriate protease (e.g., thrombin or Factor Xa). This will allow removal of the N-terminal His-tag leader sequence *in vitro* as desired. It should be noted here that noncleaved fusion proteins will be acceptable for use as antigens. However, for biochemical, cell biology or enzyme assays, the leader sequence may need to be removed because a native protein is better than a fusion protein.

7. **Multiple cloning site (MCS).** The MCS contains different restriction enzyme sites for insertion of a specific cDNA for expression of a specific protein. Usually, commercial vectors contain overlapping bases for adjacent restriction enzymes. Perhaps it is for ease of use that all three reading-frame vector series have been developed. This strategy will allow cloning of a cDNA to be expressed at different restriction enzyme sites without causing reading-frame shifts.
8. **Terminator.** A transcriptional terminator sequence [splicing and poly(A) site] follows immediately downstream from the MCS.
9. **HIS4 gene**. This gene codes for histidinol dehydrogenase that will serve as a selectable marker for isolation of recombinant *Pichia* colonies.
10. **3′AOX1**. This 3′ portion of the *AOX1* gene will facilitate plasmid construct integration at the *AOX1* locus.
11. **Plasmid backbone (Ampicillin gene, F1 ori and ColE1).** A plasmid-based expression vector should contain the plasmid backbone elements. The ampicillin-resistant gene or *Amp*^{r} will serve as an antibiotic marker for selection of positive transformed colonies. In theory, only bacteria carrying an *Amp*^{r} gene in the introduced plasmid DNA can be resistant to ampicillin killing action. The F1 ori sequence will allow rescue of single-stranded plasmid DNA by infection of F′-containing host strains with helper phage (e.g., R408). The ColE1 element makes it possible to replicate the introduced plasmid DNA in a copy number as high as possible.

Because of the unique restriction enzyme sites indicated in the figure, if recombinant plasmid DNA is digested with a unique restriction enzyme such as *Bgl* II or *Not* I, the linearized DNA can undergo gene replacement by DNA homologous recombination at the 5′-AOX1 and 3′-AOX1 sites (Figure 18.5). When that occurs, the expression cassette will integrate and replace the endogenous *AOX1* gene, generating a mutant *AOX1* gene. As a result, most of the alcohol oxidase activity of the yeast strain will be lost, producing a strain that is phenotypically methanol utilization slow or Mut^s. Therefore, the cells will grow slowly in a methanol medium. This can serve as an indicator of transformation for integration of a foreign gene. On the other hand, if the recombinant plasmid DNA is cut with a unique enzyme such as *Sac* I or *Sal* I, it will make it possible for the linearized DNA to integrate the 5′-AOX1 or His4 site into the yeast genome. Single or multiple copies of insertion may occur. The entire project is outlined in Figure 18.6.

Cloning a cDNA to Be Expressed into Expression Vectors

Invitrogen (San Diego, CA) offers excellent plasmid-based *Pichia* expression vectors, including pHIL-D2, pHIL-S1 and pPIC9. Selection of a specific vector depends on intracellular or secretory protein expression. With permission from Invitrogen, vectors may be modified or altered as diagrammed in Figure 18.4, which is suitable for rapid cleavage and purification of the proteins of interest. Detailed protocols

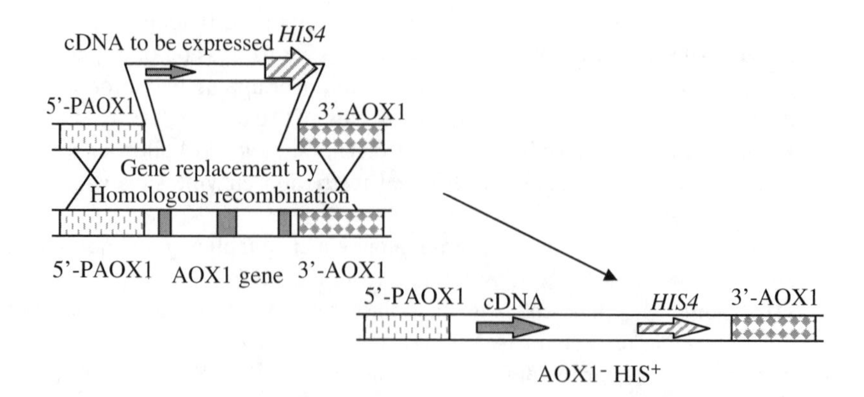

FIGURE 18.5 Replacement of *AOX1* gene with a cDNA expression cassette in *P. pastoris.*

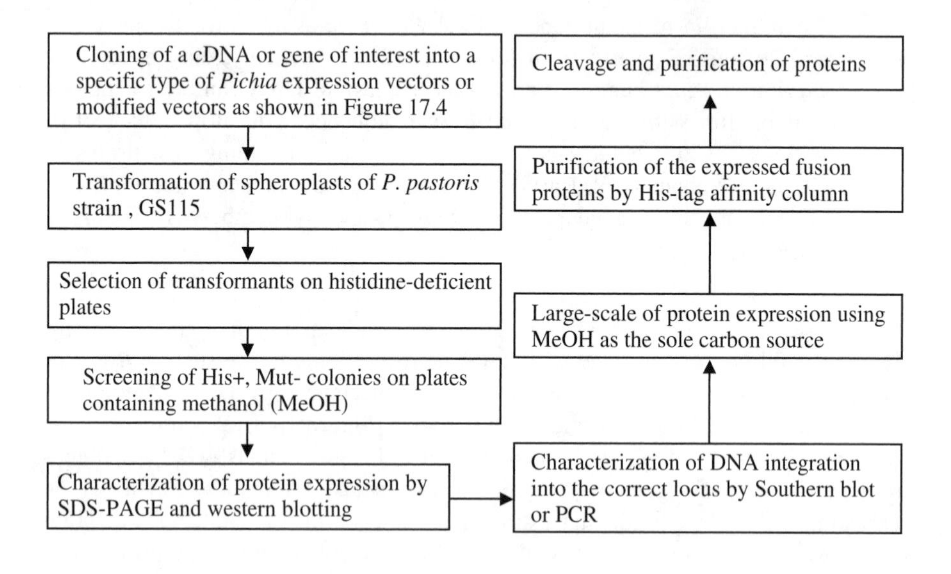

FIGURE 18.6 Flowchart of protein overexpression in yeast.

concerning DNA digestion, preparations of vector and insert DNA, ligation, transformation, selection and purification of plasmid DNA are described in Chapter 4. This section focuses primarily on some special precautions that should be taken when cloning a cDNA for expression.

1. Restriction enzymes used for DNA digestion should be unique so as to cut outside the entire coding region of the cDNA to be cloned at the MCS of the vector. To ensure success in protein expression, no enzyme cuts are allowed inside the coding region of the cDNA or outside the MCS region

of the vector DNA. Detailed restriction enzyme mapping should be carried out when preparing vector and insert DNA.

2. Carefully trim out as much as possible the 5′-UTR and 3′-UTR regions of the cDNA to be expressed using appropriate restriction enzymes or the PCR approaches. In our experience, long 5′-UTR or 3′-UTR may alter the reading frame and decrease the efficiency of protein expression.
3. Due to the orientation of the driving promoter 5′-AOX1 and the first ATG codon, which are already fixed in place in the vector, it is necessary to insert the poly(T) strand instead of the poly(A) of the cDNA in $3' \rightarrow 5'$ orientation downstream from the driving promoter. In other words, the cDNA to be expressed should be cloned in its sense orientation under the control of the driving promoter. No antisense cloning is allowed.
4. Because the first codon ATG is fixed in place in the vector, the cDNA to be expressed must be carefully oriented so that absolutely no stop codon and no reading-frame shifts are allowed. To make sure that the cDNA has been cloned at the right position with the correct reading frame starting from ATG in the vector, it is strongly recommended that the DNA constructs obtained be carefully analyzed by appropriate restriction enzyme digestion pattern analysis and by DNA sequencing of the 5′ junction site. However, the entire cDNA or construct does not need to be sequenced. In the authors' experience, as long as the 5′ portion of the cloned cDNA has the correct sequence and the correct reading frame with regard to the ATG codon and the leading sequence, the constructs should be fine without sequencing the whole cDNA.

Transformation of Yeast *Pichia* with Expression DNA Constructs

Protocol A. Linearization of Plasmid DNA

Linearization of plasmid constructs depends on particular types of transformants to be selected. We recommend that both types of HIS$^+$Mut^s and HIS$^+$Mut$^+$ GS115 transformants be isolated. As shown in Figure 18.4, to obtain HIS$^+$Mut^s, plasmid DNA should be digested with the unique enzyme, *Not* I or *Bgl* II, assuming that the cloned cDNA does not contain a *Not* I or *Bgl* II site. The linearized DNA will undergo *AOX1* gene replacement, resulting in an HIS$^+$Mut^s genotype of transformants (Figure 18.5). In this way, the HIS$^+$Mut^s phenotype of transformants will be isolated easily using His and MeOH media. *AOX1* gene replacement by DNA homologous recombination will generate one or two copies of the integrated gene to be expressed. To increase the integration copy numbers of plasmid constructs, it is recommended that the DNA be linearized with *Sal* I or *Sac* I (Figure 18.4). The merit of doing this is that the linearized DNA constructs may undergo random or site-specific (5′-AOX1 or HIS4 site) integrations, resulting in one or more copies of the cassette in the genome of *Pichia* GS115. The HIS$^+$Mut$^+$ transformants can be selected with a medium containing the His residue. Compared with gene replacement, multiple copies of integration will likely enhance the expression of foreign proteins at a high level.

Therefore, we recommend linearizing the same plasmid DNA constructs in three Eppendorf tubes as follows:

1. Linearize 15 to 25 μg DNA with *Not* I or *Bgl* II.
2. Digest 15 to 25 μg DNA with *Sal* I.
3. Linearize 15 to 25 μg DNA with *Sac* I.

Following digestion and purification of the DNA, pool the second and the third samples together, generating a DNA construct mixture that is linearized with *Sal* I or *Sac* I. The DNA mixture and the DNA linearized with *Not* I or *Bgl* II can then be separately transferred into yeast GS115. As a result, HIS^+Mut^s and multiple copies of HIS^+Mut^+ transformants can be isolated, respectively. Detailed protocols for restriction enzyme digestion, electrophoresis and DNA purification are described in Chapter 4.

Protocol B. Transformation of *Pichia* by Electroporation

Compared with other techniques, electroporation is the fastest method for transforming yeast with high efficiency of transformants; it is not necessary to prepare spheroplasts. In spite of the fact that up to 50% of cells may be killed by the high electric pulse, high efficiency of transformants will overcome such a minor drawback.

1. Culture 4 ml of *P. pastoris* in YPD medium at 30°C overnight with shaking at 250 rpm.
2. Inoculate 0.4 ml of the culture into 500 ml of YPD medium. Grow the cells at 30°C overnight with shaking at 250 rpm. The OD_{600} will be 1.2 to 1.5.
3. Centrifuge at 1600 × *g* for 5 min at 4°C, aspirate the supernatant and resuspend the cells in 250 ml ice-cold sterile water.
4. Centrifuge and remove the supernatant as in the previous step and resuspend the cells in 250 ml ice-cold sterile water.
5. Centrifuge and remove the supernatant as in step 3 and resuspend the cells in 20 ml ice-cold 1 *M* sorbitol.
6. Centrifuge and remove the supernatant as in step 3 and resuspend the cells in 1.5 ml ice-cold 1 *M* sorbitol.

Note: The extensive washing of the cells will minimize the salt concentration and reduce the killing percentage during electroporation.

7. Add 80 μl of the cells and 15 to 25 μg linearized DNA to a 0.2-cm electroporation cuvette on ice. Incubate the mixture on ice for 5 min.
8. Carry out electroporation according to the manufacturer's instructions. If the Gene Pulser from Bio-Rad is used, voltage, capacitance (μF) and resistance (Ω) may be 1500, 25 and 200, respectively. If an Electroporator II unit from Invitrogen is utilized, the three parameters should be 1500, 50 and 200, respectively, which is suitable for yeast electroporation.

9. Quickly add 1 ml of sorbitol solution to the cuvette to recover the cells and spread 0.2 to 0.5 ml of the cell suspension onto each 100-mm RD plate.
10. Invert the plates and incubate them at 30°C for 3 to 6 days. Colonies should be visible after 3 to 4 days' incubation. Proceed to phenotype screening.

Reagents Needed

20% Dextrose Solution

Dissolve 20 g dextrose (D-glucose) into a final volume of 100 ml dd.H_2O and filter-sterilize.

YPD (Yeast Peptone Dextrose) or YEPD (Yeast Extract Peptone Dextrose) Medium (for 1 l)

20 g Peptone
10 g Yeast extract
Dissolve in 800 ml of dd.H_2O and autoclave for 20 min in liquid cycle. Cool to approximately 50°C, add dextrose stock solution to the final concentration of 2% (w/v) and add sterile water to 1000 ml. Store at room temperature or 4°C until use.

1 M Sorbitol in Distilled Water

RD Plates (for 1 l)

Dissolve 186 g sorbitol and 20 g agar in 700 ml distilled water and autoclave for 20 min in liquid cycle. Allow it to cool to approximately 60°C with slow stirring and then place it in a 55°C water bath prior to use. Prepare a mixture containing 100 ml of 10X D, 100 ml of 10X YNB, 2 ml 500X B, 10 ml of 100X AA and 78 ml of sterile water. Warm the mixture to about 45°C and add it to the sorbitol solution. Pour the mixture into plates (approximately 30 ml/100 mm plate) in a laminar flow hood and allow the plates to evaporate the excess moisture at room temperature for a couple of days. Store the plates at 4°C until use.

RDH Plates (for 1 l)

Dissolve 186 g sorbitol and 20 g agar in 700 ml distilled water and autoclave for 20 min in liquid cycle. Allow it to cool to approximately 60°C with slow stirring and then place it in a 55°C water bath prior to use. Prepare a mixture containing 100 ml of 10X D, 100 ml of 10X YNB, 2 ml 500X B, 10 ml of 100X AA, 10 ml of 100X histidine and 68 ml of sterile water. Warm the mixture to about 45°C and add it to the sorbitol solution. Mix it and pour the mixture into plates (approximately 30 ml/100 mm plate) in a laminar flow hood and allow the plates to evaporate the excess moisture at room temperature for a couple of days. Store the plates at 4°C until use.

Protocol C. Preparation of Spheroplasts for Transformation

If an electroporator is not available, transformation of yeast can be carried out with spheroplasts. Compared to electroporation in Protocol B, transformation of spheroplasts is much more complicated, but it is a well-established method and is widely used.

1. Prepare yeast culture media and plates (see Reagents Needed).
2. Streak *P. pastoris* strain GS115 onto a YPD plate as shown in Figure 18.7.
3. Inoculate a single colony into 10 ml of YPD medium in a flask and culture the cells overnight at 28 to 30°C with shaking at 250 rpm. Store the cells at 4°C until use.
4. Inoculate 20 μl of the cell culture at step 3 into 200 ml of YPD medium in a 500-ml culture flask. Grow the cells at 28 to 30°C with shaking at 250 rpm. Six hours later, check OD_{600} once every hour to monitor the density of the cells. Store the cells at 4°C until use.

Note: An OD_{600} of 0.2 to 0.3 will result in high efficiency of transformation.

5. Centrifuge at 1600 × *g* at room temperature for 8 min and aspirate the supernatant.
6. Resuspend the cells in 25 ml of dd.H_2O and pellet the cells as in the preceding step. Repeat this step one more time.
7. Wash the cells by resuspending them in 25 ml of SED and pelletize the cells as described in step 5.
8. Resuspend the cells in 25 ml of 1 *M* sorbitol and centrifuge as described in step 5.
9. Resuspend the cells by swirling in 20 ml of SCE buffer and aliquot the cells into two tubes, A and B. Place the samples at room temperature.
10. Carry out a time course of Zymolyase digestion to determine the optimum conditions.
    a. Thaw the Zymolyase on ice and mix the slurry mixture well.
    b. Set a UV-Vis spectrophotometer at 800 nm and blank it with a reference tube containing 200 μl of SCE buffer and 800 μl of 5% SDS solution.

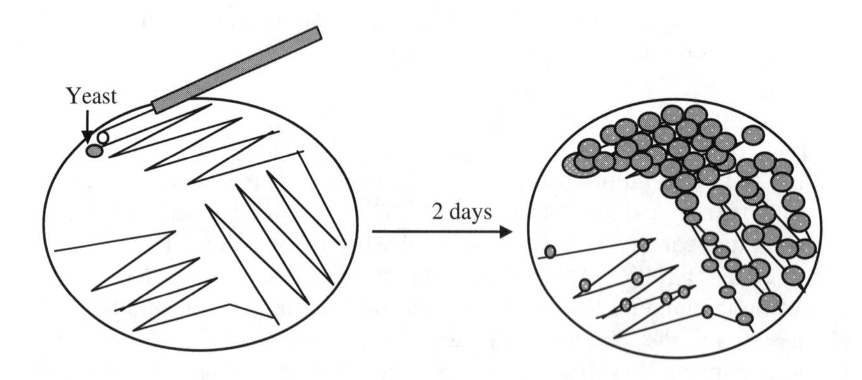

FIGURE 18.7 Streaking of yeast stock in a zig-zag pattern on a YPD plate. Two days later, well-isolated colonies will be generated in the last streaking area on the plate.

c. Label 11 microcentrifuge tubes as follows: 0, 5, 10, 15, 20, 25, 30, 35, 40, 45, and 50. To each tube, add 800 µl of 5% SDS solution.
d. Transfer 200 µl of cell suspension from tube A (step 9 above) into the tube labeled "0." Place the tube on ice until use. This tube serves as the zero time point.
e. Quickly mix Zymolyase, add 7.5 µl of it to tube A, and gently mix. Incubate the digestion at 30°C without shaking because the Zymolyase-digested cells are very weak.
f. Every 5 min, transfer 200 µl of the cells from tube A into an appropriate tube labeled at step (c). Place the tubes on ice until use.
g. Measure the OD_{600} for all samples and determine the percentage of spheroplasting for each time point using the equation below:

$$\% \text{ spheroplasting} = 100 - [(OD_{600} \text{ at time t}/OD_{600} \text{ zero time point}) \times 100]$$

For instance, if t = 0, OD_{600} is 0.2; if t = 20 min, the OD_{600} is 0.04. Then, % spheroplasting = 100 – [(0.04/0.2) × 100] = 100 – 20 = 80%. In general, 70 to 80% spheroplasting is considered an optimum digestion condition.

11. Quickly mix Zymolyase, add 7.5 µl of it to tube B at step 9 and gently mix it. Incubate the digestion at 30°C without shaking for the optimum incubation time established earlier.
12. Centrifuge at 800 × *g* for 10 min at room temperature and carefully aspirate the supernatant.
13. Gently resuspend the cells in 10 ml of 1 *M* sorbitol solution by gently tapping the tube instead of vortexing. Centrifuge and remove the supernatant as described in the previous step.

Note: Sorbitol will help the spheroplasts regenerate cell walls after transformation.

14. Wash the cells by resuspending them in 10 ml of CaS buffer and carry out centrifugation as in step 12.
15. Gently resuspend the spheroplasts in 500 to 600 µl of CaS buffer and use the spheroplasts within 30 min for high efficiency of transformation.

Reagents Needed

20% Dextrose Solution

20 g Dextrose (glucose) in final 100 ml dd.H_2O and filter-sterile

YPD (Yeast Peptone Dextrose) or YEPD (Yeast Extract Peptone Dextrose) Medium (for 1 l)

20 g Peptone
10 g Yeast extract
Dissolve in 800 ml of dd.H_2O and autoclave for 20 min in liquid cycle. Cool to approximately 50°C, add dextrose stock solution to the final concentration

of 2% (w/v) and sterile water to 1000 ml. Store at room temperature or 4°C until use.

YPD (Yeast Peptone Dextrose) or YEPD (Yeast Extract Peptone Dextrose) Plates (for 1 l)

20 g Peptone
10 g Yeast extract
20 g Difco agar
Dissolve in 800 ml of dd.H_2O and autoclave for 20 min in liquid cycle. Cool to approximately 50°C with slow stirring and add dextrose stock solution to a final concentration of 2% (w/v). Add sterile water to 1000 ml. Pour plates (approximately 30 ml/100-mm plate) in a laminar flow hood and allow the plates to evaporate the excess moisture at room temperature for a couple of days. Store the plates at 4°C until use.

The following buffers or solutions should be autoclaved or sterile-filtered.

1 M Sorbitol in distilled water

0.5 M EDTA Stock

0.5 *M* EDTA in 900 ml distilled water and warm the solution to dissolve EDTA completely. Allow it to cool to approximately 40°C and begin to adjust the pH to 8.0 while the solution is cooling to <30°C. Add sterile water to 1000 ml. Store at room temperature until use.

1 M Tris-HCl stock, pH 7.5

1 *M* Tris in 900 ml distilled water and adjust the pH to 7.5. Add sterile water to 1000 ml. Autoclave and store at room temperature until use.

0.5 M Sodium Citrate Stock, pH 5.8

1 *M* Tris in 900 ml distilled water and adjust the pH to 5.8. Add sterile water to 1000 ml. Store at room temperature until use.

SE Buffer

1 *M* Sorbitol
25 m*M* EDTA (from stock)

SCE Buffer

1 *M* Sorbitol
1 m*M* EDTA (from stock)
10 m*M* Sodium citrate (from stock)

CaS Buffer

1 *M* Sorbitol
10 m*M* Tris-HCl (from stock)
10 m*M* $CaCl_2$

Zymolyase

3 mg/ml in distilled water. Store at –20°C prior to use.

Protocol D. Transformation of Spheroplasts with Linearized DNA

1. Prepare culture media and plates (see Reagents Needed). This should be done several days prior to transformation.
2. Transfer 0.1 ml of the spheroplasts prepared in Protocol C into a sterile 10- or 15-ml Falcon plastic tube with a cap and add 15 to 25 μg linearized DNA constructs to spheroplast suspension. Gently mix and incubate at room temperature for 10 to 15 min.
3. During the incubation, prepare 1 ml of fresh PEG-CaT solution for each transformation with 1:1 (v/v) of 40% PEG:CaT solution.
4. When the incubation is complete, add 1 ml of fresh PEG-CaT solution to the mixture, gently mix and continue to incubate at room temperature for 10 to 15 min.
5. Centrifuge at 800 × *g* for 10 min at room termperature; aspirate the supernatant carefully and completely.
6. Gently resuspend the pellet in 150 μl of SOS medium and incubate it at room temperature for 20 to 25 min.
7. Add 0.85 ml of 1 *M* sorbitol solution to mixture.
8. Melt premade RD agarose using a microwave oven, cool it to approximately 60°C and then place it in a 55°C water bath prior to use.

Tip: Aliquot the top agarose medium into sterile tubes (5 ml each) and place the tubes in a water bath.

9. Add 0.1 ml of the spheroplasts–DNA mixture obtained at step 6 to 5 ml of molten RD agarose, briefly mix it and pour on a 100-mm RD plate. Quickly and carefully overlay the entire surface of the plate with the mixture. Allow the agarose to harden at room temperature for about 4 min. The surface should be very flat and without clumps. Repeat the procedure until all plates are complete.
10. Add 0.1 ml of spheroplasts without DNA constructs to 5 ml of molten RD agarose and pour on a plate.

This will serve as a negative control.

11. Invert the plates to avoid potential moisture drops and incubate the plates at 28 to 30°C for 4 to 7 days. Proceed to phenotype screening.

Transformant colonies should be visible after 4 days of incubation. However, no colonies should be formed in the negative control plate.

Reagents Needed

40% PEG

40% (w/v) PEG in distilled water

CaT Buffer

20 m*M* Tris-HCl, pH 7.5 (from stock) and 20 m*M* $CaCl_2$

SOS Solution

1 *M* Sorbitol, 0.3X YPD and 10 m*M* $CaCl_2$

10X D

200 g Dextrose or D-glucose in a final volume of 1000 ml water. Filter-sterilize it and store at room temperature.

10X YNB (for 1 l)

134 g Yeast nitrogen base (YNB) with ammonium sulfate but no amino acids in a final volume of 1 l. Briefly warm it to dissolve the chemicals and filter-sterilize. Store at 4°C.

100X Histidine

0.4 g L-histidine in 100 ml water. Briefly warm to dissolve the chemical and sterilize the solution. Store at 4°C.

100X AA

0.5 g Each of L-glutamic acid, L-methionine, L-lysine, L-leucine, and L-isoleucine in 100 ml water. Briefly warm to dissolve the chemicals and filter-sterilize. Store at 4°C.

500X B

20 mg Biotin in 100 ml water. Sterilize and store at 4°C.

RD (Regeneration Dextrose) Medium (for 1 l)

Dissolve 186 g sorbitol in 700 ml distilled water and autoclave for 20 min in liquid cycle. Allow it to cool to approximately 60°C with slow stirring and then place in a 55°C water bath prior to use. Make a mixture containing 100 ml of 10X D, 100 ml of 10X YNB, 2 ml 500X B, 10 ml of 100X AA and 88 ml of sterile water. Warm the mixture to about 45°C and add it to the sorbitol solution.

RDH (Regeneration Dextrose and Histidine) Medium (for 1 l)

Dissolve 186 g sorbitol in 700 ml distilled water and autoclave for 20 min in liquid cycle. Allow it to cool to approximately 60°C with slow stirring and then place it in a 55°C water bath prior to use. Prepare a mixture containing 100 ml of 10X D, 100 ml of 10X YNB, 2 ml 500X B, 10 ml of 100X AA, 10 ml of 100X histidine and 78 ml of sterile water. Warm the mixture to about 45°C and add it to the sorbitol solution.

RD Plates (for 1 l)

Dissolve 186 g sorbitol and 20 g agar in 700 ml distilled water and autoclave for 20 min in liquid cycle. Allow it to cool to approximately 60°C with slow stirring and then place in a 55°C water bath prior to use. Prepare a mixture containing 100 ml of 10X D, 100 ml of 10X YNB, 2 ml 500X B, 10 ml of 100X AA and 78 ml of sterile water. Warm the mixture to about

45°C and add it to the sorbitol solution. Pour the mixture into plates (approximately 30 ml/100-mm plate) in a laminar flow hood and allow the plates to evaporate the excess moisture at room temperature for a couple of days. Store the plates at 4°C until use.

RDH Plates (for 1 l)

Dissolve 186 g sorbitol and 20 g agar in 700 ml distilled water and autoclave for 20 min in liquid cycle. Allow it to cool to approximately 60°C with slow stirring and then place in a 55°C water bath prior to use. Prepare a mixture containing 100 ml of 10X D, 100 ml of 10X YNB, 2 ml 500X B, 10 ml of 100X AA, 10 ml of 100X histidine and 68 ml of sterile water. Warm the mixture to about 45°C and add it to the sorbitol solution. Mix it and pour the mixture into plates (approximately 30 ml/100-mm plate) in a laminar flow hood and allow the plates to evaporate the excess moisture at room temperature for a couple of days. Store the plates at 4°C until use.

RD Top Agarose (500 ml)

Dissolve 93 g sorbitol and 10 g agar or agarose in 350 ml distilled water and autoclave for 20 min in liquid cycle. Allow it to cool to approximately 60°C with slow stirring and then place in a 55°C water bath prior to use. Prepare a mixture containing 50 ml of 10X D, 50 ml of 10X YNB, 1 ml 500X B, 5 ml of 100X AA and 49 ml of sterile water. Warm the mixture to about 45°C and add it to the sorbitol solution. The agarose mixture can be immediately used or stored at 4°C. Premelt it in a microwave and cool to approximately 45°C prior to use.

RDH Top Agarose (500 ml)

Dissolve 93 g sorbitol and 10 g agar or agarose in 350 ml distilled water and autoclave for 20 min in liquid cycle. Allow it to cool to approximately 60°C with slow stirring and then place in a 55°C water bath prior to use. Prepare a mixture containing 50 ml of 10X D, 50 ml of 10X YNB, 1 ml 500X B, 5 ml of 100X AA, 5 ml of 100X histidine and 39 ml of sterile water. Warm the mixture to about 45°C and add it to the sorbitol solution. The agarose mixture can be immediately used or stored at 4°C. Premelt it in a microwave and cool to approximately 45°C prior to use.

SCREENING AND ISOLATION OF HIS$^+$MUT$^+$ AND HIS$^+$MUT^S PHENOTYPE

To better understand the principle of selection or screening, it is necessary to go back to the expression cassette shown in Figure 18.4. If plasmid constructs are linearized with *Not* I or *Bgl* II and introduced into *Pichia* GS115, *AOX1* gene replacement is expected to take place, resulting in His$^+$Mut^s genotype transformants. Transformant colonies will display if the transformed cells are plated onto a plate lacking histidine. In other words, if all goes well, only His$^+$ phenotype transformants will grow. Additionally, if His$^+$ phenotype transformants are inoculated on a plate containing methanol, they will grow slowly (Mut^s phenotype), compared with wild-type cells, because at least one copy of the *AOX1* gene coding for alcohol oxidase has been replaced by the protein expression cassette. Due to the decrease in alcohol oxidase activity, enzyme is not sufficient to metabolize the methanol as the sole carbon source. This provides an efficient way to distinguish positive transformants (His^+Mut^s phenotype).

If the plasmid constructs are linearized with *Sal* I or *Sac* I and then transferred into the yeast, *AOX1* gene replacement will most likely not occur. Instead, multiple copies of integration may take place. If all goes well, phenotype of His^+Mut^+ transformants will be obtained. Therefore, they can grow on plates without histidine and are capable of metabolizing methanol as with wild-type GS115. The screening procedure is as follows:

1. Pick an His^+ colony using a sterile toothpick and streak it in a single spot, first on an MM plate and then an MD plate. Use a fresh toothpick and repeat streaking until 66 to 100 colonies have been inoculated. Mark and record the position for each individual transformant on the plates. Meanwhile, wild-type GS115 colonies should be streaked in the same way as for the control.
2. Invert the plates and incubate at 30°C for 2 to 4 days.
3. Score His^+Mut^s colonies that grow normally on MD plates but very slowly on MM plates, compared to wild-type GS115 cells (Figure 18.8). On the other hand, His^+Mut^+ transformants should grow normally on both MD and MM plates.

Reagents Needed

10X M

5% Methanol in distilled water and then filter-sterilize.

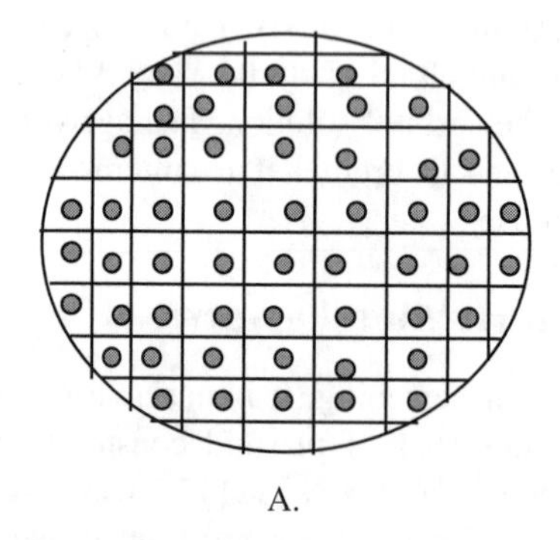

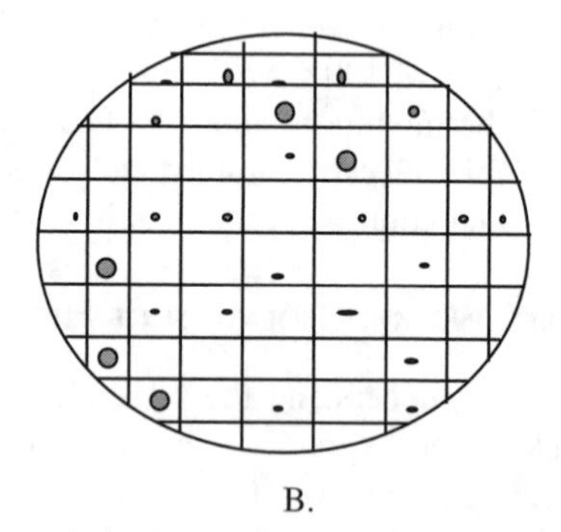

FIGURE 18.8 Screening of transformant GS115 containing *AOX1* gene replacement. (A) Colonies growing on an MD plate. (B) The same colonies are inoculated on an MM plate.

MD (Minimal Dextrose) Medium (for 1 l)

Prepare a mixture containing 100 ml of 10X D, 100 ml of 10X YNB and 2 ml 500X B. Add distilled water to 1000 ml and filter-sterilize. Store the plates at 4°C until use.

MD (Minimal Dextrose) Plates (for 1 l)

Dissolve 15 g agar in 800 ml distilled water and autoclave for 20 min using the liquid cycle. Allow solution to cool to approximately 60°C with slow stirring and then place in a 55°C water bath prior to use. Prepare a mixture containing 100 ml of 10X D, 100 ml of 10X YNB and 2 ml 500X B. Warm the mixture to about 45°C and add it to the solution. Pour the mixture into plates (approximately 30 ml/100-mm plate) in a laminar flow hood and allow the plates to evaporate the excess moisture at room temperature for a couple of days. Store the plates at 4°C until use.

MDH (Minimal Dextrose Histidine) Plates (for 1 l)

Dissolve 15 g agar in 790 ml distilled water and autoclave for 20 min in liquid cycle. Allow the solution to cool to approximately 60°C with slow stirring and then place in a 55°C water bath prior to use. Prepare a mixture containing 100 ml of 10X D, 100 ml of 10X YNB, 2 ml 500X B and 10 ml of 100X histidine stock. Warm the mixture to about 45°C and add it to the solution. Pour the mixture into plates (approximately 30 ml/100-mm plate) in a laminar flow hood and allow the plates to evaporate the excess moisture at room temperature for a couple of days. Store the plates at 4°C until use.

MM (Minimal Methanol) Medium (for 1 l)

Prepare a mixture containing 100 ml of 10X M, 100 ml of 10X YNB and 2 ml 500X B. Add distilled water to 1000 ml and filter-sterilize. Store the plates at 4°C until use.

MM (Minimal Methanol) Plates (for 1 l)

Dissolve 15 g agar in 800 ml distilled water and autoclave for 20 min in liquid cycle. Allow the solution to cool to approximately 60°C with slow stirring and then place in a 55°C water bath prior to use. Prepare a mixture containing 100 ml of 10X M, 100 ml of 10X YNB and 2 ml 500X B. Warm the mixture to about 45°C and add it to the solution. Pour the mixture into plates (approximately 30 ml/100-mm plate) in a laminar flow hood, and allow the plates to evaporate the excess moisture at room temperature for a couple of days. Store the plates at 4°C until use.

MMH (Minimal Methanol Histidine) Plates (for 1 l)

Dissolve 15 g agar in 790 ml distilled water and autoclave for 20 min in liquid cycle. Allow the solution to cool to approximately 60°C with slow stirring and then place in a 55°C water bath prior to use. Prepare a mixture containing 100 ml of 10X M, 100 ml of 10X YNB, 2 ml 500X B and 10 ml of 100X histidine stock. Warm the mixture to about 45°C and add it to the solution. Pour the mixture into plates (approximately 30 ml/100-mm plate) in a laminar flow hood and allow the plates to evaporate the excess moisture at room temperature for a couple of days. Store the plates at 4°C until use.

Identification of Transformants Containing the cDNA to Be Expressed

Based on the phenotype screening data, His^+Mut^s and His^+Mut^+ transformants may be characterized by PCR to make sure that the cDNA to be expressed is integrated into the host genome. To obtain this information, a forward primer can be designed in the 5′-P_{AOX1} region, and a reverse primer designed from the 3′-end of the cDNA. Or, the forward primer may be designed from the 5′-end of the cDNA and the reverse primer designed from the 3′-AOX1. Following PCR amplification using genomic DNA from individual colonies, analyze the PCR products by 1% agarose gel electrophoresis. Positive transformants should reveal a DNA band as expected. If not, discard the colony. Detailed PCR protocols and agarose gel electrophoresis are described in Chapter 2. In addition, the insertion copy number of the cDNA can be determined by digesting yeast genomic DNA with a restriction enzyme that cuts outside the cDNA. The digested DNA is subjected to Southern blot hybridization using the cDNA as a probe. Compared to nontransfected control DNA, the number of detected bands will most likely be the copy number of the cDNA insertion. Detailed protocols for Southern blotting are described in Chapter 7.

Expression of Proteins in Transformants

Determination of Optimum Conditions for Protein Expression

Prior to carrying out large-scale expression of proteins of interest, it is necessary to perform a time-course expression on a small-scale.

1. Inoculate individual positive GS115 transformants (His^+Mut^s or His^+Mut^+ with cDNA inserts confirmed) into 10 ml of BMG or BMGY in a 125-ml flask and grow the cells at 28 to 30°C for 12 to 16 h with shaking at 250 rpm. An OD_{600} = 2 to 6 indicates that the cells are in a log-phase state of growth.
2. Centrifuge at 3000 × *g* for 5 min at room temperature and aspirate the supernatant. To induce expression, resuspend the cells in MM, or BMM or BMMY medium to an OD_{600} = 1.0.
3. Grow 50 ml of the cell suspension cells at 28 to 30°C with shaking at 250 rpm.
4. Add 100% methanol to the final concentration of 0.5% every 24 h.
5. Label nine microcentrifuge tubes (1.5 ml) for nine time points: 0, 12, 24, 36, 48, 60, 72, 84, and 96. Transfer 1 ml of the cell culture into an appropriate time-point tube since the addition of methanol.

6. Centrifuge at 3000 × *g* for 5 min at room temperature. Transfer the supernatant into a fresh tube for secreted expression and store the supernatant and cell pellet at –80°C until use.
7. Analyze the supernatant and the cell pellets for protein expression by SDS-PAGE followed by Coomassie blue staining, or by western blot analysis if specific antibodies are available. The cells can be completely solublized under denaturation conditions that can be loaded into SDS-PAGE. The detailed protocols for protein extraction, solublization, SDS-PAGE and western blot analysis are described in great detail in Chapter 11.

Note: The size of the expressed protein can be estimated based on the amino acids encoded by the reading frame of the cloned cDNA plus the leader sequence, assuming that an average of molecular weight for each amino acid is 100 Da. However, it does not matter if it is not convenient to estimate the size of the product because, if all goes well, the band of the expressed protein should be much larger than that of an uninduced protein sample, and there should be a time-course induction pattern (Figure 18.9).

8. Following characterization of transformant colonies, the colonies that produce large amounts of foreign proteins should be frozen back in as many vials as possible using sterile glycerol to a final concentration of 15% (v/v) and stored at –80°C.

Reagents Needed

10X GY

10% (v/v) Glycerol in distilled water and sterilize by filtering

1 M Potassium Phosphate Buffer, pH 6.0

13.2% (v/v) of 1 *M* K_2HPO_4 and 86.8% (v/v) of 1 *M* KH_2PO_4. If necessary, adjust the pH with phosphoric acid or KOH.

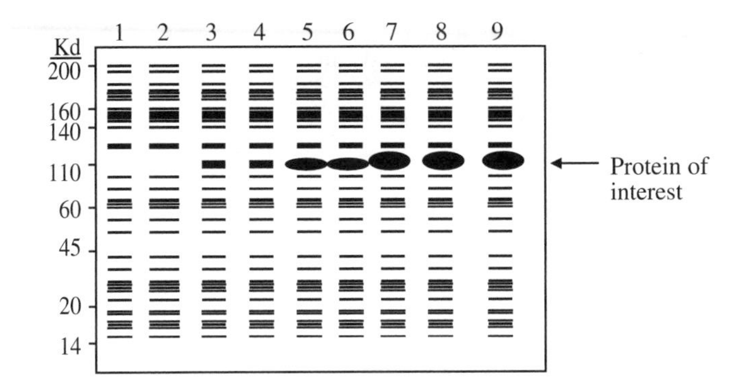

FIGURE 18.9 Analysis of time-course of protein expression yeast by SDS-PAGE and Coomassie blue staining. Lane 1 through lane 9: induction of a 110-kDa protein with methanol at 0, 12, 24, 36, 48, 60, 72, 84, and 96 h, respectively.

BMG (Buffered Minimal Glycerol)

700 ml Distilled water, 100 ml of 10X YNB, 100 ml of 1 *M* potassium phosphate buffer, pH 6.0, 2 ml of 500X B and 100 ml of 10X GY. Mix well and sterilize by filtering. Store at 4°C.

BMM (Buffered Minimal Methanol)

700 ml Distilled water, 100 ml of 10X YNB, 100 ml of 1 *M* potassium phosphate buffer, pH 6.0, 2 ml of 500X B and 100 ml of 10X M. Mix well and sterilize by filtering. Store at 4°C.

Large-Scale Expression of Proteins

Once the optimal conditions have been determined for individual colonies, expression of the protein of interest can be scaled up using large volumes as in fermentation approaches.

1. Inoculate a GS115 transformant colony (His^+Mut^s or His^+Mut^+ with cDNA inserts confirmed) into 25 ml of BMG or BMGY in a 250-ml flask and grow the cells at 28 to 30°C for 12 to 16 h with shaking at 250 rpm. An OD_{600} = 2 to 6 indicates that the cells are in a log-phase state of growth.
2. Centrifuge at 3000 × *g* for 5 min at room temperature and aspirate the supernatant. To induce expression, resuspend the cells in MM, or BMM or BMMY medium to an OD_{600} = 1.0.
3. Grow the cell suspension cells in 2 to 4 l of culture medium at 28 to 30°C with shaking at 250 rpm.
4. Add 100% methanol to a final concentration of 0.5% every 24 h until the optimal time of induction is reached.
5. Centrifuge at 3000 × *g* for 5 min at room temperature. Transfer the supernatant to a fresh tube for secreted expression and store the supernatant and cell pellet at –80°C until use. Proceed to protein extraction and purification.

Extraction of Total Proteins for Purification

Procedures are virtually the same as described for extraction of proteins from bacteria except for secretory proteins in the yeast culture medium.

1. Precipitate the secreted proteins in the harvested supernatant with 60 to 80% (w/v) $(NH4)_2SO_4$. Dissolve well and allow precipitation to take place at 4°C for 30 min.
2. Centrifuge at 15,000 rpm for 20 min at 4°C and aspirate the supernatant.
3. Dissolve the protein pellet in 3 to 5 ml of protein buffer and transfer the sample into a dialysis bag with a molecular weight cut-off of less than the molecular weight of the fusion proteins.
4. Seal the bag at both ends using appropriate clips or by making knots.
5. Place the sample bag in a beaker containing 1 l of dialysis buffer.

6. Perform dialysis in a cold room with stirring for 1 to 2 days. Change the dialysis buffer every 12 h.
7. Concentrate the protein samples using a plastic disposable microconcentrator from Amicon and centrifuge at 1000 rpm at 4°C for an appropriate period of time. An alternative and effective way is to partially freeze-dry the samples or partially lyophilize in a speed vac system under cold conditions until a minimum volume is reached.
8. Take 20 to 100 μl of the sample to measure the concentration (see Chapter 11). Store the protein sample at –20°C until use.
9. Take 2 to 3 μg protein to analyze the purity by SDS-PAGE or western blotting.

Reagents Needed

Protein Buffer

20 m*M* Tris-HCl, pH 7.5
0.5 *M* NaCl

Dialysis Buffer

20 m*M* Tris-HCl, pH 7.5
10 m*M* NaCl

Carry Out Fusion Protein Purification and Cleavage

Detailed protocols for purification and cleavage of proteins from bacteria are described in Part A.

REFERENCES

1. Studier, F.W. and Moffatt, B.A., Use of bacteriophage T7 RNA polymerase to direct selective high-level expression of cloned genes, *J. Mol. Biol.*, 189, 113–130, 1986.
2. Tabor, S. and Richardson, C.C., DNA sequence analysis with a modified bacteriophage T7 DNA polymerase, *Proc. Natl. Acad. Sci. USA,* 84, 4767–4771, 1987.
3. Rose, M. and Botstein, D., Yeast promoters and *lac* Z fusions designed to study expression of cloned genes in yeast, *Methods Enzymol.*, 101, 181–192, 1993.
4. Schein, C.H., Production of soluble recombinant proteins in bacteria, *BioTechnology,* 7, 1141–1149, 1989.
5. Smith, D.B. and Johnson, K.S., Single-step purification of polypeptides expressed in *Escherichia coli* as fusions with glutathione S-transferase, *Gene*, 67, 31–40.
6. Wu, W., Extraction and purification of proteins/enzymes, in *Handbook of Molecular and Cellular Methods in Biology and Medicine,* Kaufman, P.B., Wu, W., Kim, D., and Cseke, L., pp. 41–64, CRC Press, Boca Raton, FL, 1995.
7. Wu, W., Electrophoresis, blotting, and hybridization, in *Handbook of Molecular and Cellular Methods in Biology and Medicine,* Kaufman, P.B., Wu, W., Kim, D., and Cseke, L., pp. 87–122, CRC Press, Boca Raton, FL, 1995.
8. Wearne, S.J., Factor Xa cleavage of fusion proteins: elimination of nonspecific cleavage by reversible acylation, *FEBS Lett.*, 263, 23–26, 1990.

9. Treco, D.A., Growth and manipulate of yeast, in *Current Protocols in Molecular Biology*, Susubel, F.M., Brent, R., Kingston, R.E., Moore, D.D., Seidman, J.G., Smith, J.A., and Struhl, K., Eds., John Wiley & Sons, New York, 13.2.1.–13.2.11, 1995.
10. Treco, D.A., Manipulation of yeast genes. in *Current Protocols in Molecular Biology,* Susubel, F.M., Brent, R., Kingston, R.E., Moore, D.D., Seidman, J.G., Smith, J.A., and Struhl, K., Eds., John Wiley & Sons, New York, 13.7.1–13.7.6, 1995.
11. Reynolds, A., Yeast vetors and assays for expression of cloned genes. In *Current Protocols in Molecular Biology,* Susubel, F.M., Brent, R., Kingston, R.E., Moore, D.D., Seidman, J.G., Smith, J.A., and Struhl, K., Eds., John Wiley & Sons, New York, 13.6.1–13.6.4, 1995.

19 Quantitative Analysis of Functional Genome by Current Real-Time RT-PCR

CONTENTS

INTRODUCTION

Genomic analysis plays an important role in quantification of gene expression.[1–9] Traditionally, methods used for quantitative analysis of gene expression at the mRNA level include northern blot, RNA protection assays and non real-time RT-PCR.[1–3] The drawback, however, is that these traditional procedures are time consuming and costly, and experience significant variation. For instance, non real-time RT-PCR involves establishing linear response curves prior to analyzing and comparing a specific mRNA level among RNA samples or different treatment groups.[2–5] First, it is necessary to set up multiple experiments to establish optimal PCR cycles using the same amount of RNA as a template (e.g., 1 μg RNA per reaction) but different PCR cycles (e.g., 5, 10, 15, 20, 25) followed by 1 to 1.5% agarose gel visualization of the PCR products. The purpose is to find a PCR cycle range in which the signal intensity of PCR products or bands is in a linear range. Secondly, the amount of RNA (e.g., 0.5, 1.0, or 1.5 μg RNA/reaction) should be optimized in a linear range of PCR cycle number (e.g., 20 cycles). Again, it is necessary to identify an optimal amount of RNA (e.g., 0.5 μg per reaction with 20 cycles), which will reveal signal intensity of PCR bands or products within a linear range. Additionally, it may be necessary to optimize other conditions such as buffers or solutions, concentration of $MgCl_2$ or temperature.

Obviously, these are very time-consuming and labor-intensive activities. Why are these optimization steps necessary? Many researchers do not prefer to optimize PCR conditions. Instead, they randomly select an amount of RNA (e.g., 2 μg/reaction) and PCR cycles (e.g., 35), run RT-PCR and compare signal intensity among RNA samples. This may appear to be simple and rapid; however, many manuscripts included these types of data. Commonly, a severe critic will occur. Why? One mistake might be not to know that some signal intensity of PCR bands or products may reach a plateau instead of being within a linear range; therefore, a comparison of specific mRNA levels between samples or treatment groups is inaccurate. Nevertheless, such traditional non real-time RT-PCR is still a valuable tool for analysis of gene expression as long as it is performed in a scientific manner (Figure 19.1).

To overcome the shortcoming of non real-time RT-PCR, real-time RT-PCR technology has been developed.[1,6] The principles of this technology include:

1. A short, fluorescent probe hybridizes to the region between forward and reverse primers. The 5′ end of the probe is labeled with a fluorescent dye as a reporter (e.g., FAM or VIC). The 3′ end of the probe is a nonfluorescent quencher such as TAMRA.

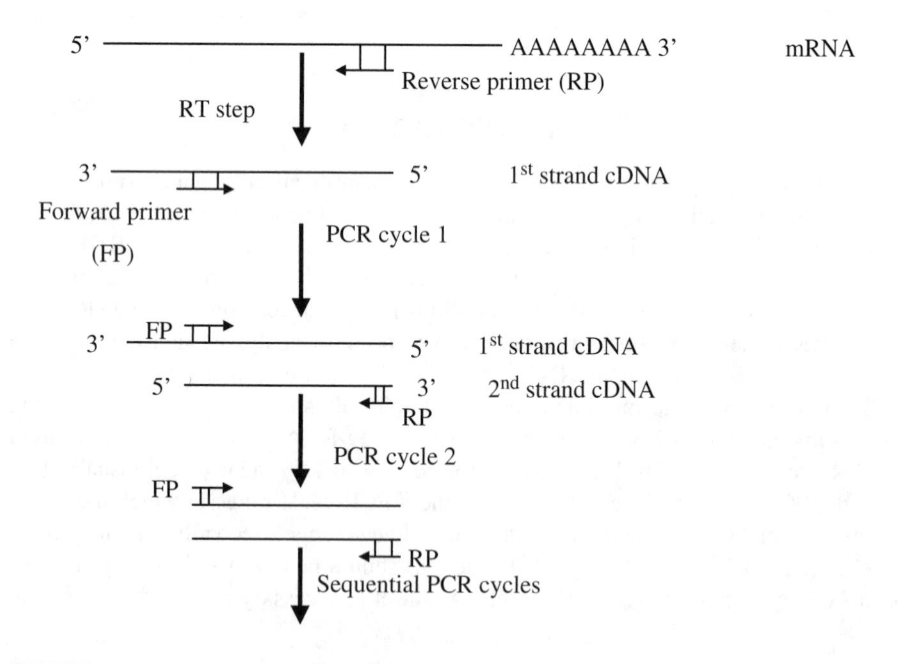

FIGURE 19.1 Schematic diagram of RT-PCR.

2. When the probe is intact, due to an energy transfer, the signal of the reporter is minimal but the signal of the quencher is strong. However, during primer extension and strand replacement, the 5′ end of the probe is sequentially cleaved by the 5′ nuclease activity of DNA *Taq* polymerase. As a result, the fluorescent signals increase but the quencher signals decrease.
3. Instruments such as ABI Prizm 7900 HT are equipped with a CCD camera that captures and records live the fluorescent fluctuations in all wells due to changes in concentration during PCR cycling. When RT-PCR is complete, the recorded data are utilized to generate a plot curve for each well, including background, geometric, linear and plateau phases.
4. To analyze and compare signal plat curves among different wells or samples (e.g., target gene or internal control gene), the Ct (cycle threshold) line can simply be placed or adjusted within the linear phase across all wells or samples for comparison. A Log_2Ct value will be automatically calculated for each well by software such as SDS 2.1.

The present chapter describes in detail how to analyze the mRNA levels or functional genomics of gene expression by real-time RT-PCR (Figure 19.2).

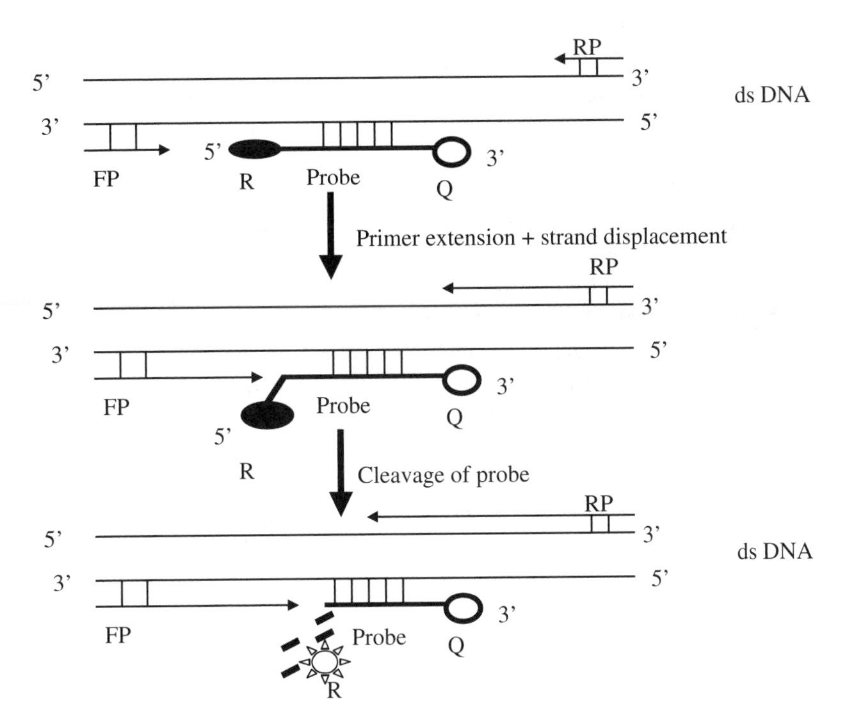

FIGURE 19.2 Procedure of fluorescence real-time PCR.

ISOLATION OF TOTAL RNA FROM TISSUES OR ORGANS OF INTEREST

1. Label one test tube (5 to 10 ml in size) for each sample and add 1 ml of Trizol reagents per 0.1 g tissue to each tube.

Warning: Trizol or phenol is very toxic and should be handled carefully. For protection, wear gloves, lab coat and safety glasses. Gloves should be changed periodically during RNA isolation to minimize RNase activities. Whenever possible, Trizol reagents or phenol should be handled in a fume hood with a glass door placed between the face and the inside of the hood.

2. Quickly cut a tiny piece of each tissue (e.g., 100 mg) on a clean plastic Petri dish or an equivalent using a pair of scissor and forceps. Immediately place the cut piece into an appropriate tube labeled at step 1.

Tips: In case of many samples, cut and process fewer than 10 samples per round because the tissue samples should be homogenized as soon as possible to minimize RNase activities. Another round of RNA isolation can be started during precipitation of the prior RNA samples at –20°C for 1 to 2 h.

3. Carefully homogenize each sample to a fine homogenate. Rinse the polytron probe quickly in dd.H_2O a couple of times prior to homogenizing the next sample.

Tips: The tube can be held with one hand and the speed of the polytron controlled with the other. Contract the probe with the tissue close to the bottom of the tube and start homogenizing for approximately 10 s. Turn on the polytron again and continue to homogenize for about 3 bursts, 5 to 10 s per burst by homogenizing up and down the homogenate. Note: To avoid breaking the tube, never let the polytron probe come in contact with the bottom.

4. Slightly shake the tubes, place them on a bench and allow lysis of cells to occur for 5 min at room temperature (RT). At this point, label clean 1.5-ml microcentrifuge tubes corresponding to the homogenate tubes.
5. Add 0.2 ml of chloroform to each homogenate, tightly cap the tubes and vigorously shake each tube for approximately 30 s to optimize the lysis of cells and shear genomic DNA. Place the tubes on a bench for 2 min.
6. Transfer the lysate to the appropriate microcentrifuge tubes.
7. Centrifuge for 10 min at 4°C. Label a set of clean microcentrifuge tubes.
8. Carefully transfer the top aqueous phase to fresh tubes.

Tips: This is a critical step in terms of RNA purity. Transfer the liquid a little at a time (e.g., 100 to 150 µl per transfer), starting from the top portion to close to the interphase. In general, 250 to 350 µl of the liquid can be transferred. Whenever possible during the transfer, the interphase containing proteins, lipids, carbohydrates, genomic DNA, etc. should not be disturbed. Once the top transferring liquid lowers approximately 2 mm to the interphase, stop. Do not to be greedy! Otherwise, the RNA will be contaminated, which will be evidenced by a low ratio of A_{260}/A_{280} *(e.g., <1.5).*

9. Add cold isopropanol to each tube (0.8 volume of the transferred liquid) and allow RNA to be precipitated for 1 to 2 h at –20°C. Longer precipitation is acceptable.
10. Centrifuge for 10 min at 4°C. Decant the liquid; the RNA pellet should be visible at the bottom of the tube. Add 1 ml of 75% ethanol to each tube to wash residual salt.
11. Centrifuge for 5 min at 4°C.
12. Dry the RNA samples in a speed vac for about 20 min. Avoid overdrying the samples.
13. Dissolve RNA in 100 to 200 µl RNase-free water or TE buffer for 40 min at RT or 20 min at 50°C.
14. Carry out determination of RNA concentration in a spectrophotometer according to the instructions.

Note: An excellent RNA sample should have a ratio of A_{260}/A_{280} >1.9 to 2.0. In contrast, a ratio of A_{260}/A_{280} <1.5 is a poor RNA preparation that contains proteins.

15. Run 5 µg RNA of each sample in 1% agarose denaturing gel with ethidium bromide and carry out electrophoresis until the first blue dye reaches approximately 2 cm from the end of the gel. Take a picture of the gel.

Tips: A good RNA preparation should display a sharp band of 28S rRNA, a sharp band of 18S rRNA and a smear in each lane. The signal intensity of 28S rRNA:18S rRNA bands should be approximately 3:1. In case of RNA degradation, no visible or very weak rRNA bands or only the 18S rRNA band is visible.

16. Store RNA samples at –80°C until use.

ISOLATION OF TOTAL RNA FROM CULTURED CELLS

1. Aspirate or decant culture medium from the culture dishes or plates and quickly examine the cells under a microscope to estimate cell confluence and density. It is not necessary to scrape the cells out of Petri dishes or plates. Instead, add an appropriate volume of RNA isolation buffer such as Trizol reagents directly to the dish or plate, approximately 1 ml per 10^{5-6} cells.

Warning: Trizol or phenol is very toxic and should be handled carefully. For protection, wear gloves, lab coat and safety glasses. Gloves should be changed periodically during RNA isolation to minimize RNase activities. Whenever possible, Trizol reagents or phenol should be handled in a fume hood with a glass door placed between the face and the inside of the hood.

2. Quickly overlay the dish or wells of the plate with extraction buffer and pat the dish or plate by hand to loosen cells. This takes a couple of minutes. Periodically, monitor cell lysis under a microscope until cells disappear.

Tips: To expedite cell lysis, use 1- to 5-ml pipette to flush the bottom of the dish or plate until cells disappear, indicating that a good lysis occurred.

3. Once cell lysis has completed, transfer the lysate to 1.5-ml microcentrifuge tubes and place them on a bench for 5 min at RT to maximize further lysis.
4. Add 0.2 ml of chloroform to each 1 ml of lysate, tightly cap the tubes and vigorously shake each tube for approximately 30 s to optimize the lysis of cells and shear genomic DNA. Place the tubes on a bench for 2 min.
5. Centrifuge for 10 min at 4°C. Label a set of clean microcentrifuge tubes.
6. Carefully transfer the top aqueous phase to fresh tubes.

Tips: This is a critical step in terms of RNA purity. Transfer the liquid a little at a time (e.g., 100 to 150 μl per transfer), starting from the top portion to close to the interphase. In general, 250 to 350 μl of the liquid can be transferred. Whenever possible during the transfer, the interphase containing proteins, lipids, carbohydrates, genomic DNA, etc. should not be disturbed. Once the top transferring liquid lowers approximately 2 mm to the interphase, stop. Do not be greedy! Otherwise, the RNA will be contaminated, which will be evidenced by a low ratio of A_{260}/A_{280} (e.g., <1.5).

7. Add cold isopropanol to each tube (0.8 volume of the transferred liquid) and allow RNA to be precipitated for 1 to 2 h at –20°C. Longer precipitation is acceptable.
8. Centrifuge for 10 min at 4°C. Decant the liquid; the RNA pellet should be visible at the bottom of the tube. Add 1 ml of 75% ethanol to each tube to wash residual salt.
9. Centrifuge for 5 min at 4°C.
10. Dry the RNA samples in a speed vac for about 20 min. Avoid overdrying the samples.
11. Dissolve RNA in 100 to 200 μl RNase-free water or TE buffer for 40 min at RT or 20 min at 50°C.
12. Determine the concentrations of RNA samples using a spectrophotometer according to the instructions.

Note: An excellent RNA sample should have a ratio of A_{260}/A_{280} >1.9 to 2.0. In contrast, a ratio of A_{260}/A_{280} <1.5 is a poor RNA preparation that contains proteins.

13. Run 5 μg RNA of each sample in 1% agarose denaturing gel with ethidium bromide and carry out electrophoresis until the first blue dye reaches approximately 2 cm from the end of the gel. Take a picture of the gel.

Tips: A good RNA preparation should display a sharp band of 28S rRNA, a sharp band of 18S rRNA and a smear in each lane. The signal intensity of 28S rRNA:18S rRNA bands should be approximately 3:1. In case of RNA degradation, no visible or very weak rRNA bands or only the 18S rRNA band is visible.

14. Store the RNA samples at –80°C until use.

PREPARATION OF RNA POOL

For drug, compound, chemical or pharmacogenomic *in vitro* or *in vivo* studies, there are many groups of samples: control vs. different treatment (e.g., 10 groups). Each group consists of many animals (e.g., N = 5 per group). In this case, 50 RNA samples are isolated from 50 animals. Because it is time consuming to analyze the mRNA level of interest in 50 individual animals, quantitative real-time RT-PCR using the RNA pool/group can be performed or the same amount of RNA (e.g., 10 μg) from each of five animals can be combined in one vial. The final volume of each group should be the same (e.g., 100 μl). Hence, the final RNA concentration will be the same among animals or groups: 0.5 μg/μl. This serves to minimize potential variations if individual animal RNA samples are pipetted instead of the RNA group pool. We recommend preparing a final concentration of 50 ng/μl RNA pool per group to be utilized for real-time RT-PCR. Store the RNA samples at –80°C until use.

DESIGNING PRIMERS AND PROBES

Design of primers and probes is a critical step in a successful real-time RT-PCR. Poor-quality primers or probes could result in failure. The authors offer the following tips to help achieve successful results. A couple of software programs are available for designing primers and probes. Because we utilized the Prizm 7900 HT system (Applied Biosystems, CA) for real-time RT-PCR, Primer Express was primarily used for this purpose. Additionally, we employed other software including Lasergene (DNA Star) and GCG program.

1. Obtain gene sequences of interest from GenBank via NCBI and save a sequence in a file folder with .seq at the end of sequence name (e.g., GFPcDNA.seq). Given a specific accession number or EST ID number, software such as Primer Express and Lasergene can directly link to and import sequences from databases.
2. Design and select a forward primer, a reverse primer and a probe between forward and reverse primers under default conditions (e.g., Primer Express) or specified parameters.

Guidelines: (1) The 3′ end of forward or reverse primer should be approximately 10 to 25 bases away from the 5′ or 3′ end of the probe between the primers. (2) The length of the primers is 15 to 22 nucleotides. (3) The length of the probe is 17 to 25 bases. (4) The forward and reverse primers should not form heterodimmers. No more than 5 bases are allowed to form intraloops of primers or probe. (5) The GC content of primers should be <60%, but the probe may have up to 80% GC content. (6) The Tm value for primers should be 58 to 60°C. However, Tm for the probe should be 68 to 70°C. (7) Avoid Gs at the 5′ end of the probe or primers.

3. Carefully check the sequences prior to sending them out for synthesis. Primers and probes do not need HPLC purification. Standard desalt works fine in our laboratories.

4. Accurately calculate the concentrations of primers or probes and briefly spin down the primer or probe powder prior to adding dd.H_2O to the vials. It is recommended that the concentrations for a primer or probe be 100 and 200 μM, respectively.

Notes: Making the correct concentration may be a problem for many people due to confusion. It is important to bear in mind that M, mM, μM, or nM refers to the concentration of 1 l. 1M = 1000 mM, 1 mM = 1000 μM, 1 μM = 1000 nM, 1 nM = 1000 pM, and 1 pM = 1000 fM. It is improper to say or write 1 μM of something per ml or μl because 1 μM of that substance refers to the concentration of 1 l. Therefore, whether using 10 ml, 1 ml or 1 μl, the concentration is 1 μM in this case. How to dissolve a primer or oligonucleotide powder in a correct volume of water to make the proper concentration? For example, if one received an oligo primer powder with a specification of 50 nmol and wanted to resuspend the primer to a concentration of 200 μM, how much water would be needed (final volume)? Here is a formula:

$$\text{Volume} \times \text{Concentration} = \text{Content (e.g., mol, mmol, } \mu\text{mol, nmol, pmol, fmol).}$$

For the present example, volume (l) × 200 μM = 50/1000 μmol, so volume = 0.05/200 = 0.00025 (liter or l) = 250 μl. One can add 250 μl of dd.H_2O to the vial containing 50 nmol of primer powder, generating a concentration of 200 μM primer stock solution.

5. Make a working solution of primers or probe by diluting the stock primer or probe solution from 200 to 10 μ*M* in RNase- and DNase-free water. Forward and reverse primers can be combined into one vial, 10 μ*M* each. Place the primer or probe solution at –20 or 4°C until use.

REAL-TIME RT-PCR

ASSEMBLY OF REACTIONS

1. Thaw primers, probes and reagents at RT and RNA samples on ice.
2. Design a real-time RT-PCR assay using a 96-well template (Figure 19.3) as a guide. We recommend a triplicate (three wells) for each RNA sample (individual or RNA group pool).
3. Carefully calculate the amount or volume of primers, probes, RNA, reagents and dd.H_2O needed. Assemble a reaction mixture for each gene of interest in a microcentrifuge tube. The following cocktail is designed for 10 reactions or wells, total of 50 μl/well:
   5X One-step RT-PCR master mixture, 100 μl
   100X RNase inhibitor (optional), 5.0 μl
   Target gene primers (forward and reverse primers, 10 μ*M* each), 25 μl
   Internal control primers (e.g., GAPDH forward and reverse primers, 10 μ*M* each), 25 μl

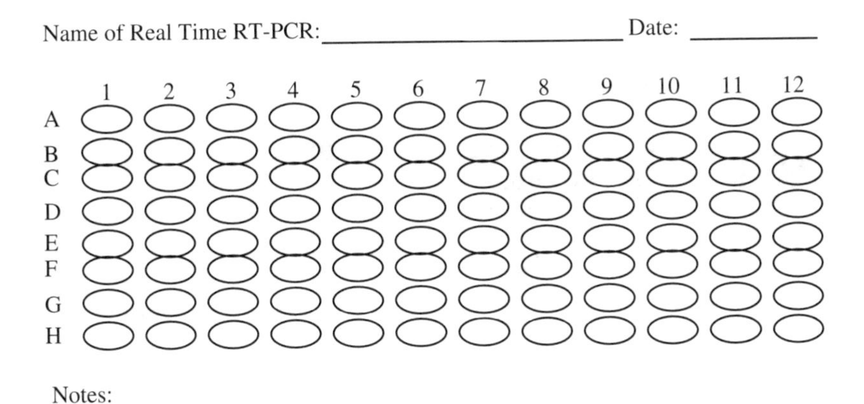

FIGURE 19.3 96-well template used for real-time RT-PCR.

Target gene probe (fluorescent FAM-labeled, 200 μ*M*), 2.5 μl
Internal control probe (fluorescent VIC-labeled, 200 μ*M*), 2.5 μl
RNA (50 ng/μl), 40 μl
dd.H_2O, 300 μl
Total volume of 500 μl

Notes: (1) The RT-PCR master mixture contains reverse transcriptase, DNA Taq polymerase, dNTPs, $MgCl_2$, buffer, etc. and is commercially available (AB Applied Biosystems, CA; AHW Biomedical Systems, MI). (2) Assembly of this cocktail or mixture should be based on the conditions of preoptimization. Initially, optimization experiments should be set up. Based on our experience, we recommend trying a couple of conditions, including final concentration of primers (e.g., 125, 250 and 500 nM), probes (2.5 and 5 μM) and RNA (50, 100, and 150 ng/well). In this case, it may be necessary to add different components individually to appropriate wells. (3) Once the conditions are optimized based on real-time RT-PCR results, it is strongly recommended to assemble the reaction cocktail prior to transferring an appropriate volume to a well (e.g., 50 μl/well), minimizing variation by potential pipetting error. In general, the fewer the pipetting steps, the less variation there will be. It is imperative to minimize variation as much as possible to quantify and compare gene expression levels among samples or treatments. (4) Whenever possible, target gene and internal control should be set up and processed in the same wells. We recommend GAPDH as an internal control. Many researchers use 18S rRNA as internal control; the drawback, however, is that 18S rRNA is overabundant. As a result, RT-PCR results may not reveal any difference in terms of 18S rRNA level between one- and twofold total RNA as template.

4. Accurately aliquot the same volume of the reaction cocktail to each well according to the predesigned 96-well plate template.
5. Firmly cover the plate with a sheet of sticky film to prevent evaporation during RT-PCR.

Programming and Performance of Real-Time RT-PCR

1. Set up a real-time RT-PCR protocol and start a run according to the instructions.
   a. Turn on the computer that controls the ABI Prizm 7900 HT System.
   b. Open the SDS 2.1 program.
   c. Turn on ABI Prizm 7900 HT. The instrument indicates red light → yellow light → green light. The instrument needs to be turned on for at least 15 min prior to RT-PCR.
   d. Click **File** of SDS 2.1 program → click **New** → select **quantitation → name → OK**.
   e. Click **Detector Manager** → create a fluorescent color for each gene of interest — for example, Name: ABC1 mRNA (target gene); Color/Detector: FAM green, Description: unknown. Name: p53 mRNA (target gene); Color/Detector: FAM yellow, Description: unknown. Name: GAPDH mRNA (internal control gene); Color/Detector: VIC pink, Description: unknown. Once all genes are set up, click **Save**, **Copy file** and then **Done**.
   f. Click **Instrument set up** → highlight specific wells according to the 96-well plate template designed in the first place → select the specific target and internal control genes and colors set up by **Detector Manager** → repeat the process for other wells.
   Notes: To obtain convenient export results to a terminal computer, the SDS 2.1 software recommends that wells in the plate be highlighted from row to row instead of from column to column. Therefore, when designing a 96-well plate template at the beginning of the experiment, set up triplicates for each gene in the same row instead of the same column.
   g. Click **Profile** → 48°C, 30 min (for RT step) → 95°C, 10 min → 40 cycles of [95°C, 15 s (denaturation) → 60°C, 1 min (annealing/extension)] → volume: 50 µl.
   h. Click **Data Collection**: 95°C and 60°C points.
   i. Click **Open** → place the plate in place → click **Close**.
2. Click **Run**. It displays the thermal profile and remaining time.

Analysis of Data

1. Analyze real-time RT-PCR data after "The run is successfully completed" is seen on the screen. Click **Open** → to remove the plate → click **Close** to close the door. Click **Analyze** → **Results** → **Amplification Plot**. At this point, the plate template and amplification plot are displayed on the screen.
2. Highlight specific wells for individual genes in a row-to-row fashion according to the initial design of the 96-well template. The amplification plot window will automatically display RT-PCR curves for a particular gene. Adjust/place the Ct (cycle threshold) line in the linear phase across

all the curves; a Ct value will be automatically displayed for each well on the same screen. Repeat the procedure until all the wells for all the genes of interest are analyzed.

Notes: (1) All the wells designed for a specific gene (target gene or internal control gene) should be simultaneously highlighted from row to row instead of column to column. (2) A Ct line should be placed in the linear phase (not geometric phase or plateau phase) and should be across all the curves for different RNA groups. Figure 19.4 to Figure 19.6 indicate the right or wrong way to place the Ct line. (3) A Ct value for a particular well should not be >40 because the maximum PCR cycle number was set at 40. Ct values of >40 seen in some wells indicate that the Ct line needs to be readjusted or the conditions of RT-PCR for those particular genes should be optimized. For statistical analysis, a triplicate should be set up for each RNA sample (e.g., treated vs. control). If all goes well, Ct values for three wells should be similar and the amplified well positions displayed in the amplification plot window almost horizontal. Any significant differences among triplicate wells indicate a large variation occurred during assembly of reactions and aliquots among wells.

3. Save the results and export the result to a terminal computer (if necessary) for further analysis. Generate a table or figure of real-time RT PCR results to compare the expression levels of genes of interest using Microsoft Excel or equivalent software (Table 19.1 and Figures 19.7 and 19.8).

VALIDATION

Validate real-time RT-PCR data by traditional northern blotting or RNA protection assays (optional).

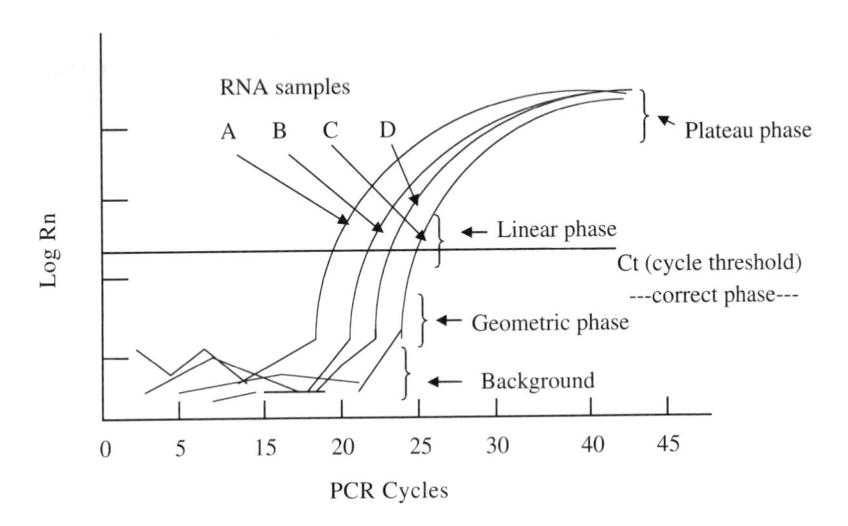

FIGURE 19.4 A diagram of correct Ct line across real-time RT-PCR curves.

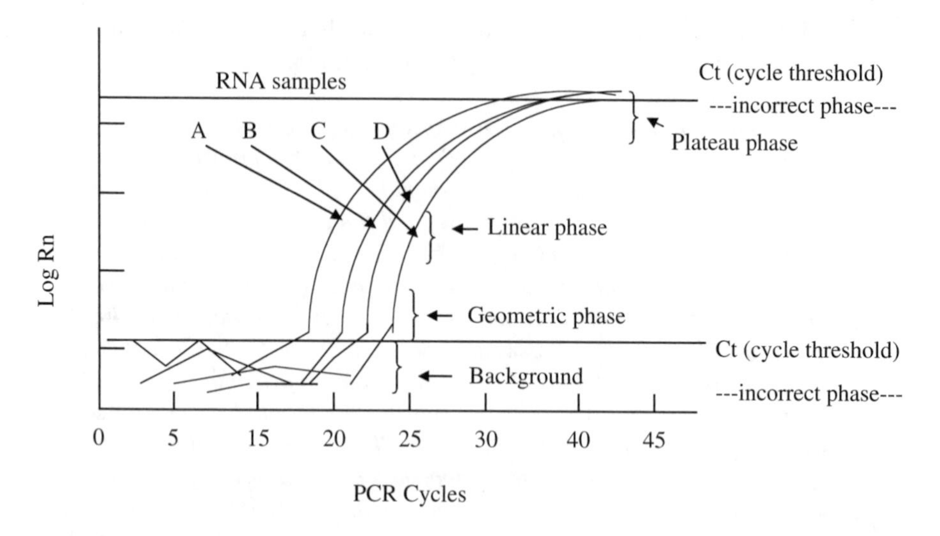

FIGURE 19.5 A diagram of incorrect adjustment of Ct line across the real-time RT-PCR curves.

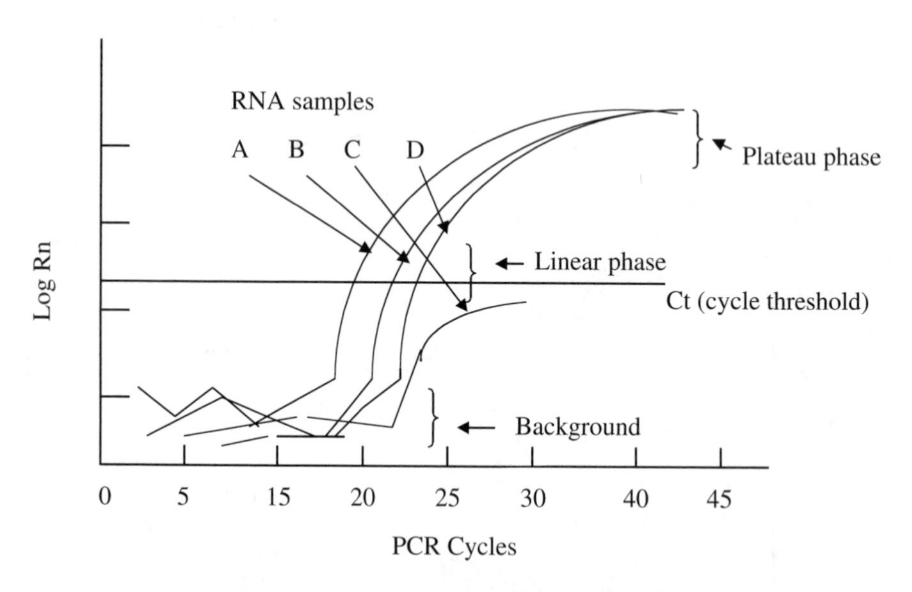

FIGURE 19.6 A diagram of incorrect amplification curve for RNA sample C. RNA samples A, B and D are fine.

TABLE 19.1
Analysis and Comparison of mRNA Levels between Control and Treated Groups

	Ct Value						Ct Mean Value						
	ABC1 Well 1	ABC1 Well 2	ABC1 Well 3	GAPDH Well 1	GAPDH Well 2	GAPDH Well 3	ABC1	GAPDH	ΔCt: ABC1 - GAPDH	ΔΔCt: Treated Group – Control (ΔCt-ΔCt)	mRNA Fold-Change vs. Control = 2^ – ΔΔCt	Actual mRNA Fold-Change vs. Control Group Using an Excel Formula: *IF (L2 < 1, –1/L2, L2), then enter	mRNA Level % Change (Treated vs. Control Group)
Control RNA	22.52	22.01	22.65	19.55	19.67	19.2	22.39	19.49	2.90	0	1	1	100%
Treated RNA 1	20.9	20.4	20.1	19.55	19.67	19.2	20.47	19.49	0.98	–1.92	3.7842	3.7842	378%
Treated RNA 2	21.6	21.2	21.7	19.55	19.67	19.2	21.5	19.49	2.01	–0.89	1.8531	1.8531	185%
Treated RNA 3	25.2	25.8	25.6	19.55	19.67	19.2	25.53	19.49	6.04	3.14	0.1134	–8.8152	–882%

Notes:

1. Ct values are derived from Log_2. Therefore, ΔCt data Log_2.values. To calculate a Log_2.value, use the formula: = Log($_{base,}$ number). For example, Log_2.value for 8 can be calculated as = Log($_2$, 8) = 3.

2. Conversion of Log_2.value to number, the formula is = 2^ΔΔCt for positive ΔCt or Log_2.value, for example, 2^3 = 8. For negative ΔCt or Log_2.value, the formula is = 2^ – ΔΔCt (e.g., 2^–3 = –8. In case that 2^–ΔΔCt or 2^ΔΔCt values are <1, a conversion is needed to generate actual fold-change using an excel formula: = IF (column row < 1, –1/column row, column row). For example, IF (L2 < 1, –1/L2, L2), then Enter key.

3. Based on the converted numbers, the mRNA fold-change or percentage changes can be calculated compared to control group. Please bear in mind that fold- or % changes are relative vs. control group.

4. For reproducibility and statistically significant data, we strongly recommend repeating real-time RT-PCR at least three times in order to calculate average fold- or percentage-change with standard deviation in the figures. Only consistent results are meaningful.

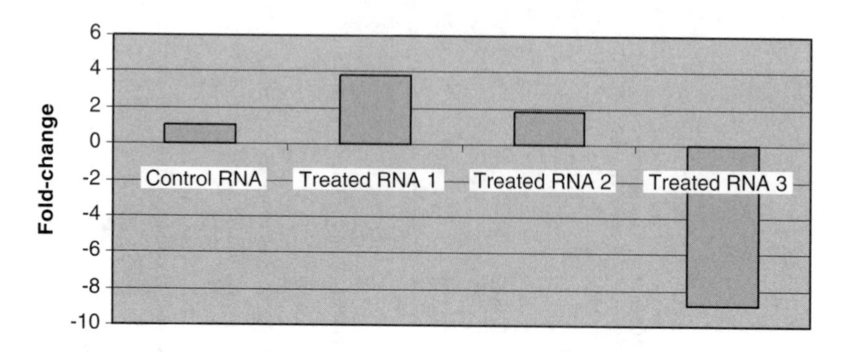

FIGURE 19.7 Fold-change of mRNA level in treated or control groups.

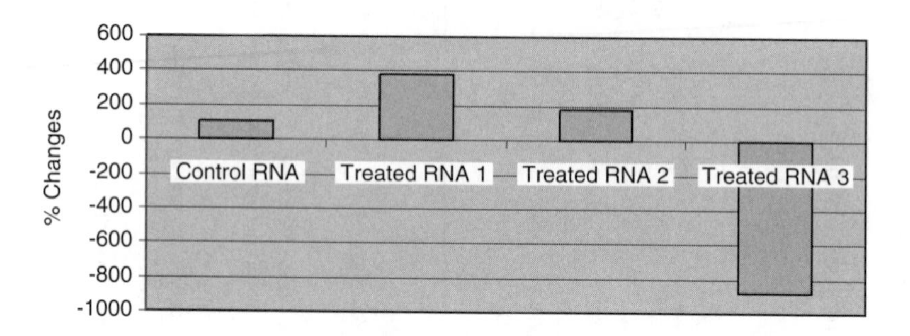

FIGURE 19.8 Percentage change in mRNA level compared to the control group.

REFERENCES

1. Innis, M.A., Myambo, K.B., Gelfand, D.H., and Brow, M.A., DNA sequencing with *Thermus aquaticus* DNA polymerase and direct sequencing of polymerase chain reaction amplified DNA, *Proc. Natl. Acad. Sci. USA*, 85, 9436, 1988.
2. Holland, P.M., Abramson, R.D., Watson, R., and Gelfand, D.H., Detection of specific polymerase chain reaction product by utilizing the 5′ → 3′ exonuclease activity of *Thermus aquaticus* DNA polymerase, *Proc. Natl. Acad. Sci. USA*, 88, 7276, 1991.
3. Srarnaki, O., Tanner, M., and Visakorpi, T., Amplification and overexpression of elongin C gene discovered in prostate cancer by cDNA microarrays, *Lab Invest.*, 82(5):629, 2002
4. Witney, A., Haynes, J.D., Moch, J.K., Carucci, D.J., and Adams, J.H., Transcripts of developmentally regulated Plasmodium falciparum genes quantified by real-time RT-PCR, *Nucl. Acids Res.,* 15, 2224, 2002.
5. Lin, C.Y., Chao, K.H., Wu, M.Y., Yang, Y.S., and Ho, H.N., Defective production of interleukin-11 by decidua and chorionic villi in human anembryonic pregnancy, *J. Clin. Endocrinol. Metab.*, 87, 2320, 2002.
6. Dotsch, J., Repp, R., Rascher, W., and Christiansen, H., Diagnostic and scientific applications of TaqMan real-time PCR in neuroblastomas, *Expert Rev. Mol. Diagn.*, 1, 233, 2001.

7. Gleissner, B. and Thiel, E., Detection of immunoglobulin heavy chain gene rearrangements in hematologic malignancies, *Expert Rev. Mol. Diagn.*, 1, 191, 2001.
8. Pfaffl, M.W., Georgieva, T.M., Georgiev, I.P., Ontasouka, E., Hageleit, M., and Blum, J.W., Real-time RT-PCR quantification of insulin-like growth factor (IGF)-1, IGF-1 receptor, IGF-2, IGF-2 receptor, insulin receptor, growth hormone receptor, IGF-binding proteins 1, 2, and 3 in the bovine species, *Domestic Anim. Endocrinol.*, 22, 91, 2002.
9. Mansson, E., Liliemark, E., Soderhall, S., Gustafsson, G., Eriksson, S., and Albertioni, F., Real-time quantitative PCR assays for deoxycytidine kinase, deoxyguanosine kinase, and 5′-nucleotidase mRNA measurement in cell lines and in patients with leukemia, *Leukemia,* 16, 386, 2002.

20 High-Throughput Analysis of Gene Expression by Cutting-Edge Technology — DNA Microarrays (Gene Chips)

CONTENTS

INTRODUCTION

We are living in an era of a revolution in genomics and biotechnology. The flood of information about the human genome is particularly overwhelming because technology is forging ahead and bringing about rapid changes day after day. With the genome sequences of several organisms now in public databases, the next logical step is to reveal and understand the biological systems and mechanisms. To achieve such goals mandates high-throughput methodologies. Certainly, microarry technology plays a major role in genomic and proteomic profiling. DNA microarray or gene chip technology makes it possible to analyze simultaneously thousands of genes combined with a dozen parameters within a single experiment, and to increase the speed and efficiency of scientific research and drug discovery significantly.[1–10] The major principle of this technology is that hundreds and thousands of genes (i.e., oligonucleotides, EST, cDNA, etc.) are immobilized or arrayed onto small slides or chips via an appropriate technology; the chip is then hybridized with complementary probes usually generated from mRNA samples. Following washing, the slide or chip is scanned into a computer and analyzed using the appropriate software.[1–5,7–12] The procedure is similar to traditional Southern or northern blot hybridization except for the high throughput. The major merit of DNA microarry technology is that a single chip in a single experiment can reveal thousands of gene expression profiles, including up-regulated, down-regulated or unchanged genes, compared to a control group. Hence, this powerful technology is widely utilized in new gene discovery, DNA–DNA interaction, DNA–RNA interaction and DNA–protein interaction.[2–6,12–15]

This chapter describes this technology and, via step-by-step protocols, helps readers to use DNA microarrays or chips for research as well as for drug discovery. (Figure 20.1).

ISOLATION OF TOTAL RNA FROM CELLS OR TISSUES

1. Add the appropriate volume of RNA isolation buffer such as Trizol reagents to the tube containing cells or tissue (approximately 1 ml per 10^{5-6} cells or 1 mg tissue).

Warning: Trizol or phenol is very toxic and should be handled carefully. For protection, wear gloves, lab coat and safety glasses. Gloves should be changed periodically during RNA isolation to minimize RNase activities. Whenever possible, Trizol reagents or phenol should be handled in a fume hood with a glass door placed between the face and the inside of the hood.

2. Suspend the cells or homogenize the tissue to a fine homogenate and allow the cells to lyse for 5 to 10 min at room temperature.

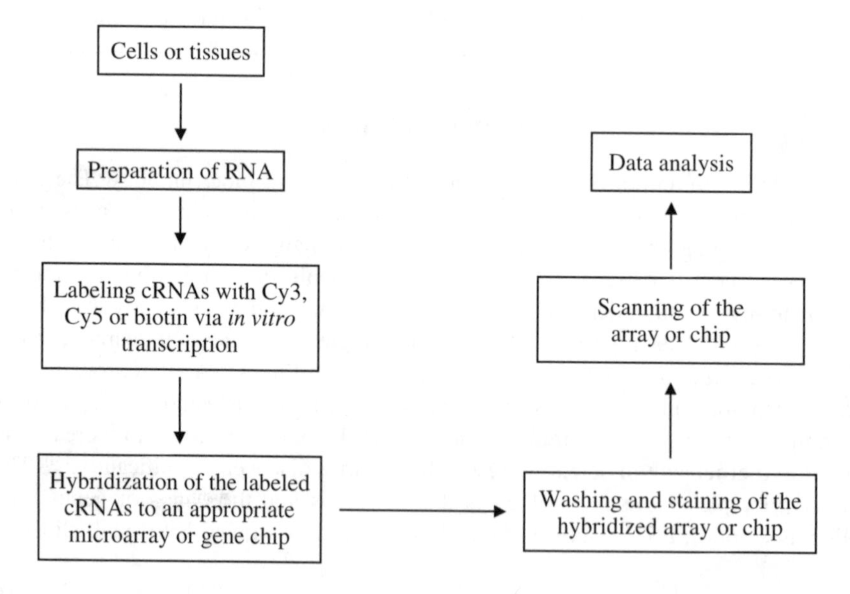

FIGURE 20.1 A general procedure of gene expression profiling via DNA microarray.

3. Add 0.2 ml of chloroform to each 1 ml of lysate, tightly cap the tubes and vigorously shake each tube for approximately 30 s to optimize the lysis of cells and shear genomic DNA. Place the tubes on a bench for 2 min.
4. Centrifuge for 10 min at 4°C. Label a set of clean microcentrifuge tubes.
5. Carefully transfer the top aqueous phase to fresh tubes.

Tips: This is a critical step in terms of RNA purity. Transfer the liquid a little at a time (e.g., 100 to 150 µl per transfer) starting from top portion to close to the interphase sequentially. In general, 250 to 350 µl of the liquid can be transferred. Whenever possible during the transfer, the interphase containing proteins, lipids, carbohydrates, genomic DNA, etc. should not be disturbed. Once the top transferring liquid lowers to approximately 2 mm to the interphase, stop. Do not be greedy! Otherwise, the RNA will be contaminated, which will be evidenced by a low ratio of A_{260}/A_{280} (e.g., <1.5).

6. Add cold isopropanol to each tube (0.8 volume of the transferred liquid) and allow RNA to be precipitated for 1 to 2 h at –20°C. Longer precipitation is acceptable.
7. Centrifuge for 10 min at 4°C. Decant the liquid; the RNA pellet should be visible at the bottom of the tube. Add 1 ml of 75% ethanol to each tube to wash residual salt.
8. Centrifuge for 5 min at 4°C.
9. Dry the RNA samples in a speed vac for about 20 min. Avoid overdrying the samples.
10. Dissolve RNA in 100 to 200 µl RNase-free water or TE buffer for 40 min at RT or 20 min at 50°C.
11. Determine the concentrations of RNA samples using a spectrophotometer according to the instructions.

Note: An excellent RNA sample should have a ratio of A_{260}/A_{280} >1.9 to 2.0. In contrast, a ratio of A_{260}/A_{280} <1.5 is a poor RNA preparation that contains proteins.

12. Run 5 µg RNA of each sample in 1% agarose denaturing gel with ethidium bromide and carry out electrophoresis until the first blue dye reaches approximately 2 cm from the end of the gel. Take a picture of the gel.

Tips: A good RNA preparation should display a sharp band of 28S rRNA, a sharp band of 18S rRNA and a smear in each lane. The signal intensity of 28S rRNA:18S rRNA bands should be approximately 3:1. In case of RNA degradation, no visible or very weak rRNA bands or only the 18S rRNA band is visible.

13. Store RNA samples at –80°C until use.

PREPARATION OF FLUORESCENT PROBES FROM mRNA

Once high-quality RNA has been obtained, synthesis of cDNA probes can be performed using mRNAs as templates. A standard reaction is total 25 µl using 0.2 to

2 μg mRNA or 5 to 10 μg total RNAs. For each additional 1 μg mRNA, increase the reaction volume by 10 μl.

1. Anneal 2 μg of mRNAs as templates with 1 μg of primer of oligo(dT) in a 1.5-ml sterile, RNase-free microcentrifuge tube. Add nuclease-free dd.H_2O to total 19 μl. Heat the tube at 65°C for 5 min to denature the secondary structures of mRNAs and allow it to cool slowly to room temperature to complete the annealing. Briefly spin down the mixture.
2. To the annealed primer–template, add the following in the order listed:
   8 μl 5X First strand buffer
   4 μl 0.1 *M* DTT
   4 μl 10X dNTP
   4 μl Cy3 or Cy5 dUTP or biotin dUTP
   1 μl RNase inhibitor
   2 μl Superscript-RT (AMV reverse transcriptase)

   Incubate at 42°C for 60 min; heat at 94°C for 5 min.

   Function of components: The 5X reaction buffer contains components for the synthesis of cDNAs. Ribonuclease inhibitor serves to inhibit RNase activity and to protect the mRNA templates. Sodium pyrophosphate suppresses the formation of the hairpins commonly generated in traditional cloning methods, thus avoiding the S1 digestion step, which is difficult to carry out. AMV reverse transcriptase catalyzes the synthesis of the first-strand cDNAs from mRNA templates according to the rule of complementary base pairing.
3. To remove RNAs completely, add RNase A (DNase free) to a final concentration of 20 μg/ml or 1 μl of RNase (Boehringer Mannheim, 500 μg/ml) per 10 μl of reaction mixture. Incubate at 37°C for 30 min, followed by inactivation of RNase A by heating at 94°C for 5 min.
4. (Optional) Extract the probes from the unlabeled reaction mixture with one volume of TE-saturated phenol/chloroform. Mix well and centrifuge at 11,000 × *g* for 5 min at room temperature.
5. Carefully transfer the top, aqueous phase to a fresh tube. Do not take any white materials at the interphase between the two phases. To the supernatant, add 0.5 volume of 7.5 *M* ammonium acetate or 0.15 volume of 3 *M* sodium acetate (pH 5.2), mix well and add 2.5 volumes of chilled (at –20°C) 100% ethanol. Gently mix and precipitate cDNAs at –20°C for 2 h.
6. Centrifuge in a microcentrifuge at 12,000 × *g* for 5 min. Carefully remove the supernatant and briefly rinse the pellet with 1 ml of cold 70% ethanol. Centrifuge for 4 min and carefully aspirate the ethanol.
7. Dry the cDNAs under vacuum for 15 min. Dissolve the probe pellet in 15 to 25 μl of TE buffer. Take 2 μl of the sample to measure the concentration of probes prior to the next reaction. Store the sample at –20°C until use.

Materials Needed

50 m*M* EDTA
7.5 *M* Ammonium acetate
Ethanol (100% and 70%)
Chloroform:isoamyl alcohol (24:1)
2 *M* NaCl
0.2 *M* EDTA
1 mg/ml Carrier DNA (e.g., salmon sperm)
Trichloroacetic acid (5% and 7%)

First-Strand 5X Buffer

50 m*M* Tris-HCl, pH 8.3 (at 42°C)
250 m*M* KCl
2.5 m*M* Spermidine
50 m*M* $MgCl_2$
50 m*M* DTT
5 m*M* Each of dATP, dCTP, dGTP and dTTP

TE Buffer

10 m*M* Tris-HCl, pH 8.0
1 m*M* EDTA

TE-Saturated Phenol/Chloroform

Thaw phenol crystals at 65°C and mix equal parts of phenol and TE buffer. Mix well and allow the phases to separate at room temperature for 30 min. Take one part of the lower, phenol phase to a clean beaker or bottle and mix with one part of chloroform:isoamyl alcohol (24:1). Mix well and allow phases to be separated; store at 4°C until use.

GENERATION OF DNA MICROARRAYS OR CHIPS

DNA chips are commercially available (Affymatrix: rat array, mouse array, human array, etc.), but can be very costly. However, it is possible to produce arrays by spotting appropriate DNA (cDNA, PCR products, oligos) onto precoated glass slides using an appropriate spotter (Figure 20.2). The spotted slides should be compatible with a commercial scanner. The authors provide the following conditions for arraying slides.

Corning CMT-GAPS silane-coated slides
Spotting buffer: 50% DMSO, 20 m*M* Tris-HCl, 50 m*M* KCl, pH 6.5
72°F (22°C) with 40 to 50% relative humidity
UV cross-linking at 90 mJ, bake at 80°C for 2 h

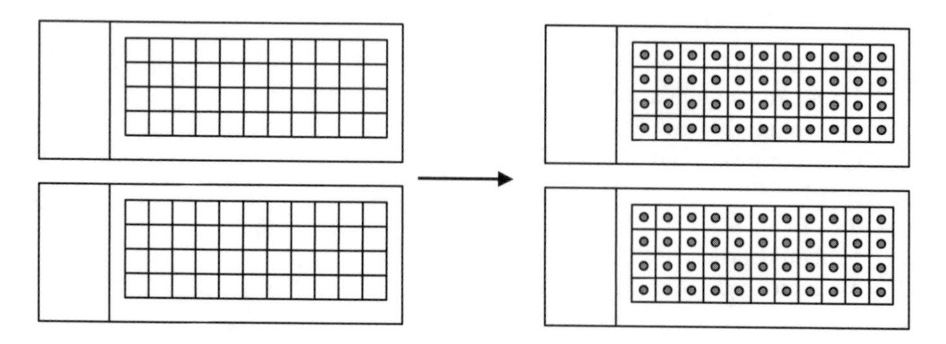

FIGURE 20.2 A diagram of arraying DNA onto slides.

PREHYBRIDIZATION OF GENE CHIPS OR ARRAYS

1. Mark the chip (array or slide) with a pencil.
2. Vapor-moisturize the chip (array side facing down) over boiling water for 1 min.
3. Quickly place the chip in a UV cross-linker (i.e., Stratalinker, 250 mJ, setting 2500, silane slides) with the array side up.
4. Remoisturize the chip over steam as described in step 2.
5. Rinse the chip or slide in 0.1% SDS for 20 s and in dd.H_2O for 20 s.
6. Heat the chip or slide (i.e., boil in water bath) at 95°C for 3 min. (This is necessary for spotted dsDNA chips; however, it is optional for ssDNA chips.)
7. Add 60 μl of prehybridization buffer to the chip and place a coverslip over the array.
8. Add 10 μl dd.H_2O to each corner of the chamber to maintain humidity.
9. Place the chip or slide in the hybridization chamber and incubate in a 50°C water bath for 1 h.

HYBRIDIZATION

1. Remove water from the array–slide.
2. Add 60 μl of hybridization solution over the array (40 μl hybridization buffer, 1 μl blocking solution, 19 μl probes).
3. Place a new coverslip on the array and place the array in the hybridization chamber.
4. Perform hybridization overnight in a 50°C water bath.

WASHING

1. Place the slide in 50 to 100 ml of wash solution containing 2X SSC and 0.1% SDS, and carefully remove the coverslip from the array.

2. Shake the array for 10 min.
3. Replace 2X SSC with 100 ml of 0.2X SSC solution, and shake the array for 15 min.
4. Place the slide in the 50-ml tube and spin for 5 min at 1000 rpm to dry it.
5. Keep the array in a dark place until scanning.

Materials Needed

1X dNTPs Solution

25µl dGTP
25 µl dATP
25 µl dCTP
10 µl dTTP
415 µl DEPC-H_20
500 µl Total

Prehybridization Buffer

7.0 ml Formamide
0.4 ml ss Salmon sperm DNA
1.0 ml 50X Denhardt's
4 ml 20X SSPE
1.0 ml 10% SDS
6.6 ml dd.H_20
20 ml Total

Hybridization Buffer

0.7 ml Formamide (3%)
50 µl 20% SDS (0.5%)
100 µl 50X Denhardt's (2.5X)
400 µl 20X SSPE (4X)
1.25 ml Total

Blocking Solution (20X Stock)

80 µl PolydA (1µg/µl)
16 µl tRNA (10µg/µl)
400 µl Human/mouse Cot1 DNA (1µg/µl)
480 µl Total
Precipitate the combined reagents with ethanol.
Resuspend in 40 µl dd.H_2O.

SCANNING GENE CHIPS USING GENEPIX 4000A MICROARRAY SCANNER

1. Turn on the scanner, and start GenePixPro 3.0 software.
2. Open the scanner door and insert the chip with the hybridized side down and the label facing toward you.

3. In the **hardware settings** window, set PMTs at 600 in 635 nm (Cy 5) and 532 nm (Cy 3) channels.
4. Click **preview scan** to determine location of scanning areas or spots and signal intensities.
5. Draw a scan area around the entire array and perform a quick visual inspection of hyb and make initial adjustment of PMTs.
6. For gene expression hybs, set the ratio over the entire scan area to be 1.0 by raising or lowering the red and green PMTs to obtain the color balance.
7. In the **hardware settings** window, change the **lines to average** to two prior to performing the data scan. This will allow the scanner to scan each pixel twice and average the collected data counts, which can reduce background noise.
8. Click **data scan** to perform a high-resolution scanning. During the scanning, click the **histogram** tab located at the top of the screen to view the relative intensities of both channels dynamically as scanning takes place. The histogram settings should be: Image: Both, X-axis: Fullscale, Y-axis: Log Axis on Fullscale. The histogram displays the percentage of normalized counts at a given intensity or quanta. The histogram only shows the pixels that can be affected by artifacts and dirt. If a lot of dirt or artifacts are present, try to zoom in on a clean portion of the array to determine accurate PMT settings.
9. Based on the histograms, adjust the PMTs so that the pixels can represent the entire intensity range. In some cases, saturated pixels (i.e., counts >70,000) will be thrown out and spots with pixel counts close to the background will be given poor data.
10. Once the PMT levels are set to make the intensity ratio near 1.00, click **data scan** over the scan area and save the results.
11. Go to the **open/save** button and select **save images** to save the images at 635 or 532 nm (type = multiimage tiff).
12. Assign spot identities and calculate results using GenePixPro software. The software can deconvolute the identities of PCR products from a 96-well text file into a 384-well format (save this information in the form of GAL file, GenepixArrayList). The 384-well files will then be changed to a grid (the GPS, GenePixSettingFile). The GPS will assign each spot on the chip its identity based on printing paramters (the number of tips used to print, spacing between spots, number of plates printed).
13. Go to the **help** section, the **image** tab and then **array list generator** to create the GAL.
14. Following creation of a GAL and GPS files, obtain and analyze results (Figure 20.3 through Figure 20.5).

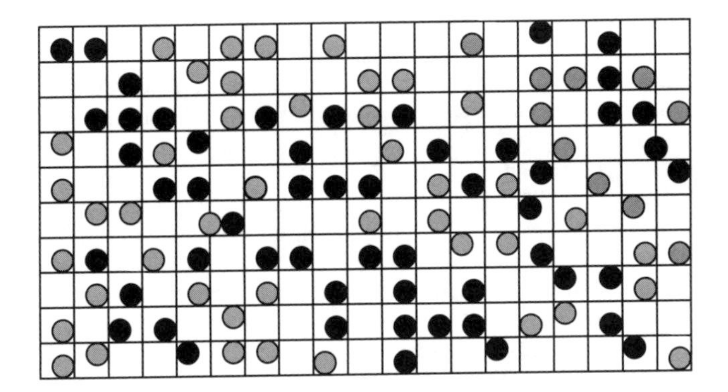

FIGURE 20.3 An image of scanned DNA microarray chips.

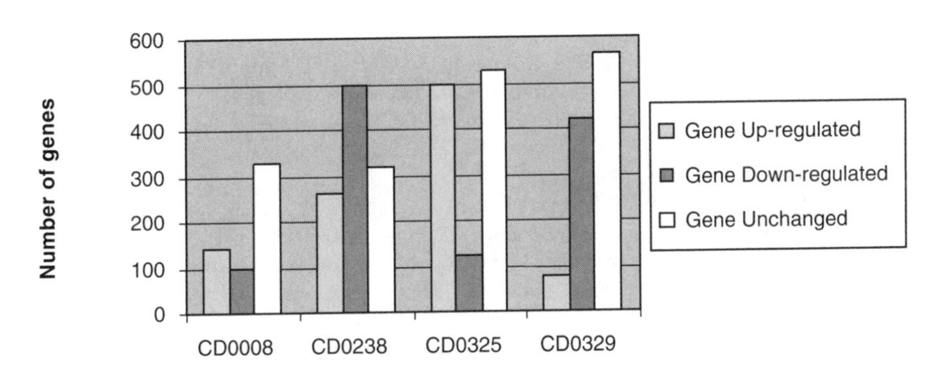

FIGURE 20.4 High-throughput analyses of effects of new drugs on gene expression profiles in mice.

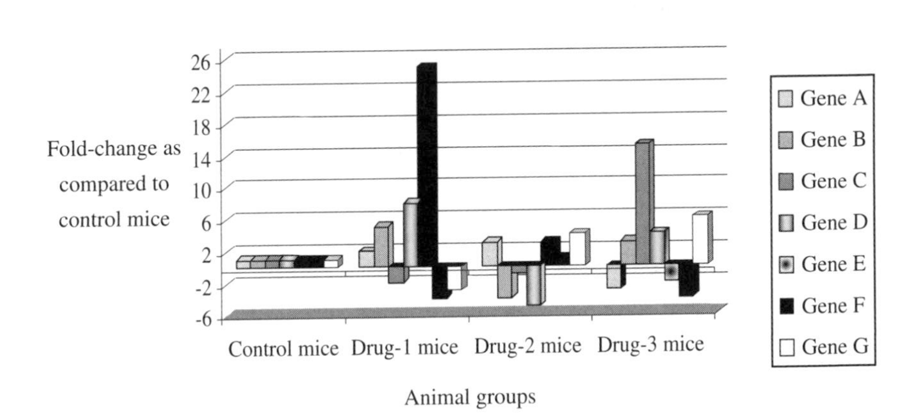

FIGURE 20.5 Fold-changes in gene expression in mice treated with new drugs.

REFERENCES

1. Wolfgang, M.C., Lee, V.T., Gilmore, M.E., and Lory, S., Coordinate regulation of bacterial virulence genes by a novel adenylate cyclase-dependent signaling pathway, *Dev. Cell*, 4(2):253–263, 2003.
2. Martinez, M.J., Aragon, A.D., Rodriguez, A.L., Weber, J.M., Timlin, J.A., Sinclair, M.B., Haaland, D.M., and Werner-Washbume, M., Identification and removal of contaminating fluorescence from commercial and in-house printed DNA microarrays, *Nucl. Acids Res.*, 31(4):e18, 2003.
3. Lopez, I.P., Marti, A., Milagro, F.I., Zulet, M.M., Moreno-Aliaga, M.J., Martinez, J.A., and De Miguel, C., DNA microarray analysis of genes differentially expressed in diet-induced (cafeteria) obese rats, *Obesity Res.*, 11(2):188–194, 2003.
4. Chen, X., Whitney, E.M., Gao, S.Y., and Yang, V.W., Transcriptional profiling of Kruppel-like factor 4 reveals a function in cell cycle regulation and epithelial differentiation, *J. Mol. Biol.,* 326(3):665–677, 2003.
5. Hatfield, G.W., Hung, S.P., and Baldi, P., Differential analysis of DNA microarray gene expression data, *Mol. Microbiol.*, 47(4):871–877, 2003.
6. Nam, J.H., Yu, C.H., Hwang, K.A., Kim, S., Ahn, S.H., Shin, J.Y., Choi, W.Y., Joo, Y.R., and Park, K.Y., Application of cDNA microarray technique to detection of gene expression in host cells infected with viruses, *Acta Virol.,* 46(3):141–146, 2002.
7. Yeatman, T.J., The future of clinical cancer management: one tumor, one chip, *Am. Surg.*, 69(1):41–44, 2003.
8. Le, B.S., Demolombe, S., Chambellan, A., Bellocq, C., Aimond, F., Toumaniantz, G., Lande, G., Siavoshian, S., Baro, I., Pond, A.L., Nerbonne, J.M., Leger, J.J., Escande, D., and Charpentier, F., Microarray analysis reveals complex remodeling of cardiac ion channel expression with altered thyroid status: relation to cellular and integrated electrophysiology, *Circ. Res.*, 92(2):234–242, 2003.
9. Hedges, J.F., Cockrell, D., Jackiw, L., Meissner, N., and Jutila, M.A., Differential mRNA expression in circulating gammadelta T lymphocyte subsets defines unique tissue-specific functions, *J. Leukoc. Biol.*, 73(2):306–314, 2003.
10. Borucki, M.K., Krug, M.J., Muraoka, W.T., and Call, D.R., Discrimination among *Listeria monocytogenes* isolates using a mixed genome DNA microarray, *Vet Microbiol.*, 92(4):351–362, 2003.
11. Andersen, P.S., Jespersgaard, C., Vuust, J., Christiansen, M., and Larsen, L.A., High-throughput single strand conformation polymorphism mutation detection by automated capillary array electrophoresis: validation of the method, *Hum Mutat.*, 21(2):116–122, 2003.
12. Baran, M.F., Fagard, R., Girard, B., Camkilleri, B.S., Zeng, F., Lenoir, G.M., Raphael, M., and Feuillard, J., Gene array identification of Epstein–Barr virus-regulated cellular genes in EBV-converted Burkitt lymphoma cell lines, *Lab Invest.*, 82(11):1463–1479, 2002.
13. Yang, I.V., Chen, E., Hasseman, J.P., Liang, W., Frank, B.C., Wang, S., Sharov, V., Saeed, A.I., White, J., Li, J., Lee, N.H., Yeatman, T.J., and Quackenbush, J., Within the fold: assessing differential expression measures and reproducibility in microarray assays, *Genome Biol.*, 3(11), 2002.
14. Goda, H., Shimada, Y., Asami, T., Fujioka, S., and Yoshida, S., Microarray analysis of brassinosteroid-regulated genes in Arabidopsis, *Plant Physiol.*, 130(3):1319–1334, 2002.
15. Haupl, T., Burmester, G.R., and Stuhlmuller, B., New aspects of molecular biology diagnosis. Array technology and expression profile for characterization of rheumatic diseases, Z *Rheumatol.*, 61(4):396–404, 2002,

21 Construction and Screening of Human Antibody Libraries: Using State-of-the-Art Technology — Phage Display

CONTENTS

INTRODUCTION

Antibodies have been widely utilized for many decades in almost all aspects of the life sciences, including novel research and development (R&D), pharmaceutical discoveries, vaccines, and treatment of human diseases.[1–3] Antibodies can be divided into two major categories: monoclonal and polyclonal. Traditionally, monoclonal antibodies are generated from mice and hybridoma cell lines. Polyclonal antibodies can be produced in rabbits, sheep, dogs, goats, etc. In terms of specifics, monoclonal antibodies have been demonstrated to be much higher than polyclonal antibodies.[2–4] However, generation of monoclonal antibodies takes longer (approximately 6 to 9 months) than polyclonal antibodies (about 1.5 to 3 months). Until recently, production of antibodies was limited to very time-consuming and laborious processes. Monoclonal antibodies, in particular, involve animal immunization schemes and generation of hybridoma cell lines, namely, monoclonal cells. Even so, such traditional methodologies can hardly produce human antibodies using animal systems. This problem has been resolved by the invention of phage display technology,[2,5] which is considered a revolutionary breakthrough in immunology. This technology involves current DNA recombination and gene amplification and makes it possible to clone and produce monoclonal antibodies including human antibodies in bacteria.[5–7]

This chapter describes in detail how to utilize phage display technology to produce monoclonal antibodies such as human antibody libraries. Successful cloning of antibody genes mandates understanding antibody structure and function, and the principles of phage display technology. Native antibodies have a tetrameric structure consisting of two identical heavy chains (55 kDa/440 amino acids) and light chains (24 kDa/220 amino acids). The N-terminal end of each chain contains the variable domain that defines the binding specificity of an antibody. The variable domains of heavy (VH) and light (VL) chains give rise to a heterodimeric molecule, namely, an Fv fragment. The C-terminal region of each chain contains constant regions, the C region. In mouse, five classes of constant heavy (CH) chain genes ($\alpha, \delta, \varepsilon, \gamma, \mu$) and two classes of constant light (CL) chain genes (κ, λ) are present. The DNA and amino acid sequences are relatively conserved within each class. Based on CH regions, antibodies can be grouped into five major classes: IgA, IgD, IgG, IgE and IgM. Furthermore, an antibody is distinguished by a light chain (κ, λ) paired with its heavy chain. The classes of antibodies can be isotyped with commercially available reagents.

Antibody display was initially demonstrated using antibody variable region fragments. Genes encoding the VH and VL can be fused together with a DNA fragment encoding a flexible linker peptide, generating a single protein with covalently linked VH and VL, termed a single-chain Fv molecule (scFv). Alternatively, antibodies can be expressed in the form of an FAb fragment composed of a whole light chain (LC) and variable heavy chain (VH), along with its first constant domain (CH1). All three forms, Fv, scFv and FAb, have been expressed on the surface of a phage.[5,6] A number of functional antibodies have been successfully expressed on the surface of the phage, including alkaline phosphatase, protein A, CD4, growth hormone, etc.[4–11]

There are conservative sequences within the variable regions of antibody genes among species. Alignment and analysis of variable domains can lead to the determination of species-specific consensus sequences, which will be used for design of PCR primers. Hence, the entire repertoires of antibody genes can be prepared by

PCR. In this way, recombinant phage antibody technology has the power and versatility to mimic the features of immune diversity and monoclonal selection.

Phage display of recombinant antibodies takes advantage of the unique features of the M13 phage life cycle and the expression of a couple of phage proteins on the surface of the M13 phage.[2–4] This phage is filamentous and approximately 895 nm long and 9 nm in diameter. Its genome is single-stranded DNA that contains 6.4 kb encoding 10 different proteins. The genome encodes a coat protein comprising approximately 2700 copies of the gene 8 protein (g8p). Also, the M13 phage expresses three to five copies of the gene 3 adsorption protein (g3p) on its tip.[2,3] Phage display technology utilizes the features of g3p and engineered recombinant gene III, generating fusion g3p and expression of recombinant proteins or antibodies on the surface of the M13 phage. When these antibodies bind to antigen, they bind the phage to the antigen as well, indicating that antigen-reactive antibodies are expressed on the tips of the M13 phage. The phage particles can be detected by an enzyme-labeled antibody against the g8 coat proteins (g8p).

M13 phage does not produce lytic infection in *E. coli* as seen in most of the other bacteriophage life cycles. M13 phage induces a state in which the infected host cells produce and secrete phage particles without undergoing lysis. Initially, infection takes place when the phage attaches to the f pilus of a male *E. coli*, which is mediated by the phage g3p. As the phage enters the cell, its coat protein is removed and deposited in the inner membrane. Meanwhile, the host DNA replication machinery changes the single-stranded (ss) phage parental DNA into the double-stranded (ds) plasmid-like replicative form (RF). The dsDNA is replicated and undergoes rolling circle replication to produce ssDNA for packaging into new phage particles. The RF also serves as the template for transcription of phage mRNA. The leader sequences of g3p and g8p direct the transport of these proteins into the inner membrane of the bacterial cell. The ssDNA extrudes through the inner membrane and is assembled into a mature phage particle that extrudes through the cell wall into the medium without disrupting host cell growth. Phage replication continues until the host cell dies.

Many types of vectors are commercially available, including those from Cambridge Antibody Technology and Novagen Corporation. In recent years, a phagemid vector system has been developed that combines the features of phage and plasmid vectors. The phagemid carries the M13 (ssDNA) and plasmid (dsDNA) origins of replication. Phagemids can be grown conveniently as plasmids or packaged as recombinant phage with the aid of a helper phage (i.e., M13K07). Despite its containing the M13 origin of replication, the phagemid lacks the phage protein genes required to generate a complete phage. Hence, cells transformed with phagemid need to be infected with a helper phage that encodes proteins for replication and package ss phagemid DNA into an M13 phage particle, namely, phage rescue.

The presence of a resistance marker such as ampicillin facilitates selection of the transformed cell using antibiotic medium. Because the M13K07 helper phage contains a defective origin of replication (IG region), the phagemid DNA is more efficiently replicated and packaged than the helper phage genome. Therefore, the majority of the phage produced contains phagemid DNA. The K07 origin is sufficient to replicate the helper phage genome in the absence of a competing phagemid. The kanamycin resistance marker on the K07 genome allows cells infected with the helper phage to be selected specifically over noninfected cells. In this way, following infection with the helper phage, the culture of phagemid-containing *E. coli* can be grown in medium containing ampicillin and kanamhycin. Ampicillin selects for cells carrying phagemid and kanamycin selects for cells infected with K07 helper phage. The detailed procedures are described next.

PREPARATION OF WHITE BLOOD CELLS (WBCs)/LYMPHOCYTES FOR RNA ISOLATION

Preparation of WBCs from Blood

1. Dilute 20 ml of heparinized blood with 20 ml of PBS.
2. Divide the diluted blood into two 20-ml fractions and overlay each onto a 15-ml ficoll plaque as a "cushion."
3. To separate red blood cells (RBCs), centrifuge for 30 min at approximately 1000 rpm at room temperature.
4. Carefully collect the peripheral blood lymphocytes from the interface by aspiration with a Pasteur pipette.
5. Dilute the cells with an equal volume of PBS and centrifuge as described in step 3.
6. Resuspend the cells in 1 ml of PBS and centrifuge as described in step 3.
7. Aspirate the supernatant; cells can be used for RNA isolation immediately or frozen at –70°C until use.

Note: In human blood, there are approximately 5 to 11 × 10^6 WBCs/ml, of which about 30% are lymphocytes.

Preparation of Lymphocytes from Tissues

1. Place spleen/lymphonote tissues in a sterile Petri dish containing 15 ml of PBS.
2. Under a scope, tease the tissue apart using sterile forceps and a disposable syringe needle to produce a single-cell suspension of lymphocytes and erythrocytes.
3. Transfer the mixture to a tube using a 22-gauge needle and syringe and allow the clumps to settle to the bottom.
4. Transfer the supernatant to a fresh, sterile container.
5. Centrifuge for 10 min at approximately 1000 rpm at room temperature.
6. Discard the supernatant, resuspend the cells in 10 ml of PBS and centrifuge as described in step 5.
7. Aspirate the supernatant; cells can be used for RNA isolation immediately or frozen at –70°C until use.

Note: There are approximately 11 × 10^8 WBCs in mouse spleen or an equivalent size of human spleen, of which about 35% are B cells.

ISOLATION OF TOTAL RNA FROM LYMPHOCYTES

1. Add an appropriate volume of RNA isolation buffer such as Trizol reagents to the tube containing WBCs, approximately 1 ml per 10^{5-6} cells.

Warning: Trizol or phenol is very toxic and should be handled carefully. For protection, wear gloves, lab coat and safety glasses. Gloves should be changed periodically during RNA isolation to minimize RNase activities. Whenever possible, Trizol reagents or phenol should be handled in a fume hood with a glass door placed between the face and the inside of the hood.

2. Suspend the cells and allow them to lyse for 5 to 10 min at room temperature.
3. Add 0.2 ml of chloroform to each 1 ml of lysate, tightly cap the tubes and vigorously shake each tube for approximately 30 s to optimize the lysis of cells and shear genomic DNA. Place the tubes on a bench for 2 min.
4. Centrifuge for 10 min at 4°C. Label a set of clean microcentrifuge tubes.
5. Carefully transfer the top aqueous phase to fresh tubes.

Tips: This is a critical step in terms of RNA purity. Transfer the liquid a little at a time (e.g., 100 to 150 µl/transfer) starting from the top portion to draw sequentially close to the interphase. In general, 250 to 350 µl of the liquid can be transferred. Whenever possible during the transfer, do not disturb the interphase containing proteins, lipids, carbohydrates, genomic DNA, etc. Once the top transferring liquid lowers approximately 2 mm to the interphase, stop. Do not to be greedy! Otherwise, the RNA will be contaminated, which will be evidenced by a low ratio of A_{260}/A_{280} (e.g., <1.5).

6. Add cold isopropanol to each tube (0.8 volume of the transferred liquid) and allow RNA to be precipitated for 1 to 2 h at –20°C. Longer precipitation is acceptable.
7. Centrifuge for 10 min at 4°C. Decant the liquid; the RNA pellet should be visible at the bottom of the tube. Add 1 ml of 75% ethanol to each tube to wash residual salt.
8. Centrifuge for 5 min at 4°C.
9. Dry the RNA samples in a speed vac for about 20 min. Avoid overdrying the samples.
10. Dissolve RNA in 100 to 200 µl RNase-free water or TE buffer for 40 min at RT or 20 min at 50°C.
11. Determine the concentrations of RNA samples according to the instructions of the spectrophotometer used.

Note: An excellent RNA sample should have a ratio of A_{260}/A_{280} of >1.9 to 2.0. In contrast, an A_{260}/A_{280} ratio of <1.5 is a poor RNA preparation that contains proteins.

12. Run 5 µg RNA of each sample in 1% agarose denaturing gel with ethidium bromide and carry out electrophoresis until the first blue dye reaches approximately 2 cm from the end of the gel. Take a picture of the gel.

Tips: A good RNA preparation should display a sharp band of 28S rRNA, a sharp band of 18S rRNA and a smear in each lane. The signal intensity of

28S rRNA:18S rRNA bands should be approximately 3:1. In case of RNA degradation, no visible or very weak rRNA bands or only an 18S rRNA band is visible.

13. Store RNA samples at –80°C until use.

SYNTHESIS OF cDNAS

SYNTHESIS OF FIRST-STRAND cDNAS FROM MRNAS

Once high-quality RNA has been obtained, synthesis of cDNAs can be performed using mRNAs as templates (Figure 21.1). Setting up two reactions in two microcentrifuge tubes that will be carried out separately all the way to the end is recommended. The benefit is that if one reaction is somehow stopped by an accident during the experiments, the other can be used as a back-up. Otherwise, it is necessary to start all over again, thus wasting time and money. A standard reaction is a total of 25 µl using 0.2 to 2 µg mRNA or 5 to 10 µg total RNAs. For each additional 1 µg mRNA, increase the reaction volume by 10 µl. It is recommended to carry out a control reaction under the same conditions whenever possible.

1. Anneal 2 µg of mRNAs as templates with 1 µg of primer of oligo(dT) in a 1.5-ml sterile, RNase-free microcentrifuge tube. Add nuclease-free dd.H_2O to 15 µl. Heat the tube at 70°C for 5 min to denature the secondary structures of mRNAs and allow it to cool slowly to room temperature to complete the annealing. Briefly spin down the mixture.
2. To the annealed primer–template, add the following in the order listed. To prevent precipitation of sodium pyrophosphate when adding it to the reaction, the buffer should be preheated at 40 to 42°C for 4 min prior to the addition of sodium pyrophosphate and AMV reverse transcriptase. Gently mix well following each addition.
   First-strand 5X buffer, 5 µl
   rRNasin ribonuclease inhibitor, 50 units (25 units/µg mRNA)
   40 mM sodium pyrophosphate, 2.5 µl
   AMV reverse transcriptase, 30 units (15 units/µg mRNA)
   Add nuclease-free dd.H_2O to a final of 25 µl.

Function of components: The 5X reaction buffer contains components for the synthesis of cDNAs. Ribonuclease inhibitor serves to inhibit RNase activity and to protect the mRNA templates. Sodium pyrophosphate suppresses the formation of the hairpins commonly generated in traditional cloning methods, thus avoiding the S1 digestion step that is difficult to carry out. AMV reverse transcriptase catalyzes the synthesis of the first-strand cDNAs from mRNA templates according to the rule of complementary base pairing.

3. As a tracer reaction, transfer 5 µl of the mixture to a fresh tube that contains 1 µl of 4 µCi of [α-^{32}P]dCTP (>400 Ci/mmol, less than 1 week old). The synthesis of the first-strand cDNAs will be measured by trichloroacetic acid (TCA) precipitation and alkaline agarose gel electrophoresis using the tracer reaction.

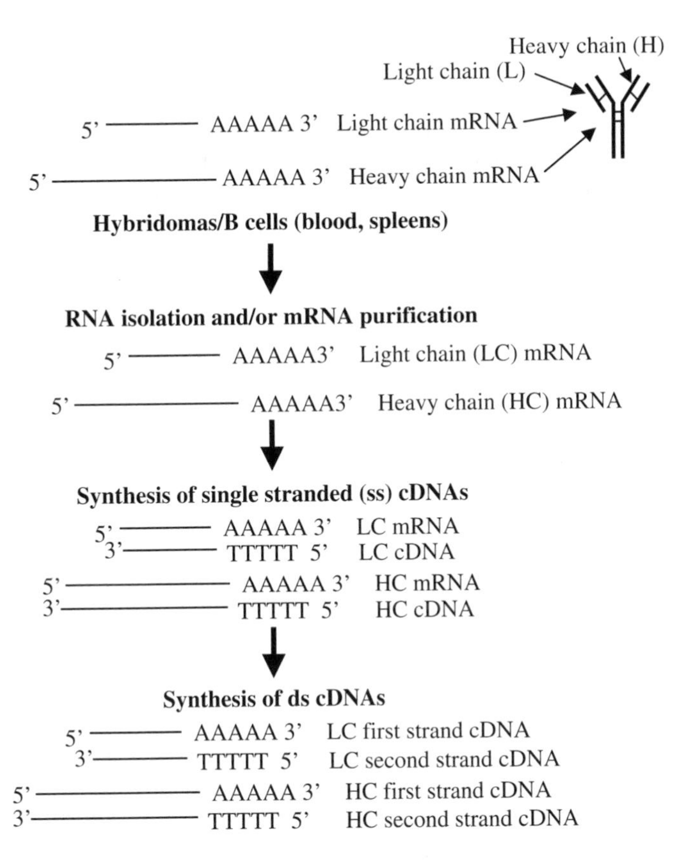

FIGURE 21.1 Isolation of mRNAs and synthesis of cDNAs containing mRNAs/cDNAs encoding the light chains and heavy chains of antibodies.

Caution: [α-^{32}P]dCTP is a toxic isotope. Gloves should be worn when carrying out the tracer reaction, TCA assay and gel electrophoresis. Waste materials such as contaminated gloves, pipette tips, solutions and filter papers should be put in special containers for disposal of waste radioactive materials.

4. Incubate both reactions at 42°C for 1 to 1.5 h and then place the tubes on ice. At this point, the synthesis of the first-strand cDNAs has been completed. To the tracer reaction mixture, add 50 m*M* EDTA solution up to a total volume of 100 μl and store on ice until use for TCA incorporation assays and analysis by alkaline agarose gel electrophoresis.
5. To remove RNAs completely, add RNase A (DNase free) to a final concentration of 20 μg/ml or 1 μl of RNase (Boehringer Mannheim, 500 μg/ml) per 10 μl of reaction mixture. Incubate at 37°C for 30 min.
6. Extract the cDNAs from the unlabeled reaction mixture with one volume of TE-saturated phenol/chloroform. Mix well and centrifuge at 11,000 × *g* for 5 min at room temperature.

7. Carefully transfer the top, aqueous phase to a fresh tube. Do not take any white materials at the interphase between the two phases. To the supernatant, add 0.5 volume of 7.5 *M* ammonium acetate or 0.15 volume of 3 *M* sodium acetate (pH 5.2), mix well and add 2.5 volumes of chilled (at –20°C) 100% ethanol. Gently mix and precipitate cDNAs at –20°C for 2 h.
8. Centrifuge in a microcentrifuge at 12,000 × *g* for 5 min. Carefully remove the supernatant, and briefly rinse the pellet with 1 ml of cold 70% ethanol. Centrifuge for 4 min and carefully aspirate the ethanol.
9. Dry the cDNAs under vacuum for 15 min. Dissolve the cDNA pellet in 15 to 25 μl of TE buffer. Take 2 μl of the sample to measure the concentration of cDNAs prior to the next reaction. Store the sample at –20°C until use.

Materials Needed

[α-^{32}P]dCTP (>400 Ci/mmol)
50 m*M* EDTA
7.5 *M* Ammonium acetate
Ethanol (100% and 70%)
Chloroform:isoamyl alcohol (24:1)
2 *M* NaCl
0.2 *M* EDTA
1 mg/ml Carrier DNA (e.g., salmon sperm)
Trichloroacetic acid (5% and 7%)

First-Strand 5X Buffer

250 m *M* Tris-HCl, pH 8.3 (at 42°C)
250 m*M* KCl
2.5 m*M* Spermidine
50 m*M* $MgCl_2$
50 m*M* DTT
5 m*M* each of dATP, dCTP, dGTP and dTTP

TE Buffer

10 m*M* Tris-HCl, pH 8.0
1 m*M* EDTA

TE-Saturated Phenol/Chloroform

Thaw phenol crystals at 65°C and mix an equal part of phenol and TE buffer. Mix well and allow the phases to separate at room temperature for 30 min. Take one part of the lower, phenol phase to a clean beaker or bottle and mix with one part of chloroform:isoamyl alcohol (24:1). Mix well, allow phases to be separated, and store at 4°C until use.

Alkaline Gel Running Buffer (Fresh)

30 m*M* NaOH
1 m*M* EDTA

2X Alkaline Buffer

20 m*M* NaOH
20% Glycerol
0.025% Bromophenol blue (use fresh each time)

Synthesis of Double-Stranded cDNAs from the Subtracted cDNAs

1. Set up the following reaction:
   Subtracted first-strand cDNAs, 1 µg
   Second-strand 10X buffer, 50 µl
   Poly(A)$_{12–20}$ primers, 0.2 µg
   E. coli DNA polymerase I, 20 units
   Add nuclease-free dd.H_2O to a final volume of 50 µl.
2. Set up a tracer reaction for the second-strand DNA by transferring 5 µl of the mixture to a fresh tube containing 1 µl 4 µCi [α-^{32}PJdCTP (>400 Ci/mmol). The tracer reaction will be used for TCA assay and alkaline agarose gel electrophoresis to monitor the quantity and quality of the synthesis of the second-strand DNA, as described previously.
3. Incubate both reactions at 14°C for 3.5 h and then add 95 µl of 50 m*M* EDTA to the 5 µl tracer reaction.
4. Heat the unlabeled (nontracer portion) double-strand DNA sample at 70°C for 10 min to stop the reaction. At this point, the double-strand DNAs should be produced. Briefly spin down to collect the contents at the bottom of the tube and place on ice until use.
5. Add four units of T4 DNA polymerase (2 u/µg input cDNA) to the mixture and incubate at 37°C for 10 min. This step functions to make the blunt-end double-stranded cDNAs by T4 polymerase. The blunt ends are required for later adapter ligations.
6. Stop the T4 polymerase reaction by adding 3 µl of 0.2 *M* EDTA and place on ice.
7. Extract the cDNAs with one volume of TE-saturated phenol/chloroform. Mix well and centrifuge at 11,000 × *g* for 4 min at room temperature.
8. Carefully transfer the top aqueous phase to a fresh tube and add 0.5 volume of 7.5 *M* ammonium acetate or 0.15 volume of 3 *M* sodium acetate (pH 5.2). Mix well and add 2.5 volumes of chilled (–20°C) 100% ethanol. Gently mix and precipitate the cDNAs at –20°C for 2 h.
9. Centrifuge at 12,000 × *g* for 5 min and carefully remove the supernatant. Briefly rinse the pellet with 1 ml of cold 70% ethanol, centrifuge at 12,000 × *g* for 5 min and aspirate the ethanol.
10. Dry the cDNA pellet under vacuum for 15 min. Dissolve the cDNA pellet in 20 µl of TE buffer. Measure the concentration of the cDNAs as described earlier. Store the sample at –20°C until use.

Note: Prior to adapter ligation, it is strongly recommended that the quantity and quality of the double-stranded (ds) cDNAs should be checked by TCA precipitation and gel electrophoresis as described previously.

Materials Needed

[α-^{32}P]dCTP (>400Ci/mmol)
50 m*M* EDTA
3 *M* Sodium acetate buffer, pH 5.2
Ethanol (100% and 70%)
Chloroform:isoamyl alcohol (24:1)
0.2 *M* EDTA
Trichloroacetic acid (5% and 7%)

Hybridization Buffer (2X)

40 m*M* Tris-HCl, pH 7.7
1.2 *M* NaCl
4 m*M* EDTA
0.4% SDS
1 µg/µl Carrier yeast tRNA (optional)

Phosphate Buffer (Stock)

0.5 *M* Monobasic sodium phosphate
Adjust the pH to 6.8 with 0.5 *M* dibasic sodium phosphate.

DESIGN OF PRIMERS FOR CONSTRUCTION OF ANTIBODY V GENE REPERTOIRE

PCR or RT-PCR amplification of antibody variable regions (V genes) is carried out using primers complementary with the ends of the VH and VL. The products of VH and VL PCR can be linked by PCR to form a recombinant, single-chain Fv fragment. Table 21.1 and Table 21.2 list a series of primers that have been tested.

PCR AMPLIFICATION OF HEAVY- AND LIGHT-CHAIN VARIABLE REGIONS (VH, VL), LINKER AND ASSEMBLY OF RECOMBINANT SINGLE-CHAIN DNA VIA PCR

Figure 21.2 illustrates the procedures for amplification and assembly of VH–Linker–VL using PCR technology. Using appropriate forward and reverse primers listed in the previous section, the variable regions of heavy chain (VH), light chain (VL) and a linker fragment can be amplified in separate PCR reactions. The PCR products for VH and VL are approximately 350 and 325 base pairs in length, respectively. A typical PCR reaction is as follows:

1. Prepare a PCR cocktail in a 0.5-ml microcentrifuge tube as follows:
   Forward primer, 5 to 8 pmol (30 to 40 ng), depending on the size of the primer
   Reverse primer, 5 to 8 pmol (30 to 40 ng), depending on the size of the primer
   cDNAs, 1 to 2 µg
   10X Amplification buffer, 4 µl

TABLE 21.1
Series of Primers for PCR Amplification of Human Antibody Variable Regions (V gene)

Primers for human V gene PCR
Human heavy chain variable region (VH) forward primers
5′ CAGGTGCAGCTGGTGCAGTCTGG 3′
5′ GAGGTGCAGCTGGTGGAGTCTGG 3′
5′ CAGGTCAACTTAAGGGAGTCTGG 3′
5′ CAGGTGCAGCTGCAGGAGTCGGG 3′
5′ CAGGTACAGCTGCAGCAGTCAGG 3′
Human join region of heavy chain (JH) reverse primers
5′ TGAAGAGACGGTGACCATTGTCCC 3′
5′ TGAGGAGACGGTGACCAGGGTGCC3′
5′ TGAGGAGACGGTGACCGTGGTCCC 3′
5′ TGAGGAGACGGTGACCAGGGTTCC 3′
Human V kappa forward primers
5′ GAAATTGTGCTGACTCAGTCTCC 3′
5′ GACATCCAGATGACCCAGTCTCC3′
5′ GATGTTGTGATGACTCAGTCTCC 3′
5′ GAAATTGTGTTGACGCAGTCTCC 3′
5′ GAAACGACACTCACGCAGTCTCC 3′
5′ GACATCGTGATGACCCAGTCTCC 3′
Human J kappa reverse primers
5′ ACGTTTGATTTCCACCTTGGTCCC 3′
5′ ACGTTTAATCTCCAGTCGTGTCCC 3′
5′ ACGTTTGATCTCCAGCTTGGTCCC 3′
5′ ACGTTTGATATCCACTTTGGTCCC 3′
5′ ACGTTTGATCTCCACCTTGGTCCC 3′
Human V lambda forward primers
5′ AATTTTATGCTGACTCAGCCCCA 3′
5′ CAGTCTGTGTTGACGCAGCCGCC3′
5′ CAGTCTGCCCTGACTCAGCCTGC 3′
5′ TCCTATGTGCTGACTCAGCCACC 3′
5′ CAGGCTGTGCTCACTCAGCCGTC 3′
5′ TCTTCTGAGCTGACTCAGGACCC 3′
Human J lambda reverse primers
5′ ACCTAAAACGGTGAGCTGGGTCCC 3′
5′ ACCTAGGACGGTGACCTTGGTCCC 3′
5′ ACCTAGGACGGTCAGCTTGGTCCC 3′

Primers for linker fragment PCR
Forward primers of JH for scFv linker
--------------Heavy chain region------ --Linker--
5′ GGACCACGGTCACCGTCTCCTCAGGTGG 3′
5′ GCACCCTGGTCACCGTCTCCTCAGGTGG 3′

-- *continued*

TABLE 21.1 (continued)
Series of Primers for PCR Amplification of Human Antibody Variable Regions (V gene)

5′ GGACAATGGTCACCGTCTCTTCAGGTGG 3′

5′ GAACCCTGGTCACCGTCTCCTCAGGTGG 3′

Reverse primers of V kappa for scFV linker

------------Light chain region----------- ----Linker----

5′ GGAGACTGAGTCAGCACAATTTCCGATCCGCC 3′

5′ GGAGACTGCGTGAGTGTCGTTTCCGATCCGCC 3′

5′ GGAGACTGGGTCATCTGGATGTCCGATCCGCC 3′

5′ GGAGACTGCGTCAACACAATTTCCGATCCGCC 3′

Reverse primers of V lambda linker

-------------Light chain region---------- ------------Linker ----------------

5′ GGCGGCTGCGTCAACACAGACTG CGATCCGCCACCGCCAGAG 3′

5′ GCAGGCTGAGTCAGAGCAGACTG CGATCCGCCACCGCCAGAG 3′

5′ GGTGGCTGAGTCAGCACATAGGA CGATCCGCCACCGCCAGAG 3′

5′ TGGGGCTGAGTCAGCATAAAATT CGATCCGCCACCGCCAGAG 3′

Primers for generation of full-length (V_H + Linker + V_L) fragment with restriction enzyme sites

Human VH forward primer with Sfi I site

-----------Heavy chain V region----------

5′ GTCCTCGCAACTGCGGCCCAGCCGGCC
ATGGCCCAGGTGCAGCTGGTGCAGTCTGG 3′

5′ GTCCTCGCAACTGCGGCCCAGCCGGCC
ATGGCCCAGGTCAACTTAAGGGAGTCTGG 3′

5′ GTCCTCGCAACTGCGGCCCAGCCGGCC
ATGGCCGAGGTGCAGCTGGTGGAGTCTGG 3′

5′ GTCCTCGCAACTGCGGCCCAGCCGGCC
ATGGCCCAGGTACAGCTGCAGCAGTCAGG 3′

Human J kappa reverse primers with Not I sites

----------------Light chain V region-------

5′ GAGTCATTCTCGACTTGCGGCCGCACGTTTGATTTCCACCTTGGTCCC 3′

5′ GAGTCATTCTCGACTTGCGGCCGCACGTTTGATCTCCAGCTTGGTCCC 3′

5′ GAGTCATTCTCGACTTGCGGCCGCACGTTTAATCTCCAGTCGTGTCCC 3′

5′ GAGTCATTCTCGACTTGCGGCCGCACGTTTGATCTCCACCTTGGTCCC 3′

Human J lambda reverse primers with Not I sites

-----------------Light chain V region-------

5′ GAGTCATTCTCGACTTGCGGCCGCACCTAGGACGGTGACCTTGGTCCC 3′

5′ GAGTCATTCTCGACTTGCGGCCGCACCTAGGACGGTCAGCTTGGTCCC 3′

5′ GAGTCATTCTCGACTTGCGGCCGCACCTAAAACGGTGAGCTGGGTCCC 3′

TABLE 21.2
Primer Series for PCR Amplification of Mouse Antibody Variable Regions (V Gene)

Primers for V gene PCR
Mouse VH forward primer
5′ AGGTGCACAAGCTGCAGGCAGTCATGG 3′
Mouse JH reverse primer
5′ TGAGGAGACGGTGACCGTGGTCCCTTGGCCCC 3′
Mouse V kappa reverse primer
5′ GACATTGAGCTCACCCAGTCTCCA 3′
Mouse V lambda reverse primers
5′ CCGTTTGATTTCCAGCTTGGTGCC 3′
5′ CCGTTTCAGCTCCAGCTTGGTCCC 3′

Primer for linker fragment PCR
Mouse JH: for scFv linker
5′ GGGACCAC GGTCACCGTCTCCTCA 3′
Mouse V kappa reverse primer
5′ TGGAGACTGGGTGAGCTCAATGTC 3′

Primers for generation of full-length fragment PCR
Mouse VH forward primer with Sfi I site
---------------Heavy chain V region---------
5′ TCCTCGCAACTGCGGCCCAGCCGGCCATGGCCCAGGTGCACAAGCTGCAGGCAGTC 3′
Mouse V lambda reverse primer
--------------Light chain V region--------
5′ GAGTCATTCTGCGGCCGCCCGTTTGATTTCCAGCTTGGTGCC 3′
5′ GAGTCATTCTGCGGCCGCCCGTTTCAGCTCCAGCTTGGTCCC 3′

dNTPs (2.5 m*M* each), 4 μl
Taq or *Tth* DNA polymerase, 5 to 10 units
Add dd.H_2O to a final volume of 40 μl.

Note: The DNA polymerase should be high fidelity and long expand, which is commercially available.

2. Carefully overlay the mixture with 30 μl of mineral oil to prevent evaporation of the samples during the PCR amplification. Place the tubes in a thermal cycler and perform PCR.

Profile	Predenaturation	Cycling (35 cycles)			Hold
		Denaturation	Annealing	Extension	
Template	94°C, 3 min	94°C, 0.5 min	55°C, 0.5 min	72°C, 1.5 min	4°C

3. Purification of PCR products using appropriate reagents (i.e., a kit from Qiagen).
4. Carry out PCR amplification and agarose gel purification of recombinant, single-chain (VH–linker–VL) using forward and reverse primers with restriction enzyme sites at the ends, as illustrated in Figure 21.3.
5. Perform restriction enzyme digestion and purification of the single-chain DNA (Figure 21.4).

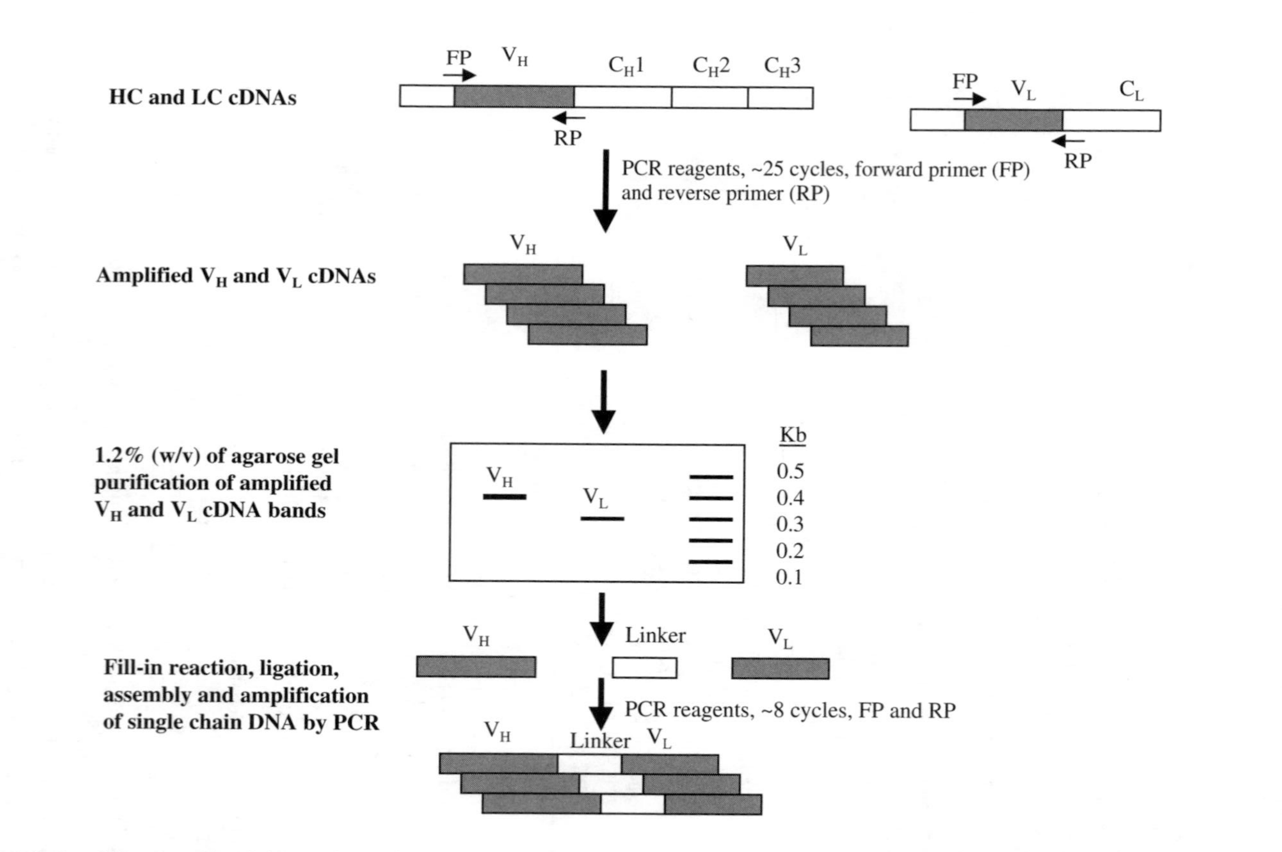

FIGURE 21.2 Amplification of variable regions of heavy chain (HC) and light chain (LC) cDNAs and assembly of recombinant single-chain cDNA.

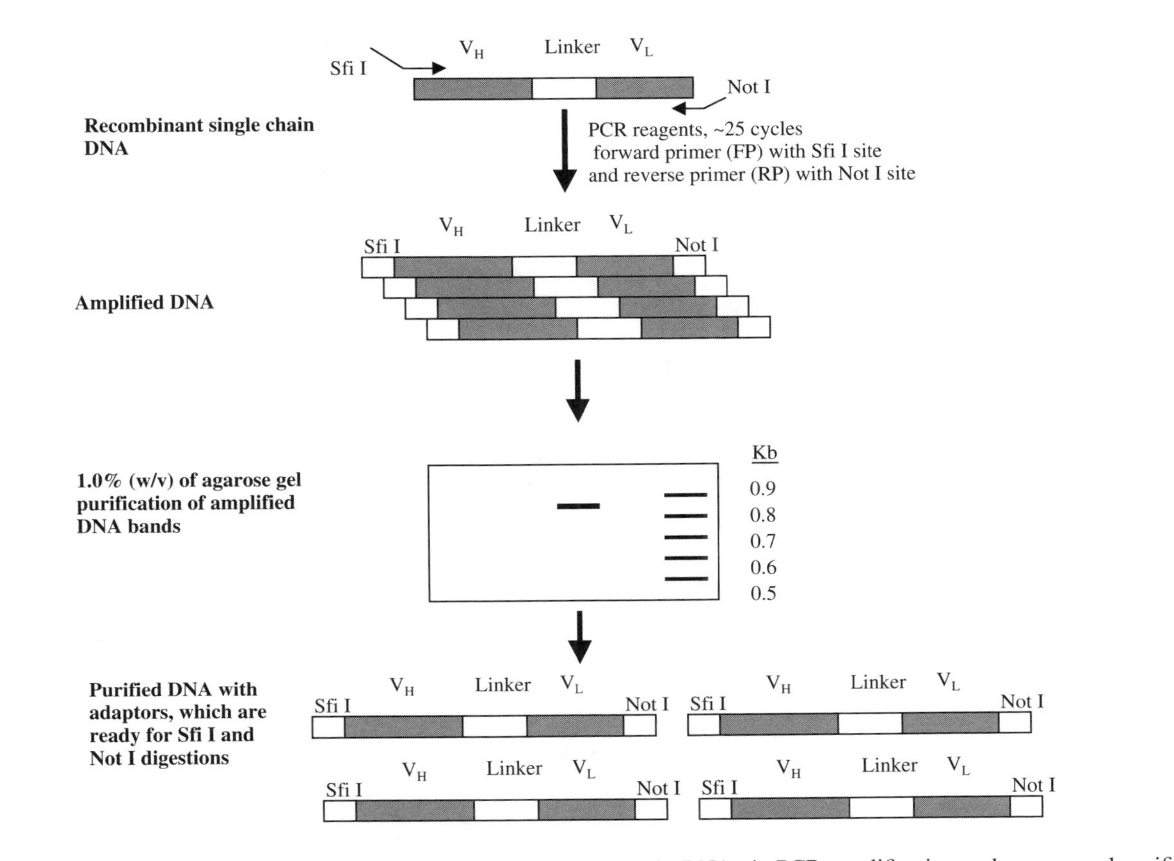

FIGURE 21.3 Addition of appropriate adapter ends to the recombinant single-chain DNA via PCR amplification and agarose gel purification.

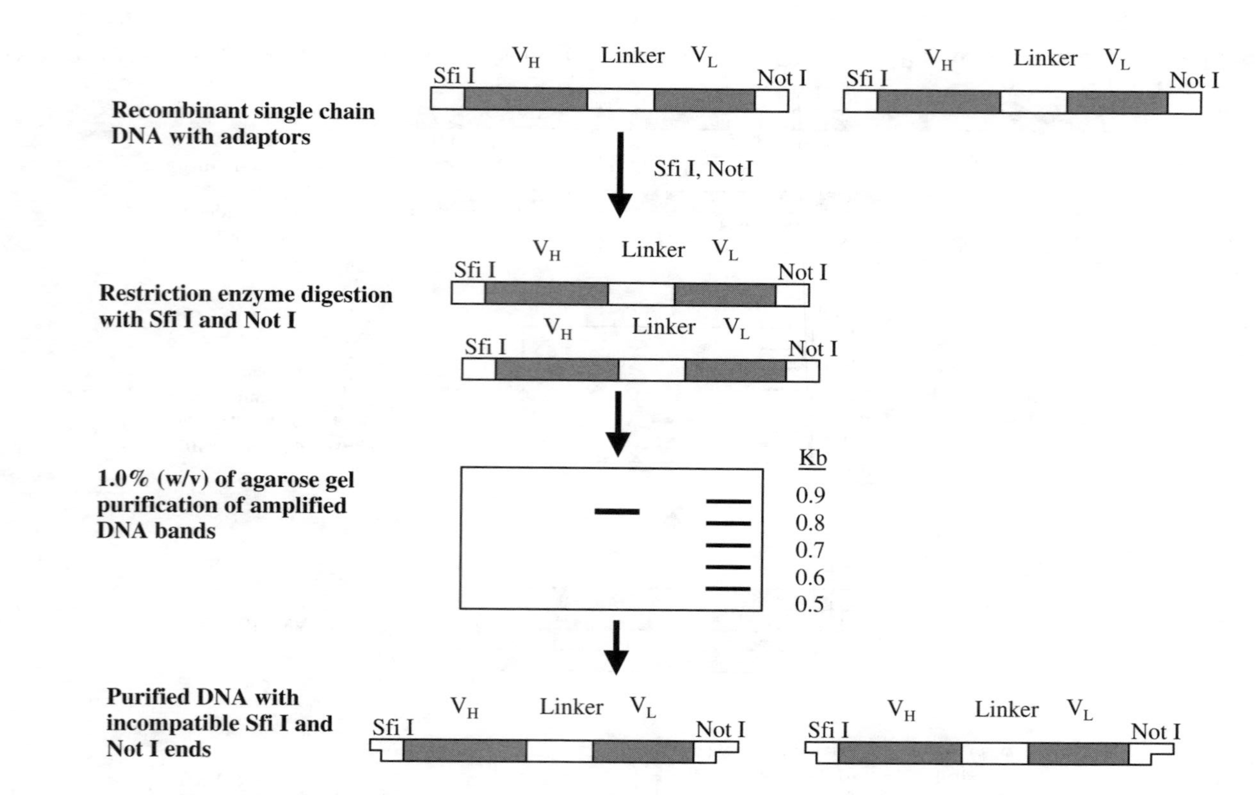

FIGURE 21.4 Generation of incompatible sticky ends by restriction enzyme digestion with *Sfi* I and *Not* I as an example.

GENERATION OF RECOMBINANT DNA CONSTRUCTS FOR EXPRESSING RECOMBINANT ANTIBODIES ON THE SURFACE OF PHAGE

As shown in Figure 21.5, following restriction enzyme digestion of recombinant, single-chain DNA (VH–linker–VL), the DNA should be ligated with appropriate vectors that are commercially available. For detailed protocols (ligation, bacterial transformation, isolation of plasmid DNA and validation of cloned sites), please refer to Chapter 4 of this book.

EXPRESSION, BIOPANNING AND CHARACTERIZATION OF PHAGE THAT PRODUCES MONOCLONAL ANTIBODIES ON ITS TIP

This general procedure is illustrated in Figure 21.6. A vector carrying an inducible *lac* promoter should be utilized for cloning and expression of recombinant antibodies. The *lac* promoter is in turn regulated by the *lac* I^q gene that encodes the *lac* repressor. Under appropriate conditions, the transcription of cloned gene under the *lac* promoter by the *E. coli* DNA-dependent RNA polymerase is inhibited by the *lac* repressor, blocking the expression of the g3p fusion protein or recombinant antibodies. Given that accumulation of g3p is toxic to the host cell, the *lac* promoter needs to be controlled tightly prior to infection with K07 helper phage to block the expression of g3p. Bacterial strains such as *E. coli* TG1 and HB2125 should be employed because they contain *lac* repressor gene.

Additionally, low temperature (i.e., 30°C) and culture medium containing 5% glucose can be used to minimize the "leaky" status of the *lac* promoter. The glucose forces the transformed bacteria to shut down alternate metabolic pathways, repressing the *lac* operon. During K07 helper phage infection, the glucose medium is removed and the *lac* promoter is allowed to express the ScFv-g3p fusion protein. Such residual, low levels of g3p will be sufficient for phage assembly. Alternatively, some commercial vectors have the *lac* promoter and the *lac* repressor gene. In this case, expression of ScFv-g3p fusion protein can be induced with 1 m*M* IPTG. It should be noted that such IPTG induction needs to be controlled because overexpression of fusion proteins can kill the infected cells, producing no phage particles.

INFECTION OF BACTERIA CARRYING RECOMBINANT DNA CONSTRUCTS AND RESCUING PHAGEMID LIBRARIES

1. Inoculate a putative bacterial colony in 50 ml of LB medium containing 3 to 5% (w/v) glucose and 100 μg/ml of ampicillin in a 250 ml of sterile flask.
2. Grow the cells for approximately 6 to 12 h at 30°C until the OD_{600} is about 0.5.

Notes: Bacteria are in midlog phase and express the F pilus. If the OD is over 0.5, dilution can be done.

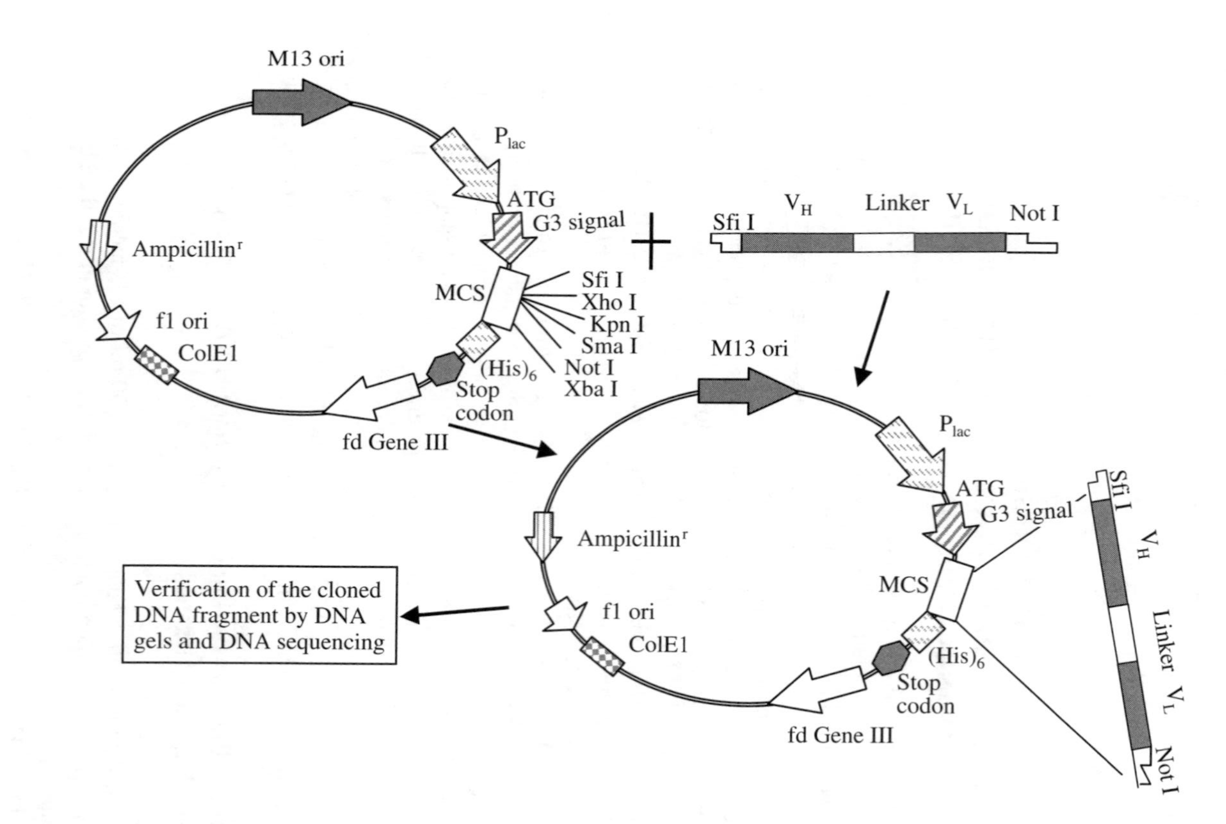

FIGURE 21.5 Cloning and generation of recombinant DNA construct for expressing recombinant antibodies and proteins of interest.

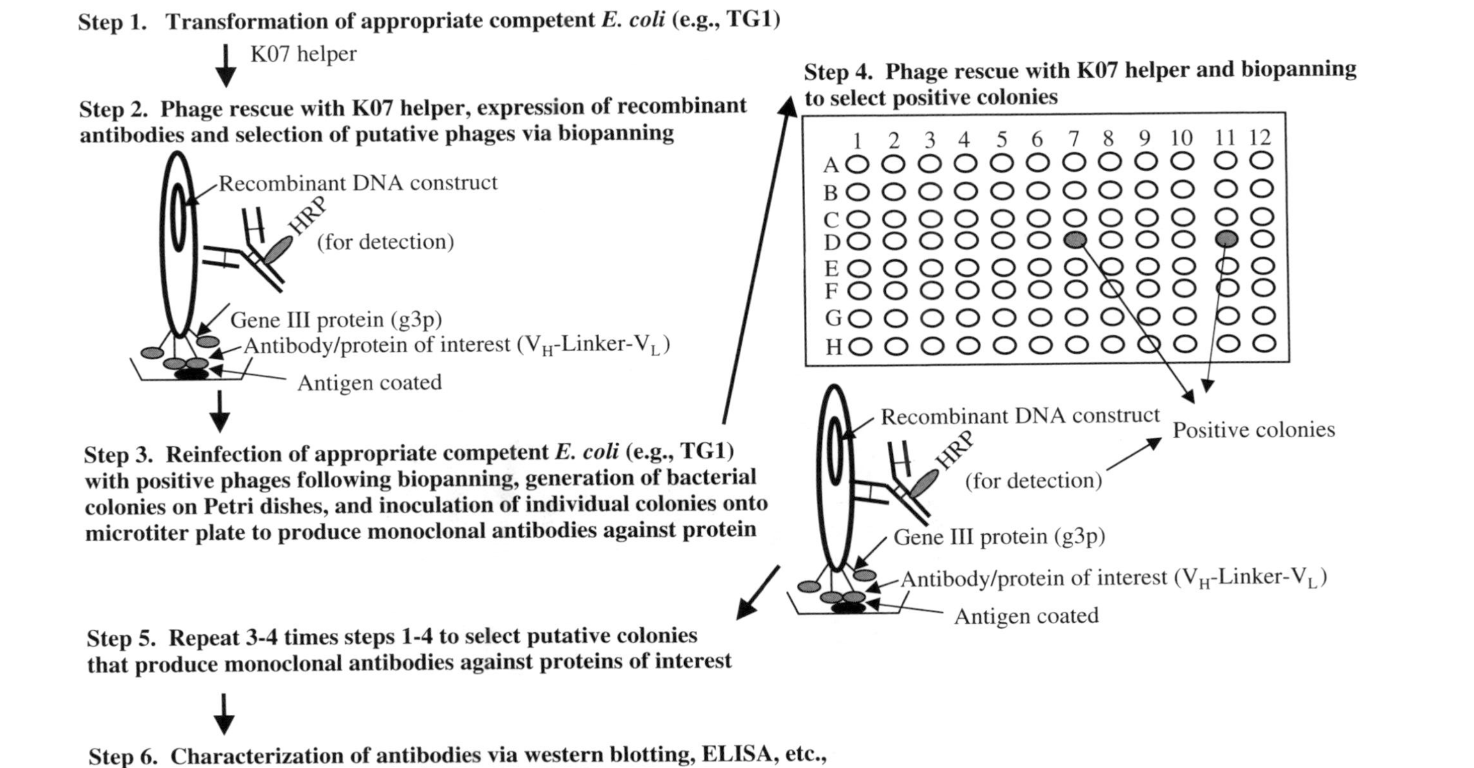

FIGURE 21.6 Expression and purification of phages that produce monoclonal antibodies against proteins of interest.

3. Perform infection by adding 2×10^{10} pfu M13K07 helper phage to the 50 ml of LB culture and grow the cells at 37°C for 1 h.

Note: The ratio of helper phage to bacteria is approximately 1:1.

4. Centrifuge at $3000 \times g$ for 10 min and resuspend the cells in 500 ml of LB medium containing 100 μg/ml of ampicillin and 50 μg/ml of kanamycin in a 2-l flask.

Note: The culture medium at this point should not contain glucose. Removing glucose serves to induce intact fusion g3p.

5. Incubate at 30°C overnight with vigorous shaking.
6. Harvest the phage–antibody particles by centrifugation and PEG precipitation. Add 0.2 volume of precipitation solution (20% PEG, 2.5 *M* NaCl) and allow to precipitate for 30 to 60 min at 4°C, followed by centrifugation at $5000 \times g$ for 10 min.
7. Resuspend the particles in 0.5 ml of PBS or TE buffer.

Note: This usually gives rise to 5×10^{11} pfu/ml of infectious particles.

Selection of Phage–Antibody Particles by Biopanning

1. Coat a 96-well microtiter/ELISA plate overnight with 150 μl/well of antigen solution at 4°C or for 2 to 3 h at room temperature.

Note: The antigen of interest can be dissolved in 20 mM Tris buffer (pH 7.5) or 50 mM sodium hydrogen carbonate, pH 9.5. The concentration of antigen can be 0.1 to 1mg/ml.)

2. Briefly rinse the plate with PBS and block wells with 0.35 ml/well of PBS containing 5% (w/v) nonfat dry milk for 2 h at room temperature.
3. Add approximately 1 x 10^6 pfu phage per well and allow incubation to occur for 1 to 2 h at room temperature with shaking.
4. Wash the plate 3 x 5 min with PBS containing 0.1% Tween-20.

Note: This step serves to wash away nonspecific phage particles. Particles with recombinant antibodies on their surface should bind to the coated antigen.

5. Add 150 μl of detection antibody solution (i.e., labeled antibodies against g8p on the surface of phage diluted at 1:5000 in PBS. The antibodies can be labeled with alkaline phosphatase, AP or horseradish peroxidase, HRP) to each well. Incubate at room temperature for 2 h with agitation.
6. Wash the plate as described in step 4.
7. Detect positive wells by adding 100 μl of substrate solution for AP or HRP for 20 min at room temperature. Colored wells indicate presence of antibody–phage particles.
8. Elute the phage from the positive wells by adding 0.2 ml/well of elution buffer (100 m*M* triethylamikne or 100 m*M* glycine (pH 3.0). Elute for 15 min at room temperature with agitation.

Note: Once the phage particles are eluted, the solution should be neutralized immediately using 0.1 ml of 1 M *Tris buffer (pH 7.5). Store the viral particles at 4°C until use.*

9. Reinfect bacteria with the eluted phage particles as described in the previous section and further selection of putative phage particles as described in step 1 to step 8 in this section.
10. Characterize recombinant antibodies via western blotting or ELISA.
11. Following putative phage (monoclonal) identification, production and validation of monoclonal antibodies can be scaled up against specific proteins of interest.

REFERENCES

1. McCafferty, J., Grffiths, A.D., Winter, G., and Chiswell, D.J., Phage antibodies: filamentous phage displaying antibody variable domains, *Nature*, 348, 552–554, 1990.
2. Hoogenboom, H.R. and Winter, G., By-passing immunization: human antibodies from synthetic repertoires of germline VH gene segments rearranged *in vitro*, *J. Mol. Biol.*, 227, 381–388, 1992.
3. McCafferty, J. and Johnson, K.S., Construction and screening of antibody display libraries, in *Phage Display of Peptides and Proteins,* Kay, B.K., Winter, J., and McCaffey, J. (Eds.). Academic Press, San Diego, CA, 1996, 79–110.
4. Huston, J.S., Levinson, D., Mudgett, H.M., Tai, M.S., Novotny, J., Margolies, M.N., Ridge, R.J., Bruccoleri, R.E., Haber, E., Crea, R., and Opperman, H., Protein engineering of antibody binding sites: recovery of specific activity in an antidigoxin single-chain Fv analogue produced in *E. coli, Proc. Natl. Acad. Sci. USA*, 85, 5879–5883, 1988.
5. Willats, W.G., Phage displaly: practicalities and prospects, *Plant Mol. Biol.*, 50(6):837–854, 2002.
6. Steinhauer, C., Wingren, C., Hager, A.C., and Borrebaeck, C.A., Single framework recombinant antibody fragments designed for protein chip applications. *Biotechniques*, 38–45, 2002.
7. Wu, W., Construction and screening of subtracted and complete expression cDNA libraries, in *Methods in Gene Biotechnology,* 1st ed., Wu, W., Welsh, M.J., Kaufman, P.B., and Zhang, H., (Eds.), CRC Press, Boca Raton, FL, 1997, 29–65.
8. Wu, W. and Zhang, H., Subcloning of gene or DNA fragments, in *Methods in Gene Biotechnology,* 1st ed., Wu, W., Welsh, M.J., Kaufman, P.B., and Zhang, H., (Eds.,) CRC Press, Boca Raton, FL, 1997, 67–88.
9. Sidhu, S.S., Fairbrother, W.J., and Deshayes, K., Exploring protein–protein interaction with phage display, *Chembiochemistry,* 4(1): 14–25, 2003.
10. Krichevsky, A., Graessmann, A., Nissim, A., Piller, S.C., Zakai, N., and Loyter, A., Antibody fragments selected by phage display against the nuclear localization signal of the HIV-1 Vpr protein inhibit nuclear import in permeabilized and intact cultured cells, *Virology,* 305(1):77–92, 2003.
11. Hartley, O., The use of phage display in the study of receptors and their ligands, *J. Recept. Signal. Transduct. Res.*, 22(1):373–392, 2002.

22 Down-Regulation of Gene Expression in Mammalian Systems via Current siRNA Technology

CONTENTS

INTRODUCTION

Short-interfering RNA (siRNA) — double-stranded RNA (dsRNA) with approximately 21 nt in length — has been a powerful tool widely used for down-regulation of expression of specific genes.[1–5] Introduction of siRNAs to mammalian cells leads to the sequence-specific destruction of endogenous mRNA molecules complementary to siRNAs (Figure 22.1). The cellular process that undergoes degradation of target mRNA via siRNA is termed RNA interference (RNAi). During RNAi, siRNAs appear to serve as guides for enzymatic cleavage of complementary RNAs. In addition, in some systems, siRNAs can function as primers for an RNA-dependent RNA polymerase that can synthesize additional siRNAs, which can enhance the effects of RNAi.[1,4–8] Nevertheless, the mechanism by which complementary RNAs can be destroyed by siRNAs is not well understood and the specific enzymes that mediate siRNA functions remain to be identified.

It has been demonstrated that relatively long siRNAs (>30 nt in length) do not work well in mammalian cells in terms of RNAi due to the nonspecificity of siRNAs. However, siRNAs with 19 to 21 nt in length are very effective because they are too short to trigger nonspecific dsRNA response.[5,9–11] Therefore, 19- to 21-nt double-stranded siRNAs have been widely utilized in down-regulation or knocking-down of expression of genes of interest.

Initially, siRNAs were synthesized *in vitro* and then applied to specific cells to inhibit mRNA of interest. Although this technique has been successful, it is time consuming, costly and less effective. Recently, development of vectors has overcome the limitations of traditional methodology. Double-stranded siRNA oligonucleotides can be cloned downstream of the U2, U6 or Hi RNA polymerase III promoter that leads to expression of siRNAs *in vivo*. This chapter provides a guide to utilize siRNA technology to inhibit the expression of genes of interest.

DESIGN OF OLIGONUCLEOTIDES FOR HAIRPIN siRNA

SELECTION OF siRNA TARGET SITE

It is imperative to select an optimal target site of specific mRNA. The criteria and recommendations can be summarized as follows:

1. Target region to be considered:
   a. The upstream nucleotide sequence from the initiation codon AUG and the first codon AUG should be selected as the most effective target region when designing an oligomer for siRNA. The hybridization of the oligomers may prevent the small subunit of the ribosome from interacting with the upstream region of the initiating AUG codon, thus blocking the initiation of translation.
   b. SiRNA oligonucleotides can also be designed to be complementary to the coding regions to block the chain elongation of the specific polypeptide. However, this is less effective than the selection described in (a).
   c. The terminal region of translation is a good target to block the termination of translation. The partial-length polypeptide may not be folded properly and will be rapidly degraded.
   d. Unlike design of primers for PCR, which involves the *Tm* value, the siRNA oligomers for inhibition of gene expression do not need computer analysis of *Tm*.

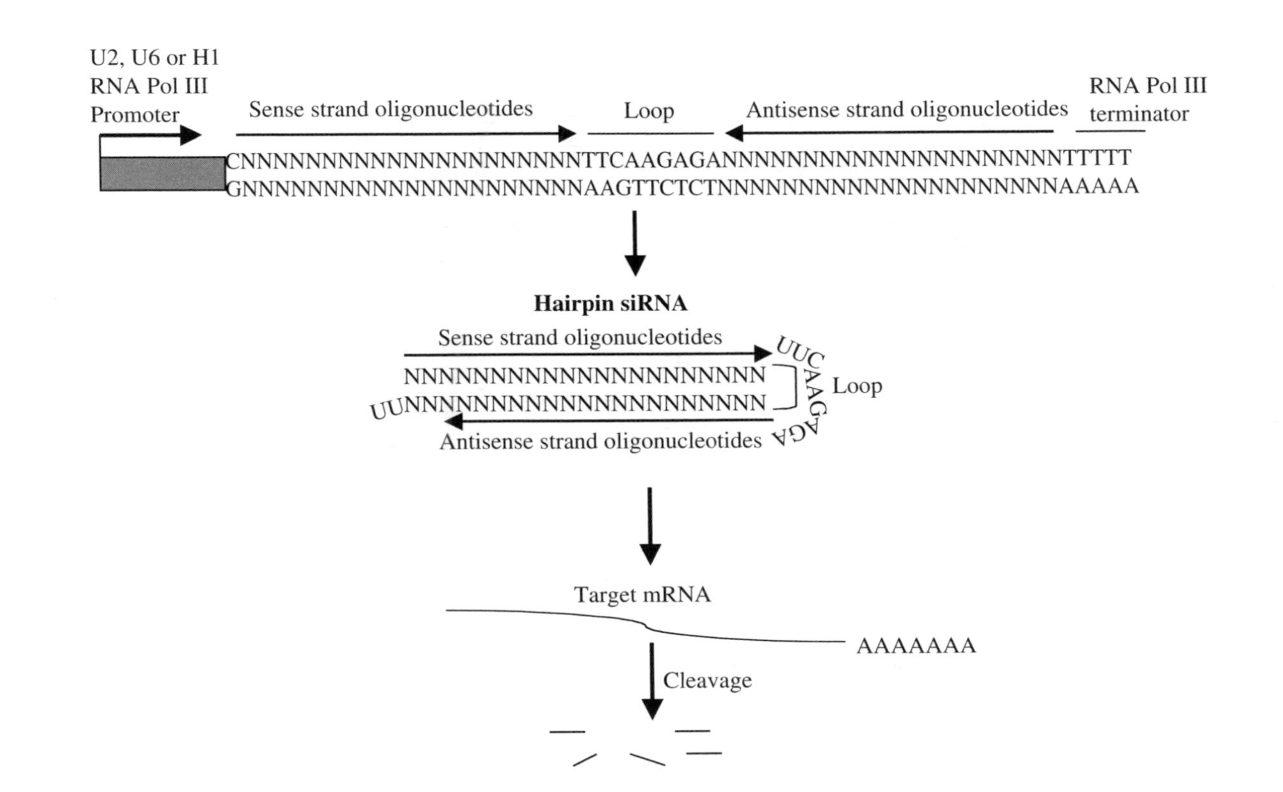

FIGURE 22.1 Down-regulation of mRNA expression via siRNA.

2. Target sequences: For targeting a single mRNA, it is necessary to analyze the full-length cDNA among other cDNAs of the family using appropriate DNA software such as Lasergene. To avoid inhibiting nontargeting genes, it is important to identify a unique sequence as the target site of the specific cDNA of interest. It would be a big mistake to select highly homologous sequences.

Selection of Oligomer Size and Orientation for siRNAs

1. To down-regulate the expression of a specific gene effectively, it is recommended to design 53 to 55 nt oligonucleotide fragment for generation of hairpin siRNA, which is composed of 19- to 21-nt target sequence (sense orientation) + 9-nt loop + 19- to 21-nt target sequence (antisense orientation) + 3- to 4-nt 3′ terminal thymidine tract (Figure 22.1). An example would be as follows:

   5′ GCTGGACCTGACGTAACCGTA (sense target sequence) TTCAAGAGA (a loop) TGACCTGGACTGCATTGGCAC (antisense target sequence) TTTT (terminal sequence) 3′.

   Additionally, specific restriction enzyme sites may be added at 5′ and 5′ ends for cloning into the appropriate vector.
2. Given that it is more effective for RNA polymerase III to initiate transcription of siRNAs if the 5′ end of oligonucleotides is a purine, it is recommended that a G or A be designed at the 5′ end of the sense target sequence.
3. Design another oligonucleotide fragment that is complementary to the sequence listed in (a). For instance, 3′ CGACCTGGACTGCATTGGCAT (sense target sequence) AAGTTCTCT (loop) ACTGGACCTGACGTAACCGTC (antisense target orientation) AAAA 5′. This oligonucleotide fragment will hybridized to the fragment in (a), generating a double-stranded DNA fragment to be cloned into an expression vector (Figure 22.1).

Synthesis of Hairpin siRNA-Encoding Oligonucleotides

Based on the design of the oligonucleotides, an appropriate scale of oligos can be ordered and chemically synthesized; no extra purification is necessary.

PREPARATION OF DOUBLE-STRANDED OLIGONUCLEOTIDE FRAGMENT FOR CLONING

1. Dissolve the oligonucleotides synthesized in the previous section in TE buffer (10 m*M* Tris, pH 8.0, 1 m*M* EDTA) to final concentration of 1 μg/μl.
2. Anneal both strands of hairpin siRNA oligonucleotides as follows:

   Annealing solution, 25 μl
   Sense strand oligonucleotide fragment solution, 1 μl
   Antisense strand oligonucleotide fragment solution, 1 μl
   Total volume = 25 μl

3. Heat the mixture to 92°C for 5 min and slowly cool to RT and incubate for 1 h at room temperature. Store the annealed double-stranded oligonucleotide fragment or hairpin siRNA oligo insert at 4°C until use.

CLONING HAIRPIN siRNA OLIGO INSERT INTO APPROPRIATE VECTORS

As illustrated in Figure 22.2, a double-stranded hairpin siRNA oligo insert can be ligated to an appropriate vector for expressing hairpin siRNA. The procedure is as follows:

1. Assemble a ligation reaction in an Eppendorf tube
   Restriction enzyme digested vector DNA, 1 μg
   Hairpin siRNA oligoinsert DNA, 20 to 50 ng
   10X Ligase buffer, 2 μl
   T4 DNA ligase (Weiss units), 10 units
   Add dd.H_2O to a final volume of 20 μl.
2. Incubate the ligation for 2 h to overnight at 4°C. The ligated mixture is ready for transformation to bacteria.

Notes: Double-stranded hairpin siRNA oligonucleotides can be fused with a reporter gene system so that inhibition of reporter gene activity can be rapidly and accurately measured or observed under a fluorescent microscope (Figure 22.3).

3. Thaw an aliquot of 0.2 ml of frozen $CaCl_2$-treated competent cells on ice.
4. Assemble the transformation as follows:

Components	Tube 1	Tube 2
Cells	50 μl	50 μl
DNA	50 ng in 5 μl	200 ng in 5 μl

5. Incubate on ice for 50 to 60 min.
6. Heat-shock at 42°C for 1 to 1.5 min and quickly place the tubes on ice for 1 min.

Tip: Heat-shocking and chilling on ice should not extend beyond the time indicated. Otherwise, the efficiency of transformation will decrease 10- to 20-fold.

7. Transfer the cell suspension to sterile culture tubes containing 1 ml of LB medium without antibiotics. Incubate at 37°C for 1 to 2 h with shaking at 140 rpm to recover the cells.
8. Transfer 50 to 150 μl of the cultured cells per plate onto the center of LB plates containing 50 μg/ml ampicillin (or 0.5 m*M* IPTG and 40 μg/ml X-Gal for color selection). Immediately spread the cells over the entire surface of the LB plates using a sterile, bent glass rod.
9. Invert the plates and incubate them at 37°C overnight. Selected colonies should be visible 14 h later.

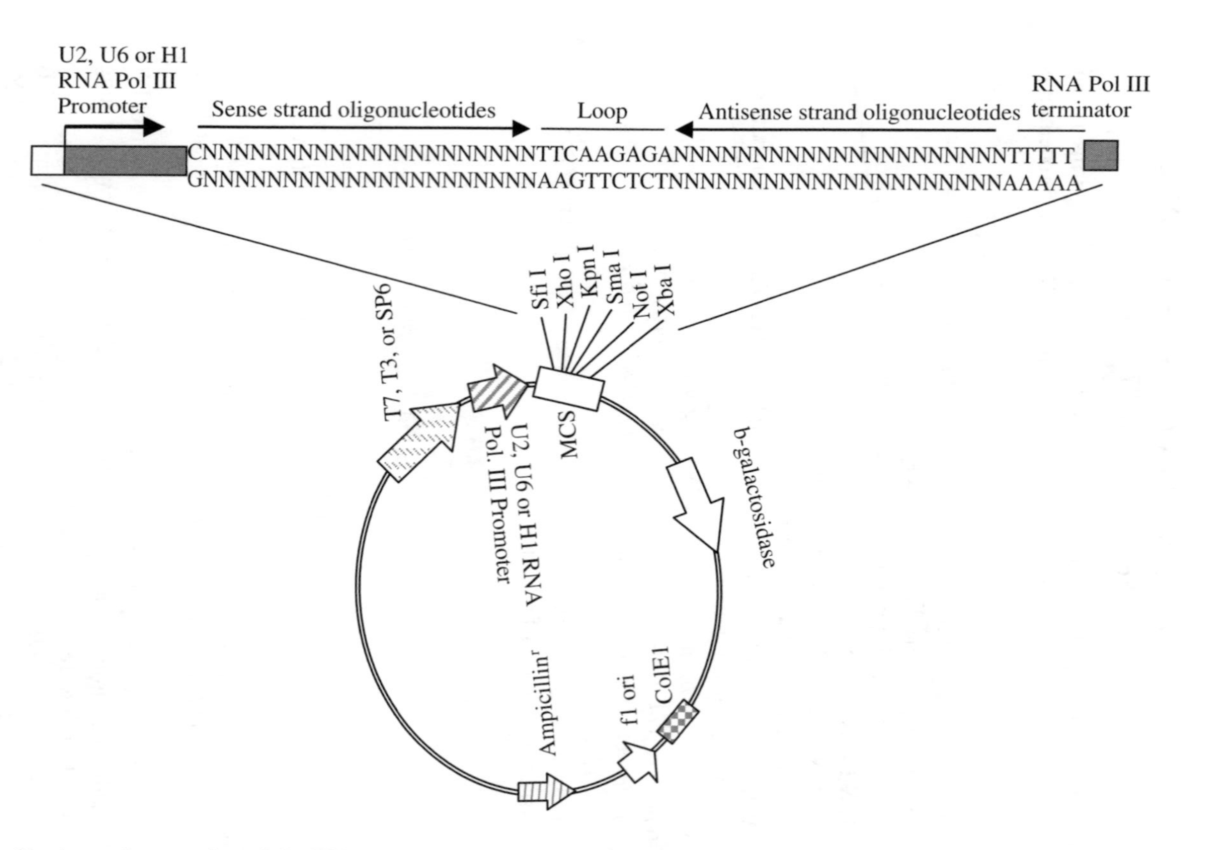

FIGURE 22.2 Cloning and generation of the DNA construct for expression of siRNA.

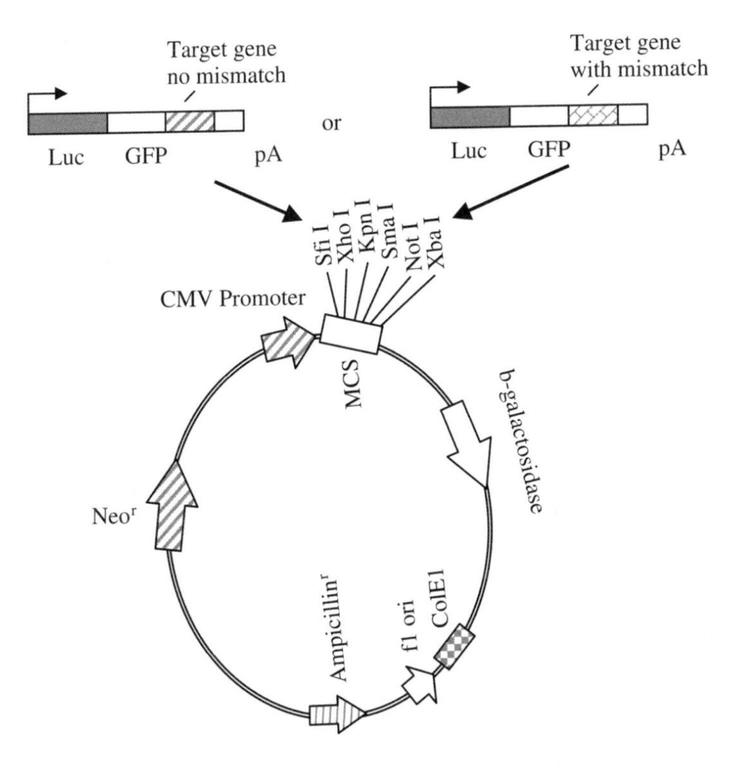

FIGURE 22.3. Cloning and generation of reporter gene constructs for siRNA/fluorescent assays.

PURIFICATION OF PLASMID DNA CARRYING HAIRPIN siRNA INSERT

1. Inoculate a single colony in 5 ml of LB medium and 50 μg/ml of appropriate antibiotics (e.g., ampicillin), depending on the specific antibiotic-resistant gene carried by the specific plasmid. Culture the bacteria at 37°C overnight with shaking at 200 rpm.
2. Transfer the overnight culture into microcentrifuge tubes (1.5 ml/tube) and centrifuge at 6000 rpm for 4 min. Carefully aspirate the supernatant.

Note: If the plasmid carried by bacteria is a high-copy number plasmid, 1.5 ml of cell culture usually gives a yield of approximately 40 μg plasmid DNA using the present method. In this way, 10 tubes of preparation can generate approximately 400 μg of plasmid DNA, which is equivalent to a large-scale preparation. Therefore, the volume of bacterial culture used for purification of plasmid DNA depends on a particular experiment.

3. Add 0.1 ml of ice-cold lysis buffer to each tube and gently vortex for 1 min. Incubate at room temperature for 5 min.

Note: The function of this step is to lyse the bacteria by hyperlytic osmosis, releasing DNA and other contents.

4 Add 0.2 ml of a freshly prepared alkaline solution to each tube and mix by inversion. **Never VORTEX.** Place the tubes on ice for 5 min.

Note: This step denatures chromosomal DNA and proteins.

5. Add 0.15 ml of ice-cold potassium acetate solution to each tube. Mix by inversion for 10 s and incubate on ice for 5 min.

Note: The principle of this step is to precipitate proteins, polysaccharides and genomic DNA selectively. Under this particular buffer condition, plasmid DNA still remains in the supernatant without being precipitated.

6. Centrifuge at 12,000 × *g* for 5 min and carefully transfer the supernatant to fresh tubes.
7. To degrade total RNAs from the sample, add RNase (DNase free, Boehringer Mannheim) into the supernatant at 1.5 μl/10 μl supernatant. Incubate the tubes at 37°C for 30 min.
8. Transfer the RNase-treated supernatant into a minifilter cup (0.4 ml/cup) placed in an Eppendorf tube. Cap the tube and centrifuge at 12,000 × *g* for 30 s. Remove the filter cup and the collected supernatant in the Eppendorf tube is ready for DNA precipitation.

Note: The membrane filter at the bottom of the cup is specially designed to retain proteins and allow nucleic acids to pass through. One cup can be reused four to five times if it is used for the same supernatant pool. However, it is not wise to reuse it for different plasmid samples due to potential contamination. The author routinely utilizes this type of filter for plasmid DNA preparation. The DNA obtained is pure and high yield with a ratio (A_{260}/A_{280}) of approximately 2.0.

9. Add 2 to 2.5 volumes of 100% ethanol to each tube. Mix by inversion and allow the plasmid DNA to precipitate for 10 min at –80°C or 30 min at –20°C.
10. Centrifuge at 12,000 × *g* for 10 min. Carefully decant the supernatant, add 1 ml of prechilled 70% ethanol and centrifuge at 12,000 × *g* for 5 min. Carefully aspirate the ethanol and dry the plasmid DNA under vacuum for 15 min.
11. Dissolve the plasmid DNA in 15 μl of TE buffer or sterile deionized water. Take 1 μl to measure the concentration and store the sample at –20°C until use.

Note: At this point, the DNA is ready for transfection into cells of interest.

Reagents Needed

LB (Luria-Bertaini) Medium (per Liter)

10 g Bacto-tryptone
5 g Bacto-yeast extract
5 g NaCl
Adjust pH to 7.5 with 2 *N* NaOH and autoclave. If needed, add antibiotics after the autoclaved solution has cooled to less than 50°C.

TE Buffer

10 m*M* Tris-HCl, pH 8.0
1 m*M* EDTA

Lysis Buffer

25 m*M* Tris-HCl, pH 8.0
10 m*M* EDTA
50 m*M* Glucose

Alkaline Solution

0.2 *N* NaOH
1% SDS

Potassium Acetate Solution (pH 4.8)

Prepare 60 ml of 5 *M* potassium acetate. Add 11.5 ml of glacial acetic acid and 28.5 ml of H_2O to total volume of 100 ml. This solution is 3 *M* with respect to potassium and 5 *M* with respect to acetate. Store on ice prior to use.

RNase (DNase free)

Boehringer Mannheim, Cat. No: 119915, 500 μg/ml.

TRANSIENT TRANSFECTION OF CELLS OF INTEREST WITH PURIFIED PLASMID DNA TO EXPRESS siRNAs FOR INHIBITING TARGET mRNA

Mammalian cells can be transfected with commercially available reagents and kits (i.e., Roche–Molecular Chemicals, Invitrogen, etc.). Calcium phosphate methods may also be used. The principle of calcium phosphate transfection is that the DNA to be transferred is mixed with $CaCl_2$ and phosphate buffer to form a fine calcium phosphate precipitate. The precipitated complexes or particles are then placed on a cell monolayer. They bind or attach to the plasma membrane and are taken into the cell by endocytosis. This is one of the most widely used methods for transient and stable transfection.

1. Perform trypsinizing to remove adherent cells for subculture or cell counting as well as for monolayer cell preparation:
   a. Aspirate the medium from the cell culture or the tissue culture dish and wash the cells twice with 10 to 15 ml of PBS buffer or other calcium- and magnesium-free salt solutions.
   b. Remove the washing solution and add 1 ml of trypsin solution per 100-mm dish. Quickly overlay the cells with the solution and incubate at 37°C for 5 to 10 min, depending on the cell type (e.g., NIH 3T3, 2 min; CHO, 2 min; HeLa, 2.5 min; COS, 3 min).

c. Closely monitor the cells. Once they begin to detach, immediately remove the trypsin solution and tap the bottom and sides of the culture flask or dish to loosen the remaining adherent cells.

d. Add culture medium containing 10% serum (e.g., FBSor FCS) to the cells to inactivate the trypsin. Gently shake the culture dish. The cells can be used for cell subculturing or counting.

2. Prepare cell monolayers the day prior to the transfection experiment. The cell monolayers should be at 60 to 70% confluence. Aspirate the old culture medium and feed the cells with 5 to 10 ml of fresh culture medium.

Notes: (1) It is recommended that this procedure be carried out in a sterile laminar flow hood. Gloves should be worn to prevent contamination of the cells. (2) The common cell lines used for transfection are NIH 3T3, CHO, HeLa, COS and mouse embryonic stem (ES) cells. Other mammalian cells of interest can also be transfected in this way. (3) The density of the cells to be seeded depends on the particular cell line. A general guideline is 1 to 5 × 10^6 cells per 100-mm culture dish. (4) All of the solutions and buffers should be sterilized by autoclaving or by sterile-filtration.

3. Prepare a transfection mixture for each 100-mm dish as follows:

   2.5 M $CaCl_2$ solution, 50 µl
   DNA construct or constructs (cotransfection), 15 to 30 µg
   Add dd.H_2O to a final volume of 0.5 ml.

Mix well, and slowly add the DNA mixture dropwise to 0.5 ml of well-suspended 2X HBS buffer with continuous mixing by vortexing. Incubate the mixture at room temperature for 30 to 45 min to allow coprecipitation to occur.

Notes: After the DNA mixture has been added to the 2X HBS buffer, the mixture should be slightly opaque due to the formation of fine coprecipitate particles of DNA with calcium phosphate.

4. Slowly add the precipitated mixture dropwise to the dish containing the cells while continuously swirling the plate. For each 100-mm dish add 1 ml of the DNA mixture, or add 0.5 ml of the DNA mixture to each 60-mm dish.
5. Place the dishes in a culture incubator overnight at 37°C and provide 5% CO_2.
6. Aspirate the medium and wash the monolayer once with 10 ml of serum-free DMEM medium and once with 10 ml of PBS. Add fresh culture medium and return the dishes to the incubator. In general, cells can be harvested 36 to 48 h after transfection for transient expression analysis.

Reagents Needed

10X HBS Buffer

5.94% (w/v) HEPES [4-(2-hydroxyethyl)-1-piperazine ethanesulfonic acid], pH 7.2
8.18% (w/v) NaCl
0.2% (w/v) Na_2HPO_4

2X HBS Buffer

Dilute the 10X HBS in dd.H_2O.
Adjust the pH to 7.2 with 1 *N* NaOH solution.

DNA to Be Transferred

Circular or linear recombinant DNA constructs in 50 µl of TE buffer, pH 7.5

2.5 M $CaCl_2$

DMEM Culture Medium

Dissolve one bottle of Dulbecco's modified eagle's medium (DMEM) powder in 800 ml dd.H_2O; add 3.7 g $NaHCO_3$; adjust the pH to 7.2 with 1 *N* HCl. Add dd.H_2O to 1 l. Sterile-filter the medium (do not autoclave) in a laminar flow hood. Aliquot the medium into 200- to 500-ml portions in sterile bottles. Add 10 to 20% FBS, depending on the particular cell type, to the culture medium. Store at 4°C until use.

PBS Buffer

0.137 *M* NaCl
4.3 m*M* Na_2HPO_4
2.7 m*M* KCl
1.47 m*M* KH_2PO_4
The final pH is 7.2 to 7.4.

1X Trypsin–EDTA Solution

0.05% (w/v) Trypsin
0.53 m*M* EDTA, pH 7.6
Dissolve in PBS buffer, pH 7.4 to 7.6.

ISOLATION OF TOTAL RNA FROM TRANSFECTED CELLS

1. Add an appropriate volume of RNA isolation buffer such as Trizol reagents to the tube containing approximately 1 ml per 10^{5-6} cells.

Warning: Trizol or phenol is very toxic and should be handled carefully. For protection, wear gloves, lab coat and safety glasses. Gloves should be changed periodically during RNA isolation to minimize RNase activities. Whenever possible, Trizol reagents or phenol should be handled in a fume hood with a glass door placed between the face and the inside of the hood.

2. Suspend the cells and allow them to lyse for 5 to 10 min at room temperature.
3. Add 0.2 ml of chloroform to each 1 ml of lysate, tightly cap the tubes and vigorously shake each tube for approximately 30 s to optimize the lysis of cells and shear genomic DNA. Place the tubes on a bench for 2 min.

4. Centrifuge for 10 min at 4°C. Label a set of clean microcentrifuge tubes.
5. Carefully transfer the top aqueous phase to fresh tubes.

Tips: This is a critical step in terms of RNA purity. The liquid should be transferred a little at a time (e.g., 100- to 150-µl/transfer) starting from the top portion to sequentially close to the interphase. In general, 250 to 350 µl of the liquid can be transferred. Whenever possible during the transfer, do not disturb the interphase containing proteins, lipids, carbohydrates, genomic DNA, etc. Once the top transferring liquid lowers to approximately 2 mm to the interphase, stop. Do not be greedy! Otherwise, the RNA will be contaminated, which will be evidenced by a low ratio of A_{260}/A_{280} (e.g., <1.5).

6. Add cold isopropanol to each tube (0.8 volume of the transferred liquid) and allow RNA to be precipitated for 1 to 2 h at –20°C. Precipitation for a longer period is acceptable.
7. Centrifuge for 10 min at 4°C. Decant the liquid; the RNA pellet should be visible at the bottom of the tube. Add 1 ml of 75% of ethanol to each tube to wash residual salt.
8. Centrifuge for 5 min at 4°C.
9. Dry the RNA samples in a speed vac for about 20 min. Avoid overdrying the samples.
10. Dissolve RNA in 100 to 200 µl RNase-free water or TE buffer for 40 min at RT or 20 min at 50°C.
11. Determine the concentrations of RNA samples using a spectrophotometer according to the instructions.

Note: An excellent RNA sample should have a ratio of A_{260}/A_{280} of >1.9 to 2.0. In contrast, an A_{260}/A_{280} ratio of <1.5 is a poor RNA preparation that contains proteins.

12. Run 5 µg RNA of each sample in 1% agarose denaturing gel with ethidium bromide and carry out electrophoresis until the first blue dye reaches approximately 2 cm from the end of the gel. Take a picture of the gel.

Tips: A good RNA preparation should display a sharp band of 28S rRNA, a sharp band of 18S rRNA and a smear in each lane. The signal intensity of 28S rRNA:18S rRNA bands should be approximately 3:1. In case of RNA degradation, no visible or very weak rRNA bands, or only an 18S rRNA band, is visible.

13. Store RNA samples at –80°C until use.

ANALYSIS OF INHIBITION OF SPECIFIC mRNA VIA siRNAs BY NORTHERN BLOTTING

Once RNAs are isolated from the transfected or nontransfected cells, analysis of inhibition of target mRNA can be performed by northern blot hybridization or other methodologies. Detailed protocols for northern blot hybridization are described in Chapter 10 of this book. As shown in Figure 22.4, if the siRNA experiment goes well, the target mRNA level in cells transfected with the vector without a hairpin siRNA oligonucleotide fragment should be similar to that in nontransfected cells serving as a control group. In contrast, the mRNA level in cells transfected with the DNA construct containing a hairpin siRNA oligonucleotide fragment should be significantly inhibited (approximately 70 to 90% reduction) compared with non-transfected cells used as control. It should be noted that if the reduction level of the targeting mRNA is less than 30% that of the control group, it may be necessary to optimize conditions so that siRNA effects can be over 50%. It is recommended to hybridize the membrane simultaneously with the probe for target mRNA and the probe for internal control, such as actin or GAPDH, that can serve to normalize the mRNA level of the target gene.

EXTRACTION OF PROTEINS FOR PROTEOMIC OR REPORTER GENE ANALYSIS

Isolation of Proteins for Western Blotting

Following aspiration of culture medium, cells can be lysed with 1 ml/100-mm Petri dish of lysis buffer (100 m*M* Tris, pH 6.8, 1% SDS, 5% β-mercaptoethanol). Overlay the entire dish and allow cells to lyse for approximately 5 min. Transfer the protein solution to a tube ready for protein concentration determination (see Chapter 11 of this book for detailed procedures[7,8]).

Extraction of Proteins for Reporter Activity Assays

For reporter gene activity assays, using a mild lysis buffer without detergents, such as SDS and mercaptoethanol, is recommended to preserve enzymatic activities. Following aspiration of culture medium, add 1 ml/100-mm Petri dish of lysis buffer (e.g., 100 m*M* Tris, pH 7.0, 1.5% Triton 100-X, 05 *M* urea) to the dish. Overlay the entire surface of the dish with buffer, scrape the cells from the dish and transfer the mixture into a tube. Allow the cells to lyse for approximately 1 to 2 h with shaking at 4°C. Proceed to determine the protein concentration (see Chapter 11 for detailed protocols).

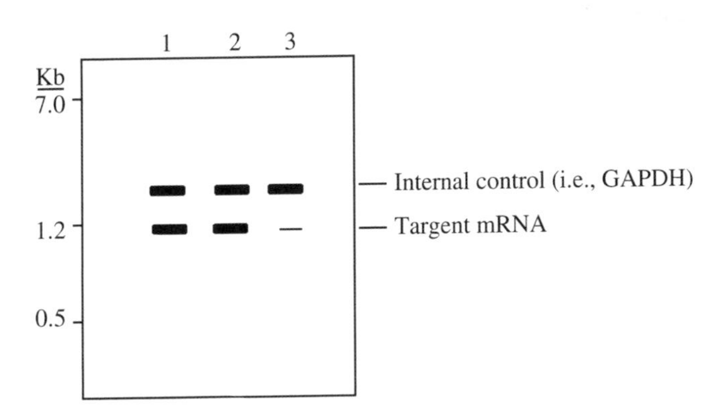

FIGURE 22.4 Northern blot analysis of inhibition of the target mRNA via siRNA. Lane 1: nontransfected cells. Lane 2: cells transfected with vector without siRNA construct. Lane 3: cells transfected with siRNA constructs.

ANALYSIS OF siRNA-MEDIATED INHIBITION OF GENE EXPRESSION AT PROTEOMIC LEVEL BY WESTERN BLOTTING

To characterize down-regulation of gene expression via siRNAs, western blot hybridization is recommended, which not only provides information about the expression level of the target protein, but also reveals the molecular weight of the protein. Chapter 11 describes in detail the protocols for performing western blotting. Figure 22.5 illustrates the effects of siRNAs on a target protein. The protein levels in nontransfected cells and cells transfected with the vector without siRNA-encoding oligonucleotides should be similar to each other. However, there is a significant reduction of protein in cells transfected with DNA constructs containing the hairpin siRNA oligonucleotide fragment, indicating that siRNAs caused significant inhibition of target mRNA, which in turn led to reduction of translation of the protein of interest due to insufficient availability of the mRNA template.

REPORTER GENE ASSAYS OF DOWN-REGULATION OF GENE EXPRESSION VIA siRNAs

As illustrated in Figure 22.3, DNA constructs may be generated, which contain a reporter gene such as luciferase, green fluorescent protein (GFP) fused with the 3′-UTR of the luciferase gene and hairpin siRNA-encoding oligonucleotide fragment. Moreover, the hairpin siRNA oligonucleotide fragment may be manipulated because one or two bases mismatch as control. If all goes well, the mismatch control will not facilitate mRNA degradation via siRNAs. The strategy and the power of these types of constructs are to provide rapid quantification of target protein using the

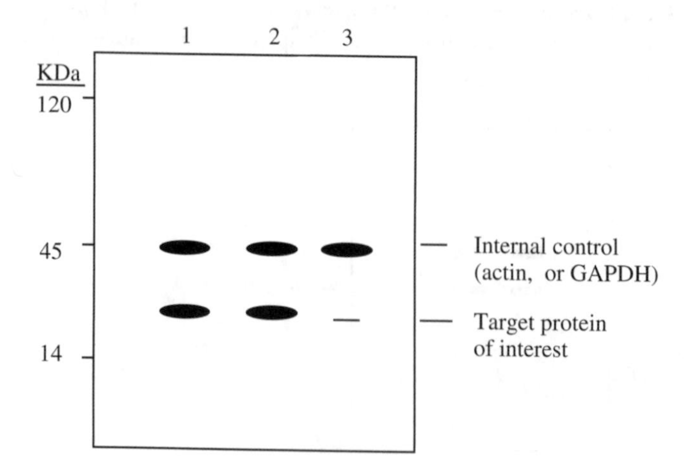

FIGURE 22.5 Western blot analysis of inhibition of the protein of interest via siRNA. Lane 1: nontransfected cells. Lane 2: cells transfected with vector without siRNA construct. Lane 3: cells transfected with siRNA constructs.

reporter gene assay or fix the cells for confocal, fluorescent localization of the target protein in cells.

Additionally, a β-galactosidase is included in the vector, which will provide a tool for normalization of reporter gene assays downstream. For luciferase assays, commercial kits and the dual-light system (Applied Biosystem) may be used. As seen in Figure 22.6, luciferase activities in control groups (A and B) are very similar. However, cells cotransfected with the control plasmid DNA and the construct containing the hairpin siRNA-encoding oligonucleotide fragment caused significant reduction of luciferase (group C). Cells cotransfected with the control plasmid with mismatch in the target region and the DNA construct containing the hairpin siRNA-encoding oligonucleotide fragment did not alter luciferase activity compared to group A and group B.

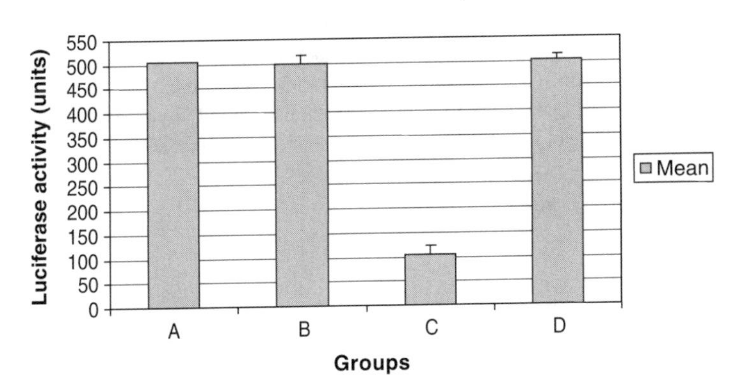

FIGURE 22.6 Analysis of gene expression by siRNA/reporter gene assays. Group A: cells transfected with control DNA construct (Luc-GFP-target gene without mismatch bases). Group B: cells transfected with control DNA construct (Luc-GFP-target gene with mismatch bases). Group C: cells cotransfected with control A construct (Luc-GFP-target gene without mismatch bases) plus siRNA constructs as illustrated in Figure 22.2. Group D: cells cotransfected with siRNA construct (Figure 22.2) plus control B construct (Luc-GFP-target gene 1 to 2 bases mismatch).

REFERENCES

1. Amarzguioui, M., Holen, T., Babaie, E., and Prydz, H., Tolerance for mutations and chemical modification in siRNA, *Nucleic Acids Res.*, 31(2):589–595, 2003.
2. Czauderna, F., Fechtner, M., Aygun, H., Arnold, W., Klippel, A., Giese, K., and Kaufmann, J., Functional studies of the PI(3)-kinase signalling pathway employing synthetic and expressed siRNA, *Nucleic Acids Res.*, 31(2):670–682, 2003.
3. Randall, G., Grakoui, A., and Rice C.M., Clearance of replicating hepatitis C virus replicon RNAs in cell culture by small interfering RNAs, *Proc. Natl. Acad. Sci. USA*, 100(1):235–240, 2003.
4. Qin, X.F., An, D.S., Chen, I.S., and Baltimore, D., Inhibiting HIV-1 infection in human T cells by lentiviral-mediated delivery of small interfering RNA against CCR5, *Proc. Natl. Acad. Sci. USA*, 100(1):183–188, 2003.

5. Vickers, T.A., Koo, S., Bennett C.F., Crooke, S.T., Dean, N.M., and Baker, B.F., Efficient reduction of target RNAs by siRNA and RNase H dependent antisense agents: a comparative analysis, *J. Biol. Chem.,* 278(9):7108–7118, Dec. 23, 2002.
6. Couzin, J., Breakthrough of the year, *Science,* 298(5602):2283, 2002.
7. Wu, W., Analysis of gene expression at the RNA level, in *Methods in Gene Biotechnology,* 1st ed., Wu, W., Welsh, M.J., Kaufman, P.B., and Zhang, H., (Eds.), CRC Press, Boca Raton, FL, 1997, 225–240.
8. Wu, W., Analysis of gene expression at the protein level, in *Methods in Gene Biotechnology,* 1st ed., Wu, W., Welsh, M.J., Kaufman, P.B., and Zhang, H., (Eds.), CRC Press, Boca Raton, FL, 1997, 241–258.
9. Siegmund, D., Hadwiger, P., Pfizenmaier K., Vornlocher, H.P., and Wajant, H., Selective inhibition of FLICE-like inhibitory protein expression with small interfering RNA oligonucleotides is sufficient to sensitize tumor cells for TRAIL-induced apoptosis, *Mol. Med.,* 8(11):725–732, 2002.
10. Xu, Z., Kukekov, N.V., and Greene, L.A., POSH acts as a scaffold for a multiprotein complex that mediates JNK activation in apoptosis, *EMBO J.*, 22(2):252–261, 2003.
11. Robert, M.F., Morin, S., Beaulieu, N., Gauthier, F., Chute, I.C., Barsalou, A., and MacLeod, A.R., DNMT1 is required to maintain CpG methylation and aberrant gene silencing in human cancer cells, *Nat Genet.,* 33(1):61–65, 2003.

Index

A

B

C

E

F

G

H

I

J

K

L

M

N

O

P

R

S

T

U

V

W

X